Molecular Mechanisms of Synaptic Plasticity 2.0: Dynamic Changes in Neurons Functions, Physiological and Pathological Process

Molecular Mechanisms of Synaptic Plasticity 2.0: Dynamic Changes in Neurons Functions, Physiological and Pathological Process

Editor

Giuseppina Martella

MDPI

Basel • Beijing • Wuhan • Barcelona • Belgrade • Novi Sad • Cluj • Manchester

Editor
Giuseppina Martella
Department of Humanistic Sciences
Faculty of Motor Sciences
Pegaso University
Naples
Italy

Editorial Office
MDPI
St. Alban-Anlage 66
4052 Basel, Switzerland

This is a reprint of articles from the Special Issue published online in the open access journal *International Journal of Molecular Sciences* (ISSN 1422-0067) (available at: www.mdpi.com/journal/ijms/special_issues/synaptic_plasticity_2).

For citation purposes, cite each article independently as indicated on the article page online and as indicated below:

Lastname, A.A.; Lastname, B.B. Article Title. *Journal Name* **Year**, *Volume Number*, Page Range.

ISBN 978-3-0365-9045-5 (Hbk)
ISBN 978-3-0365-9044-8 (PDF)
doi.org/10.3390/books978-3-0365-9044-8

Contents

About the Editor

Giuseppina Martella

Giuseppina Martella is an associate professor of Physiology at the Department of Human Sciences of Pegaso Telematic University. She also continues her research at the Laboratory of Physiology and Plasticity, Fondazione Santa Lucia, IRCCS, Rome. She holds a Master's Degree in Biology specializing in Pathophysiology from the University of Rome Tor Vergata. She furthered her training in the Neurophysiology labs of Professor Giorgio Bernardi.

She has been interested in neurological pathologies since 2000. She then obtained her specialization in Clinical Pathology in 2007, continuing to work on movement disorders under the guidance of Professors Stefani, Calabresi, and Pisani. In 2012, she received her doctorate in Neuroscience. She has been the recipient of numerous fellowships and research grants.

Since 2014, she has been devoted to studying Synaptic Plasticity phenomena and focused on Parkinson's Disease and Genetic Dystonias. She is the author of more than 70 papers in international journals and is the recipient of several Research Awards and Grants. Her primary goal is the knowledge of physiological mechanisms and their alterations in Neurological disorders; to achieve this goal, she is currently collaborating with other researchers on translational projects that start from the molecular and continue in the patient.

Preface

Synaptic plasticity is a complex and crucial neuronal mechanism linked to principal memory and motor functions. During the developmental period into old age, the neural frame is subject to structural and functional modifications in response to external stimuli. This essential skill of neuronal cells underpins the ability to learn about mammalian organisms (Glanzman et al., 2010).

Synaptic plasticity phenomena include microscopic changes such as spine pruning and macroscopic changes such as cortical remapping in response to injury (Citri and Malenka 2008; Hofer et al., 2009). The increase in neurological and neuropsychiatric disorders in the current century—although this increase did not occur among the aging population—has resulted in a greater urgency to understand the aberrant processes connected to these diseases (Martella et al., 2016; 2018; Bonsi et al. 2018).

In the last decades, it has been highlighted that de novo protein synthesis (mRNA transcription, mRNA and protein degradation, histone acetylation, DNA methylation, and miRNA regulation), as well as a new set of signaling molecules (endogenously generated cannabinoids, peptides, Neurotrophins, protein kinases, and ubiquitindash proteasome system), have been implicated in synaptic transmission and plasticity.

Giuseppina Martella
Editor

International Journal of
Molecular Sciences

MDPI

Editorial

Molecular Mechanisms of Synaptic Plasticity 2.0: Dynamic Changes in Neurons Functions, Physiological and Pathological Process

Giuseppina Martella [1,2]

[1] Laboratory of Neurophysiology and Plasticity, IRCCS Fondazione Santa Lucia, 00143 Rome, Italy; martella@med.uniroma2.it or g.martella@hsantalucia.it

[2] Department of Humanistic Sciences, Faculty of Motor Sciences, Pegaso University, 80143 Naples, Italy

Due to the success of the first Special Issue on synaptic plasticity, I endeavored to promote a new Special Issue with an emphasis on dynamic changes in neuronal functions and physiological and pathological processes.

I have endorsed this Special Issue with the aim of collecting scientific research capable of intensifying readers' interest in phenomena related to synaptic plasticity.

Synaptic plasticity is a crucial molecular mechanism whose actions are carried out from the developmental period to old age. Described briefly, our brain is subject to structural and functional modifications in response to external stimuli. Synaptic plasticity phenomena include macroscopic changes and cortical remapping in response to injury, microscopic changes, and spine pruning [1,2]. The emergency resulting from the increase in neuropsychiatric and neurological disorders over the last few years has stressed the urgency of understanding the aberrant processes connected to synaptic plasticity failure [3–5].

The authors invited to contribute to this Special Issue have provided important contributions consisting of the translational and clinical studies that I am pleased to promote in these few lines. This Special Issue contains two original clinical articles, two literature reviews, and five original translation articles. I hope that I have provided you with an extensive thematic collection and that you enjoy your reading.

check for updates

Citation: Martella, G. Molecular Mechanisms of Synaptic Plasticity 2.0: Dynamic Changes in Neurons Functions, Physiological and Pathological Process. *Int. J. Mol. Sci.* **2023**, *24*, 12685. https://doi.org/10.3390/ijms241612685

Received: 2 August 2023
Accepted: 7 August 2023
Published: 11 August 2023

Patnaik and co-workers examined 29 patients with certified bipolar disorder, 32 patients affected by cerebellar neurodegenerative pathologies, and 37 healthy subjects using the 3T-MRI technique in order to determine the similarities and differences in cerebellar grey matter loss. They found a pattern of grey matter cerebellar alterations in both the bipolar and cerebellar groups that involved the anterior and posterior cerebellar regions, demonstrating the involvement of the cerebellum in the synaptic plasticity of patients with bipolar disorder [6].

In multiple sclerosis, inflammation can modify synaptic transmission and plasticity. Professor Centonze's group explored the influence of proinflammatory cytokines on associative Hebbian synaptic plasticity. In their cohort, they found that IL-1β levels were associated with synaptic hyperexcitability and inversely related to LTP-like synaptic plasticity.

These findings support the evidence that anti-IL-1β drugs represent a new potential therapeutic target. Considering the identification of IL-1β as a marker of inflammatory synaptopathy, antagonistic drugs may also represent a specific target during the different phases of the progression of multiple sclerosis [7].

In their systematic review, Eltokhi and coworkers underline the correlation between neuropsychiatric disorders and deficits in the glutamatergic system, and they also consider the psychiatric features of neurodevelopmental disorder as well as autism. Alterations in synaptic plasticity, accompanied by structural modifications of excitatory synapses, were observed in schizophrenia and autism spectrum disorders using EM-imaging methods. In addition, it was revealed that the expression of glutamatergic receptors is differentially affected in various brain regions, thus revealing an undeniable link between altered synaptic plasticity and psychiatric illness [8].

D'Angelo and colleagues investigated the A2A receptor's role in different areas inside of a DYT1 mouse model of dystonia. They showed that A2A receptors are significantly improved in the striatal and globus pallidus nuclei and reduced in the entopeduncular nucleus. These opposite modifications may suggest that the pathophysiology of dystonia is correlated with an imbalance of the direct or indirect pathway [9].

The motor thalamus (MTh) is involved in the basal ganglia cortical loop and acts on the codifying of motor information. Di Giovanni and his co-workers showed for the first time that acute dopamine depletion caused by tetrodotoxin (TTX) results in an increase in GABA concentration in the MTh along with unchanged glutamate levels. Chronic dopamine denervation via 6-hydroxydopamine (6-OHDA) in anesthetized rats affects the coupling of MTh cortical activity in relation to the TTX-induced acute depletion state. The authors' findings demonstrate that the MTh, among other areas in the basal ganglia, is influenced by DA chronic deprivation and makes alterations in the basal ganglia network in relation to counterbalancing the profound alteration arising after the onset of the acute DA depletion state [10].

La Recchiuta aimed to emphasize the role of the amygdala and medial prefrontal cortex in functional plasticity and synaptic wiring in conditions of fear extinction. Her results demonstrated that the optogenetic activation of pyramidal neurons in mice conditioned by induced fear extinction deficits causes an increase in cellular excitability, excitatory neurotransmission, and spinogenesis and is also associated with modifications of the transcriptome of amygdala pyramidal neurons [11].

The brain-derived neurotrophic factor (BDNF) drives brain development and maturation. Altered BDNF levels have been observed in many neurological diseases to such an extent that new therapeutic strategies are being developed to increase the level of BDNF.

Fingolimod-phosphate (FTY720-P) can modulate BDNF levels. However, the mechanisms by which the FTY720-P operates are still unclear. Patnaik and colleagues have shown that FTY720-P can regulate dendritic architecture, increase dendritic spine density, and modify the morphology of mature primary hippocampal neuron cultures. This study confirms that BDNF-dependent therapies may represent a new goal for many neurological diseases [12].

The Ghiglieri group, on the other hand, carried out scientific work demonstrating that food restriction can improve the lifespan of different species.

According to their analyses, food is a natural reward, and where this reformulation is restricted, some neuroadaptive responses are necessary to maintain physiological homeostasis.

They examined the AMPA receptor subunit composition in the dorsal striatal neurons of mice that had been acutely food deprived and showed that even moderate food deprivation in experimental animal models reflects a series of neuroadaptations and the remodeling of striatal synaptic plasticity [13].

Stroke is a great enemy of modern medicine. Usually, deficiencies of the neurovascular unit caused by reperfusion lesions, or inflammatory processes, constitute the main field of study. In their literature review, De Luca and collaborators provided a road map that could help to improve both therapy and rehabilitation through the knowledge of translational studies. They also suggest that further research should involve the cellular capacity to avoid neuroinflammatory phenomena and the capacity of cells during reperfusion to actively reshape the matrix [14].

Conflicts of Interest: The author declares no conflict of interest.

References

1. Citri, A.; Malenka, R.C. Synaptic Plasticity: Multiple Forms, Functions, and Mechanisms. *Neuropsychopharmacology* **2008**, *33*, 18–41. [CrossRef] [PubMed]
2. Hofer, S.B.; Mrsic-Flogel, T.D.; Bonhoeffer, T.; Hübener, M. Experience Leaves a Lasting Structural Trace in Cortical Circuits. *Nature* **2009**, *457*, 313–317. [CrossRef] [PubMed]

3. Martella, G.; Madeo, G.; Maltese, M.; Vanni, V.; Puglisi, F.; Ferraro, E.; Schirinzi, T.; Valente, E.M.; Bonanni, L.; Shen, J.; et al. Exposure to Low-Dose Rotenone Precipitates Synaptic Plasticity Alterations in PINK1 Heterozygous Knockout Mice. *Neurobiol. Dis.* **2016**, *91*, 21–36. [CrossRef] [PubMed]
4. Martella, G.; Bonsi, P.; Johnson, S.W.; Quartarone, A. Synaptic Plasticity Changes: Hallmark for Neurological and Psychiatric Disorders. *Neural Plast.* **2018**, *2018*, 9230704. [CrossRef] [PubMed]
5. Bonsi, P.; Ponterio, G.; Vanni, V.; Tassone, A.; Sciamanna, G.; Migliarini, S.; Martella, G.; Meringolo, M.; Dehay, B.; Doudnikoff, E.; et al. RGS9-2 Rescues Dopamine D2 Receptor Levels and Signaling in DYT1 Dystonia Mouse Models. *EMBO Mol. Med.* **2019**, *11*. [CrossRef] [PubMed]
6. Saundh, S.L.; Patnaik, D.; Gagné, S.; Bishop, J.A.; Lipsit, S.; Amat, S.; Pujari, N.; Nambisan, A.K.; Bigsby, R.; Murphy, M.; et al. Identification and Mechanistic Characterization of a Peptide Inhibitor of Glycogen Synthase Kinase (GSK3β) Derived from the Disrupted in Schizophrenia 1 (DISC1) Protein. *ACS Chem. Neurosci.* **2020**, *11*, 4128–4138. [CrossRef] [PubMed]
7. Stampanoni Bassi, M.; Buttari, F.; Nicoletti, C.G.; Mori, F.; Gilio, L.; Simonelli, I.; De Paolis, N.; Marfia, G.A.; Furlan, R.; Finardi, A.; et al. Interleukin-1β Alters Hebbian Synaptic Plasticity in Multiple Sclerosis. *Int. J. Mol. Sci.* **2020**, *21*, 6982. [CrossRef] [PubMed]
8. Eltokhi, A.; Santuy, A.; Merchan-Perez, A.; Sprengel, R. Glutamatergic Dysfunction and Synaptic Ultrastructural Alterations in Schizophrenia and Autism Spectrum Disorder: Evidence from Human and Rodent Studies. *Int. J. Mol. Sci.* **2020**, *22*, 59. [CrossRef] [PubMed]
9. D'angelo, V.; Giorgi, M.; Paldino, E.; Cardarelli, S.; Fusco, F.R.; Saverioni, I.; Sorge, R.; Martella, G.; Biagioni, S.; Mercuri, N.B.; et al. A2a Receptor Dysregulation in Dystonia Dyt1 Knock-out Mice. *Int. J. Mol. Sci.* **2021**, *22*, 2691. [CrossRef] [PubMed]
10. Di Giovanni, G.; Grandi, L.C.; Fedele, E.; Orban, G.; Salvadè, A.; Song, W.; Cuboni, E.; Stefani, A.; Kaelin-Lang, A.; Galati, S. Acute and Chronic Dopaminergic Depletion Differently Affect Motor Thalamic Function. *Int. J. Mol. Sci.* **2020**, *21*, 2734. [CrossRef] [PubMed]
11. Laricchiuta, D.; Balsamo, F.; Fabrizio, C.; Panuccio, A.; Termine, A.; Petrosini, L. CB1 Activity Drives the Selection of Navigational Strategies: A Behavioral and c-Fos Immunoreactivity Study. *Int. J. Mol. Sci.* **2020**, *21*, 1072. [CrossRef] [PubMed]
12. Patnaik, A.; Spiombi, E.; Frasca, A.; Landsberger, N.; Zagrebelsky, M.; Korte, M. Fingolimod Modulates Dendritic Architecture in a BDNF-Dependent Manner. *Int. J. Mol. Sci.* **2020**, *21*, 3079. [CrossRef] [PubMed]
13. Campanelli, F.; Laricchiuta, D.; Natale, G.; Marino, G.; Calabrese, V.; Picconi, B.; Petrosini, L.; Calabresi, P.; Ghiglieri, V. Long-Term Shaping of Corticostriatal Synaptic Activity by Acute Fasting. *Int. J. Mol. Sci.* **2021**, *22*, 1916. [CrossRef] [PubMed]
14. De Luca, C.; Colangelo, A.M.; Virtuoso, A.; Alberghina, L.; Papa, M. Neurons, Glia, Extracellular Matrix and Neurovascular Unit: A Systems Biology Approach to the Complexity of Synaptic Plasticity in Health and Disease. *Int. J. Mol. Sci.* **2020**, *21*, 1539. [CrossRef]

International Journal of *Molecular Sciences*

MDPI

Article

miRNA-132/212 Deficiency Disrupts Selective Corticosterone Modulation of Dorsal vs. Ventral Hippocampal Metaplasticity

Shima Kouhnavardi [1], Maureen Cabatic [1,†], M. Carmen Mañas-Padilla [2,†], Marife-Astrid Malabanan [1], Tarik Smani [3], Ana Cicvaric [1,4], Edison Alejandro Muñoz Aranzalez [1], Xaver Koenig [1], Ernst Urban [5], Gert Lubec [6], Estela Castilla-Ortega [2] and Francisco J. Monje [1,*]

1 Center for Physiology and Pharmacology, Department of Neurophysiology and Neuropharmacology, Medical University of Vienna, 1090 Vienna, Austria
2 Instituto de Investigación Biomédica de Málaga-IBIMA, 29590 Málaga, Spain
3 Department of Medical Physiology and Biophysics, University of Seville, 41013 Seville, Spain
4 Dominick P. Purpura Department of Neuroscience, Albert Einstein College of Medicine, New York, NY 10461, USA
5 Department for Pharmaceutical Sciences, Josef-Holaubek-Platz 2, 2D 303, 1090 Vienna, Austria
6 Programme for Proteomics, Paracelsus Medical University, 5020 Salzburg, Austria
* Correspondence: francisco.monje@meduniwien.ac.at
† These authors contributed equally to this work.

check for updates

Citation: Kouhnavardi, S.; Cabatic, M.; Mañas-Padilla, M.C.; Malabanan, M.-A.; Smani, T.; Cicvaric, A.; Muñoz Aranzalez, E.A.; Koenig, X.; Urban, E.; Lubec, G.; et al. miRNA-132/212 Deficiency Disrupts Selective Corticosterone Modulation of Dorsal vs. Ventral Hippocampal Metaplasticity. *Int. J. Mol. Sci.* **2023**, *24*, 9565. https://doi.org/10.3390/ijms24119565

Academic Editor: Giuseppina Martella

Received: 8 May 2023
Revised: 21 May 2023
Accepted: 25 May 2023
Published: 31 May 2023

Abstract: Cortisol is a potent human steroid hormone that plays key roles in the central nervous system, influencing processes such as brain neuronal synaptic plasticity and regulating the expression of emotional and behavioral responses. The relevance of cortisol stands out in the disease, as its dysregulation is associated with debilitating conditions such as Alzheimer's Disease, chronic stress, anxiety and depression. Among other brain regions, cortisol importantly influences the function of the hippocampus, a structure central for memory and emotional information processing. The mechanisms fine-tuning the different synaptic responses of the hippocampus to steroid hormone signaling remain, however, poorly understood. Using ex vivo electrophysiology and wild type (WT) and miR-132/miR-212 microRNAs knockout (miRNA-132/212$^{-/-}$) mice, we examined the effects of corticosterone (the rodent's equivalent to cortisol in humans) on the synaptic properties of the dorsal and ventral hippocampus. In WT mice, corticosterone predominantly inhibited metaplasticity in the dorsal WT hippocampi, whereas it significantly dysregulated both synaptic transmission and metaplasticity at dorsal and ventral regions of miR–132/212$^{-/-}$ hippocampi. Western blotting further revealed significantly augmented levels of endogenous CREB and a significant CREB reduction in response to corticosterone only in miR–132/212$^{-/-}$ hippocampi. Sirt1 levels were also endogenously enhanced in the miR–132/212$^{-/-}$ hippocampi but unaltered by corticosterone, whereas the levels of phospo-MSK1 were only reduced by corticosterone in WT, not in miR–132/212$^{-/-}$ hippocampi. In behavioral studies using the elevated plus maze, miRNA-132/212$^{-/-}$ mice further showed reduced anxiety-like behavior. These observations propose miRNA-132/212 as potential region-selective regulators of the effects of steroid hormones on hippocampal functions, thus likely fine-tuning hippocampus-dependent memory and emotional processing.

Keywords: microRNA; miR-132/212; synaptic plasticity; dorsal hippocampus; ventral hippocampus; corticosterone; emotional behavior; anxiety-like behavior

1. Introduction

The steroid hormone cortisol (also known as hydrocortisone) is a highly potent human hormone produced by the adrenal glands and whose secretion into the blood stream is triggered by the corticotropin-releasing hormone originating from the hypothalamus in the brain. Cortisol is central for the regulation of many biological functions, including the modulation of the body responses to stress, the reduction of inflammatory processes, the regulation of blood pressure and metabolic functions, and the influences on the immune system and circadian sleep-wake cycles [1,2]. Cortisol is also a powerful regulator of brain function, influencing learning, memory and emotional behaviors in both humans and animals [2–5]. Moreover, dysregulated cortisol levels have been identified during the onset and progression of a variety of conditions associated with memory decline and emotional dysregulation, including post-traumatic stress disorder, depression, stress and anxiety [6–15]. Several studies have also shown that altered levels of corticosterone (the molecular counterpart of cortisol in mice and rats) result in profound changes in neuronal synaptic plasticity as well as altered memory-related emotional behaviors [16–22]. The molecular mechanisms by which cortisol can distinctly affect the different regions of the brain to influence memory and emotion-related behaviors remain, however, poorly understood.

The hippocampus, a brain structure critical for learning and memory processing [23–26], is involved in the regulation of emotional memories [27–29] and is importantly associated with the modulation of the hypothalamic–pituitary–adrenal axis function in health and disease [30–33]. Interestingly, exposure to early-life stress has been shown to correlate with a reduction in the anatomical features of hippocampal structures as well as alterations in cortisol levels [30,33]. Moreover, while the hippocampus is the brain region with perhaps the highest concentration of receptor target sites for adrenocortical steroids [29], and whereas several studies have shown that corticosterone can directly influence hippocampal synaptic potentiation [34,35], little continues to be known about how cortisol differentially affects the different regions of the hippocampus to influence memory-related emotional behaviors [28]. Here, we hypothesized that microRNAs (miRNAs) are key molecular elements participating in the modulation of the region-specific effects of glucocorticoid hormones on hippocampal synaptic transmission and metaplasticity (the newly generated synaptic changes that happen after synaptic plasticity has occurred [36–38]).

miRNAs, are non-encoding short transcript RNAs that participate in the posttranscriptional modulation of gene expression [39,40]; that serve in a variety of pivotal functions of the central nervous systems, including the modulation of synaptic activity [41–46]; and which are importantly involved in the brain neuropathology [47–51]. Elevated levels of some miRNAs (e.g., miR-455-3p), have been found in serum samples from Alzheimer's disease patients relative to those obtained from healthy controls, thus suggesting the possibility that miR-455-3p and other miRNAs [49] could even be used as biomarkers in the contexts of very severe neuropathologies, as is the case of Alzheimer's disease [48]. In line with this, specific studies of the levels of the miRNA miR-132 have been conducted using quantitative real time PCR in postmortem brain tissue samples obtained from deceased patients that had being diagnosed with Alzheimer's disease and with mild cognitive-decline. These studies showed significantly enhanced levels of miR-132 in the samples from both Alzheimer's disease and cognitive-decline groups as compared with the data obtained from their normal (control) samples, suggesting that miR-132 might be a critical player in the pathogenesis of Alzheimer's disease [47]. Interestingly, other authors have described not elevated but rather markedly reduced levels of miR-132 at middle and advanced stages of Alzheimer's disease [50], and that the reduction in this specific miRNA aggravates amyloid and TAU pathological features [52]. For many years, miR-132 has been jointly examined together with miR-212 in a variety of different multidisciplinary studies (see for example [53–58]), as both these molecules are described to belong to the same highly conserved cluster family of miRNAs derived from a shared phylogenetic ancestor; further having their gene chromosomic organization arrayed as a tandem [59]. Similarly, in experiments using genetically modified mice, the combined genetic deletion of the genes encoding for miR-132 and miR-212 has been shown to result in an enhancement in the levels of amyloid beta peptides as well as in augmented amyloid-related plaque formation [51]. All these observations had clearly identified the microRNA212/132 cluster family as highly interesting candidates in studies about the brain neuronal function in health and disease.

Several groups, including ours, have previously described some of the functional effects of the double-deletion of the genes encoding for the miRNAs 132 and 212 in the regulation of plasticity-related functions in the mouse hippocampus [43,57,60,61]. However, although both steroid-hormones, including cortisol and corticosterone, as well as the miRNAs 132 and 212, have been independently implicated in the regulation of synaptic plasticity and emotion-related functions ([62,63]), the potential functional crosslink between steroid hormones; brain region-specific regulation of synaptic plasticity; and the miRNAs 132 and 212, remained -to our knowledge- unexplored. Therefore, the main objective of this work was to use a previously described double miR-132 and miR-212 miRNAs knockout mouse line [56] (here referred as miRNA-132/212$^{-/-}$, or KO), in order to electrophysiologically study the synaptic responses to corticosterone stimulation in hippocampal slices, and to examine whether these responses are functionally homogeneous or not when measurements at the dorsal and ventral hippocampi are compared. We have obtained experimental data suggesting a potential role for the miR-132/12 in hippocampal neuroendocrine signaling and emotion-related behaviors.

2. Results

2.1. Comparable Basal Synaptic Transmission in Dorsal and Ventral Hippocampi of WT and miR–132/212$^{-/-}$ Mice

In order to examine the physiological relevance of the microRNAs 132 and 212 in mediating the hippocampal synaptic responses to corticosterone stimulation, we explored the effect of 1 μM corticosterone on basal synaptic transmission and memory-related synaptic plasticity in the dorsal and ventral hippocampus of wild type and miR–132/212$^{-/-}$ mice. To this aim, we conducted recordings of extracellular field potentials ex vivo in acutely-dissociated hippocampal slices following standardized electrophysiological protocols previously described by our group and others [43,60,61,64–70].

As illustrated in Figure 1A, for these experiments, we extracted the hippocampus and split it into two main pieces lengthwise the septo-temporal axis, relative to the morphological transverse middle, and then slices for electrophysiological recordings were obtained from about 30–40% of the section comprising the defined center towards either the dorsal or the ventral ending regions (longitudinally). These areas are known to exhibit classical LTP responses as induced by electrical stimulation [71,72]. Subsequently, field potential recordings were conducted at the CA1 region upon delivering electrical stimulation at the Schaffer's collaterals from the CA3 region for both dorsal and ventral regions (Figure 1B). For plasticity experiments, recordings were conducted in the presence and absence of corticosterone in slices from both WT and miR–132/212$^{-/-}$ mice. Corticosterone was applied in the bath for the time specified in the figures and then removed using gravity-based-perfusion and peristaltic pump-driven solution exchange. Before examining the effect of 1 μM corticosterone on the properties of synaptic plasticity, we verified the functional integrity of the synaptic circuits by measuring input/output (I/O) field responses (see Materials and Methods) in untreated dorsal and ventral hippocampi of WT and miR–132/212$^{-/-}$ mice. As shown in Figure 1CD, no statistically significant differences were found in basal synaptic transmission when comparing the data obtained in slices derived from WT and miR–132/212$^{-/-}$ mice. Two-way RM-ANOVA with Bonferroni's and Geisser–Greenhouse's corrections, and alpha set to 0.05 (n = 21 animals per group), showed no significance for the effects of the 1 μM corticosterone treatment for WT vs. miR–132/212$^{-/-}$ at dorsal (p = 0.7450; $F_{(1, 40)}$ = 0.1073) or ventral (p = 0.1748; $F_{(1, 40)}$ = 1.908) hippocampi.

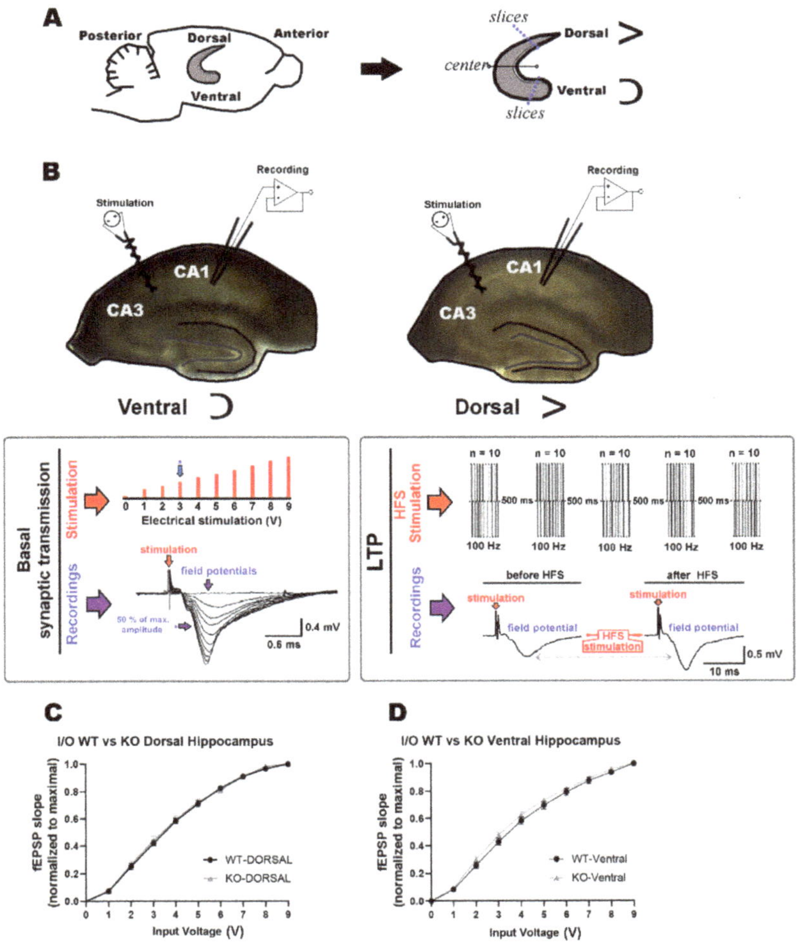

Figure 1. Description of the electrophysiological approach and study of basal synaptic transmission. (**A**) The cartoon on the left represents a sagittal section of the mouse brain with the relative localization of one of the two hippocampi and Anterior–Posterior and Dorsal–Ventral references indicated. To the right, the black arrow points towards the extracted hippocampus, which was divided into two parts using its anatomical middle as the nominal center. Slices were subsequently obtained from the ending dorsal (represented by a left-90°-rotated "V" shape) and ventral (represented by a horizontally flipped "C" shape) regions that comprise circa 30–40% (blue dotted lines) of the portion located longitudinally from the center towards each respective end. (**B**) Two representative microphotographs of hippocampal slices from the ventral (**left**) and dorsal (**right**) regions were used to hereby illustrate the synaptic regions examined electrophysiologically and the positioning of the recording and stimulating electrodes. The tossed "V"- and "C"-like black lines partially surrounding the dentate gyrus (gray line inside the slices) illustrate the morphology typically observed ventrally and dorsally under the microscope. The CA3 and CA1 regions are indicated. In the large, boxed inset below (**left**): Schematic representation of the protocol used to generate input/output (I/O) curves used to assess basal synaptic transmission. The large, red-filled arrow at the top left points towards the delivered electrical stimulation pulses, which consisted of 10 discrete 200 μs voltage steps (in red) from 0–9 V delivered with 15 s intervals. Below, the large purple-filled arrow on the left points towards representative traces of the 10 elicited field potential recordings. The small blue arrow

preceded by an asterisk indicates a field recording with an amplitude of approximately 50% of the maximal achievable amplitude, which, in this diagram, and as an illustration, would have been generated by the step of voltage stimulation of 3 V, as indicated in the upper panel by a small blue arrow preceded by an asterisk. In the large, boxed inset below (**right**): Schematic representation of the high-frequency stimulation (HFS) protocol used to generate long-term potentiation (LTP). In the top panel, the large, red-filled arrow on the left points towards the 5 delivered bursts of electrical stimulation pulses, each burst consisting of 10 biphasic voltage steps (100 μs/phase, 100 Hz, with 500 ms intervals) given at voltage intensities eliciting about 50% of the maximal inducible amplitude. In the lower panel, the large, purple-filled arrow on the left points towards representative traces illustrating elicited field potential recordings during the baseline recordings (before HFS) and after having delivered the LTP-inducing protocol (after HFS). During experiments, the "before" and "after" field recordings are obtained upon delivering one biphasic pule (100 μs/phase) with inter-stimulus intervals of 30 s, using the same voltage stimulation that elicited about 50% of the maximal inducible amplitude. (**C**,**D**) The line charts represent the data from the changes in the field slopes versus the different values of voltage delivered (normalized to maximal slope) for slices from the dorsal and ventral regions, respectively, obtained from WT and miRNA-132/212$^{-/-}$ mice (n = 21 animals per group). No statistically significant differences were observed between the different groups (details in the main text). Data are shown as mean ± SEM.

2.2. miRNA-132/212 Gene-Depletion Differentially Affects Short-Term Plasticity in Ventral and Dorsal Hippocampus

In order to further characterize the functional activity at the dorsal and ventral hippocampi of WT and miR–132/212$^{-/-}$ mice, we sought to investigate the properties of presynaptic-dependent short-term plasticity. To this aim, we implemented electrophysiological protocols of Paired-Pulse-induced Facilitation (PPF) in brain slices following methods previously described by our group and others [60,61,73,74] (see also Materials and Methods). Through PPF protocols, it is possible to experimentally evoke a form of short-term plasticity that has been functionally associated with the modulation of exocytosis [75] and which is further proposed to influence the properties of long-term forms of synaptic plasticity [76].

We therefore examined PPF in dorsal and ventral hippocampi in slices from WT and KO mice. As depicted in Figure 2, analysis of the changes in the relative amplitudes (Figure 2A) and field slope ratios (Figure 2B) measured at all the different interpulse intervals examined (see also Material and Methods) showed significant differences between the groups indicating a marked impact of the miRNA-132/212 gene deletion on the properties of dorsal vs. ventral responses during short-term synaptic plasticity. Mixed-effects model three-way ANOVA with Tukey multiple comparisons correction showed significant (**) effect of the interpulse interval (p = 0.0012), as well as highly significant (****) difference for WT dorsal–WT ventral vs. KO dorsal–KO ventral (p < 0.0001) and high significance (****) for WT dorsal–KO dorsal vs. WT ventral–KO ventral (p < 0.0001) for the analyses of the data from the raw amplitudes (with Chi-square = 20.98; df = 1; p < 0.0001; (****)) (Figure 2A). Accordingly, mixed-effects model three-way ANOVA with Tukey multiple comparisons correction showed significant (**) effect of the interpulse interval (p = 0.0033), as well as both significant (*) difference for WT dorsal–WT ventral vs. KO dorsal–KO ventral (p = 0.0327) and high significance (****) for WT dorsal–KO dorsal vs. WT ventral–KO ventral (p = 0.0001) for the analyses of the slopes (Chi-square = 14.71; df = 1; p = 0.0001; (***)) (Figure 2B).

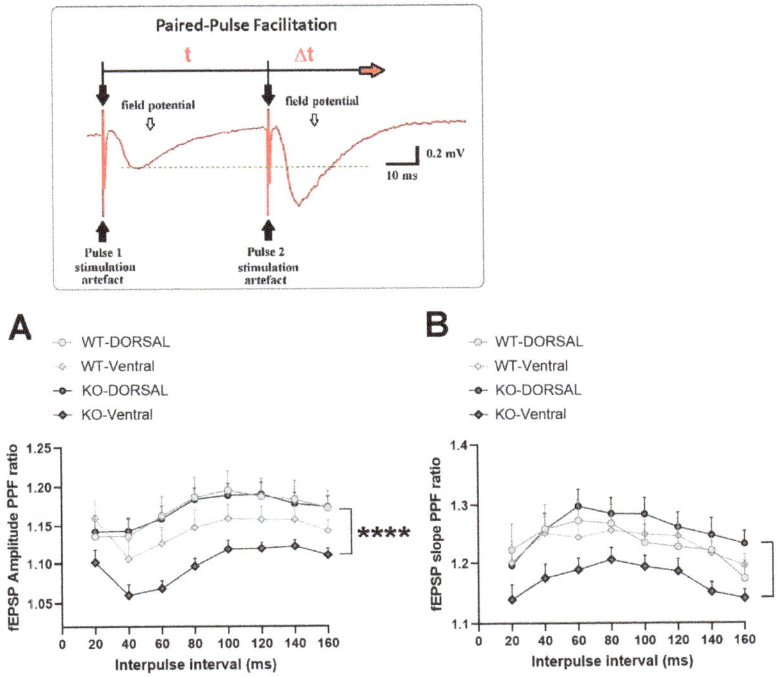

Figure 2. miRNA-132/212 gene-deletion influences presynaptic-dependent short-term facilitation in the mouse hippocampus. The cartoon in the upper inset illustrates the paired-pulse facilitation protocol used to induce short-term facilitation. Two consecutive pulses of electrical stimulation (indicated by black filled arrows) were delivered with an initial 20 ms interpulse interval (t), followed by consecutive increments in 20 ms (Δt) of duration until reaching 160 ms. The red-filled arrow pointing towards the right indicates the advance in time. Ratios for the values of raw amplitude or initial decay slope of the field potential responses ($EPSP_2/EPSP_1$) were used to quantify the power of paired-pulse-induced facilitation. No major differences are detected in the PPF amplitude (**A**) and field-slope ratios (**B**) when recordings from the dorsal and ventral hippocampi are examined in response to the different interpulse time intervals in slices from WT animals. Conversely, a marked difference is observed in the properties of the PPF amplitude, and field-slope ratios are examined in slices derived from miR–132/212$^{-/-}$ mice. $p < 0.05$ was considered significant. * $p < 0.05$, **** $p < 0.0001$. Data are shown as mean ± SEM. A total of 21–22 animals per group were examined. Statistical values are described in the main text.

2.3. miRNAs-132/212 Regulate the Region-Specific Effects of Corticosterone on Metaplasticity in the Mouse Hippocampus

We next analyzed the effect of the 1 μM corticosterone treatment on dorsal hippocampal metaplasticity in slices obtained from WT animals. To this aim, two consecutive LTP-inducing stimulation protocols known to induce metaplasticity [36,77] were delivered before and after corticosterone treatment, which generated the transient post-tetanic potentiation peaks PTP1 and PTP2 (see also Materials and Methods). Figure 3A shows the development in time of the averaged field-slope responses (normalized to the initial baseline). No differences between the untreated and the corticosterone-treated group were apparent for PTP1, and in the treated group, corticosterone treatment did not result in salient differences in the subsequent development of the potentiated field responses compared to the untreated group. However, the PTP2 response obtained after corticosterone treatment presented a statistically significant reduction in its initial amplitude, whereas the subsequent field slopes developed in time in a pattern that did resemble that of the untreated group (Figure 3A). Two-way RM-ANOVA with Bonferroni's and Geisser–Greenhouse's correction, and alpha set to 0.05 (with n = 10–11 animals per group), showed no significance for the effect of the treatment ($F_{(1, 19)} = 0.01610$; $p = 0.9004$) but a significant (****) time × treatment interaction ($F_{(159, 3021)} = 2.482$; $p < 0.0001$). Unpaired t-test (two-tailed) was also used to determine significance for three individual times: 10.5 min, corresponding to the first PTP response (Figure 3B); 25 min, corresponding to the early effect of corticosterone on the synaptic responses (Figure 3C); and 50.5 min, corresponding to the second PTP response (Figure 3D). Significant differences were found only for the PTP2 response (50.5 min: $p = 0.0078$; Welch-correction t = 3.149, df = 12.80; 10.5 min: $p = 0.4654$; Welch-correction t = 0.7529, df = 12.50; 25 min: $p = 0.7769$, Welch-correction t = 0.2876, df = 18.25).

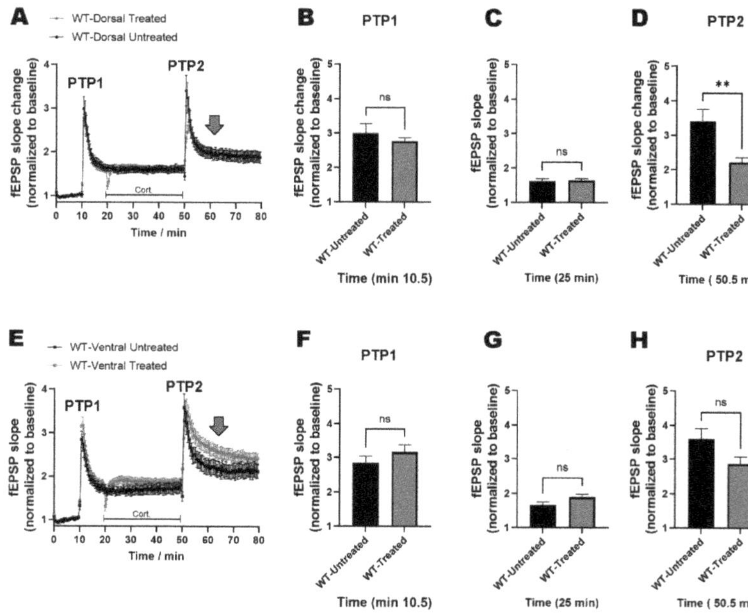

Figure 3. Corticosterone affects hippocampal metaplasticity in a region-specific manner. (**A**) Double high-frequency stimulation protocols, known to induce LTP (see also Materials and Methods) and generate robust peaks of post-synaptic-potentiation (PTP) responses (PTP1 and PTP2), were implemented in slices in order to examine the effect of 1 μM corticosterone on facilitated synaptic transmission and plasticity in the dorsal hippocampus of WT animals. No salient differences were apparent for the PTP1 response or the succeeding development of synaptic transmission during the exposure to corticosterone, as compared to the untreated control group. However, a very pronounced, statistically significant reduction of the PTP2 response was observed in the corticosterone-treated group, with the following field responses developing analogously to those of the untreated control group. Independent statistical analyses (details in the main text) were conducted for the PTP1 response measured at 10.5 min (**B**), which showed no significant differences; as well as for the development of synaptic transmission in the presence of corticosterone at 25 min (**C**), also showing no differences; and for the PTP2 response at 50.5 min (**D**), which showed significant (**) differences (statistical values in the main text). Corresponding analyses were done for the ventral hippocampus, which showed no differences for PTP1 and PTP2 (**E**), and also no significant differences in the field responses of 10.5 min (**F**), 25 min (**G**) and 50.5 min (**H**). Gray-filled large arrows show effects on metaplasticity. ns = not significant. $p < 0.05$ was considered significant. ** $p < 0.01$. Data are shown as mean ± SEM. Statistical values and the number of subjects are described in the main text.

An equivalent examination was conducted in order to study the effects of corticosterone on ventral hippocampal metaplasticity using the double LTP-inducing protocol in the WT group. As depicted in Figure 3E, the PTP1 responses showed no differences when the untreated and corticosterone-treated groups were compared. Two-way RM-ANOVA with Bonferroni's and Geisser–Greenhouse's correction, and alpha set to 0.05 (with n = 10–11 per group), showed no significance for the effect of the treatment ($F_{(1, 19)} = 2.375$; $p = 0.1398$) but a significant (****) time × treatment interaction (F (159, 3021) = 2.884; $p < 0.0001$). As conducted for dorsal hippocampal slices, unpaired t-test (two-tailed) was also used to examine differences in the ventral hippocampus for 10.5 min, 25 min and 50.5 min. As illustrated in Figure 3F–H, no significant differences were found in any of the three examined time points. (10.5 min: $p = 0.2748$; Welch-correction t = 1.124, df = 19; 25 min: $p = 0.0814$, Welch-correction t = 1.840, df = 19; 50.5 min: $p = 0.0703$; Welch-correction t = 1.917, df = 19), but a mixed-effects model (REML) without sphericity assumption, used to independently verify the behavior of the data right after the PTP2 (min 52–79), found a highly significant time × treatment interaction ($p < 0.0001$ (****), $F_{(54, 1026)} = 1.959$), a phenomenon not observed in dorsal hippocampal slices (see large down-pointing gray-filled arrows after PTP2 in Figure 3A,E).

2.4. *miRNA–132/212 Gene Deletion Disrupts the Region-Specific Effect of Corticosterone on Hippocampal Metaplasticity*

In order to assess the relevance of the microRNAs 132 and 212 as regulators of the effects of corticosterone in hippocampal synaptic transmission and metaplasticity, we next investigated the acute impact of corticosterone stimulation on the dorsal and ventral hippocampus using slices from miR–132/212$^{-/-}$ mice. We first examined the effects of corticosterone on metaplasticity using the paired LTP-inducing protocol in the dorsal hippocampus. As shown in Figure 4A, slices from miR–132/212$^{-/-}$ mice presented with metaplasticity responses are indistinguishable from those observed in slices from their related WT counterparts after corticosterone treatment. That is, corticosterone induced a reduction in the amplitude of the PTP2 response in miR–132/212$^{-/-}$ slices but did not affect the temporal development of the subsequent field responses relative to the untreated miR–132/212$^{-/-}$ control group. Two-way RM-ANOVA with Bonferroni's and Geisser–Greenhouse's correction, and alpha set to 0.05 (with n = 10–11 per group), showed no significance for the effect of the treatment ($F_{(1, 19)} = 0.002677$; $p = 0.9593$) but a significant (****) time × treatment interaction ($F_{(159, 3021)} = 5.025$; $p < 0.0001$). Unpaired t-test (two-tailed), used to examine differences at 10.5, 25 and 50.5 min, returned non-significant values of $p = 0.2252$ and Welch-correction t = 1.280 with df = 11.75 for 10.5 min (Figure 4B); $p = 0.0549$ and Welch-correction t = 2.048 with df = 18.62 for 25 min (Figure 4C); and significant values of $p = 0.0083$ (****) with Welch-correction t = 3.011 and df = 16.01 only for 50.5 min (Figure 4C).

We subsequently analyzed the impact of corticosterone on metaplasticity in the ventral hippocampus of slices derived from miR–132/212$^{-/-}$ mice. Figure 4E shows comparable metaplasticity-related LTP responses between the untreated and the corticosterone-treated groups for the PTP1 response and no statistically significant differences at 10.5 min timepoint (Figure 4F). However, corticosterone produced a statistically significant increase in the synaptic responses in the presence of corticosterone as recorded at 25 min, which is indicated by a large gray-filled down-pointing arrow in Figure 4E, a phenotype also revealed at the 25 min timepoint in Figure 4G (**). Moreover, corticosterone also produced a significant reduction in the PTP2 response in ventral hippocampal slices from miR–132/212$^{-/-}$ mice (Figure 4H), an effect not detected in corticosterone-treated ventral hippocampal slices derived from WT mice (Figure 3H). Two-way RM ANOVA with stacked matching and no sphericity assumption (done with Geisser–Greenhouse's and Bonferroni's corrections and alpha set to 0.05) for data depicted in Figure 4E reported no significance for the effect of treatment ($p = 0.0991$) and a highly statistically significant (****) time × treatment interaction ($p < 0.0001$; $F_{(159, 3180)} = 3.899$). Independent two-tailed unpaired t-test yielded values of $p = 0.6237$ with Welch-corrections of $t = 0.4998$ and df = 16.76 for datapoints at 10.5 min (Figure 4F); $p = 0.0064$ (**) with Welch-corrections of $t = 3.232$ and df = 13.26 for 25 min (Figure 4G); and $p = 0.0492$ (*) with Welch-corrections of $t = 2.107$ and df = 18.27 for 50.5 min (Figure 4H).

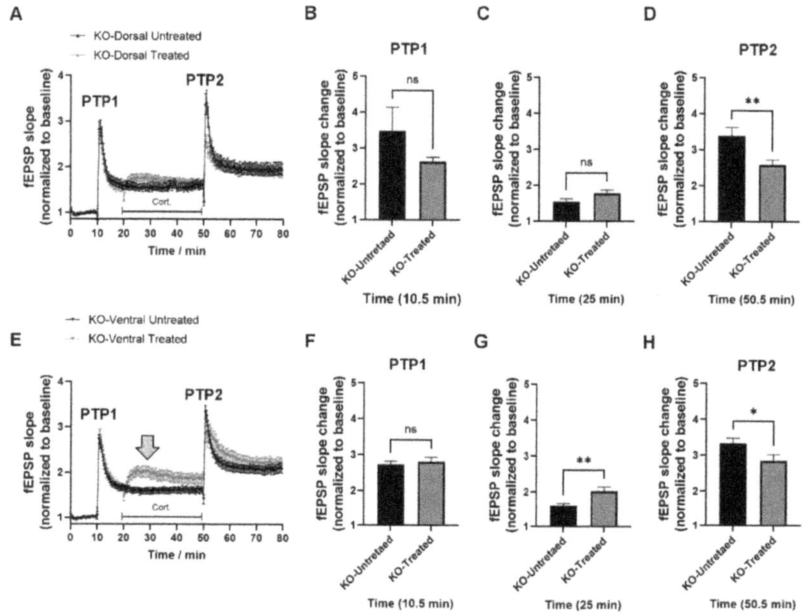

Figure 4. miRNAs–132/212 gene deletion impairs the region-selective effect of corticosterone on hippocampal synaptic plasticity. The properties of synaptic transmission and plasticity examined in the dorsal hippocampus of untreated and corticosterone-treated slices derived from miR–132/212$^{-/-}$ mice were comparable with the WT group described above. That is, no major differences were observed for PTP1, but a significant reduction in the PTP2 response (**A**) was apparent, also with no detectable differences for 10.5 min (**B**) and 25 min (**C**), and a statistically significant (**) difference between untreated and treated groups for 50.5 min (**D**). However, for the ventral hippocampus, while slices from miR–132/212$^{-/-}$ mice exhibited similar PTP1 responses for the untreated and corticosterone-treated groups (**E**), the facilitated postsynaptic responses measured in the presence of corticosterone presented an enhanced amplitude (large gray-filled arrow) as well as a significant reduction of the PTP2 response, to a degree that was not observed in the corticosterone-treated WT group. Statistical analyses conducted for responses measured at 10.5 min showed no significant differences (**F**); whereas both recordings at 25 min (**G**) and 50.5 min (**H**) showed statistically significant differences (** and *, respectively). ns = not significant. * $p < 0.05$, ** $p < 0.01$. Data are shown as mean ± SEM. Statistical values and number of subjects are described in the main text.

Taken together, all these observations point towards a possible involvement of the miRNA 132/212 cluster family in the region-specific regulation of the effects of glucocorticoid hormones on hippocampal synaptic metaplasticity functions and memory-related emotional behaviors. These data further suggested that molecular elements known to mediate in the regulation of both synaptic functions and glucocorticoid hormone activity could be differentially affected by corticosterone in WT and miR–132/212$^{-/-}$ mice hippocampi. In an effort to launch a first experimental verification of this hypothesis, we used the technique of western blot (Materials and Methods) and examined the expression levels of CREB, Sirt1, MSK1, CDK5 and PTEN in the hippocampi of WT and miR–132/212$^{-/-}$ mice.

2.5. Deletion of miR–132/212$^{-/-}$ Prevents Enhanced Expression of CREB in the Hippocampi of Corticosterone Treated Slices

The participation of CREB in the regulation of synaptic plasticity and hippocampus-dependent learning and memory functions has been extensively described, and functional crosslinks between microRNA-mediated regulation (including the 132/212 cluster) and CREB expression have also been proposed [44,78–83]. Moreover, our group has recently described significantly enhanced levels of CREB in the hippocampus of miR–132/212$^{-/-}$ mice [61]. However, to the best of our knowledge, the acute effects of corticosterone on the levels of CREB has not been examined in hippocampi lacking miR-132/212. We therefore prepared acutely-dissociated hippocampal slices from both WT and miR–132/212$^{-/-}$ mice and subsequently stimulated them for 1 h with corticosterone. Two-way ANOVA (Alpha 0.05) reported a significant effect of the genotype (F (1, 12) = 6.504; p = 0.0255), high significance for the effects of the treatment ($F_{(1, 12)}$ = 45.16; $p < 0.0001$), and a significant interaction between factors ($F_{(1, 12)}$ = 80.05; $p < 0.0001$).

As shown in Figure 5A,C, further post hoc analyses revealed that untreated hippocampi derived from miR–132/212$^{-/-}$ mice presented enhanced expression levels of CREB compared to untreated WT controls, thus corroborating our previously reported observations [61]. However, both the levels of CREB (Figure 5B,C) were considerably reduced in the corticosterone-treated hippocampi of miR–132/212$^{-/-}$ mice when compared to both untreated KO miR–132/212$^{-/-}$ slices and corticosterone-treated hippocampi from the WT animals. Phospho-CREB was also downregulated in KO compared to WT slices when treated with corticosterone (Figure 5B,D), as shown by unpaired t-test ($p < 0.0001$; t = 8.834, df = 6).

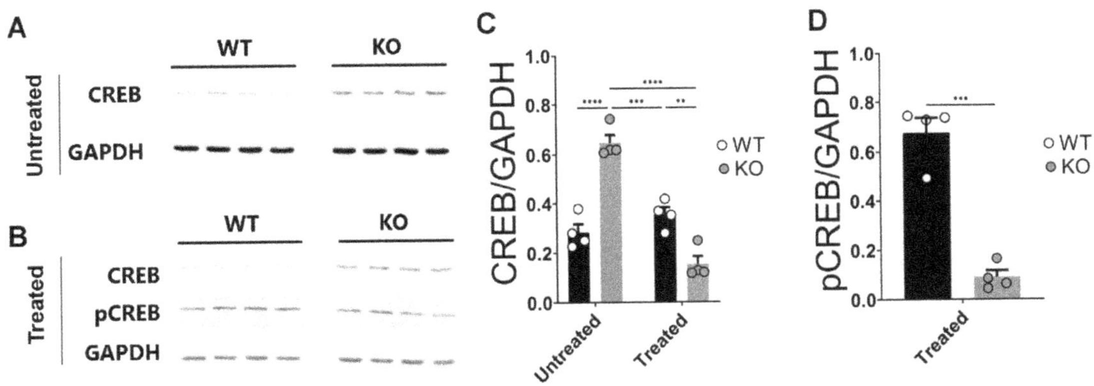

Figure 5. Corticosterone induces reduction in the levels of CREB in miR–132/212$^{-/-}$ mice hippocampus. (**A**,**B**) Representative photographs of blotting membranes containing transferred proteins obtained from untreated and corticosterone-treated hippocampal tissue from WT and miRNA-132/212$^{-/-}$ (KO) mice, respectively. The membranes were incubated with antibodies for CREB, pCREB and GAPDH. (**C**) The chart shows the averaged data for the levels of CREB normalized to those of the GAPDH as derived from densitometric analysis of the blots. The levels of CREB are significantly augmented in the hippocampus of the miRNA-132/212 KO mice and significantly reduced in response to corticosterone treatment. (**D**) Charts of the blots averaged densitometry data for the levels of pCREB relative to GAPDH, as examined using corticosterone-treated hippocampal tissue from WT and miRNA-132/212$^{-/-}$ mice. Note the significant reduction in the levels of the two forms of CREB in response to corticosterone. The results in all charts are shown in each case as a fold change relative to the detected levels of the enzyme GAPDH. ** $p < 0.01$, *** $p < 0.001$, **** $p < 0.0001$. Data are shown as mean ± SEM (n = 4 animals per group).

2.6. Sirt1 Protein Levels Are Enhanced and Insensitive to Corticosterone Stimulation in the miR–132/212$^{-/-}$ Mice Hippocampi

A functional link has been very recently established between the brain neuronal microRNAs miR-132 and miR-212 and their target-regulated protein Sirtuin 1 (Silent Information Regulator 1; in short, Sirt1) in the context of Alzheimer's Disease [84]. Sirt1 has indeed been shown to regulate endocrine activity and participates in the regulation of memory-related neuronal functions through its capability to induce axonal and dendritic morphological rearrangements [84,85], see also [86]. Moreover, corticosterone (the brain stress-signaling molecule in rodents equivalent to cortisol in humans) has also been shown to dose-dependently regulate the levels of Sirt1 in cellular stress models [87]. While a crosslink between Sirt1 and the miRNAs 132/-212 had been previously examined [88], the acute impact of corticosterone on the levels of hippocampal Sirt1 and its relation to the levels of the miRNAs 132 and 212 remained, to our knowledge, uncharacterized. We, therefore, next sought out to examine the acute effects of corticosterone on the levels of Sirt1 in the hippocampi of both WT and miR–132/212$^{-/-}$ mice (see also Materials and Methods). Two-way ANOVA (Alpha 0.05) reported statistically significant differences (***) for genotype ($F_{(1, 12)} = 31.52$; $p = 0.0001$); also, significance (*) for treatment ($F_{(1, 12)} = 5.293$; $p = 0.0401$) with no differences for interactions ($F_{(1, 12)} = 2.287$; $p = 0.1564$). Subsequent post hoc analysis showed that the hippocampi of miR–132/212$^{-/-}$ mice presented a significant enhancement in the levels of Sirt1, while the corticosterone treatment did not significantly revert this phenotype (Figure 6).

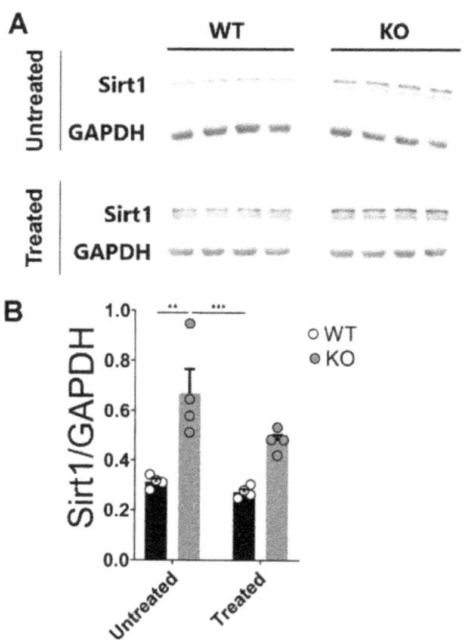

Figure 6. Corticosterone does not affect the otherwise enhanced levels of Sirt1 detected in the miR–132/212$^{-/-}$ mice hippocampus. (**A**) Representative photographs of blot membranes derived from WB studies using untreated (upper set) and corticosterone-treated (lower set) hippocampal tissue from WT and miRNA-132/212$^{-/-}$ (KO) mice. The membranes were incubated with antibodies for Sirt1 and GAPDH. (**B**) Averaged levels of Sirt1 under untreated and corticosterone-treated conditions, normalized to GAPDH values, as derived from densitometric analysis of blots. The levels of Sirt1 appeared significantly enhanced in the untreated hippocampi from miRNA-132/212$^{-/-}$ (KO) mice. The levels of hippocampal Sirt1 continued to be significantly enhanced upon corticosterone-treatment in miRNA-132/212$^{-/-}$ (KO) mice compared to those of the WT controls. Results in charts represent fold changes relative to GAPDH. ** $p < 0.01$, *** $p < 0.001$. Data are shown as mean ± SEM ($n = 4$ animals per group).

2.7. Corticosterone Effects on MSK1 Levels in WT and miR–132/212$^{-/-}$ Mice Hippocampi

The Mitogen- and Stress-activated Kinase 1 (MSK1) enzyme is a nuclear serine/ threonine protein kinase whose in vivo activity becomes triggered upon activation of either the Mitogen-Activated Protein Kinases (MAPK) ERK1/2 or p38, which participate in the structural/functional modifications of the histones during the regulation of gene transcription that is associated with emotional processing and hippocampus-dependent learning and memory functions [89–93]. Moreover, it has also been shown that MSK1 is associated with the regulation of transcription of miR-132/212 [56]. However, how glucocorticoid stimulation could impact the levels of MSK1 in a miRNA-132/212$^{-/-}$-dependent manner remained unknown. We therefore investigated the effects of acute corticosterone stimulation on the hippocampal levels of MSK1 in WT and miR–132/212$^{-/-}$ mice. For total MSK1 analyses, two-way ANOVA (Alpha 0.05) reported no differences for genotype ($F_{(1, 12)} = 0.1503$; $p = 0.7050$), significant (****) differences for the treatment ($F_{(1, 12)} = 63.25$; $p < 0.0001$) and significant (**) differences for the interaction ($F_{(1, 12)} = 9.618$; $p = 0.0092$). For pMSK1 analyses, two-way ANOVA (Alpha 0.05) reported no significant differences for genotype ($F_{(1, 12)} = 3.486$; $p = 0.0865$), significant (***) differences for the treatment ($F_{(1, 12)} = 30.38$; $p = 0.0001$) and no significant differences for the interaction ($F_{(1, 12)} = 2.473$; $p = 0.1418$). Further post hoc analysis revealed that the untreated hippocampi from miR–132/212$^{-/-}$ and WT showed comparable levels of MSK1 and phospho-MSK1 (pMSK1) (Figure 7A–D). Additionally, the corticosterone treatment resulted in an overall pronounced elevation in the detected levels of MSK1 in the hippocampi of both WT and miR–132/212$^{-/-}$ animals, indicating that corticosterone has the ability to modulate the levels of total MSK1 in a manner independent of the levels of miRNAs 132/212 (Figure 7C). However, an inhibitory effect of corticosterone was observed for the levels of p-MSK1 in WT hippocampi that was not present in the hippocampi from miR–132/212$^{-/-}$ mice (Figure 7D), indicating the possible role of pMSK1 in the corticosterone response that is missing in the KO mouse.

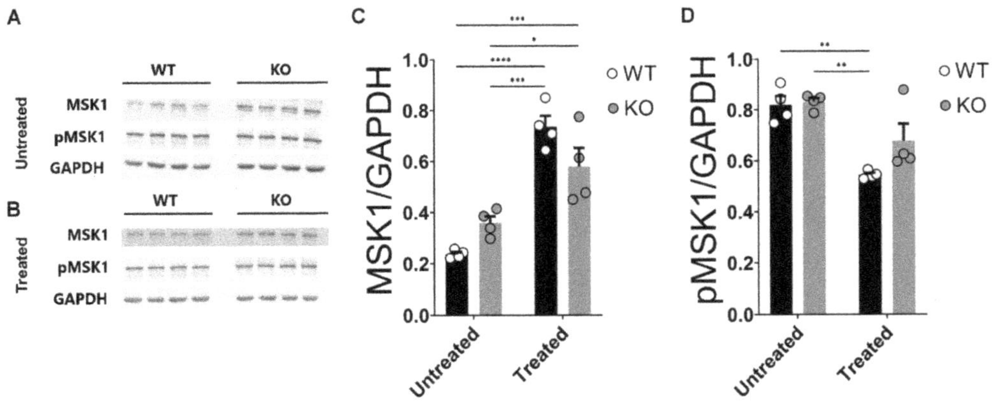

Figure 7. The levels of hippocampal MSK1 in WT and miR–132/212$^{-/-}$ are comparable and enhanced by corticosterone. (**A**,**B**) Representative blots for untreated and corticosterone-treated hippocampal tissue from WT and miRNA-132/212$^{-/-}$ (KO, in the figure) mice in membranes incubated with antibodies for MSK1, phospho-MSK1 (pMSK1) and GAPDH. (**C**) Averaged MSK1 levels, relative to GAPDH, as from densitometric analysis of blots. WT and miRNA-132/212$^{-/-}$ presented with enhanced MSK1 levels in the corticosterone-treated groups. (**D**) Averaged pMSK1 levels, relative to GAPDH, as from densitometric analysis of blots. While the levels of pMSK1 were comparable between miRNA-132/212$^{-/-}$ and WT hippocampi, corticosterone significantly reduced the levels of pMSK1 in the WT group. $p < 0.05$ was considered significant. * $p < 0.05$, ** $p < 0.01$, *** $p < 0.001$, **** $p < 0.0001$. Results represent fold changes relative to GAPDH. Data are shown as mean ± SEM (n = 4 animals per group).

2.8. Unaltered Levels of CDK5 and PTEN in the Hippocampi of miRNA-132/212$^{-/-}$ Mice

Both CDK5 and PTEN are brain neuronal proteins expressed in the hippocampus that are involved in the regulation of memory-related synaptic functions [94–103]. Additionally, both CDK5 and PTEN have been also associated with the stress-related effects of corticosterone and the hippocampal function (see also [104–106] and [94,107,108], respectively). Similarly, both CDK5 and PTEN have been shown to be regulated by microRNAs (see e.g., [109,110] and [111,112], respectively). However, the impact of corticosterone on the hippocampal levels of CDK5 and PTEN and its relation to miRNAs 132 and 212 remained, to our knowledge, uncharacterized. We therefore examined the impact of corticosterone on the levels of both CDK5 and PTEN in the hippocampus of WT and miRNA-132/212$^{-/-}$. Figure 8 shows data from WB experiments indicating that neither the deletion of the genes encoding for the miRNAs 132 and 212 nor the treatment with corticosterone affects the expression levels of CDK5 or PTEN in either WT or miRNA-132/212$^{-/-}$ mice hippocampi. For CDK5 analyses, two-way ANOVA (Alpha 0.05) reported no significant differences for genotype ($F_{(1, 12)} = 1.303$; $p = 0.2759$), interaction ($F_{(1, 12)} = 0.2628$; $p = 0.6175$) or treatment ($F_{(1, 12)} = 0.08639$; $p = 0.7738$). For PTEN analyses, two-way ANOVA (Alpha 0.05) reported no significant differences for genotype ($F_{(1, 12)} = 0.2911$; $p = 0.5994$), no differences for the treatment ($F_{(1, 12)} = 3.248$; $p = 0.0967$) and no differences for the interaction ($F_{(1, 12)} = 0.2734$; $p = 0.6106$).

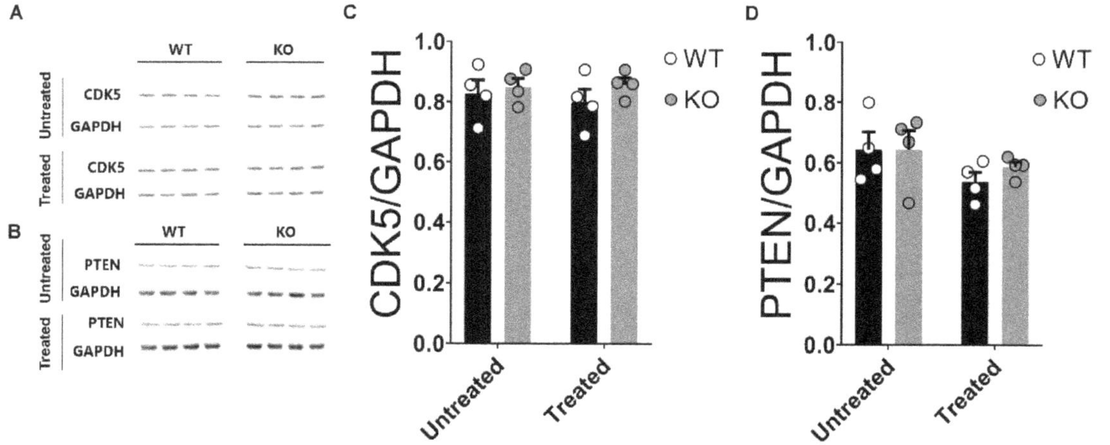

Figure 8. Corticosterone does not affect the otherwise comparable levels of CDK5 or PTEN in WT and miRNA-132/212$^{-/-}$ mice hippocampi. (**A**,**B**) Blots for untreated and corticosterone-treated (treated) hippocampal tissue from WT and miRNA-132/212$^{-/-}$ (KO) mice in membranes incubated with antibodies for CDK5, PTEN and GAPDH. (**C**,**D**) Averaged CDK5 and PTEN levels, respectively, relative to GAPDH, as from densitometric analysis of blots for untreated and corticosterone-treated tissue. No significant differences in the levels of CDK5 or PTEN were found between the untreated or corticosterone-treated hippocampi of WT and miRNA-132/212$^{-/-}$ mice. Results represent fold changes relative to GAPDH. Data are shown as mean ± SEM ($n = 4$ animals per group).

2.9. Reduced Anxiety-Like Behavior in miRNA-132/212$^{-/-}$ Mice

Our data from the pharmacological treatment with corticosterone in hippocampal electrophysiology, as well as the western blot analyses of changes in the levels of molecules involved in the physiological responses to stress described here in response to corticosterone treatment, suggested that the miRNA 132/212 cluster was implicated in the regulation of emotional behaviors. Consequently, changes in anxiety-like behaviors were predicted to be detectable in miRNA-132/212$^{-/-}$ mice. In order to test this hypothesis, we conducted behavioral experiments using the open field test (OFT) and the elevated plus maze [EPM

(Materials and Methods)]. As shown in Figure 9, no significant differences in the behavior of WT and miRNA-132/212$^{-/-}$ mice were observed in the OFT. In the OFT, unpaired two-tailed *t*-test yielded values of $p = 0.064$ (t = 1.949, df = 22) for total distance traveled comparisons (Figure 9A) with no between-group differences found in the cumulative duration in the center of the open field (Figure 9B). In the EPM, however, statistically significant differences in features associated to anxiety-like behaviours became apparent in miRNA-132/212$^{-/-}$ subjects when compared to WT mice. Unpaired two-tailed *t*-test yielded values of $p = 0.0094$ (t = 2.874, df = 20) for the absolute time spent in the open arm (Figure 9C), with no differences detected for the latency to the first access to the open arm ($p = 0.4491$; t = 0.7721, df = 20) in the EPM (Figure 9D). Moreover, when the time spent in the open arms was analysed relative to the time spent in the closed arms (here referred to as analysis factor score), miRNA-132/212$^{-/-}$ animals exhibited significantly (**) enhanced time spent in the open arms compared to their relative WT control counterparts ($p = 0.0094$; t = 2.872, df = 20) (Figure 9E). Taken together, all these observations encourage further research in order to independently verify the herein proposed potential involvement of the 132 and 212 miRNA cluster in the modulation of stress-related emotional responses mediated by steroid hormones belonging to the pituitary adrenocortical axis.

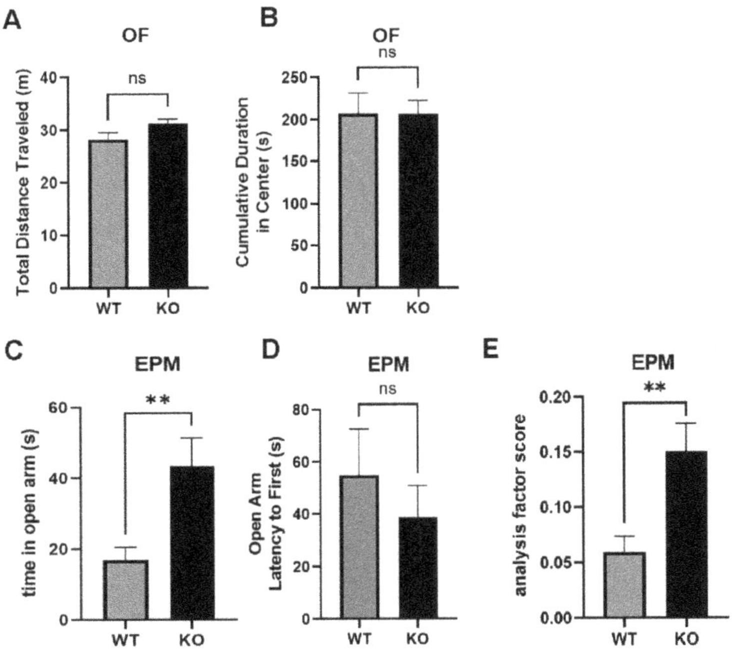

Figure 9. Performance of WT and miRNA-132/212$^{-/-}$ mice in the open field (OF) and EPM. In the OF, no differences in the behavior of the animals were apparent after examinations of the total distance travelled (**A**) and cumulative time duration that the animals spent in the center (**B**). An $n = 12$ animals per group was used for the OF studies. In the EPM, miRNA-132/212$^{-/-}$ (KO) mice presented significantly increased time spent in the open arms (**C**), whereas no differences were detectable in the latency to the first enter to the open arms (**D**) compared to their WT littermates used as controls. (**E**) Analysis of the time spent in the open arms relative to closed arms showed enhanced time in the open arms for miRNA-132/212$^{-/-}$ mice compared to WT controls. An $n = 10$–12 animals per group was used for EPM studies, ns = not significantly different. $p < 0.05$ was considered significant. ** $p < 0.01$. Data are expressed as mean ± SEM.

3. Discussion

Steroid-hormones, including cortisol, have been proposed as modulators of the hippocampal functions in the context of mood-related disorders in humans [113,114]. Corticosterone is a very potent steroid-hormone influencing several neuronal functions (including neuronal morphology and synaptic plasticity in rodents), and also affecting learning and memory and emotional behaviors in a manner analogue to that of human cortisol [16–19]. In humans, the physiological relevance of neurohormonal regulators becomes even more apparent for its relation to conditions such as stress, depression and anxiety; and for its crosslink with other maladies and psychiatric disorders, some of which have been widely studied in experimental animal models [6–14]. In the amygdala, a brain region critical for the regulation of learning, memory, and emotional behaviors [115,116], regulation by steroid-hormone-mediated signaling has been previously described [63,117–119].

The hippocampus had also been shown to play pivotal roles not only in spatial learning and memory, but also in the processing of emotion-related memory storage [23–29,120]. Circa 123 years ago, Ramon y Cajal had indeed provided some of the first anatomical, pictographic renditions highlighting some of the distinctive morphological differences between the dorsal and the ventral hippocampus (see also [121]). However, experimental/unequivocal demonstrations about the relevance of specific regions of the hippocampus as selected areas important for the regulation of emotional behaviors started to emerge only about 15-20 years ago [121–127], and, notwithstanding of this, little continue to be known about the molecular mechanisms determining the region-specific effects of steroid-hormones on the hippocampal function [30–33].

3.1. miRNAs 132/212 in the Brain Neuronal Function

Using a mouse model of learned safety, Ronovsky et. al., proposed that, in the amygdala, miR-132/212 regulate fear-inhibitory mechanisms as well as emotional-responses and plasticity-related synaptic functions [62]. Ronovsky et. al., also identified candidate proteins in the amygdala as potential gene targets for miRNA-132/212 which could participate in the miR-132/212-mediated regulation of mood-related behaviors [62]. We here expanded that line of research, by addressing the impact of miRNA-132/212 gene deletion on hippocampal synaptic functions and the effects to emotion-related steroid-hormone stimulation. Our data proposes the first functional description, to the best of our knowledge, for an involvement of miR-132/212 in the region-specific regulation of the effects of steroid-hormones on synaptic transmission and plasticity in the hippocampus.

Interestingly, other authors have found a long-term enhancement of corticosterone as well as impaired hippocampal synaptic plasticity in trauma-susceptible mice [128]. In line with these studies, and in agreement with our data, other reports have also shown a dysregulation of miRNAs in the hippocampus of trauma-susceptible mice [129]. Our behavioral examinations showing altered anxiety-like behavior in miRNA-132/212$^{-/-}$ animals provide further support to the possible involvement of the miRNAs 132/212 in mood-related behaviors, as they reveal the existence of basal differences in behavioral responses that are known to be influenced by neuroendocrine signaling. Discrepancies between data from different tests examining anxiety-like behaviors (e.g., using the open field) have precedents in the scientific literature. These differences can be due to factors such as the size of the experimental arenas (many different laboratories use open fields of different dimensions), differences in the different intensities of illumination used in different laboratories, or the use of partially reflective or even transparent materials in some arenas (see e.g., [130]). All these different subtle factors are known to influence the outcome in different experimental settings depending on different animal strains; depending on the type of pharmacological treatments; or depending on the genetic modifications introduced into the experimental subjects (see also [131–137]). Consequently, additional examinations (implementing other behavioral tests (e.g., Novelty suppressed feeding; Light/dark box, or Social Interaction Tests), are therefore encouraged.

3.2. Proteins Associated with Mood and Steroid Hormone Signaling

We here provide a basic biochemical analysis of proteins implicated in the regulation of steroid-hormone signaling, which propose the potential participation of the miRNAs 132/212 in the regulation of hippocampal steroid hormone signaling and emotion-related behaviors. Our results are in line with previous observations from our group and others [53,112], which pointed towards a possible involvement of miRNA-132/212 in the processing of emotion-related functions. We had previously described that miRNA-132/212 deletion influences the expression levels of alpha7-nAChR hippocampal receptors [43], which, in the basolateral amygdala, are proposed to mediate emotional behaviors [138,139]. We had also described that miRNA-132/212 deletion influenced the effects of nicotine (a modulator of mood and anxiety [140]) on hippocampal synaptic plasticity [61]. Previous reports had also associated the function of SIRT1 with miR-132- and miR-212-mediated regulation in the context of aging and Alzheimer's Disease [84]. Moreover, the levels of miR-132 and miR-212 have been shown to become primarily augmented in the hippocampus in response to short-term (5 h) but not to long-lasting (15 days) exposure to stress, and depletion of the miR-132 has been shown to result in enhanced anxiety-like behaviors [112]. In this last report ([112]), the authors used single miR-132 and combined miRNA-132/212 depletion and showed, in both cases, altered expression levels for hippocampal SIRT1 and PTEN-2; two proteins known as targets of miR-132 that are involved in the modulation of anxiety-like behaviors. Interestingly, while the pioneering results from Aten, et.al., 2019 [112] already showed a marked, yet not significant, trend towards enhanced levels of SIRT1 in miRNA-132/212$^{-/-}$ mice hippocampi, their results are perfectly in line with our observations here showing (with different methods) significantly enhanced levels of SIRT1 in the hippocampus of miRNA-132/212$^{-/-}$ mice. Taken together, data derived from the pioneering studies from the group of Dr. Obrietan [112], together with our data here describing enhanced levels of SIRT1 in the hippocampi of miRNA-132/212$^{-/-}$ mice (as compared to their related WT counterparts), and also the absence of significant effects of corticosterone in the two groups showed by us, indicate that whereas the presence of miRNAs 132/212 might negatively regulate the levels of SIRT1, this action is not affected by corticosterone treatment.

Additionally, previous reports have proposed a role for the proteins CDK5 and PTEN in the regulation of brain neuronal synaptic functions. For example, PTEN has been associated to the regulation of the cognitive function (including social behaviors) likely via its capability to regulate brain growth and to modulate neuronal circuit formation and synaptic functions, as examined in cortico-amygdala synapses [94]. Other groups have established PTEN is involved in the regulation of emotion-related fear memories as well as spatial memory, and that PTEN is also a critical regulator of hippocampal LTP via its functional association with the protein CaMKII [95] (see also [96]). CDK5, on the other hand, has been functionally involved in the regulation of hippocampal dendrite morphology [101], in the regulation of both hippocampal neurotransmitter release and amplitude of hippocampal field EPSP slope [102], and CDK5 has been also proposed as a possible molecular regulator of amyloid beta production and in the mechanisms associated to the pathogenesis of Alzheimer's disease [98,103].

Our collaborative groups had also described PTEN as a potential target for miRNAs 132 and 212 in the amygdala [62]. Moreover, previous reports have described that the enhancement in the levels of miR-132 can induce augmented expression of CDK5 [141]. However, our findings here show comparable levels of PTEN and of CDK5 in WT and miRNA-132/212$^{-/-}$ mice hippocampi without detectable effects in both cases of corticosterone. These observations thus suggest that miRNA-132/212$^{-/-}$ might not play major roles in the regulation of the levels of PTEN and CDK5 in the mouse hippocampus. Nevertheless, given that total hippocampal tissue was examined here, further studies need to be conducted examining, separately, whether differences in the levels of these proteins or their transcripts might still exist when dorsal vs. ventral hippocampal regions are examined and compared using higher resolution techniques.

Here, we also show that whereas MSK1 shows comparable levels in WT and miRNA-132/212$^{-/-}$ hippocampi (and in both cases, their levels become strikingly enhanced by corticosterone), the levels of phospho-MSK1, on the contrary, become significantly reduced only in WT hippocampi. Interestingly, and in line with our results here, previous observations have proposed a BDNF-related regulation in the levels of the miRNA-132/212 clusters via MSK activity [56], all together thus accumulating experimental evidence in support for the existence of a possible functional hippocampal crosslink between miRNA-132/212$^{-/-}$ and MSK1 likely associated to the regulation of the effects of steroid hormones and, possibly, associated emotional and behavioral effects.

Additionally, we observed significantly enhanced levels of hippocampal CREB in miRNA-132/212$^{-/-}$ mice, together with a significantly stronger reduction of CREB levels in response to the corticosterone treatment (an effect of corticosterone also observed for pCREB) compared to WT mice hippocampi. Future research addressing the levels of pCREB under untreated condition might provide a larger picture of the impact of the miRNA 132/212 gene deletion on the hippocampal response to corticosterone and how this, comparatively, affects the levels of CREB phosphorylation (see also Figure 10). Abundant literature has linked CREB to hippocampal function in the context of stress ([92,119,142–145]) and anxiety ([80,146–149]). The link between CREB and members of the miRNA-132/-212 gene cluster has also been described before (see for example [150,151]). Similarly, an involvement of miR-132/212 in stress/anxiety-related behaviors as well as in the regulation of the levels of Sirt1 and PTEN have been also recently described [112]. Our data here showed that miRNA-132/212$^{-/-}$ deletion significantly changes the effects of corticosterone on the levels of CREB, which is a critical regulator of memory [152], but not of other proteins involved in the responses to stress (e.g., PTEN).

Our work thus provides unprecedent data suggesting that miRNA-132/212 might participate, in vivo, in the regulation of synaptic plasticity and mood-related behaviors by fine-tuning the effects of steroid hormones via regulation of the levels of specific gene-product targets in an inter- and intra-brain-region selective manner (Figure 10). Further experiments are thus required in order to elucidate whether the levels of proteins that mediate in steroid hormone regulation are influenced by miRNA-132/212 in a subregion-specific manner. In the light of recent reports linking the hippocampus to learned safety [153,154], our findings also propose miRNA-132/212 as a potential modulator of learned safety through their capability to influence the levels of molecular targets responsive to neuroendocrine signals in both the hippocampus and in the amygdala (see also [62,155,156]). These observations encourage further verification in female subjects, as the existence of gender-specific effects of steroid-hormones on behavior, cognition, as well as on neuronal morphology and plasticity-related functions, have been established [157–165], and also brain microRNAs are regulated in a sex-dependent manner (see also [166–168]).

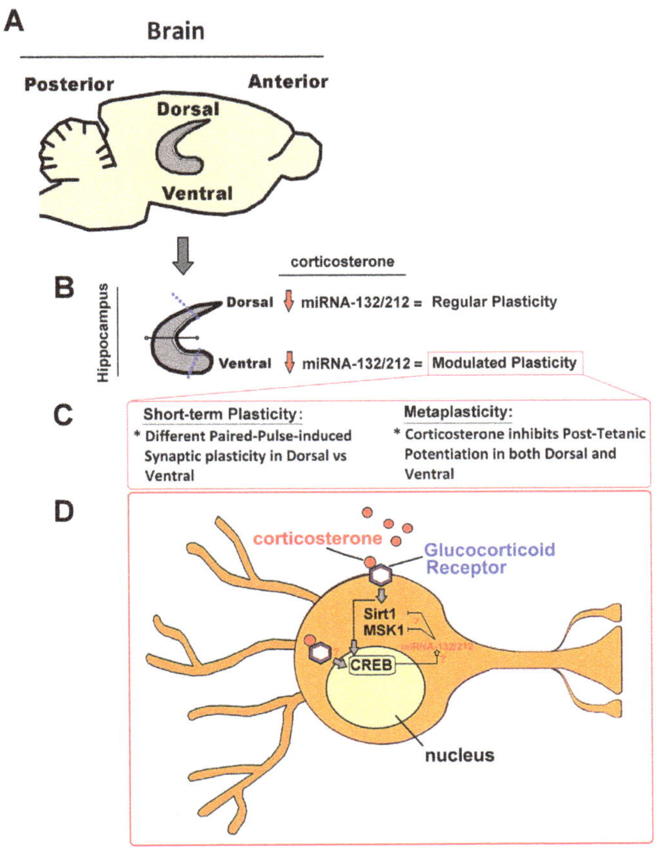

Figure 10. A possible role of the miRNA-132/212 family in the regulation of the effects of corticosterone on hippocampal synaptic plasticity. The hippocampus (**A**,**B**) is not a homogenous structure, as it presents several anatomical and functional differences between its dorsal and ventral regions. For example, the dorsal region has been shown to be primarily involved in spatial learning and memory functions, whereas the ventral region has been predominantly implicated in emotion-related processing. Here, we postulate that changes in the levels of miRNA-132/212 (e.g., its downregulation, as illustrated by red arrows in (**C**), could result in changes in the properties of both short-term and long-term forms of synaptic plasticity as well as in modulated metaplasticity (**C**). Alterations in the levels of miRNA-132/212 could also influence the synaptic responses to neuromodulators, such as corticosterone (**D**), which could, in a loop-like manner, impact the levels of miRNA-132/212 via activation of CREB, thus regulating the expression of proteins known to be downstream components of the glucocorticoid receptor (such as Sirt1 and MSK1). This work therefore proposes that miRNA-132/212 might contribute to increasing the in vivo mechanisms responsible for the functional heterogenicity between hippocampal regions, which might be critical to differentially distribute the effects of corticosterone on brain areas important for different cognitive functions.

4. Materials and Methods

4.1. Animals

All the experiments reported here were performed using male adult (8–10 weeks old) wild type (WT) C57Bl/6 (substrain N) mice as well as knockout (KO) miRNA-132/212 (miRNA-132/212$^{-/-}$) mice. These knockout mice, originally produced and described by the group of Dr. Pankratov [57], were engineered via the insertion of LoxP sites at non-coding regions (1st intron and exon 2) of the RNA gene encoding for the miRNAs 132 and 212, and generated in a C57Bl/6 background [57]. Here, as controls, only WT littermate animals (as verified by PCR-based genotyping), were used. All experiments followed ethic directives of the Bundesministerium für Wissenschaft und Forschung of Austria (BMWF-66.009/0200-WF/V/3b/2016). ARRIVE and U.K. Animals usage guidelines (Scientific Procedures Act, 1986 and associated guidelines, EU Directive 2010/63/EU for animal experiments) were implemented. Animals were kept in standard Thoren Plexiglas cages housed in a colony-room with a temperature of (22 ± 2) °C. Since animal grouping has been reported to reduce stress and aggression [169], in agreement with reports examining the importance of cage size for animal wellbeing [170], 3–5 mice were grouped together per Thoren Plexiglass cage (with cages of ≈ 22 × 31 × 16 cm). Cages were located in a Thoren mouse vent rack with maximizer (Thoren, Hazleton, PA, USA). The housing room had an automatically controlled illumination system programmed on a 12 h light/dark cycle (with light switched on at 6:00 a.m. to deliver (200 ± 20) lux). All cages were provided with aspen wood bedding (ABEDD-LAB & VET Service, Vienna, Austria), and every cage was equipped with 2 layers of fragrance-free TORK Advance Soft paper (Tork AT, Essity Austria GmbH, Storchengasse 1, Vienna (1150) Austria) which animals consistently shredded and used as nesting material. Both bedding and nesting materials were entirely changed once a week. All animals had both food and water available ad libitum. All the animals were handled by expert technical personnel, and an expert veterinarian supervised animal management. While in this work we are only including male animals, our group is also currently conducting parallel experiments using female experimental subjects in order to examine the possible influence of sex as a critical factor mediating the effects of both corticosterone treatment and miRNA-132/212 gene deletions on the biophysical properties of hippocampal circuits (see also Discussion).

4.2. Animal Grouping and Work Plan

Experimental animals were randomly assigned to one of the different experimental groups studied, which were organized as follows: 2 groups were used for untreated controls (untreated WT vs. untreated KO), and 2 groups were used for corticosterone-treated (corticosterone-treated WT vs. corticosterone-treated KO). For the electrophysiological studies (see below) and all the four groups referred, the experiments were organized by sub-dividing the recordings conducted in the hippocampus into recordings obtained from the dorsal and ventral regions. For the western blot analysis, the experimental groups studied were organized as follows: 2 groups were used as untreated controls (untreated WT vs. untreated KO), and 2 groups were used for corticosterone-treated (corticosterone-treated WT vs. corticosterone-treated KO). It must be noted that for the WB experiments, the entire hippocampus was used; that is, the hippocampal tissue was not sub-divided into dorsal or ventral regions (see also the sections Discussion and Limitations of this work). For behavioral studies, animals were separated into two major groups (WT vs. KO), and they were not subjected to any pharmacological treatments. The number of subjects used in each experiment has been noted in the main text and/or in Figures Legends.

4.3. Slice Electrophysiology

For the preparation of hippocampal slices, the animals were subjected to mild sedation with low CO_2 inhalation followed by quick cervical dislocation and swift decapitation using a sharp-blade guillotining (DCAP-M, World Precision Instruments, Inc., Sarasota, FL, USA). Brains were extracted and placed on an ice-cold artificial Cerebrospinal Fluid (aCSF) solution containing (in mM): 125 NaCl, 2.5 $CaCl_2$, 2.5 KCl, 1 $MgCl_2$, 20 $NaHCO_3$, 25 D-glucose, 1 NaH_2PO_4 (pH 7.4). Hippocampi were transversally sliced using a McIlwain tissue chopper (TC752, Campden Instruments Ltd., Loughborough, UK) into 400 μm thick sections. Slices were transferred into an aCSF-filled recovery chamber submerged in a bath filled with water at 30 °C, where they recovered for at least 1 h before electrophysiology measurements. All the aCSF solutions were continuously supplied with a carbogen gas mixture (95% medical O_2 + 5% medical CO_2). The slice electrophysiology measurements were conducted with slices placed in a submerged recording chamber that was continuously supplied with 3–4 mL/min of a pre-carbogenated aCSF that had been pre-warmed at (30 $\pm$ 2) °C. Field excitatory postsynaptic potentials (fEPSPs) were obtained through heat-pulled glass capillary pipettes (capillary glass from Harvard Apparatus, GmbH; Hugo Sachs Elektronik, Germany). Capillary tubes were pulled using a horizontal P-87 puller from Sutter Instrument (Model P-87, Novato, CA (94949), USA). Heat-pulled capillary pipettes were then back-filled with aCSF, yielding series resistances of (3 $\pm$ 1) MΩ. fEPSPs were registered from the hippocampal CA1 region, with recording electrodes positioned at the stratum–radiatum layer. fEPSPs were evoked by electrically stimulating the Schaffer's collateral projections that originated from the CA3 region, as described before [43,68]. To induce the fEPSPs, biphasic-square pulses of voltage were delivered through bipolar electrodes made of tungsten wire isolated to the tip with a Teflon coating layer (~50 µm diameter tips). The voltage pulses were generated from an ISO-STIM 01D stimulator (NPI Electronics, Tamm, Germany).

To examine the properties of basal synaptic transmission, input/output (I/O) curves were generated by plotting the raw amplitudes (and first-decaying phase slopes (normalized to maximum)) of the recorded output fEPSPs against the different values of delivered input voltages, consisting of increasing voltage steps of 200 µs, with pulses of 0–9 V delivered (in 1 V increments) with interpulse intervals of 15 s (see Figure 1).

In order to induce long-term potentiation (LTP), 5 separate sets of electrical stimulation were applied (500 ms apart), each comprising a total of 10 biphasic voltage pulses (100 µs/phase) delivered at 100 Hz (Figure 1; see also [66–68,171,172]). For LTP experiments, 10 min baseline field recordings were followed by the delivery of the LTP-inducing protocol; subsequently, an additional 40 min of field recordings were obtained, and a second LTP-inducing electrical stimulation protocol was delivered, followed by 30 min of field recordings. This protocol, thus, comprises an experimental way to induce hippocampal metaplasticity (see also [36,77]). Changes across time in the values of the slopes of the initial field decay (obtained by offline linear fittings of the recorded traces and normalized to baseline) were used as a measure of synaptic plasticity. LTP measurements were averaged from the values obtained from all slices within each animal, thus generating a single value per subject. For the electrophysiological recordings, the corticosterone (1 µM) treatment was conducted only for LTP experiments as follows: 10 min of baseline recordings > 1st high-frequency stimulation step > 10 min of recordings in ACSF solution > 30 min recordings in either ACSF solution (for untreated controls) or ACSF solution + 1 µM corticosterone applied in the bath (for the treatment groups) > 2nd high-frequency stimulation step and immediately start completely washing out the bath solution using only ACSF solution without corticosterone for the rest of the recordings. Recordings were obtained using an AxoClamp-2B amplifier, digitalized using the Digidata-1440 interface, and acquired and analyzed using the pClamp-10 (version 11.1) software (all from Axon Instruments, Molecular Devices, 660-665 Eskdale Rd, Winnersh, Triangle, Wokingham RG41 5TS, UK). Paired-pulse-induced plasticity examinations were also conducted by delivering voltages evoking ~50% of the maximum inducible field amplitude, as described before [43].

4.4. Western Blotting

In order to conduct the Western blot (WB) examinations, hippocampi from both WT and miR–132/212$^{-/-}$ mice (8–10 weeks old) were extracted and hippocampal slices prepared, allowed to recover for 1 h, and subsequently stimulated with 1 µM corticosterone or vehicle (Cat.Nr.27840, Sigma-Aldrich, Zimbagasse 5, Vienna (1140) Austria) for 1 h. Immediately after, the tissue was carefully transferred into Eppendorf® tubes and snap-frozen in liquid nitrogen. For WB preparations, the tissue was treated with a freshly prepared homogenizing protein lysis buffer with the following composition (in mM): 150 NaCl, 1 EDTA, 10 Tris-HCl, 10 NaF, 10 Na_3VO_4, 5 $Na_4P_7O_2$, 0.5% Triton $\times$100, 1% SDS and the protease inhibitor cocktail cOmplete™ (Roche Diagnostics, Mannheim, Germany). Samples of protein-containing tissue extracts were subsequently weight-separated by 10%

SDS-PAGE gel electrophoresis. Separated protein contents were then transferred from the gels into polyvinylidene fluoride (PVDF) membranes, which were then subjected to 1 h of blocking treatment (5% BSA in TBST) at room temperature. The membranes were then exposed to an overnight bath treatment at 4 °C with the respective primary antibodies. The next day, the membranes were washed out and exposed to a 1 h bath treatment (room temperature) with the corresponding secondary antibodies. Indirect detection of the hippocampal protein levels was determined by chemiluminescence-reporter labeling of secondary antibodies and fluorescent image acquisition using a FluorChem HD2 system (Alpha Innotech, San Leandro, CA, USA). Obtained images were examined using the open-source image analysis software ImageJ (Version 1.53t) [173]. Protein levels from detected bands were quantified by densitometrical analysis with data normalized to values of GAPDH levels. The antibodies used were: CREB (Cell Signaling Technology (Danvers, MA, USA); USA, Cat. Nr. 9197); p-CREB (Cell Signaling Technology; USA, Cat. Nr. 9196s); Sirt1(Cell Signaling Technology; USA, Cat. Nr. 2028); MSK1 (Cell Signaling Technology; USA, Cat. Nr. 3489); p-MSK1 (Cell Signaling Technology; USA, Cat. Nr. 9595); CDK5 (Cell Signaling Technology; USA, Cat. Nr. 2506); Pten (Abcam (Biomedical Campus, Discovery Dr, Trumpington, Cambridge CB2 0AX, United Kingdom); UK, Cat. Nr. ab154812); GAPDH (Thermo Fisher Scientific. 168 Third Avenue. Waltham, MA USA 02451); USA, Cat. Nr. MA5-15738).

4.5. Behavioral Testing

All the behavioral experiments described here were conducted at morning hours, throughout the light phase of the imposed light/dark cycle, in a noise-isolated room. Before the start of the experiments, all the animals went through a period of daily handling for 5 min, conducted by the experimenter and implemented to familiarize the animals with the experimenter and the handling, thus reducing stress of the animals derived from the manipulations required for the experiments. On the testing days and before the beginning of the experiments, the animals were allowed to rest for 1 h in their home cages so that they could become habituated to the testing room.

4.6. Open Field Test

The Open Field test (OFT) is commonly used in experiments used with rodents in order to monitor basic locomotor activity as well as some aspects of anxiety-related behaviors [174]. Experiments in the OFT were conducted following protocols previously described by our group [65–67,175,176]. In brief, mice were placed in the center of an open field arena, illuminated at ~300 lux, consisting of a four white-mate plastic-walled box (30 × 30 × 30 cm). Animals were allowed to freely occupy the arena for 10 min, and were then returned to their home cages. The OFT box was then thoroughly cleaned using 70% ethanol and allowed to dry for several minutes before reuse. In order to examine the animals' behavior in the open field, the cumulative time that animals spent in the center of the arena as well as the total distance traveled were considered (Figure 9; see also [177]). All the OFT routines were digitalized using a zenithal high-definition digital video camera. Spatial/Temporal behavioral factors were evaluated using the XT12 version of the Ethovision software package (Noldus, Wageningen, The Netherlands).

4.7. Elevated Plus Maze

Behavioral measurements using the Elevated Plus Maze (EPM) (see Figure 9) were conducted as previously described [65,178]. This test continues to be widely used for the examination of anxiety-like behaviors in experimental animal models, including mice and rats, and relies on the instinctive fear response that animals display to open, elevated zones. In the case of mice and rats with healthy/natural self-preservation instincts, they will initially have the generalized tendence to avoid entering to open/elevated potentially dangerous areas, and would remain in or move towards enclosed zone in search for safety [179]. The experiments were conducted using a customized plus-shaped platform made of white-mate Plexiglas (elevated 50 cm above ground) having two of its opposing arms enclosed by 40 cm high walls. Time in the open arms as well as open arm latency to first were examined in the EPM. All arms had 50 cm in length and were 10 cm wide. The open arms were exposed to an illumination of ~110 lux, whereas ~15 lux illuminated the enclosed arms. The testing sessions (5 min each) begin immediately after having placed the animal at the central square area, with animals positioned facing towards one of the open arms, and animals were allowed to freely move across the maze settings. The EPM behavioral performance was digitalized using a zenithal high-definition digital video camera. Spatial/Temporal behavioral factors were evaluated using the XT12 version of the Ethovision software package (Noldus, Wageningen, The Netherlands). Behavioral parameters in the EPM were examined following protocols previously described [180].

4.8. Statistical Analysis

All data analyses were conducted using the GraphPad-Prism-9 software package (version 9.5.1(733), GraphPad Software, 225 Franklin Street. Fl. 26, Boston, MA 02110, USA). In order to comply with the 3Rs regulation and reduce as much as possible the number of animals, for the electrophysiological and behavioral analyses, statistical calculations were based on power analysis to estimate the minimum sample size required, using the G*Power software (version 3.1.9.7). For biochemical studies, since the normalized data obtained from examinations using western blot are relative/semi-quantitative, we used a small number of animals in agreement with related published studies by other groups [181]. Unpaired two-tailed *t*-tests (with confidence level set to 95%) was used to examine differences of means between groups in behavioral experiments. Two-way ANOVA (Alpha 0.05) with Tukey's multiple comparisons test was used to examine data from WB experiments. Three- and two-way repeated measures (RM)-ANOVA with Tukey multiple comparisons, and/or Bonferroni's and Geisser–Greenhouse's corrections (and alpha set to 0.05), were used for the data derived from the electrophysiological analyses as in each case indicated in the main text. For bias control, data were examined by researchers who were blind to the experimental groups.

Author Contributions: S.K. conducted the electrophysiological recordings, analyzed data and contributed to the preparations of the final figures. M.C. conducted biochemical experiments and handled animals with assistance from M.-A.M. and M.C.M.-P. and E.C.-O. led behavioral experiments. T.S. and A.C. supervised data analyses of biochemical studies. E.A.M.A. helped with slice preparations. X.K., E.U. and G.L. provided crucial scientific advice. F.J.M. conceived, directed and funded the project and supervised the work of S.K., M.C.M.-P. and E.A.M.A. contributed to data analyses and wrote the manuscript. All authors have read and agreed to the published version of the manuscript.

Funding: F.J.M., S.K. and E.A.M.A. received support from the Austrian Science Fund/FWF (Project Nr. P_31004 to F.J.M.).

Institutional Review Board Statement: Animal studies followed directives from the Bundesministerium für Wissenschaft und Forschung of Austria (BMWF-66.009/0200-WF/V/3b/2016).

Informed Consent Statement: Not applicable.

Data Availability Statement: All the data derived from this work are included in the article.

Conflicts of Interest: Authors are not aware of any competing financial interests or other associations influencing this work.

Abbreviations

aCSF = artificial cerebrospinal fluid; EMP = elevated plus maze; fEPSPs = field excitatory post-synaptic potentials; HFS = high frequency stimulation; HPA = hypothalamic–pituitary–adrenal; KO = knockout; LTP = long-term potentiation; PTP = post-tetanic potentiation; I/O = input/output; miRNA = microRNA; OFT = open field test; WB = western blot; WT = wild type.

References

1. Thau, L.; Gandhi, J.; Sharma, S. Physiology, Cortisol. In *StatPearls*; StatPearls Publishing: Treasure Island, FL, USA, 2023.
2. Oakley, R.H.; Cidlowski, J.A. The biology of the glucocorticoid receptor: New signaling mechanisms in health and disease. *J. Allergy Clin. Immunol.* **2013**, *132*, 1033–1044. [CrossRef]
3. Payne, J.D.; Jackson, E.D.; Hoscheidt, S.; Ryan, L.; Jacobs, W.J.; Nadel, L. Stress administered prior to encoding impairs neutral but enhances emotional long-term episodic memories. *Learn. Mem. (Cold Spring Harb. N. Y.)* **2007**, *14*, 861–868. [CrossRef]
4. Roozendaal, B.; Okuda, S.; Van der Zee, E.A.; McGaugh, J.L. Glucocorticoid enhancement of memory requires arousal-induced noradrenergic activation in the basolateral amygdala. *Proc. Natl. Acad. Sci. USA* **2006**, *103*, 6741–6746. [CrossRef] [PubMed]
5. Lupien, S.J.; Maheu, F.; Tu, M.; Fiocco, A.; Schramek, T.E. The effects of stress and stress hormones on human cognition: Implications for the field of brain and cognition. *Brain Cogn.* **2007**, *65*, 209–237. [CrossRef]
6. Hayley, S.; Poulter, M.O.; Merali, Z.; Anisman, H. The pathogenesis of clinical depression: Stressor- and cytokine-induced alterations of neuroplasticity. *Neuroscience* **2005**, *135*, 659–678. [CrossRef] [PubMed]
7. Pariante, C.M.; Lightman, S.L. The HPA axis in major depression: Classical theories and new developments. *Trends Neurosci.* **2008**, *31*, 464–468. [CrossRef]
8. Silverman, M.N.; Sternberg, E.M. Glucocorticoid regulation of inflammation and its functional correlates: From HPA axis to glucocorticoid receptor dysfunction. *Ann. N. Y. Acad. Sci.* **2012**, *1261*, 55–63. [CrossRef]
9. Munck, A.; Guyre, P.M.; Holbrook, N.J. Physiological functions of glucocorticoids in stress and their relation to pharmacological actions. *Endocr. Rev.* **1984**, *5*, 25–44. [CrossRef]
10. Aguilera, G. Regulation of pituitary ACTH secretion during chronic stress. *Front. Neuroendocr.* **1994**, *15*, 321–350. [CrossRef]
11. Varghese, F.P.; Brown, E.S. The Hypothalamic-Pituitary-Adrenal Axis in Major Depressive Disorder: A Brief Primer for Primary Care Physicians. *Prim. Care Companion J. Clin. Psychiatry* **2001**, *3*, 151–155. [CrossRef]
12. Owens, M.J.; Nemeroff, C.B. The role of corticotropin-releasing factor in the pathophysiology of affective and anxiety disorders: Laboratory and clinical studies. *Ciba Found. Symp.* **1993**, *172*, 296–308. [CrossRef] [PubMed]
13. Young, E.A.; Haskett, R.F.; Murphy-Weinberg, V.; Watson, S.J.; Akil, H. Loss of glucocorticoid fast feedback in depression. *Arch. Gen. Psychiatry* **1991**, *48*, 693–699. [CrossRef] [PubMed]
14. Lipov, E.; Kelzenberg, B.; Rothfeld, C.; Abdi, S. Modulation of NGF by cortisol and the Stellate Ganglion Block—Is this the missing link between memory consolidation and PTSD? *Med. Hypotheses* **2012**, *79*, 750–753. [CrossRef] [PubMed]
15. Sagmeister, M.S.; Harper, L.; Hardy, R.S. Cortisol excess in chronic kidney disease—A review of changes and impact on mortality. *Front. Endocrinol.* **2022**, *13*, 1075809. [CrossRef] [PubMed]
16. Brown, E.S.; Rush, A.J.; McEwen, B.S. Hippocampal remodeling and damage by corticosteroids: Implications for mood disorders. *Neuropsychopharmacology* **1999**, *21*, 474–484. [CrossRef]
17. Lupien, S.J.; McEwen, B.S. The acute effects of corticosteroids on cognition: Integration of animal and human model studies. *Brain Res.* **1997**, *24*, 1–27. [CrossRef]
18. Charmandari, E.; Tsigos, C.; Chrousos, G. Endocrinology of the stress response. *Annu. Rev. Physiol.* **2005**, *67*, 259–284. [CrossRef]
19. Sapolsky, R.M.; Romero, L.M.; Munck, A.U. How do glucocorticoids influence stress responses? Integrating permissive, suppressive, stimulatory, and preparative actions. *Endocr. Rev.* **2000**, *21*, 55–89. [CrossRef]
20. Brinks, V.; van der Mark, M.; de Kloet, R.; Oitzl, M. Emotion and cognition in high and low stress sensitive mouse strains: A combined neuroendocrine and behavioral study in BALB/c and C57BL/6J mice. *Front. Behav. Neurosci.* **2007**, *1*, 8. [CrossRef]
21. Eadie, B.D.; Zhang, W.N.; Boehme, F.; Gil-Mohapel, J.; Kainer, L.; Simpson, J.M.; Christie, B.R. Fmr1 knockout mice show reduced anxiety and alterations in neurogenesis that are specific to the ventral dentate gyrus. *Neurobiol. Dis.* **2009**, *36*, 361–373. [CrossRef]
22. Zhao, S.; Xu, X.; Xie, G.; Zhang, T. Chronic corticosterone exposure impairs emotional regulation and cognitive function through disturbing neural oscillations in mice. *Behav. Brain Res.* **2022**, *434*, 114030. [CrossRef]
23. Scoville, W.B.; Milner, B. Loss of recent memory after bilateral hippocampal lesions. *J. Neurol. Neurosurg. Psychiatry* **1957**, *20*, 11–21. [CrossRef] [PubMed]
24. Scoville, W.B.; Milner, B. Loss of recent memory after bilateral hippocampal lesions. *J. Neuropsychiatry Clin. Neurosci.* **2000**, *12*, 103–113. [CrossRef] [PubMed]
25. Milner, B. The medial temporal-lobe amnesic syndrome. *Psychiatr. Clin. N. Am.* **2005**, *28*, 599–611, 609. [CrossRef]
26. Milner, B.; Klein, D. Loss of recent memory after bilateral hippocampal lesions: Memory and memories-looking back and looking forward. *J. Neurol. Neurosurg. Psychiatry* **2016**, *87*, 230. [CrossRef] [PubMed]
27. Izquierdo, I.; Furini, C.R.; Myskiw, J.C. Fear Memory. *Physiol. Rev.* **2016**, *96*, 695–750. [CrossRef]
28. Patel, P.D.; Katz, M.; Karssen, A.M.; Lyons, D.M. Stress-induced changes in corticosteroid receptor expression in primate hippocampus and prefrontal cortex. *Psychoneuroendocrinology* **2008**, *33*, 360–367. [CrossRef]
29. Sapolsky, R.M.; Krey, L.C.; McEwen, B.S. Glucocorticoid-sensitive hippocampal neurons are involved in terminating the adrenocortical stress response. *Proc. Natl. Acad. Sci. USA* **1984**, *81*, 6174–6177. [CrossRef]
30. Dahmen, B.; Puetz, V.B.; Scharke, W.; von Polier, G.G.; Herpertz-Dahlmann, B.; Konrad, K. Effects of Early-Life Adversity on Hippocampal Structures and Associated HPA Axis Functions. *Dev. Neurosci.* **2018**, *40*, 13–22. [CrossRef]
31. Dai, S.; Mo, Y.; Wang, Y.; Xiang, B.; Liao, Q.; Zhou, M.; Li, X.; Li, Y.; Xiong, W.; Li, G.; et al. Chronic Stress Promotes Cancer Development. *Front. Oncol.* **2020**, *10*, 1492. [CrossRef]

32. Clow, A.; Thorn, L.; Evans, P.; Hucklebridge, F. The awakening cortisol response: Methodological issues and significance. *Stress* **2004**, *7*, 29–37. [CrossRef] [PubMed]
33. Frodl, T.; O'Keane, V. How does the brain deal with cumulative stress? A review with focus on developmental stress, HPA axis function and hippocampal structure in humans. *Neurobiol. Dis.* **2013**, *52*, 24–37. [CrossRef] [PubMed]
34. Alfarez, D.N.; Wiegert, O.; Joels, M.; Krugers, H.J. Corticosterone and stress reduce synaptic potentiation in mouse hippocampal slices with mild stimulation. *Neuroscience* **2002**, *115*, 1119–1126. [CrossRef] [PubMed]
35. Rey, M.; Carlier, E.; Talmi, M.; Soumireu-Mourat, B. Corticosterone effects on long-term potentiation in mouse hippocampal slices. *Neuroendocrinology* **1994**, *60*, 36–41. [CrossRef]
36. Abraham, W.C.; Bear, M.F. Metaplasticity: The plasticity of synaptic plasticity. *Trends Neurosci.* **1996**, *19*, 126–130. [CrossRef]
37. Caliskan, G.; Stork, O. Hippocampal network oscillations as mediators of behavioural metaplasticity: Insights from emotional learning. *Neurobiol. Learn. Mem.* **2018**, *154*, 37–53. [CrossRef]
38. Holland, L.L.; Wagner, J.J. Primed facilitation of homosynaptic long-term depression and depotentiation in rat hippocampus. *J. Neurosci.* **1998**, *18*, 887–894. [CrossRef]
39. Ambros, V. microRNAs: Tiny regulators with great potential. *Cell.* **2001**, *107*, 823–826. [CrossRef]
40. Bartel, D.P. MicroRNAs: Genomics, biogenesis, mechanism, and function. *Cell.* **2004**, *116*, 281–297. [CrossRef]
41. Edbauer, D.; Neilson, J.R.; Foster, K.A.; Wang, C.F.; Seeburg, D.P.; Batterton, M.N.; Tada, T.; Dolan, B.M.; Sharp, P.A.; Sheng, M. Regulation of synaptic structure and function by FMRP-associated microRNAs miR-125b and miR-132. *Neuron* **2010**, *65*, 373–384. [CrossRef]
42. Oliver, R.J.; Mandyam, C.D. Regulation of Adult Neurogenesis by Non-coding RNAs: Implications for Substance Use Disorders. *Front. Neurosci.* **2018**, *12*, 849. [CrossRef] [PubMed]
43. Stojanovic, T.; Benes, H.; Awad, A.; Bormann, D.; Monje, F.J. Nicotine abolishes memory-related synaptic strengthening and promotes synaptic depression in the neurogenic dentate gyrus of miR-132/212 knockout mice. *Addict. Biol.* **2020**, e12905. [CrossRef]
44. Baby, N.; Alagappan, N.; Dheen, S.T.; Sajikumar, S. MicroRNA-134-5p inhibition rescues long-term plasticity and synaptic tagging/capture in an Abeta(1-42)-induced model of Alzheimer's disease. *Aging Cell.* **2020**, *19*, e13046. [CrossRef] [PubMed]
45. Berentsen, B.; Patil, S.; Ronnestad, K.; Goff, K.M.; Pajak, M.; Simpson, T.I.; Wibrand, K.; Bramham, C.R. MicroRNA-34a Acutely Regulates Synaptic Efficacy in the Adult Dentate Gyrus In Vivo. *Mol. Neurobiol.* **2020**, *57*, 1432–1445. [CrossRef] [PubMed]
46. Zhang, H.P.; Liu, X.L.; Chen, J.J.; Cheng, K.; Bai, S.J.; Zheng, P.; Zhou, C.J.; Wang, W.; Wang, H.Y.; Zhong, L.M.; et al. Circulating microRNA 134 sheds light on the diagnosis of major depressive disorder. *Transl. Psychiatry* **2020**, *10*, 95. [CrossRef]
47. Liu, D.Y.; Zhang, L. MicroRNA-132 promotes neurons cell apoptosis and activates Tau phosphorylation by targeting GTDC-1 in Alzheimer's disease. *Eur. Rev. Med. Pharm. Sci.* **2019**, *23*, 8523–8532. [CrossRef]
48. Kumar, S.; Reddy, P.H. A New Discovery of MicroRNA-455-3p in Alzheimer's Disease. *J. Alzheimers Dis.* **2019**, *72*, S117–S130. [CrossRef]
49. Herrera-Espejo, S.; Santos-Zorrozua, B.; Alvarez-Gonzalez, P.; Lopez-Lopez, E.; Garcia-Orad, A. A Systematic Review of MicroRNA Expression as Biomarker of Late-Onset Alzheimer's Disease. *Mol. Neurobiol.* **2019**, *56*, 8376–8391. [CrossRef]
50. Salta, E.; De Strooper, B. microRNA-132: A key noncoding RNA operating in the cellular phase of Alzheimer's disease. *FASEB J.* **2017**, *31*, 424–433. [CrossRef]
51. Hernandez-Rapp, J.; Rainone, S.; Goupil, C.; Dorval, V.; Smith, P.Y.; Saint-Pierre, M.; Vallee, M.; Planel, E.; Droit, A.; Calon, F.; et al. microRNA-132/212 deficiency enhances Abeta production and senile plaque deposition in Alzheimer's disease triple transgenic mice. *Sci. Rep.* **2016**, *6*, 30953. [CrossRef]
52. Salta, E.; Sierksma, A.; Vanden Eynden, E.; De Strooper, B. miR-132 loss de-represses ITPKB and aggravates amyloid and TAU pathology in Alzheimer's brain. *EMBO Mol. Med.* **2016**, *8*, 1005–1018. [CrossRef] [PubMed]
53. Aten, S.; Hansen, K.F.; Hoyt, K.R.; Obrietan, K. The miR-132/212 locus: A complex regulator of neuronal plasticity, gene expression and cognition. *RNA Dis.* **2016**, *3*.
54. Hansen, K.F.; Sakamoto, K.; Aten, S.; Snider, K.H.; Loeser, J.; Hesse, A.M.; Page, C.E.; Pelz, C.; Arthur, J.S.; Impey, S.; et al. Targeted deletion of miR-132/-212 impairs memory and alters the hippocampal transcriptome. *Learn. Mem.* **2016**, *23*, 61–71. [CrossRef] [PubMed]
55. Mendoza-Viveros, L.; Chiang, C.K.; Ong, J.L.K.; Hegazi, S.; Cheng, A.H.; Bouchard-Cannon, P.; Fana, M.; Lowden, C.; Zhang, P.; Bothorel, B.; et al. miR-132/212 Modulates Seasonal Adaptation and Dendritic Morphology of the Central Circadian Clock. *Cell. Rep.* **2017**, *19*, 505–520. [CrossRef]
56. Remenyi, J.; Hunter, C.J.; Cole, C.; Ando, H.; Impey, S.; Monk, C.E.; Martin, K.J.; Barton, G.J.; Hutvagner, G.; Arthur, J.S. Regulation of the miR-212/132 locus by MSK1 and CREB in response to neurotrophins. *Biochem. J.* **2010**, *428*, 281–291. [CrossRef]
57. Remenyi, J.; van den Bosch, M.W.; Palygin, O.; Mistry, R.B.; McKenzie, C.; Macdonald, A.; Hutvagner, G.; Arthur, J.S.; Frenguelli, B.G.; Pankratov, Y. miR-132/212 knockout mice reveal roles for these miRNAs in regulating cortical synaptic transmission and plasticity. *PLoS ONE* **2013**, *8*, e62509. [CrossRef]
58. Wanet, A.; Tacheny, A.; Arnould, T.; Renard, P. miR-212/132 expression and functions: Within and beyond the neuronal compartment. *Nucleic Acids Res.* **2012**, *40*, 4742–4753. [CrossRef]
59. Tognini, P.; Pizzorusso, T. MicroRNA212/132 family: Molecular transducer of neuronal function and plasticity. *Int. J. Biochem. Cell. Biol.* **2012**, *44*, 6–10. [CrossRef]

60. Bormann, D.; Stojanovic, T.; Cicvaric, A.; Schuld, G.J.; Cabatic, M.; Ankersmit, H.J.; Monje, F.J. miRNA-132/212 Gene-Deletion Aggravates the Effect of Oxygen-Glucose Deprivation on Synaptic Functions in the Female Mouse Hippocampus. *Cells* **2021**, *10*, 1709. [CrossRef]
61. Stojanovic, T.; Velarde Gamez, D.; Schuld, G.J.; Bormann, D.; Cabatic, M.; Uhrin, P.; Lubec, G.; Monje, F.J. Age-Dependent and Pathway-Specific Bimodal Action of Nicotine on Synaptic Plasticity in the Hippocampus of Mice Lacking the miR-132/212 Genes. *Cells* **2022**, *11*, 261. [CrossRef]
62. Ronovsky, M.; Zambon, A.; Cicvaric, A.; Boehm, V.; Hoesel, B.; Moser, B.A.; Yang, J.; Schmid, J.A.; Haubensak, W.E.; Monje, F.J.; et al. A role for miR-132 in learned safety. *Sci. Rep.* **2019**, *9*, 528. [CrossRef] [PubMed]
63. Pu, Z.; Krugers, H.J.; Joels, M. Beta-adrenergic facilitation of synaptic plasticity in the rat basolateral amygdala in vitro is gradually reversed by corticosterone. *Learn. Mem. (Cold Spring Harb. N. Y.)* **2009**, *16*, 155–160. [CrossRef] [PubMed]
64. Kouhnavardi, S.; Ecevitoglu, A.; Dragacevic, V.; Sanna, F.; Arias-Sandoval, E.; Kalaba, P.; Kirchhofer, M.; Lubec, J.; Niello, M.; Holy, M.; et al. A Novel and Selective Dopamine Transporter Inhibitor, (S)-MK-26, Promotes Hippocampal Synaptic Plasticity and Restores Effort-Related Motivational Dysfunctions. *Biomolecules* **2022**, *12*, 881. [CrossRef] [PubMed]
65. Cicvaric, A.; Sachernegg, H.M.; Stojanovic, T.; Symmank, D.; Smani, T.; Moeslinger, T.; Uhrin, P.; Monje, F.J. Podoplanin Gene Disruption in Mice Promotes in vivo Neural Progenitor Cells Proliferation, Selectively Impairs Dentate Gyrus Synaptic Depression and Induces Anxiety-Like Behaviors. *Front. Cell. Neurosci.* **2019**, *13*, 561. [CrossRef] [PubMed]
66. Cicvaric, A.; Yang, J.; Bulat, T.; Zambon, A.; Dominguez-Rodriguez, M.; Kuhn, R.; Sadowicz, M.G.; Siwert, A.; Egea, J.; Pollak, D.D.; et al. Enhanced synaptic plasticity and spatial memory in female but not male FLRT2-haplodeficient mice. *Sci. Rep.* **2018**, *8*, 3703. [CrossRef] [PubMed]
67. Cicvaric, A.; Bulat, T.; Bormann, D.; Yang, J.; Auer, B.; Milenkovic, I.; Cabatic, M.; Milicevic, R.; Monje, F.J. Sustained consumption of cocoa-based dark chocolate enhances seizure-like events in the mouse hippocampus. *Food Funct.* **2018**, *9*, 1532–1544. [CrossRef]
68. Cicvaric, A.; Yang, J.; Krieger, S.; Khan, D.; Kim, E.J.; Dominguez-Rodriguez, M.; Cabatic, M.; Molz, B.; Acevedo Aguilar, J.P.; Milicevic, R.; et al. The brain-tumor related protein podoplanin regulates synaptic plasticity and hippocampus-dependent learning and memory. *Ann. Med.* **2016**, 1–17. [CrossRef]
69. Kim, E.J.; Monje, F.J.; Li, L.; Hoger, H.; Pollak, D.D.; Lubec, G. Alzheimer's disease risk factor lymphocyte-specific protein tyrosine kinase regulates long-term synaptic strengthening, spatial learning and memory. *Cell. Mol. Life Sci.* **2013**, *70*, 743–759. [CrossRef]
70. Monje, F.J.; Kim, E.J.; Pollak, D.D.; Cabatic, M.; Li, L.; Baston, A.; Lubec, G. Focal adhesion kinase regulates neuronal growth, synaptic plasticity and hippocampus-dependent spatial learning and memory. *Neuro-Signals* **2012**, *20*, 1–14. [CrossRef]
71. Maggio, N.; Segal, M. Striking variations in corticosteroid modulation of long-term potentiation along the septotemporal axis of the hippocampus. *J. Neurosci.* **2007**, *27*, 5757–5765. [CrossRef]
72. Maggio, N.; Segal, M. Unique regulation of long term potentiation in the rat ventral hippocampus. *Hippocampus* **2007**, *17*, 10–25. [CrossRef] [PubMed]
73. Nagy, V.; Hollstein, R.; Pai, T.P.; Herde, M.K.; Buphamalai, P.; Moeseneder, P.; Lenartowicz, E.; Kavirayani, A.; Korenke, G.C.; Kozieradzki, I.; et al. HACE1 deficiency leads to structural and functional neurodevelopmental defects. *Neurol. Genet.* **2019**, *5*, e330. [CrossRef] [PubMed]
74. Fell, C.W.; Hagelkruys, A.; Cicvaric, A.; Horrer, M.; Liu, L.; Li, J.S.S.; Stadlmann, J.; Polyansky, A.A.; Mereiter, S.; Tejada, M.A.; et al. FIBCD1 is an endocytic GAG receptor associated with a novel neurodevelopmental disorder. *EMBO Mol. Med.* **2022**, *14*, e15829. [CrossRef] [PubMed]
75. Nicoll, R.A.; Malenka, R.C. Expression mechanisms underlying NMDA receptor-dependent long-term potentiation. *Ann. N. Y. Acad. Sci.* **1999**, *868*, 515–525. [CrossRef]
76. Schulz, P.E.; Cook, E.P.; Johnston, D. Changes in paired-pulse facilitation suggest presynaptic involvement in long-term potentiation. *J. Neurosci.* **1994**, *14*, 5325–5337. [CrossRef]
77. Bortolotto, Z.A.; Collingridge, G.L. A role for protein kinase C in a form of metaplasticity that regulates the induction of long-term potentiation at CA1 synapses of the adult rat hippocampus. *Eur. J. Neurosci.* **2000**, *12*, 4055–4062. [CrossRef] [PubMed]
78. Nestler, E.J. Common molecular and cellular substrates of addiction and memory. *Neurobiol. Learn. Mem.* **2002**, *78*, 637–647. [CrossRef] [PubMed]
79. Hansen, K.F.; Karelina, K.; Sakamoto, K.; Wayman, G.A.; Impey, S.; Obrietan, K. miRNA-132: A dynamic regulator of cognitive capacity. *Brain Struct. Funct.* **2013**, *218*, 817–831. [CrossRef]
80. Fisher, M.L.; LeMalefant, R.M.; Zhou, L.; Huang, G.; Turner, J.R. Distinct Roles of CREB Within the Ventral and Dorsal Hippocampus in Mediating Nicotine Withdrawal Phenotypes. *Neuropsychopharmacology* **2017**, *42*, 1599–1609. [CrossRef]
81. Barco, A.; Patterson, S.; Alarcon, J.M.; Gromova, P.; Mata-Roig, M.; Morozov, A.; Kandel, E.R. Gene expression profiling of facilitated L-LTP in VP16-CREB mice reveals that BDNF is critical for the maintenance of LTP and its synaptic capture. *Neuron* **2005**, *48*, 123–137. [CrossRef]
82. Magill, S.T.; Cambronne, X.A.; Luikart, B.W.; Lioy, D.T.; Leighton, B.H.; Westbrook, G.L.; Mandel, G.; Goodman, R.H. microRNA-132 regulates dendritic growth and arborization of newborn neurons in the adult hippocampus. *Proc. Natl. Acad. Sci. USA* **2010**, *107*, 20382–20387. [CrossRef] [PubMed]
83. Haghparast, A.; Taslimi, Z.; Ramin, M.; Azizi, P.; Khodagholi, F.; Hassanpour-Ezatti, M. Changes in phosphorylation of CREB, ERK, and c-fos induction in rat ventral tegmental area, hippocampus and prefrontal cortex after conditioned place preference induced by chemical stimulation of lateral hypothalamus. *Behav. Brain Res.* **2011**, *220*, 112–118. [CrossRef] [PubMed]

84. Hadar, A.; Milanesi, E.; Walczak, M.; Puzianowska-Kuznicka, M.; Kuznicki, J.; Squassina, A.; Niola, P.; Chillotti, C.; Attems, J.; Gozes, I.; et al. SIRT1, miR-132 and miR-212 link human longevity to Alzheimer's Disease. *Sci. Rep.* **2018**, *8*, 8465. [CrossRef] [PubMed]
85. Herskovits, A.Z.; Guarente, L. SIRT1 in neurodevelopment and brain senescence. *Neuron* **2014**, *81*, 471–483. [CrossRef]
86. Luikart, B.W.; Bensen, A.L.; Washburn, E.K.; Perederiy, J.V.; Su, K.G.; Li, Y.; Kernie, S.G.; Parada, L.F.; Westbrook, G.L. miR-132 mediates the integration of newborn neurons into the adult dentate gyrus. *PLoS ONE* **2011**, *6*, e19077. [CrossRef]
87. Jiang, Y.; Botchway, B.O.A.; Hu, Z.; Fang, M. Overexpression of SIRT1 Inhibits Corticosterone-Induced Autophagy. *Neuroscience* **2019**, *411*, 11–22. [CrossRef]
88. Aten, S.; Page, C.E.; Kalidindi, A.; Wheaton, K.L.; Niraula, A.; Godbout, J.P.; Hoyt, K.R.; Obrietan, K. Data highlighting the expression of two miR-132/212 target genes-Sirt1 and Pten-after chronic stress. *Data Brief.* **2018**, *21*, 2323–2329. [CrossRef]
89. Arthur, J.S. MSK activation and physiological roles. *Front. Biosci.* **2008**, *13*, 5866–5879. [CrossRef]
90. Hauge, C.; Frodin, M. RSK and MSK in MAP kinase signalling. *J. Cell. Sci.* **2006**, *119*, 3021–3023. [CrossRef]
91. Chandramohan, Y.; Droste, S.K.; Arthur, J.S.; Reul, J.M. The forced swimming-induced behavioural immobility response involves histone H3 phospho-acetylation and c-Fos induction in dentate gyrus granule neurons via activation of the N-methyl-D-aspartate/extracellular signal-regulated kinase/mitogen- and stress-activated kinase signalling pathway. *Eur. J. Neurosci.* **2008**, *27*, 2701–2713. [CrossRef]
92. Chwang, W.B.; Arthur, J.S.; Schumacher, A.; Sweatt, J.D. The nuclear kinase mitogen- and stress-activated protein kinase 1 regulates hippocampal chromatin remodeling in memory formation. *J. Neurosci.* **2007**, *27*, 12732–12742. [CrossRef] [PubMed]
93. Gutierrez-Mecinas, M.; Trollope, A.F.; Collins, A.; Morfett, H.; Hesketh, S.A.; Kersante, F.; Reul, J.M. Long-lasting behavioral responses to stress involve a direct interaction of glucocorticoid receptors with ERK1/2-MSK1-Elk-1 signaling. *Proc. Natl. Acad. Sci. USA* **2011**, *108*, 13806–13811. [CrossRef] [PubMed]
94. Sanchez-Puelles, C.; Calleja-Felipe, M.; Ouro, A.; Bougamra, G.; Arroyo, A.; Diez, I.; Erramuzpe, A.; Cortes, J.; Martinez-Hernandez, J.; Lujan, R.; et al. PTEN Activity Defines an Axis for Plasticity at Cortico-Amygdala Synapses and Influences Social Behavior. *Cereb. Cortex* **2020**, *30*, 505–524. [CrossRef] [PubMed]
95. Wang, P.; Mei, F.; Hu, J.; Zhu, M.; Qi, H.; Chen, X.; Li, R.; McNutt, M.A.; Yin, Y. PTENalpha Modulates CaMKII Signaling and Controls Contextual Fear Memory and Spatial Learning. *Cell. Rep.* **2017**, *19*, 2627–2641. [CrossRef] [PubMed]
96. Sperow, M.; Berry, R.B.; Bayazitov, I.T.; Zhu, G.; Baker, S.J.; Zakharenko, S.S. Phosphatase and tensin homologue (PTEN) regulates synaptic plasticity independently of its effect on neuronal morphology and migration. *J. Physiol.* **2012**, *590*, 777–792. [CrossRef] [PubMed]
97. Silva, A.R.; Santos, A.C.; Farfel, J.M.; Grinberg, L.T.; Ferretti, R.E.; Campos, A.H.; Cunha, I.W.; Begnami, M.D.; Rocha, R.M.; Carraro, D.M.; et al. Repair of oxidative DNA damage, cell-cycle regulation and neuronal death may influence the clinical manifestation of Alzheimer's disease. *PLoS ONE* **2014**, *9*, e99897. [CrossRef]
98. Quan, Q.; Qian, Y.; Li, X.; Li, M. CDK5 Participates in Amyloid-beta Production by Regulating PPARgamma Phosphorylation in Primary Rat Hippocampal Neurons. *J. Alzheimers Dis.* **2019**, *71*, 443–460. [CrossRef]
99. Liu, W.; Zhou, Y.; Liang, R.; Zhang, Y. Inhibition of cyclin-dependent kinase 5 activity alleviates diabetes-related cognitive deficits. *FASEB J.* **2019**, *33*, 14506–14515. [CrossRef]
100. Brossaud, J.; Roumes, H.; Helbling, J.C.; Moisan, M.P.; Pallet, V.; Ferreira, G.; Biyong, E.F.; Redonnet, A.; Corcuff, J.B. Retinoic acid increases glucocorticoid receptor phosphorylation via cyclin-dependent kinase 5. *Mol. Cell. Neurosci.* **2017**, *82*, 96–104. [CrossRef]
101. Jin, X.; Sasamoto, K.; Nagai, J.; Yamazaki, Y.; Saito, K.; Goshima, Y.; Inoue, T.; Ohshima, T. Phosphorylation of CRMP2 by Cdk5 Regulates Dendritic Spine Development of Cortical Neuron in the Mouse Hippocampus. *Neural Plast.* **2016**, *2016*, 6790743. [CrossRef]
102. Tomizawa, K.; Ohta, J.; Matsushita, M.; Moriwaki, A.; Li, S.T.; Takei, K.; Matsui, H. Cdk5/p35 regulates neurotransmitter release through phosphorylation and downregulation of P/Q-type voltage-dependent calcium channel activity. *J. Neurosci.* **2002**, *22*, 2590–2597. [CrossRef] [PubMed]
103. Iijima, K.; Ando, K.; Takeda, S.; Satoh, Y.; Seki, T.; Itohara, S.; Greengard, P.; Kirino, Y.; Nairn, A.C.; Suzuki, T. Neuron-specific phosphorylation of Alzheimer's beta-amyloid precursor protein by cyclin-dependent kinase 5. *J. Neurochem.* **2000**, *75*, 1085–1091. [CrossRef] [PubMed]
104. Agasse, F.; Mendez-David, I.; Christaller, W.; Carpentier, R.; Braz, B.Y.; David, D.J.; Saudou, F.; Humbert, S. Chronic Corticosterone Elevation Suppresses Adult Hippocampal Neurogenesis by Hyperphosphorylating Huntingtin. *Cell. Rep.* **2020**, *32*, 107865. [CrossRef]
105. Mitic, M.; Simic, I.; Djordjevic, J.; Radojcic, M.B.; Adzic, M. Gender-specific effects of fluoxetine on hippocampal glucocorticoid receptor phosphorylation and behavior in chronically stressed rats. *Neuropharmacology* **2013**, *70*, 100–111. [CrossRef] [PubMed]
106. Adzic, M.; Djordjevic, J.; Djordjevic, A.; Niciforovic, A.; Demonacos, C.; Radojcic, M.; Krstic-Demonacos, M. Acute or chronic stress induce cell compartment-specific phosphorylation of glucocorticoid receptor and alter its transcriptional activity in Wistar rat brain. *J. Endocrinol.* **2009**, *202*, 87–97. [CrossRef] [PubMed]
107. Shen, Y.; Chen, L.; Zhang, Y.; Du, J.; Hu, J.; Bao, H.; Xing, Y.; Si, Y. Phosphatase and Tensin Homolog Deleted on Chromosome Ten Knockdown Attenuates Cognitive Deficits by Inhibiting Neuroinflammation in a Mouse Model of Perioperative Neurocognitive Disorder. *Neuroscience* **2021**, *468*, 199–210. [CrossRef] [PubMed]

108. Choi, G.E.; Lee, H.J.; Chae, C.W.; Cho, J.H.; Jung, Y.H.; Kim, J.S.; Kim, S.Y.; Lim, J.R.; Han, H.J. BNIP3L/NIX-mediated mitophagy protects against glucocorticoid-induced synapse defects. *Nat. Commun.* **2021**, *12*, 487. [CrossRef]
109. Chen, D.; Lan, G.; Li, R.; Mei, Y.; Shui, X.; Gu, X.; Wang, L.; Zhang, T.; Gan, C.L.; Xia, Y.; et al. Melatonin ameliorates tau-related pathology via the miR-504-3p and CDK5 axis in Alzheimer's disease. *Transl. Neurodegener.* **2022**, *11*, 27. [CrossRef]
110. Inouye, M.O.; Colameo, D.; Ammann, I.; Winterer, J.; Schratt, G. miR-329- and miR-495-mediated Prr7 down-regulation is required for homeostatic synaptic depression in rat hippocampal neurons. *Life Sci. Alliance* **2022**, *5*. [CrossRef]
111. Li, Y.; Fan, C.; Wang, L.; Lan, T.; Gao, R.; Wang, W.; Yu, S.Y. MicroRNA-26a-3p rescues depression-like behaviors in male rats via preventing hippocampal neuronal anomalies. *J. Clin. Investig.* **2021**, *131*. [CrossRef]
112. Aten, S.; Page, C.E.; Kalidindi, A.; Wheaton, K.; Niraula, A.; Godbout, J.P.; Hoyt, K.R.; Obrietan, K. miR-132/212 is induced by stress and its dysregulation triggers anxiety-related behavior. *Neuropharmacology* **2019**, *144*, 256–270. [CrossRef] [PubMed]
113. Travis, S.G.; Coupland, N.J.; Hegadoren, K.; Silverstone, P.H.; Huang, Y.; Carter, R.; Fujiwara, E.; Seres, P.; Malykhin, N.V. Effects of cortisol on hippocampal subfields volumes and memory performance in healthy control subjects and patients with major depressive disorder. *J. Affect. Disord.* **2016**, *201*, 34–41. [CrossRef] [PubMed]
114. Keller, J.; Gomez, R.; Williams, G.; Lembke, A.; Lazzeroni, L.; Murphy, G.M., Jr.; Schatzberg, A.F. HPA axis in major depression: Cortisol, clinical symptomatology and genetic variation predict cognition. *Mol. Psychiatry* **2017**, *22*, 527–536. [CrossRef] [PubMed]
115. Dominguez-Borras, J.; Vuilleumier, P. Amygdala function in emotion, cognition, and behavior. *Handb. Clin. Neurol.* **2022**, *187*, 359–380. [CrossRef]
116. Simic, G.; Tkalcic, M.; Vukic, V.; Mulc, D.; Spanic, E.; Sagud, M.; Olucha-Bordonau, F.E.; Vuksic, M.; P, R.H. Understanding Emotions: Origins and Roles of the Amygdala. *Biomolecules* **2021**, *11*, 823. [CrossRef]
117. Kim, J.J.; Lee, H.J.; Han, J.S.; Packard, M.G. Amygdala is critical for stress-induced modulation of hippocampal long-term potentiation and learning. *J. Neurosci.* **2001**, *21*, 5222–5228. [CrossRef]
118. Segall, L.A.; Milet, A.; Tronche, F.; Amir, S. Brain glucocorticoid receptors are necessary for the rhythmic expression of the clock protein, PERIOD2, in the central extended amygdala in mice. *Neurosci. Lett.* **2009**, *457*, 58–60. [CrossRef]
119. Li, Y.; He, Y.; Fan, H.; Wang, Z.; Huang, J.; Wen, G.; Wang, X.; Xie, Q.; Qiu, P. Brain-derived neurotrophic factor upregulates synaptic GluA1 in the amygdala to promote depression in response to psychological stress. *Biochem. Pharm.* **2021**, *192*, 114740. [CrossRef]
120. Mackiewicz, K.L.; Sarinopoulos, I.; Cleven, K.L.; Nitschke, J.B. The effect of anticipation and the specificity of sex differences for amygdala and hippocampus function in emotional memory. *Proc. Natl. Acad. Sci. USA* **2006**, *103*, 14200–14205. [CrossRef]
121. Fanselow, M.S.; Dong, H.W. Are the dorsal and ventral hippocampus functionally distinct structures? *Neuron* **2010**, *65*, 7–19. [CrossRef]
122. Floriou-Servou, A.; von Ziegler, L.; Stalder, L.; Sturman, O.; Privitera, M.; Rassi, A.; Cremonesi, A.; Thony, B.; Bohacek, J. Distinct Proteomic, Transcriptomic, and Epigenetic Stress Responses in Dorsal and Ventral Hippocampus. *Biol. Psychiatry* **2018**, *84*, 531–541. [CrossRef] [PubMed]
123. Ohara, S.; Sato, S.; Tsutsui, K.; Witter, M.P.; Iijima, T. Organization of multisynaptic inputs to the dorsal and ventral dentate gyrus: Retrograde trans-synaptic tracing with rabies virus vector in the rat. *PLoS ONE* **2013**, *8*, e78928. [CrossRef] [PubMed]
124. Steullet, P.; Cabungcal, J.H.; Kulak, A.; Kraftsik, R.; Chen, Y.; Dalton, T.P.; Cuenod, M.; Do, K.Q. Redox dysregulation affects the ventral but not dorsal hippocampus: Impairment of parvalbumin neurons, gamma oscillations, and related behaviors. *J. Neurosci.* **2010**, *30*, 2547–2558. [CrossRef] [PubMed]
125. Igarashi, K.M.; Ito, H.T.; Moser, E.I.; Moser, M.B. Functional diversity along the transverse axis of hippocampal area CA1. *FEBS Lett.* **2014**, *588*, 2470–2476. [CrossRef]
126. Moser, M.B.; Moser, E.I. Functional differentiation in the hippocampus. *Hippocampus* **1998**, *8*, 608–619. [CrossRef]
127. Strange, B.A.; Witter, M.P.; Lein, E.S.; Moser, E.I. Functional organization of the hippocampal longitudinal axis. *Nat. Rev.* **2014**, *15*, 655–669. [CrossRef]
128. Torrisi, S.A.; Lavanco, G.; Maurel, O.M.; Gulisano, W.; Laudani, S.; Geraci, F.; Grasso, M.; Barbagallo, C.; Caraci, F.; Bucolo, C.; et al. A novel arousal-based individual screening reveals susceptibility and resilience to PTSD-like phenotypes in mice. *Neurobiol. Stress.* **2021**, *14*, 100286. [CrossRef]
129. Maurel, O.M.; Torrisi, S.A.; Barbagallo, C.; Purrello, M.; Salomone, S.; Drago, F.; Ragusa, M.; Leggio, G.M. Dysregulation of miR-15a-5p, miR-497a-5p and miR-511-5p Is Associated with Modulation of BDNF and FKBP5 in Brain Areas of PTSD-Related Susceptible and Resilient Mice. *Int. J. Mol. Sci.* **2021**, *22*, 5157. [CrossRef]
130. Belovicova, K.; Bogi, E.; Csatlosova, K.; Dubovicky, M. Animal tests for anxiety-like and depression-like behavior in rats. *Interdiscip. Toxicol.* **2017**, *10*, 40–43. [CrossRef]
131. Hogg, S. A review of the validity and variability of the elevated plus-maze as an animal model of anxiety. *Pharmacol. Biochem. Behav.* **1996**, *54*, 21–30. [CrossRef]
132. Bourin, M.; Hascoet, M. The mouse light/dark box test. *Eur. J. Pharmacol.* **2003**, *463*, 55–65. [PubMed]
133. Hagenbuch, N.; Feldon, J.; Yee, B.K. Use of the elevated plus-maze test with opaque or transparent walls in the detection of mouse strain differences and the anxiolytic effects of diazepam. *Behav. Pharmacol.* **2006**, *17*, 31–41. [PubMed]
134. Ramos, A. Animal models of anxiety: Do I need multiple tests? *Trends Pharmacol. Sci.* **2008**, *29*, 493–498. [CrossRef] [PubMed]

135. Violle, N.; Balandras, F.; Le Roux, Y.; Desor, D.; Schroeder, H. Variations in illumination, closed wall transparency and/or extramaze space influence both baseline anxiety and response to diazepam in the rat elevated plus-maze. *Behav. Brain Res.* **2009**, *203*, 35–42. [CrossRef] [PubMed]
136. Miller, S.M.; Piasecki, C.C.; Lonstein, J.S. Use of the light-dark box to compare the anxiety-related behavior of virgin and postpartum female rats. *Pharmacol. Biochem. Behav.* **2011**, *100*, 130–137. [CrossRef] [PubMed]
137. Steimer, T. Animal models of anxiety disorders in rats and mice: Some conceptual issues. *Dialogues Clin. Neurosci.* **2011**, *13*, 495–506. [CrossRef]
138. Mineur, Y.S.; Fote, G.M.; Blakeman, S.; Cahuzac, E.L.; Newbold, S.A.; Picciotto, M.R. Multiple Nicotinic Acetylcholine Receptor Subtypes in the Mouse Amygdala Regulate Affective Behaviors and Response to Social Stress. *Neuropsychopharmacology* **2016**, *41*, 1579–1587. [CrossRef]
139. Pidoplichko, V.I.; Prager, E.M.; Aroniadou-Anderjaska, V.; Braga, M.F. alpha7-Containing nicotinic acetylcholine receptors on interneurons of the basolateral amygdala and their role in the regulation of the network excitability. *J. Neurophysiol.* **2013**, *110*, 2358–2369. [CrossRef]
140. Laviolette, S.R. Molecular and neuronal mechanisms underlying the effects of adolescent nicotine exposure on anxiety and mood disorders. *Neuropharmacology* **2021**, *184*, 108411. [CrossRef]
141. Zhang, M.; Bian, Z. Alzheimer's Disease and microRNA-132: A Widespread Pathological Factor and Potential Therapeutic Target. *Front. Neurosci.* **2021**, *15*, 687973. [CrossRef]
142. Alkadhi, K.A.; Alzoubi, K.H.; Srivareerat, M.; Tran, T.T. Chronic psychosocial stress exacerbates impairment of synaptic plasticity in beta-amyloid rat model of Alzheimer's disease: Prevention by nicotine. *Curr. Alzheimer Res.* **2011**, *8*, 718–731. [CrossRef]
143. Alkadhi, K.A.; Srivareerat, M.; Tran, T.T. Intensification of long-term memory deficit by chronic stress and prevention by nicotine in a rat model of Alzheimer's disease. *Mol. Cell. Neurosci.* **2010**, *45*, 289–296. [CrossRef] [PubMed]
144. Musazzi, L.; Tornese, P.; Sala, N.; Lee, F.S.; Popoli, M.; Ieraci, A. Acute stress induces an aberrant increase of presynaptic release of glutamate and cellular activation in the hippocampus of BDNF(Val/Met) mice. *J. Cell. Physiol.* **2022**, *237*, 3834–3844. [CrossRef] [PubMed]
145. Tan, Y.; Rouse, J.; Zhang, A.; Cariati, S.; Cohen, P.; Comb, M.J. FGF and stress regulate CREB and ATF-1 via a pathway involving p38 MAP kinase and MAPKAP kinase-2. *EMBO J.* **1996**, *15*, 4629–4642. [CrossRef] [PubMed]
146. Hernandez-Rapp, J.; Smith, P.Y.; Filali, M.; Goupil, C.; Planel, E.; Magill, S.T.; Goodman, R.H.; Hebert, S.S. Memory formation and retention are affected in adult miR-132/212 knockout mice. *Behav. Brain Res.* **2015**, *287*, 15–26. [CrossRef]
147. Lee, Y.J.; Kim, H.R.; Lee, C.Y.; Hyun, S.A.; Ko, M.Y.; Lee, B.S.; Hwang, D.Y.; Ka, M. 2-Phenylethylamine (PEA) Ameliorates Corticosterone-Induced Depression-Like Phenotype via the BDNF/TrkB/CREB Signaling Pathway. *Int. J. Mol. Sci.* **2020**, *21*, 9103. [CrossRef]
148. Lu, Y.; Sareddy, G.R.; Wang, J.; Wang, R.; Li, Y.; Dong, Y.; Zhang, Q.; Liu, J.; O'Connor, J.C.; Xu, J.; et al. Neuron-Derived Estrogen Regulates Synaptic Plasticity and Memory. *J. Neurosci.* **2019**, *39*, 2792–2809. [CrossRef]
149. Okun, E.; Griffioen, K.; Barak, B.; Roberts, N.J.; Castro, K.; Pita, M.A.; Cheng, A.; Mughal, M.R.; Wan, R.; Ashery, U.; et al. Toll-like receptor 3 inhibits memory retention and constrains adult hippocampal neurogenesis. *Proc. Natl. Acad. Sci. USA* **2010**, *107*, 15625–15630. [CrossRef]
150. Hansen, K.F.; Sakamoto, K.; Wayman, G.A.; Impey, S.; Obrietan, K. Transgenic miR132 alters neuronal spine density and impairs novel object recognition memory. *PLoS ONE* **2010**, *5*, e15497. [CrossRef]
151. Lambert, T.J.; Storm, D.R.; Sullivan, J.M. MicroRNA132 modulates short-term synaptic plasticity but not basal release probability in hippocampal neurons. *PLoS ONE* **2010**, *5*, e15182. [CrossRef]
152. Silva, A.J.; Kogan, J.H.; Frankland, P.W.; Kida, S. CREB and memory. *Annu. Rev. Neurosci.* **1998**, *21*, 127–148. [CrossRef]
153. Meyer, H.C.; Odriozola, P.; Cohodes, E.M.; Mandell, J.D.; Li, A.; Yang, R.; Hall, B.S.; Haberman, J.T.; Zacharek, S.J.; Liston, C.; et al. Ventral hippocampus interacts with prelimbic cortex during inhibition of threat response via learned safety in both mice and humans. *Proc. Natl. Acad. Sci. USA* **2019**, *116*, 26970–26979. [CrossRef] [PubMed]
154. Connor, D.A.; Kutlu, M.G.; Gould, T.J. Nicotine disrupts safety learning by enhancing fear associated with a safety cue via the dorsal hippocampus. *J. Psychopharmacol.* **2017**, *31*, 934–944. [CrossRef] [PubMed]
155. Roozendaal, B. Stress and memory: Opposing effects of glucocorticoids on memory consolidation and memory retrieval. *Neurobiol. Learn. Mem.* **2002**, *78*, 578–595. [CrossRef] [PubMed]
156. Vreugdenhil, E.; Verissimo, C.S.; Mariman, R.; Kamphorst, J.T.; Barbosa, J.S.; Zweers, T.; Champagne, D.L.; Schouten, T.; Meijer, O.C.; de Kloet, E.R.; et al. MicroRNA 18 and 124a down-regulate the glucocorticoid receptor: Implications for glucocorticoid responsiveness in the brain. *Endocrinology* **2009**, *150*, 2220–2228. [CrossRef]
157. Hillerer, K.M.; Slattery, D.A.; Pletzer, B. Neurobiological mechanisms underlying sex-related differences in stress-related disorders: Effects of neuroactive steroids on the hippocampus. *Front. Neuroendocr.* **2019**, *55*, 100796. [CrossRef]
158. Gurvich, C.; Thomas, N.; Kulkarni, J. Sex differences in cognition and aging and the influence of sex hormones. *Handb. Clin. Neurol.* **2020**, *175*, 103–115. [CrossRef]
159. Hajali, V.; Andersen, M.L.; Negah, S.S.; Sheibani, V. Sex differences in sleep and sleep loss-induced cognitive deficits: The influence of gonadal hormones. *Horm. Behav.* **2019**, *108*, 50–61. [CrossRef]
160. Lambert, K.G.; Kinsley, C.H. Sex differences and gonadal hormones influence susceptibility to the activity-stress paradigm. *Physiol. Behav.* **1993**, *53*, 1085–1090. [CrossRef]

161. McEwen, B.S.; Milner, T.A. Understanding the broad influence of sex hormones and sex differences in the brain. *J. Neurosci. Res.* **2017**, *95*, 24–39. [CrossRef]
162. Kong, F.; Zhen, Z.; Li, J.; Huang, L.; Wang, X.; Song, Y.; Liu, J. Sex-related neuroanatomical basis of emotion regulation ability. *PLoS ONE* **2014**, *9*, e97071. [CrossRef] [PubMed]
163. Mahmoud, R.; Wainwright, S.R.; Galea, L.A. Sex hormones and adult hippocampal neurogenesis: Regulation, implications, and potential mechanisms. *Front. Neuroendocr.* **2016**, *41*, 129–152. [CrossRef] [PubMed]
164. Koss, W.A.; Frick, K.M. Sex differences in hippocampal function. *J. Neurosci. Res.* **2017**, *95*, 539–562. [CrossRef] [PubMed]
165. Scharfman, H.E.; MacLusky, N.J. Sex differences in hippocampal area CA3 pyramidal cells. *J. Neurosci. Res.* **2017**, *95*, 563–575. [CrossRef]
166. Morgan, C.P.; Bale, T.L. Sex differences in microRNA-mRNA networks: Examination of novel epigenetic programming mechanisms in the sexually dimorphic neonatal hypothalamus. *Biol. Sex. Differ.* **2017**, *8*, 27. [CrossRef]
167. Murphy, S.J.; Lusardi, T.A.; Phillips, J.I.; Saugstad, J.A. Sex differences in microRNA expression during development in rat cortex. *Neurochem. Int.* **2014**, *77*, 24–32. [CrossRef]
168. Sheinerman, K.; Tsivinsky, V.; Mathur, A.; Kessler, D.; Shaz, B.; Umansky, S. Age- and sex-dependent changes in levels of circulating brain-enriched microRNAs during normal aging. *Aging (Albany N. Y.)* **2018**, *10*, 3017–3041. [CrossRef]
169. Jirkof, P.; Bratcher, N.; Medina, L.; Strasburg, D.; Ebert, P.; Gaskill, B.N. The effect of group size, age and handling frequency on inter-male aggression in CD 1 mice. *Sci. Rep.* **2020**, *10*, 2253. [CrossRef]
170. Svenson, K.L.; Paigen, B. Recommended housing densities for research mice: Filling the gap in data-driven alternatives. *FASEB J.* **2019**, *33*, 3097–3111. [CrossRef]
171. Nguyen, P.V.; Abel, T.; Kandel, E.R. Requirement of a critical period of transcription for induction of a late phase of LTP. *Sci. (N. Y.)* **1994**, *265*, 1104–1107. [CrossRef]
172. Nguyen, P.V.; Kandel, E.R. Brief theta-burst stimulation induces a transcription-dependent late phase of LTP requiring cAMP in area CA1 of the mouse hippocampus. *Learn. Mem. (Cold Spring Harb. N. Y.)* **1997**, *4*, 230–243. [CrossRef] [PubMed]
173. Schneider, C.A.; Rasband, W.S.; Eliceiri, K.W. NIH Image to ImageJ: 25 years of image analysis. *Nat. Methods* **2012**, *9*, 671–675. [CrossRef] [PubMed]
174. Prut, L.; Belzung, C. The open field as a paradigm to measure the effects of drugs on anxiety-like behaviors: A review. *Eur. J. Pharmacol.* **2003**, *463*, 3–33. [CrossRef] [PubMed]
175. Ladron de Guevara-Miranda, D.; Millon, C.; Rosell-Valle, C.; Perez-Fernandez, M.; Missiroli, M.; Serrano, A.; Pavon, F.J.; Rodriguez de Fonseca, F.; Martinez-Losa, M.; Alvarez-Dolado, M.; et al. Long-lasting memory deficits in mice withdrawn from cocaine are concomitant with neuroadaptations in hippocampal basal activity, GABAergic interneurons and adult neurogenesis. *Dis. Model. Mech.* **2017**, *10*, 323–336. [CrossRef] [PubMed]
176. Manas-Padilla, M.C.; Avila-Gamiz, F.; Gil-Rodriguez, S.; Ladron de Guevara-Miranda, D.; Rodriguez de Fonseca, F.; Santin, L.J.; Castilla-Ortega, E. Persistent changes in exploration and hyperactivity coexist with cognitive impairment in mice withdrawn from chronic cocaine. *Physiol. Behav.* **2021**, *240*, 113542. [CrossRef]
177. Bailey, K.R.; Crawley, J.N. Anxiety-Related Behaviors in Mice. In *Methods of Behavior Analysis in Neuroscience*, 2nd ed.; Buccafusco, J.J., Ed.; Frontiers in Neuroscience; CRC Press/Taylor & Francis: Boca Raton, FL, USA, 2009.
178. Manas-Padilla, M.C.; Gil-Rodriguez, S.; Sampedro-Piquero, P.; Avila-Gamiz, F.; Rodriguez de Fonseca, F.; Santin, L.J.; Castilla-Ortega, E. Remote memory of drug experiences coexists with cognitive decline and abnormal adult neurogenesis in an animal model of cocaine-altered cognition. *Addict. Biol.* **2021**, *26*, e12886. [CrossRef]
179. Pinheiro, S.H.; Zangrossi, H., Jr.; Del-Ben, C.M.; Graeff, F.G. Elevated mazes as animal models of anxiety: Effects of serotonergic agents. *Acad. Bras. Cienc.* **2007**, *79*, 71–85. [CrossRef]
180. Walf, A.A.; Frye, C.A. The use of the elevated plus maze as an assay of anxiety-related behavior in rodents. *Nat. Protoc.* **2007**, *2*, 322–328. [CrossRef]
181. Li, H.; Zhang, C.; Shen, H.; Shen, Z.; Wu, L.; Mo, F.; Li, M. Physiological stress-induced corticosterone increases heme uptake via KLF4-HCP1 signaling pathway in hippocampus neurons. *Sci. Rep.* **2017**, *7*, 5745. [CrossRef]

International Journal of
Molecular Sciences

MDPI

Article

Loss of CDKL5 Causes Synaptic GABAergic Defects That Can Be Restored with the Neuroactive Steroid Pregnenolone-Methyl-Ether

Roberta De Rosa [1], Serena Valastro [1], Clara Cambria [2], Isabella Barbiero [1], Carolina Puricelli [1], Marco Tramarin [1], Silvia Randi [1], Massimiliano Bianchi [3,4], Flavia Antonucci [2] and Charlotte Kilstrup-Nielsen [1,*]

1 Department of Biotechnology and Life Sciences (DBSV), Centre of NeuroScience, University of Insubria, 21052 Busto Arsizio, Italy
2 Department of Medical Biotechnology and Translational Medicine (BIOMETRA), University of Milan, 20129 Milan, Italy
3 Ulysses Neuroscience Ltd., Trinity College Dublin, D02 PN40 Dublin, Ireland
4 Institute of Neuroscience, Trinity College Dublin, D02 PN40 Dublin, Ireland
* Correspondence: c.kilstrup-nielsen@uninsubria.it; Tel.: +39-033-133-9430

Abstract: CDKL5 deficiency disorder (CDD) is an X-linked neurodevelopmental disorder characterised by early-onset drug-resistant epilepsy and impaired cognitive and motor skills. CDD is caused by mutations in cyclin-dependent kinase-like 5 (CDKL5), which plays a well-known role in regulating excitatory neurotransmission, while its effect on neuronal inhibition has been poorly investigated. We explored the potential role of CDKL5 in the inhibitory compartment in *Cdkl5*-KO male mice and primary hippocampal neurons and found that CDKL5 interacts with gephyrin and collybistin, two crucial organisers of the inhibitory postsynaptic sites. Through molecular and electrophysiological approaches, we demonstrated that CDKL5 loss causes a reduced number of gephyrin puncta and surface exposed γ_2 subunit-containing $GABA_A$ receptors, impacting the frequency of miniature inhibitory postsynaptic currents, which we ascribe to a postsynaptic function of CDKL5. In line with previous data showing that CDKL5 loss impacts microtubule (MT) dynamics, we showed that treatment with pregnenolone-methyl-ether (PME), which promotes MT dynamics, rescues the above defects. The impact of CDKL5 deficiency on inhibitory neurotransmission might explain the presence of drug-resistant epilepsy and cognitive defects in CDD patients. Moreover, our results may pave the way for drug-based therapies that could bypass the need for CDKL5 and provide effective therapeutic strategies for CDD patients.

Keywords: CDKL5; inhibitory synapse; gephyrin; collybistin; GABAA receptor; pregnenolone-methyl-ether

check for updates

Citation: De Rosa, R.; Valastro, S.; Cambria, C.; Barbiero, I.; Puricelli, C.; Tramarin, M.; Randi, S.; Bianchi, M.; Antonucci, F.; Kilstrup-Nielsen, C. Loss of CDKL5 Causes Synaptic GABAergic Defects That Can Be Restored with the Neuroactive Steroid Pregnenolone-Methyl-Ether. *Int. J. Mol. Sci.* **2023**, *24*, 68. https://doi.org/10.3390/ijms24010068

Academic Editor: Giuseppina Martella

Received: 2 November 2022
Revised: 9 December 2022
Accepted: 16 December 2022
Published: 21 December 2022

1. Introduction

Mutations in the X-linked cyclin-dependent kinase-like 5 (*CDKL5*) gene cause a severe neurodevelopmental disorder (Early Infantile Epileptic Encephalopahty, OMIM 300672), commonly referred to as CDKL5 deficiency disorder (CDD). CDD patients are characterised by intellectual disability, autistic features and drug-resistant epilepsy that normally manifest within the first three months of age [1]. *CDKL5* is mutated in approximately 1:50.000 live births, making CDD one of the most frequent causes of genetic epilepsy.

CDKL5 is a serine-threonine kinase that is highly abundant in the brain and whose expression peaks in the first postnatal weeks [2]. The constitutive loss of CDKL5 in CDD mouse models causes impaired learning and memory, altered locomotion and autistic-like features [3–6]. Spontaneous epilepsy has only recently been observed in aged heterozygous female mice [7,8] and in male mice harbouring the conditional knock-out (KO) of *Cdkl5* in glutamatergic forebrain neurons [9].

Various studies of *Cdkl5* mouse models and CDKL5 deficient primary neurons converge on the role of CDKL5 in regulating excitatory neurotransmission [6]. Indeed, CDKL5 interacts with the scaffolding protein PSD95, and CDKL5 deficient neurons are characterised by morphological and molecular alterations linked to excitatory synapses [10–12]. These defects depend in part on altered microtubule (MT) dynamics leading to the impaired invasion of MTs into dendritic spines and reduced spine maturation. Indeed, we previously demonstrated that CDKL5 regulates the MT-binding of the plus-end tracking protein CLIP170, thus impacting MT dynamics [13–15]. Importantly, the modulation of MT dynamics in vitro and in vivo, mediated by treatment with the pregnenolone analogue pregnenolone-methyl-ether (PME), which promotes CLIP170 functioning, can restore several CDKL5-related defects, including spine maturation, expression of synaptic proteins and hippocampal-dependent learning and memory [13–15].

Notwithstanding the importance of maintaining a proper excitation/inhibition balance, the possible role of CDKL5 in regulating inhibitory neurotransmission has 'til now been rather neglected. Mice with the conditional inactivation of *Cdkl5* in their glutamatergic forebrain neurons display altered miniature inhibitory synaptic currents (mIPSCs) [16], but a molecular basis for such a defect is still unknown.

Here, we report that CDKL5 interacts with the inhibitory postsynaptic scaffolding complex containing gephyrin and collybistin (CB) and show that CDKL5 loss leads to reduced levels and functioning of the γ_2 subunit-containing γ-aminobutyric acid type A receptors ($GABA_AR$) in vitro and in vivo. $GABA_ARs$ are heteropentameric Cl^- permeable channels, generally composed of two α- and two β-subunits and a single γ- or δ-subunit [17]. Typically, synaptic $GABA_ARs$, which mediate phasic inhibition upon presynaptic GABA release, contain a γ-subunit; contrariwise, the extrasynaptic receptor complexes that mediate tonic inhibition as a response to low ambient GABA levels contain the δ-subunit [18]. Surface levels of synaptic $GABA_ARs$ depend on a complex and dynamic regulation including transport, recycling and stabilisation, of which the latter relies on a scaffolding complex containing gephyrin and CB [19].

Alterations in any of the components linked to inhibitory neurotransmission are associated with various neurodevelopmental disorders characterised by cognitive defects, autism-like features and epilepsy [20]. The pharmacological targeting of the GABAergic system, based both on the allosteric modulation of $GABA_AR$ subtypes and on the functional control of $GABA_AR$-associated proteins, represents an important interventional strategy against epilepsy and other clinical manifestations linked to the altered $GABA_AR$ levels or functioning [21].

We here show that treatment with PME can restore $GABA_AR$ γ_2 expression and functioning in primary *Cdkl5*-KO neurons. Importantly, we also find that the synaptic $GABA_AR$ subunit γ_2 is significantly reduced in hippocampi of *Cdkl5*-KO mice, but that its levels are normalised to those of WT animals upon treatment with PME. Altogether, these data suggest that CDKL5, through its interaction with the inhibitory scaffolding complex, regulates synaptic $GABA_AR$ levels and, importantly, that altered membrane insertion of $GABA_ARs$ can be restored by the targeting of MT dynamics.

2. Results

2.1. CDKL5 Interacts with the Gephyrin-Collybistin Complex and Regulates the Number of Postsynaptic Gephyrin Clusters

To investigate the possible role of CDKL5 at the inhibitory synapses, we examined its interaction with gephyrin and CB, two proteins playing a fundamental role in the organisation of the postsynaptic sites of these synapses. To this aim, brain lysates of young mice at postnatal day 20–30 (PND20-30) were used to immunoprecipitate CDKL5 using IgGs as negative control; through the subsequent western blotting (WB), both gephyrin and CB could be detected in the CDKL5 immunocomplexes (Figure 1A,B). We further confirmed the interaction in a heterologous system expressing Flag-tagged CDKL5 in HEK293T cells together with the Myc-tagged CB2 isoform (Figure 1C). Upon immunoprecipitation of CDKL5, allowing the precipitation of both exogenous and endogenous CDKL5, overexpressed CB2 could easily be detected through WB analysis. Interestingly, the weakly expressed endogenous CB could also be detected as interacting with endogenous CDKL5 that was visible upon a higher exposure of the membrane. These results confirm a recent report by Uezu et al. [22] in which CB was identified as a direct interactor of CDKL5 through a chemical-genetic proximity-labelling approach.

Gephyrin is a key protein in the organisation of inhibitory synapses through the formation of submembranous clusters that regulate the accumulation of $GABA_A$Rs at the postsynaptic sites [23]. The accumulation of gephyrin under the cell membrane depends on the guanine nucleotide exchange factor CB that can recruit gephyrin to the postsynaptic sites. CB exists in a closed inactive conformation, which depends on its N-terminal SH3 domain [24]. When GFP-tagged gephyrin is expressed together with the SH3-containing CB2 derivative (CB2-$SH3^+$) in a heterologous system, it therefore accumulates in cytoplasmic deposits (Figure 1D), whereas deletion of the SH3 domain (CB-ΔSH3) renders CB constitutively active, leading to the translocation of gephyrin to submembranous microaggregates. Interestingly, the exogenous expression of Flag-tagged CDKL5 in cells expressing CB2-$SH3^+$ causes GFP-gephyrin to accumulate under the cell membrane, suggesting that CDKL5 is capable of activating CB through the interaction of the two proteins (Figure 1E).

We further evaluated whether CDKL5 loss influences the total levels of gephyrin or CB in neurons. WB analyses on neuronal lysates from primary hippocampal cultures of *Cdkl5*-WT and -KO neurons, cultured for 14 days in vitro (DIV), did not reveal any changes in the expression of either gephyrin or CB (Figure 1F–I). We next examined whether CDKL5 loss affects the capacity of gephyrin to form the typical submembranous clusters in primary hippocampal neurons. Interestingly, we observed a significantly reduced number of gephyrin puncta in the dendritic segments of *Cdkl5*-KO neurons (Figure 1J,K). The interaction of CDKL5 with the gephyrin-CB complex, together with the reduced number of gephyrin clusters in *Cdkl5*-KO neurons, indicate that CDKL5 might play a hitherto undescribed role in the organisation of the postsynaptic sites of inhibitory synapses.

2.2. CDKL5 Loss Affects the Membrane Levels of γ_2 Subunit-Containing $GABA_A$Rs and Impairs mIPSCs

Gephyrin deficiency impacts the surface accumulation of γ_2-containing $GABA_A$Rs [25–27]. We therefore speculated that the reduced number of gephyrin clusters in *Cdkl5*-KO neurons might influence the cell-surface expression of the $GABA_A$R subunit γ_2. To address this, we performed a cell-surface biotinylation assay. Briefly, upon biotinylation of hippocampal neurons, labelled proteins were affinity purified, and the amount of the synaptic $GABA_A$R subunit γ_2 was analysed through WB in parallel with a fraction of the total cell extract (Figure 2A). The proportion of the total receptor pool that resides in the neuronal surface was determined by quantifying the ratio of the biotinylated fraction (surface) over the amount in the total lysate (total). As control of the biotinylation procedure, we verified that the cytoplasmic protein GAPDH was barely present in the pool of affinity-purified proteins. As illustrated in the graph in Figure 2B, we found that *Cdkl5*-KO neurons displayed reduced surface levels of the γ_2 subunit, whereas its total levels were unaltered (Figure 2C,D). We further examined the surface expression of the γ_2 subunit through the immunofluorescence staining of hippocampal *Cdkl5*-WT and -KO neurons performed under non-permeabilising conditions. In accordance with the biotinylation experiment, *Cdkl5*-KO neurons displayed significantly reduced surface levels of the γ_2 subunit (Figure 2E,F).

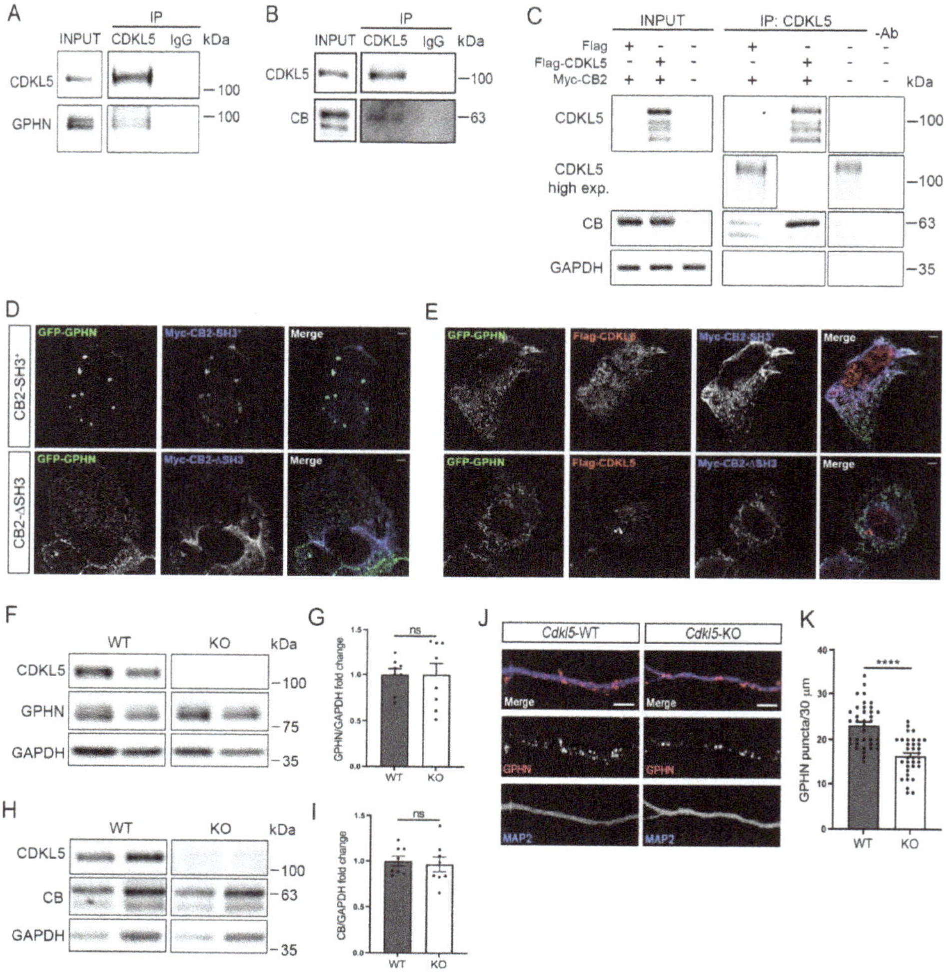

Figure 1. CDKL5 interacts with the gephyrin-collybistin complex. (**A**,**B**) Representative WBs showing the coimmuniprecipitation of CDKL5 and gephyrin (GPHN; A) or collybistin (CB; B) from PND20-30 mouse brain lysates. Unrelated IgGs were used as negative control. Whole brain lysate (input; 2.5%) and immunocomplexes were analysed through WB with antibodies against CDKL5, GPHN (**A**) and CB (**B**). n = 3. (**C**) Flag-CDKL5 was expressed in HEK293T cells together with Myc-CB2 and immunoprecipitated with a monoclonal anti-CDKL5 antibody. Whole cell lysates (input; 3%) and immunoprecipitated proteins were analysed through WB using antibodies against CDKL5, CB and, as loading control, GAPDH. n = 3. (**D**,**E**) Representative confocal images of COS7 cells expressing GFP-tagged gephyrin (GFP-GPHN, green) and Myc-CB2-SH3$^+$ or Myc-CB2-ΔSH3 (blue) together with Flag-CDKL5 (red). Scale bar: 5 μm. (**F**,**H**) Representative WB of whole cell lysates of *Cdkl5*-WT/KO primary hippocampal neurons at DIV14. Antibodies against GPHN and CB were used together with anti-CDKL5 and, as loading control, GAPDH. (**G**,**I**) The graphs show the quantification of normalised GPHN (**G**) and CB (**I**) levels. n = 8 biological replicates. Mean ± SEM. Not significant (ns), $p > 0.05$. Unpaired Student's *t*-test. (**J**) Representative images of DIV14 *Cdkl5*-WT/KO hippocampal neurons stained with antibodies against GPHN (red) and the dendritic marker MAP2 (blue). Scale bar, 5 μm. (**K**) Graph showing the quantification of GPHN puncta along 30 μm long segments. n = 7 biological replicates, N = 37/35 WT/KO neurons. Mean ± SEM. **** $p < 0.0001$. Unpaired Student's *t*-test.

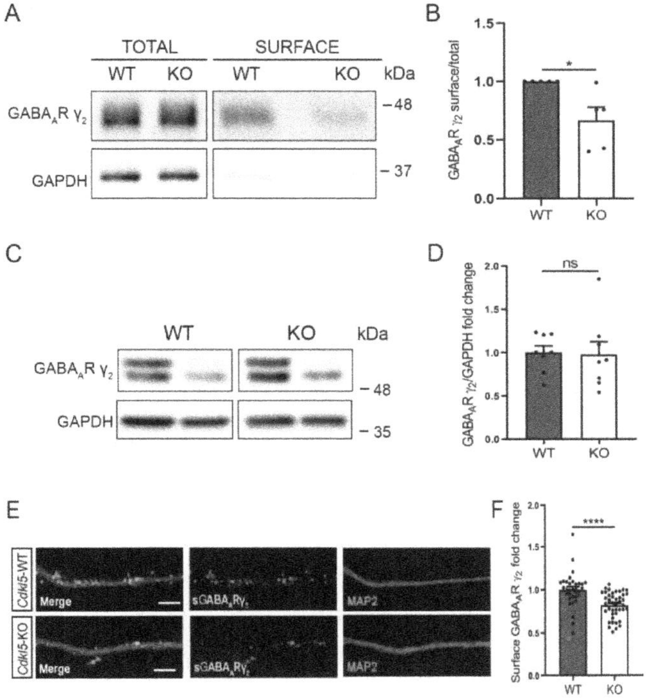

Figure 2. CDKL5 loss affects surface expression of synaptic $GABA_A$Rs in primary hippocampal neurons. (**A**) Representative WB of a biotinylation experiment on *Cdkl5*-WT/KO primary hippocampal neurons at DIV14. Levels of $GABA_A$R subunit γ_2 were analysed together with GAPDH in the surface fraction, obtained from 500 μg of lysate, and in 30 μg of whole cell lysate. (**B**) Graph showing the ratio of surface/total levels of $GABA_A$R γ_2. N = 5 biological replicates. Mean ± SEM * $p < 0.05$. Unpaired Student's *t*-test. (**C**) Representative WB analysis of whole cell lysates of *Cdkl5*-WT/KO primary hippocampal neurons at DIV14. GAPDH was used as loading control. (**D**) Graph showing the quantification of normalised $GABA_A$R γ_2 levels. N = 8 biological replicates. Mean ± SEM ns, $p > 0.05$. Unpaired Student's *t*-test. (**E**) Surface exposed $GABA_A$R γ_2 (green, s$GABA_A$R γ_2) was detected through immunostaining of *Cdkl5*-WT/KO primary hippocampal neurons at DIV14 under non-permeabilising conditions. MAP2 (blue) was used as dendritic marker. Scale bar, 5 μm. (**F**) Graph showing the quantification of the fluorescence intensity of s$GABA_A$R γ_2 staining of 30 μm long dendritic segments. n = 6 biological replicates, N = 33/42 WT/KO neurons. Mean ± SEM **** $p < 0.0001$. Unpaired Student's *t*-test.

To evaluate the functional consequences of the reduced levels of the γ_2-containing $GABA_A$Rs on the neuronal surface, we measured mIPSCs from *Cdkl5*-WT and -KO hippocampal neurons at DIV14. Notably, in *Cdkl5*-KO neurons, mIPSCs were less frequent, while the amplitude was unaltered (Figure 3A–C). The reduced frequency of mIPSCs points to possible presynaptic deficits in the *Cdkl5*-KO neurons, and we therefore analysed the presynaptic markers bassoon and vesicular GABA transporter (VGAT). Bassoon is localised at the active zone of both excitatory and inhibitory presynaptic terminals, while VGAT is specific for the inhibitory presynaptic sites. Immunofluorescence staining showed reduced numbers of both bassoon and VGAT puncta in *Cdkl5*-KO hippocampal neurons (Figure 3D–G).

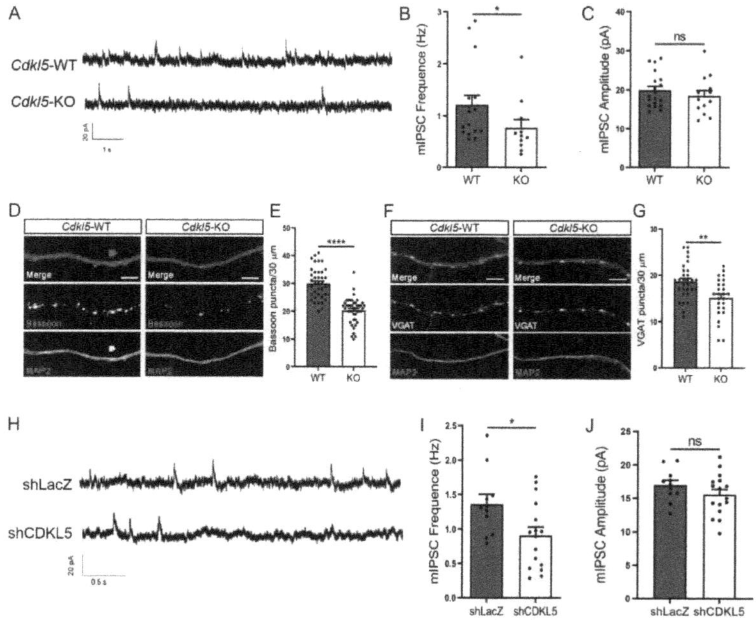

Figure 3. CDKL5 loss leads to a reduction in the frequency of mIPSCs and in presynaptic inhibitory markers. (**A**) Representative traces of mIPSCs recorded from *Cdkl5*-WT/KO primary hippocampal neurons at DIV14 patching at +10 mV. (**B**,**C**) Graphs showing mIPSC frequency (**B**) and amplitude (**C**) in *Cdkl5*-WT/KO cultures. n = 3 biological replicates; frequency: N = 16/11 WT/KO neurons. Mean ± SEM * $p < 0.05$, not significant (ns), $p > 0.05$. mIPSC frequency: Mann Whitney U test; mIPSC amplitude: Unpaired Student's *t*-test. (**D**,**F**) Representative images of *Cdkl5*-WT/KO neurons at DIV14 stained for Bassoon (red, D), VGAT (green, F) and MAP2 (blue). Scale bar, 5 μm. (**E**,**G**) Graphs showing the quantification of bassoon and VGAT puncta along 30 μm long dendritic segments. Bassoon: n = 7 biological replicates, N = 40/44 neurons. VGAT: n = 3 biological replicates, N = 30/26 neurons. Mean ± SEM ** $p < 0.01$, **** $p < 0.0001$. Unpaired Student's *t*-test. (**H**) Representative traces of mIPSCs recorded from CDKL5-silenced neurons (shCDKL5) and controls (shLacZ). Neurons were transfected with GFP-expressing shRNA vectors at DIV11 and mIPSCs were recorded at DIV14 patching at +10 mV. (**I**,**J**) Graphs showing mIPSC frequency (**I**) and amplitude (**J**) upon acute CDKL5-silencing. n = 3 biological replicates; N = 11/16 shLacZ/shCDKL5 neurons. Mean ± SEM * $p < 0.05$; ns, $p > 0.05$. Unpaired Student's *t*-test.

The interaction of CDKL5 with the postsynaptic scaffolding proteins led us to test whether the ablation of CDKL5 at the postsynaptic site would be sufficient to generate the observed reduction in mIPSC frequency. We therefore recorded the mIPSCs in neurons transfected with a construct expressing a short-hairpin RNA specific for CDKL5 (shCDKL5) or, as control, against LacZ (shLacZ). Transfected neurons are easily detectable thanks to the concomitant expression of GFP from these vectors; moreover, due to the low transfection efficiency (below 5%), the synaptic input to transfected neurons is generated from non-silenced cells. Interestingly, the mIPSC frequency was reduced in shCDKL5 neurons as compared to the shLacZ controls (Figure 3H–J), thus supporting a postsynaptic CDKL5-dependent effect on presynaptic inputs.

2.3. GABAergic Defects in Cdkl5-KO Neurons Are Normalised upon Treatment with PME

The presence of $GABA_A$Rs on the neuronal membrane is determined by the dynamic balance of delivery, recycling and degradation, which altogether depends on complex regulatory mechanisms involving the interaction of various proteins, including gephyrin, with MTs [28,29]. We previously showed that various neuronal defects linked to CDKL5 deficiency in vitro and in vivo can be restored through treatment with PME, which promotes MT dynamics [13–15]. We therefore found it intriguing to analyse whether the treatment of *Cdkl5*-KO neurons with PME could restore the observed defects at the inhibitory synapse. *Cdkl5*-WT and -KO neurons were treated with 0.3 and 1 µM PME at DIV11 and stained at DIV14 for gephyrin, $GABA_AR\ \gamma_2$, bassoon or VGAT (Figure 4A–H). Whereas treatment with PME affected neither of the pre- and postsynaptic markers in WT neurons, we observed a significant effect already with 0.3 µM of PME in *Cdkl5*-KO neurons. Indeed, at the postsynaptic site, both the gephyrin and the $GABA_AR\ \gamma_2$ levels were restored to those in WT neurons (Figure 4A–D); a similar positive effect was observed also with the presynaptic markers bassoon and VGAT (Figure 4E–H). In line with this, treatment with PME could also normalise the frequency of mIPSCs in *Cdkl5*-KO neurons, whereas no effect was observed on the amplitude (Figure 4I–K).

By treating symptomatic *Cdkl5*-KO mice with PME, we previously found that CDKL5-related hippocampal-dependent behavioural and excitatory synaptic defects benefit from increased MT dynamics in vivo [15]. Intrigued by the positive effect of PME on inhibitory synapses in vitro, we proceeded to evaluate its effect in vivo, also. With this aim, we subjected hippocampal sections of *Cdkl5*-WT and -KO mice treated with 10 mg/kg of PME for 7 days (starting at PND60) to immunofluorescence staining against $GABA_AR\ \gamma_2$. As shown in Figure 5 we observed a dramatic decrease in the density of $GABA_AR\ \gamma_2$ clusters in the dentate gyrus of the vehicle-treated *Cdkl5*-KO mice with respect to the WT mice; interestingly, upon treatment with PME, the number of $GABA_AR\ \gamma_2$ clusters was similar between the two genotypes.

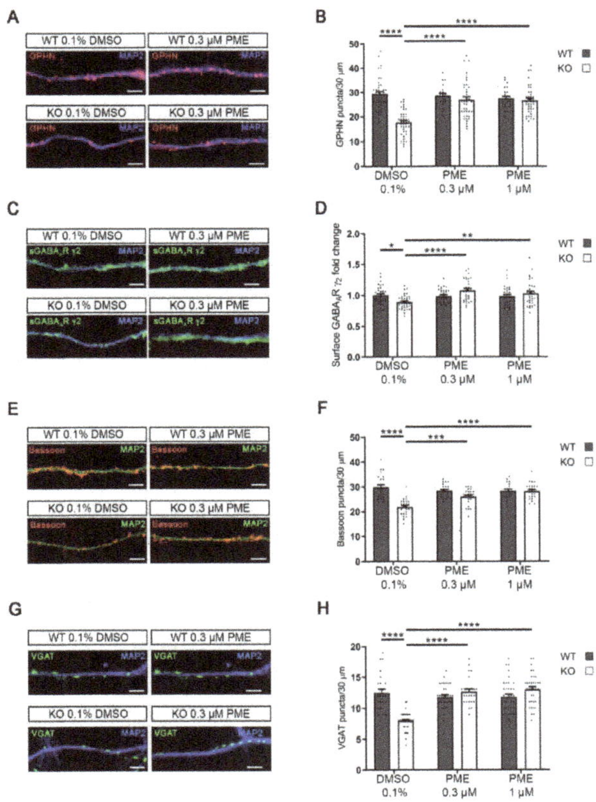

Figure 4. *Cont.*

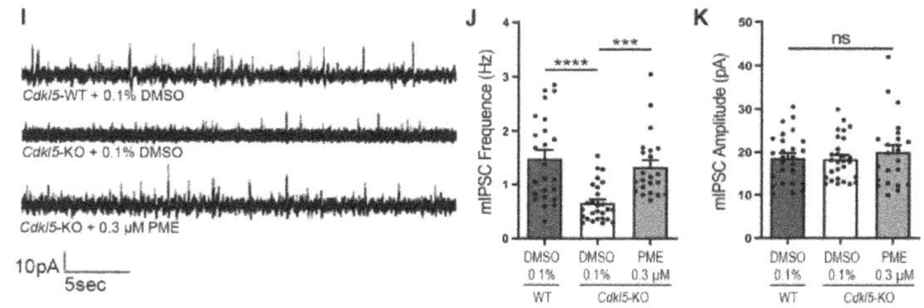

Figure 4. Treatment with PME normalises CDKL5-dependent inhibitory synaptic defects. (**A**,**C**,**E**,**G**) Representative images of *Cdkl5*-WT/KO primary hippocampal neurons stained for MAP2 together with gephyrin (GPHN, red; A), $GABA_AR$ γ_2 (green, C), bassoon (red, E) or VGAT (green, G) upon treatment with 0.3 or 1 μM of PME or vehicle (0.1% DMSO) at DIV11 for 72 h. Scale bar, 5 μm. (**B**,**D**,**F**,**H**) Graphs showing the quantification of GPHN puncta (**B**), $sGABA_AR$ γ_2 fluorescence intensity (**D**), bassoon puncta (**F**), VGAT puncta (**H**) along 30 μm long dendritic segments. GPHN: n = 6 biological replicates, N $\geq$ 19 neurons; $sGABA_AR$ γ_2: n $\geq$ 5 biological replicates, N $\geq$ 32 neurons; bassoon: n = 3 biological replicates, N $\geq$ 20 neurons; VGAT: n = 3 biological replicates, N $\geq$ 25 neurons. Mean ± SEM * $p < 0.05$; ** $p < 0.01$; *** $p < 0.001$; **** $p < 0.0001$. Two-way ANOVA, followed by Tukey's post-hoc. (**I**) Representative traces of mIPSCs in *Cdkl5*-WT and -KO primary hippocampal neurons treated as indicated with either vehicle (0.1% DMSO) or 0.3 μM PME for 72 h starting at DIV11. (**J**,**K**) Graphs showing mIPSC frequency (**J**) and amplitude (**K**) of *Cdkl5*-WT and -KO neurons treated as indicated. n = 3 biological replicates, N $\geq$ 22. Mean ± SEM. mIPSC frequency: *** $p < 0.001$, **** $p < 0.0001$. Kruskal-Wallis test, followed by Dunn's multiple comparisons test. mIPSC amplitude: ns, $p > 0.05$. One-way ANOVA, followed by Tukey's post-hoc.

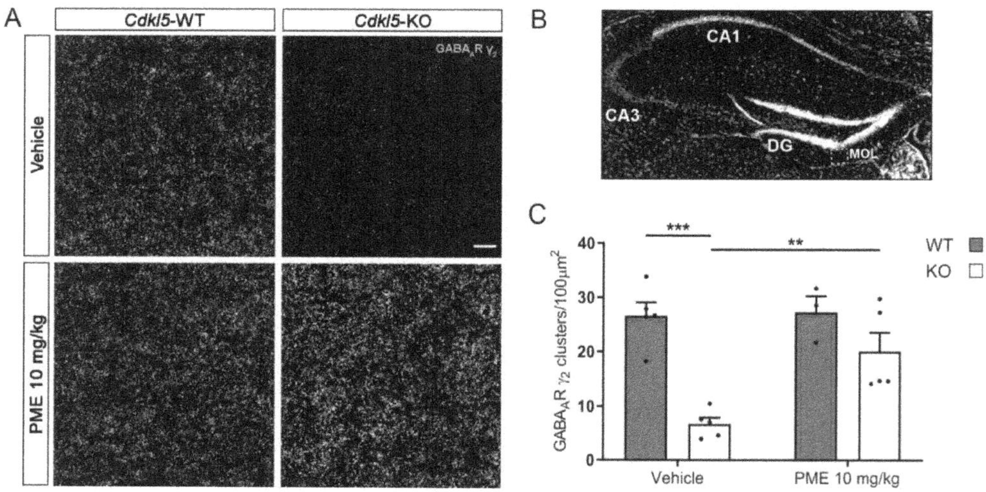

Figure 5. Reduced levels of synaptic $GABA_AR$ γ_2 in hippocampi of *Cdkl5*-KO mice are normalised upon treatment with PME. (**A**) Hippocampal slices (dentate gyrus) from *Cdkl5*-WT/KO male mice treated with PME (10 mg/kg, s.c. for seven days starting from PND60) or sesame oil (vehicle) were stained for $GABA_AR$ γ_2. Scale bar, 5 μm. (**B**) Representative image of a hippocampal section stained with the nuclear dye DAPI. Images from the molecular layer (MOL) of the dentate gyrus (DG) were used for the analyses shown in panels A and C. (**C**) Graph showing the quantification of $GABA_AR$ γ_2 clusters/100 μm^2. n $\geq$ 3. Mean±SEM, ** $p < 0.01$, *** $p < 0.001$. Two-way ANOVA followed by Tukey's post-hoc.

3. Discussion

In this study, we investigated the possible role of CDKL5 in the inhibitory synapse. Beyond the well-established role of CDKL5 at glutamatergic synapses, the results of this study reveal a hitherto undescribed function of CDKL5 in the inhibitory compartment. Our results suggest that CDKL5, through its interaction with the inhibitory scaffolding complex containing gephyrin and CB, regulates membrane levels of synaptic γ_2-containing $GABA_A$Rs. Deranged $GABA_A$R levels are frequently linked to cognitive deficits and epilepsy, and we speculate that our results may help in explaining the seizure phenotype observed in CDD patients.

3.1. CDKL5 Deficiency Leads to Dysfunctions in the Inhibitory Synapse

$GABA_A$Rs are ion channels permeable to chloride and bicarbonate ions and are, in the mammalian CNS, localised at postsynaptic inhibitory specialisations or at extrasynaptic sites, where they mediate inhibitory neurotransmission [29,30].

Herein, we analysed the synaptic $GABA_A$Rs, focusing our attention on the γ_2 subunit, which is the most abundant $GABA_A$R subunit in the rat brain [17]. Our immunofluorescence analyses showed a significant reduction of $GABA_A$R γ_2 surface expression in both the hippocampal neurons and hippocampal slices of *Cdkl5*-KO mice. This defect was corroborated by biochemical approaches. Indeed, through the highly sensitive cell-surface biotinylation assay, we showed that the absence of CDKL5 influenced surface levels of γ_2-containing $GABA_A$Rs in primary cultures of *Cdkl5*-KO neurons. Since the absence of CDKL5 did not affect the total levels of the $GABA_A$R γ_2 subunit, the reduced surface levels might be a consequence of an altered transport or recycling.

The cell membrane distribution of synaptic $GABA_A$Rs is dynamically regulated through various mechanisms, including the subsynaptic scaffolding factor gephyrin, which binds and clusters synaptic $GABA_A$Rs at sites directly opposite to GABA-releasing axon terminals [23]. Through in vitro and ex vivo immunoprecipitation experiments, we found that CDKL5 forms a complex with both CB and gephyrin. Furthermore, the reduction of surface expressed synaptic $GABA_A$Rs was accompanied by a reduction in the number of gephyrin-positive puncta in the *Cdkl5*-KO primary cultures.

The molecular interaction between CDKL5 and the cytoplasmic CB-gephyrin complex is likely a key mechanism through which CDKL5 exerts its control on synaptic $GABA_A$Rs. In particular, gephyrin takes part in the aggregation, but not in the surface insertion or stabilisation, of α_2 and γ_2 subunit-containing $GABA_A$Rs [31]. CB is a brain-specific GDP/GTP-exchange factor, which interacts with gephyrin and regulates its recruitment from intracellular deposits to postsynaptic membranes [32]. The loss of CB leads to a strong reduction in gephyrin and synaptic $GABA_A$R clusters in several regions of the forebrain, including the hippocampus, amygdala and cerebellum [33,34]. Normally, CB is present in an auto-inhibited conformation and depends on other neuronal factors such as neuroligin 2 (NL2), $GABA_A$R subunit α_2 or the Rho-like GTPase TC-10 for its activation [35,36]. Our results, showing that exogenous CDKL5 expression is sufficient for localising gephyrin under the cell membrane when coexpressed with full-length CB, suggest that CDKL5 may also be capable of relieving CB from the inhibited conformation. Our data also support a previous study by Uezu et al. [22] that identified CDKL5 as a direct interactor of CB. The presence of gephyrin in the immunocomplexes is likely caused by an indirect interaction mediated by CB. Since CB is required for inhibitory receptor clustering and function, via the recruitment of gephyrin [37], we speculate that CDKL5 plays a direct role in the stabilisation of the key components of the inhibitory synapse by the above interaction. Future studies will be performed to address how CDKL5 loss affects the gephyrin-CB complex and which $GABA_A$R subunits are affected.

3.2. CDKL5 Deficiency Impairs the Functional GABAergic Synapse

The postsynaptic inhibitory defects were accompanied by a reduction in the frequency of mIPSCs in *Cdkl5*-KO neurons. Various mechanisms can underlie the reduced frequency, such as reduced number of synapses at the pre- and postsynaptic levels, a reduced neurotransmitter release and a reduced number of presynaptic vesicles. Even if we cannot rule out a direct role for CDKL5 at the presynaptic level, we find it relevant to consider that a similar result was observed when CDKL5 expression was acutely silenced through the transfection of a CDKL5-silencing construct also expressing GFP. Our experimental settings allowed the recording of the mIPSCs of CDKL5-silenced neurons that were innervated by non-silenced cells. Therefore, the altered inhibitory neurotransmission could be ascribed to a direct role for CDKL5 at the postsynaptic site, given that the presynaptic compartment is normal. The major phenotype of CDKL5 deficiency in both *Cdkl5*-KO and silenced neurons was a remarkable decrease in mIPSC frequency, hence reflecting a strong reduction in the number of functional GABAergic synapses. We speculate that the loss of CDKL5 at the postsynaptic site influences GABAergic innervation in *Cdkl5*-KO cultures similar to what has been reported for γ_2-containing $GABA_AR$ clusters in cortical neurons silenced for GODZ, which is implicated in trafficking and postsynaptic accumulation of γ_2 subunit-containing $GABA_ARs$ [38]. In support of our hypothesis, we observed a significantly reduced number of both bassoon- and VGAT-positive puncta in the primary cultures of *Cdkl5*-KO neurons, which indicates a presynaptic defect.

3.3. PME Treatment Ameliorates CDKL5-Related Defects

Various recent data have shown that CDKL5 is involved in regulating MT dynamics [13,39–41]. Interestingly, the possibility of targeting MT interacting proteins, the function of which is impaired in the absence of CDKL5, seems to represent an interesting disease modifying therapeutic strategy for CDD [14,15,42].

Here we show that treatment in vitro with the neuroactive synthetic steroid PME restores CDKL5-dependent $GABA_AR$ defects both molecularly and functionally. Indeed, the decreased frequency of mIPSCs in primary *Cdkl5*-KO cultures was normalised, together with the surface exposure of synaptic $GABA_ARs$ and the number of gephyrin, bassoon and VGAT puncta. Intriguingly, we also found that the number of γ_2 subunit-containing $GABA_AR$ clusters was restored in the hippocampal slices of *Cdkl5*-KO mice treated with subcutaneous injections of PME (10 mg/kg) for seven consecutive days starting from PND60. Of relevance, this treatment schedule restored hippocampal-dependent learning and memory defects in *Cdkl5*-KO mice [15].

At present, we can only speculate about the precise mechanism through which PME can promote synaptic $GABA_AR$ accumulation. Our previous data showed that PME activates the +TIP CLIP170 by inducing its open conformation, thus promoting MT dynamics [15]. CLIP170 was found to be involved in the efficient loading of dynein-bound cargoes for retrograde transport in axons [43]. Dynein is central for minus-end directed transport of various cargoes and, by activating CLIP170, PME may therefore promote dynein-dependent transport. Indeed, gephyrin interacts directly with dynein and transports glycine receptor-containing vesicles in dendrites [28]. Moreover, forward trafficking of γ_2-containing $GABA_ARs$ is mediated by GABARAP-containing complexes, which interact with dynein [44]. Future studies will allow us to reveal the effect of PME on dynein-dependent transport.

In conclusion, we have demonstrated for the first time that CDKL5 plays a direct role in the expression of functional $GABA_ARs$ at synaptic sites, in part through its interaction with the cytoplasmic CB-gephyrin complex. Importantly, treatment with the synthetic neuroactive steroid PME can bypass the need for CDKL5 and restore $GABA_AR$ expression and $GABA_AR$ functioning. Currently, there are no approved therapies for CDD and any pharmacological strategies that reduce the frequency, duration or severity of seizures may positively impact the quality of life for CDD patients. PME might represent an important breakthrough in the CDD field, as the restoration of $GABA_AR$ expression might be beneficial also for the cognitive defects and autistic-like features present in CDD patients.

4. Materials and Methods

4.1. Mice

Protocols and use of animals were approved by the Animal Ethics Committee of the University of Insubria and in accordance with the guidelines released by the Italian Ministry of Health (D.L. 2014/26) and the European Community directives regulating animal research (2010/63/EU). Adult mice were euthanised by cervical dislocation, while neonates were sacrificed by exposure to CO_2 followed by decapitation.

Male *Cdkl5*-KO mice [4] on the genetic background CD1 were used. Littermate controls were used for all experiments. The day of birth was designated as postnatal day (PND) zero. After weaning, three to five animals belonging to the same litter were housed in activity enriched cages on a 12 h light/dark cycle in a temperature-controlled environment with food and water provided *ad libitum* and checked daily for general health conditions. Genotyping was performed through PCR on genomic DNA from tail biopsies using the Rapid Extract PCR Kit (PCRBIO, [15]).

4.2. Plasmids

pEGFP-GPHN, pRK5 Myc-CB2-SH3$^+$, and pRK5 Myc-CB2-ΔSH3 were kindly provided by Dr. Theophilos Papadopoulos's laboratory (Max-Planck Institute for Brain Research, Göttingen) and were generated as described previously [45]. pFlag-CDKL5 encoding the 107 kDa splice variant was generated as described elsewhere [46].

4.3. Cell Cultures and Transfections

African green monkey kidney cells (COS7) and human embryonic kidney 293T (HEK293T) cells were maintained in DMEM (Euroclone) supplemented with 10% fetal bovine serum (Euroclone), 2 mM L-glutamine (Euroclone) and penicillin/streptomycin (100 units/mL and 100 μg/mL respectively, Euroclone) at 37 °C with 5% CO_2. Cells were transfected with Lipofectamine^{TM} 3000 (Life Technologies Incorporated).

Primary hippocampal cultures were prepared from embryonic day 17 (E17) mouse embryos considering the day of the vaginal plug as E0, as described previously [47], and plated on poly-L-lysine (Sigma-Aldrich, Sant Louis, MO 63103, USA) coated plates.

CDKL5 expression was abruptly silenced in primary hippocampal neurons at DIV11 through the transfection of a shCDKL5 targeting the sequence GCAGAGTCGGCACAGCTAT, using as negative control a shRNA against LacZ (shLacZ). Plasmid transfection was performed with Lipofectamine^{TM} 2000 transfection reagent (Life Technologies Incorporated).

4.4. Pharmacological Treatment

For treatment in vitro, 0.3 or 1 μM of PME or 0.1% DMSO (vehicle) was added to the primary cultures at DIV11, and neurons were harvested after 72 h. For treatment in vivo, mice received daily subcutaneous injections from PND60 to PND66 of 10 mg/kg of PME or sesame oil (vehicle) between 9:00–11:00 as described in Barbiero et al. [15]. For neuroanatomical experiments, mice were sacrificed 24 h after ended treatment and processed as described below. PME was suspended in sesame oil and freshly prepared each day.

4.5. Antibodies

The following primary antibodies were used in immunofluorescence and western blotting experiments: mouse anti-bassoon (Santa Cruz, sc-58509), rabbit anti-CDKL5 (Sigma, HPA002847), mouse anti-CDKL5 (Santa Cruz, sc-376314), rabbit anti-collybistin (SYSY, 261003), rabbit anti-GABA$_A$R γ_2 (SYSY, 224003), goat anti-GABA$_A$R γ_2 (Invitrogen, PA5-19299), rabbit anti-GAPDH (Sigma, G9545), mouse anti-gephyrin (Santa Cruz, sc-25311), anti-GFP (chicken, Molecular Probes, A10262), chicken anti-MAP2 (SYSY, 188006), mouse anti-c-myc (clone 9E10), rabbit anti-VGAT (SYSY, 131003). Secondary Alexa Fluor anti-rabbit, -mouse and -chicken were purchased from Abcam or Invitrogen. HRP-conjugated goat anti-mouse, goat anti-rabbit and donkey anti-goat secondary antibodies for western blottings were purchased from Jackson Immunoresearch.

4.6. Immunofluorescence

COS7 cells and hippocampal neurons: neurons were grown on poly-L-lysine (1 mg/mL; Sigma-Aldrich) coated coverslips (300 cells/mm^2) until DIV14. After fixation in 4% formaldehyde (Pierce™) with 4% sucrose (Sigma-Aldrich, Sant Louis, MO 63103, USA), cells were blocked in 1X PBS, 5% goat serum (Euroclone) and 0.2% Triton X-100. Surface exposed γ_2 subunit-containing $GABA_AR$ were immunostained under non-permeabilising conditions with blocking in 1X PBS, 5% goat serum. Incubation with the primary antibody was performed overnight at 4 °C and with the secondary antibody for 1 h at room temperature. Slides were mounted with ProLong Gold antifade reagent (Life Technologies).

To quantify gephyrin, bassoon and VGAT puncta as well as surface expressed $GABA_AR$ γ_2 along MAP2-positive dendrites, images were captured with a 60X objective coupled to an Olympus BX51 fluorescence microscope equipped with Retiga R1 (QImaging) CCD camera. The number of gephyrin, bassoon and VGAT puncta, along with the fluorescence intensity of $GABA_AR$ γ_2 staining, were measured along 30 µm long segments (proximal part of secondary branches) using the software Fiji ImageJ (function: analyse particles or measure). Primary antibodies were used: bassoon, 1:50; CDKL5, 1:50; $GABA_AR$ γ_2, 1:150; gephyrin, 1:50; GFP, 1:100; MAP2 chicken, 1:500; MAP2 rabbit and mouse, 1:1000; Myc, 1:200; VGAT, 1:1000.

Hippocampal slices: 24 h after treatment, mice were decapitated (upon dislocation) and brain hemispheres were rapidly excised and frozen in liquid nitrogen. Cryosections (30 µm) were cut and mounted onto coated slides (SuperFrost® Plus, Thermo Scientific, 38116 Braunschweig, Germany) and stored at −80 °C. The samples were fixed in 2% paraformaldehyde (4 °C) for 90 s, rinsed thrice with 1X PBS and blocked for 1 h in blocking solution (0.05% goat serum, 3% Triton X-100 in 1X PBS). Upon incubation with anti-$GABA_AR$ γ_2 antibody (1:1000) in a humid chamber overnight at 4 °C, the slices were incubated with the secondary antibody (Alexa Fluor goat anti-rabbit 488 nm) in blocking solution for 1 h at room temperature, rinsed thrice with 1X PBS and mounted with ProLong Gold antifade reagent (Life Technologies). As negative control, a sample was incubated with only the secondary antibody.

Images from the molecular layer of the dentate gyrus were acquired with a LEICA TCS SL confocal microscope (LEITZ; Leica Mycrosystems, Wetzlar, Germany) with objective 63X (NA 1.32; zoom factor = 8) and the pinhole set at 1 Airy unit. Four slices per animal were analysed, and the number of $GABA_AR$ γ_2 clusters was quantified using the software Fiji ImageJ (plugin: analyse particles). Optimised threshold values and size filters were applied for all the images to identify $GABA_AR$ γ_2 clusters. The number of $GABA_AR$ γ_2 puncta was calculated in four identical sections for each slice (to have a mean of four separate zones of the dentate gyrus per single slice) and expressed per μm^2.

4.7. Western Blotting and Immunoprecipitation

Primary hippocampal neurons were lysed in 3X Laemmli buffer, and samples were separated by 10% SDS-PAGE, transferred to nitrocellulose membranes and blocked in 5% non-fat milk in TBS-T (20 mM of Tris-HCl pH 7.4, 150 mM of NaCl, 0.2% Tween-20). Blots were incubated with primary antibodies overnight at 4 °C, washed in TBS-T and incubated with appropriate secondary antibodies for 1 h at room temperature. Blots were developed with protein detection system-ECL (Genespin) coupled to G:BOX Chemi Imaging System (Syngene). Densitometric expression analyses were performed using ImageJ software.

CDKL5 was immunoprecipitated from 400 µg of HEK293T cells or 1 mg of a mouse brain extract (PND20-30) lysed in lysis buffer [mM: 50 Tris-HCl pH 7.4, 150 NaCl, 1 EDTA, 1 EGTA, 1% Triton X-100, 1X protease inhibitor cocktail (PIC, Sigma-Aldrich, Sant Louis, MO 63103, USA) and 1X PhosSTOP (Roche)] and incubated overnight at 4 °C with 1 µg of anti-CDKL5 or unrelated IgGs as control. The immunocomplexes were precipitated with protein-G agarose (Life Technologies), washed several times with lysis buffer and analysed by SDS-PAGE and western blotting.

4.8. Biotinylation Assays

Biotinylation assays were performed according to previously described protocols [48,49] with slight modifications. Primary hippocampal neurons were plated in 6-well plates coated with 0.5 mg/mL poly-L-lysine (Sigma-Aldrich, 400,000 neurons/well) and used at DIV14. Neurons were washed twice with HBSS/Ca^{2+}/Mg^{2+}, followed by incubation with 0.5 mg/mL of Sulfo-NHS-SS-Biotin (Cyanagen). Quenching was performed using HBSS/Ca^{2+}/Mg^{2+} supplemented with glycine (50 mM) and BSA (0.5%), after which cells were lysed with standard radio-immunoprecipitation assay (RIPA) buffer (mM: 50 Tris-HCl pH 7.4, 150 NaCl, 1 EDTA, 2 EGTA, 1% NP40, 0.1% SDS, 0.5% SDC, 1X PhosSTOP, 1X PIC). After correction for protein content using a BCA protein assay kit (Pierce™), biotinylated proteins were purified on StreptAvidin UltraLink Resin (Pierce™) and resolved by SDS-PAGE and western blotting. Surface expression was evaluated as the ratio between the biotinylated fraction (surface) and the total cell lysate normalised with GAPDH.

4.9. In Vitro Electrophysiological Recordings

GABA-mediated inhibitory postsynaptic currents in miniature (mIPSCs) were recorded using the patch-clamp technique in the whole cell voltage-clamp configuration in the presence of 1 μM tetrodotoxin (TTX, Tocris) to block the generation of action potentials. Recordings were obtained from primary hippocampal neurons at DIV14 (plated on poly-L-lysine coated coverslips at the density of 234 cells/mm^2) using the Axopatch 200B amplifier and the pClamp-10 software (Axon Instruments). Recordings were performed in Krebs'-Ringer's-HEPES (KRH) external solution (mM: 125 NaCl, 5 KCl, 1.2 $MgSO_4$, 1.2 KH_2PO_4, 2 $CaCl_2$, 6 glucose, 25 HEPES-NaOH pH 7.4). Recording pipettes were pulled from glass capillary (World Precision Instrument) using a two-stage puller (Narishige) and had tip resistances of 3–5 MΩ when filled with the intracellular solution (mM: 130 Cs-gluconate, 8 CsCl, 2 NaCl, 10 HEPES, 4 EGTA, 4 MgATP, 0.3 GTP pH 7.3). Voltage-clamp recordings were performed at holding potentials of +10 mV. The recorded traces were analysed using Clapfit-pClamp 10 software, after choosing an appropriate threshold. In particular, events that exceeded at least twice the standard deviation of the baseline noise were considered as mIPSCs and included in our analyses.

4.10. Statistical Analyses

All experiments and analyses were performed knowing the respective genotypes of the animals. Data were analysed with Prism software (GraphPad), and all values were expressed as the mean ± SEM. Data that were identified as significant outliers by the software were removed from the datasets. The significance of the western blotting and immunofluorescence results was evaluated using Unpaired Student's *t*-test, One-way ANOVA followed by Tukey's multiple comparisons test and Two-way ANOVA followed by Tukey's multiple comparisons test. The significance of in vitro electrophysiological studies was evaluated by Mann Whitney U test, Kruskal-Wallis test followed by Dunn's multiple comparisons test, One-way ANOVA followed by Tukey's multiple comparisons test and Unpaired Student's *t*-test. Probability values of $p < 0.05$ were considered as statistically significant.

Author Contributions: Conceptualization, R.D.R., M.T. and C.K.-N.; methodology, R.D.R., F.A. and C.K.-N.; investigation, R.D.R., S.V., C.C., I.B., C.P., M.T. and S.R.; resources, M.B.; formal analysis, R.D.R., C.P., M.T., F.A. and I.B.; writing—original draft preparation, R.D.R.; writing—review and editing, C.K.-N.; supervision, C.K.-N.; project administration, R.D.R. and C.K.-N.; funding acquisition, C.K.-N. All authors have read and agreed to the published version of the manuscript.

Funding: This research was funded by Fondazione Telethon [GGP20024], AIRETT onlus and the Italian parents' association Albero di Greta (to CKN).

Institutional Review Board Statement: The animal study protocol was approved by the Ethics Committee of the University of Insubria and by the Italian Council on Animal Care, the Italian Government decree No. 28/2019.

Informed Consent Statement: Not applicable.

Data Availability Statement: Not applicable.

Acknowledgments: We thank Theophilos Papadopoulos (Max-Planck Institute for Brain Research, Göttingen) for the CB and gephyrin constructs.

Conflicts of Interest: The authors declare no conflict of interest.

References

1. Olson, H.E.; Demarest, S.T.; Pestana-Knight, E.M.; Swanson, L.C.; Iqbal, S.; Lal, D.; Leonard, H.; Cross, J.H.; Devinsky, O.; Benke, T.A. Cyclin-dependent kinase-like 5 deficiency disorder: Clinical review. *Pediatr. Neurol.* **2019**, *97*, 18–25. [CrossRef] [PubMed]
2. Rusconi, L.; Salvatoni, L.; Giudici, L.; Bertani, I.; Kilstrup-Nielsen, C.; Broccoli, V.; Landsberger, N. CDKL5 expression is modulated during neuronal development and its subcellular distribution is tightly regulated by the C-terminal tail. *J. Biol. Chem.* **2008**, *283*, 30101–30111. [CrossRef]
3. Wang, I.T.J.; Allen, M.; Goffin, D.; Zhu, X.; Fairless, A.H.; Brodkin, E.S.; Siegel, S.J.; Marsh, E.D.; Blendy, J.A.; Zhou, Z. Loss of CDKL5 disrupts kinome profile and event-related potentials leading to autistic-like phenotypes in mice. *Proc. Natl. Acad. Sci. USA* **2012**, *109*, 21516–21521. [CrossRef] [PubMed]
4. Amendola, E.; Zhan, Y.; Mattucci, C.; Castroflorio, E.; Calcagno, E.; Fuchs, C.; Lonetti, G.; Silingardi, D.; Vyssotski, A.L.; Farley, D.; et al. Mapping pathological phenotypes in a mouse model of CDKL5 disorder. *PLoS ONE* **2014**, *9*, e91613. [CrossRef] [PubMed]
5. Okuda, K.; Kobayashi, S.; Fukaya, M.; Watanabe, A.; Murakami, T.; Hagiwara, M.; Sato, T.; Ueno, H.; Ogonuki, N.; Komano-Inoue, S.; et al. CDKL5 controls postsynaptic localization of GluN2B-containing NMDA receptors in the hippocampus and regulates seizure susceptibility. *Neurobiol. Dis.* **2017**, *106*, 158–170. [CrossRef] [PubMed]
6. Zhou, A.; Han, S.; Zhou, Z.J. Molecular and genetic insights into an infantile epileptic encephalopathy—CDKL5 disorder. *Front. Biol.* **2017**, *12*, 1–6. [CrossRef]
7. Mulcahey, P.J.; Tang, S.; Takano, H.; White, A.; Portillo, D.R.D.; Kane, O.M.; Marsh, E.D.; Zhou, Z.; Coulter, D.A. Aged heterozygous Cdkl5 mutant mice exhibit spontaneous epileptic spasms. *Exp. Neurol.* **2020**, *332*, 113388. [CrossRef]
8. Terzic, B.; Cui, Y.; Edmondson, A.C.; Tang, S.; Sarmiento, N.; Zaitseva, D.; Marsh, E.D.; Coulter, D.A.; Zhou, Z. X-linked cellular mosaicism underlies age-dependent occurrence of seizure-like events in mouse models of CDKL5 deficiency disorder. *Neurobiol. Dis.* **2021**, *148*, 105176. [CrossRef]
9. Wang, H.T.; Zhu, Z.A.; Li, Y.Y.; Lou, S.S.; Yang, G.; Feng, X.; Xu, W.; Huang, Z.L.; Cheng, X.; Xiong, Z.Q. CDKL5 deficiency in forebrain glutamatergic neurons results in recurrent spontaneous seizures. *Epilepsia* **2021**, *62*, 517–528. [CrossRef]
10. Ricciardi, S.; Ungaro, F.; Hambrock, M.; Rademacher, N.; Stefanelli, G.; Brambilla, D.; Sessa, A.; Magagnotti, C.; Bachi, A.; Giarda, E.; et al. CDKL5 ensures excitatory synapse stability by reinforcing NGL-1-PSD95 interaction in the postsynaptic compartment and is impaired in patient iPSC-derived neurons. *Nat. Cell Biol.* **2012**, *14*, 911–923. [CrossRef]
11. Sala, G.D.; Putignano, E.; Chelini, G.; Melani, R.; Calcagno, E.; Ratto, G.M.; Amendola, E.; Gross, C.T.; Giustetto, M.; Pizzorusso, T. Dendritic spine instability in a mouse model of CDKL5 disorder is rescued by insulin-like growth factor 1. *Biol. Psychiatry* **2016**, *80*, 302–311. [CrossRef]
12. Tramarin, M.; Rusconi, L.; Pizzamiglio, L.; Barbiero, I.; Peroni, D.; Scaramuzza, L.; Guilliams, T.; Cavalla, D.; Antonucci, F.; Kilstrup-Nielsen, C. The antidepressant tianeptine reverts synaptic AMPA receptor defects caused by deficiency of CDKL5. *Hum. Mol. Genet.* **2018**, *27*, 2052–2063. [CrossRef]
13. Barbiero, I.; Peroni, D.; Tramarin, M.; Chandola, C.; Rusconi, L.; Landsberger, N.; Kilstrup-Nielsen, C. The neurosteroid pregnenolone reverts microtubule derangement induced by the loss of a functional CDKL5-IQGAP1 complex. *Hum. Mol. Genet.* **2017**, *26*, 3520–3530. [CrossRef] [PubMed]
14. Barbiero, I.; Peroni, D.; Siniscalchi, P.; Rusconi, L.; Tramarin, M.; De Rosa, R.; Motta, P.; Bianchi, M.; Kilstrup-Nielsen, C. Pregnenolone and pregnenolone-methyl-ether rescue neuronal defects caused by dysfunctional CLIP170 in a neuronal model of CDKL5 Deficiency Disorder. *Neuropharmacology* **2020**, *164*, 107897. [CrossRef] [PubMed]

15. Barbiero, I.; Zamberletti, E.; Tramarin, M.; Gabaglio, M.; Peroni, D.; De Rosa, R.; Baldin, S.; Bianchi, M.; Rubino, T.; Kilstrup-Nielsen, C. Pregnenolone-methyl-ether enhances CLIP170 and microtubule functions improving spine maturation and hippocampal deficits related to CDKL5 deficiency. *Hum. Mol. Genet.* **2022**, *31*, 2738–2750. [CrossRef] [PubMed]
16. Tang, S.; Wang, I.T.J.; Yue, C.; Takano, H.; Terzic, B.; Pance, K.; Lee, J.Y.; Cui, Y.; Coulter, D.A.; Zhou, Z. Loss of CDKL5 in glutamatergic neurons disrupts hippocampal microcircuitry and leads to memory impairment in mice. *J. Neurosci.* **2017**, *37*, 7420–7437. [CrossRef]
17. Olsen, R.W.; Sieghart, W. GABAA receptors: Subtypes provide diversity of function and pharmacology. *Neuropharmacology* **2009**, *56*, 141–148. [CrossRef] [PubMed]
18. Tretter, V.; Moss, S.J. GABAA receptor dynamics and constructing GABAergic synapses. *Front. Mol. Neurosci.* **2008**, *1*, 7. [CrossRef] [PubMed]
19. Krueger-Burg, D.; Papadopoulos, T.; Brose, N. Organizers of inhibitory synapses come of age. *Curr. Opin. Neurobiol.* **2017**, *45*, 66–77. [CrossRef]
20. Rodriguez, R.A.; Joya, C.; Hines, R.M. Common ribs of inhibitory synaptic dysfunction in the umbrella of neurodevelopmental disorders. *Front. Mol. Neurosci.* **2018**, *11*, 132. [CrossRef]
21. Schulte, C.; Maric, H.M. Expanding GABAAR pharmacology via receptor-associated proteins. *Curr. Opin. Pharmacol.* **2021**, *57*, 98–106. [CrossRef] [PubMed]
22. Uezu, A.; Kanac, J.; Bradshaw, T.W.A.; Soderblom, E.J.; Catavero, C.M.; Burette, A.C.; Weinberg, R.J.; Soderling, S.H. Identification of an elaborate complex mediating postsynaptic inhibition. *Science* **2016**, *353*, 1123–1129. [CrossRef]
23. Tyagarajan, S.K.; Fritschy, J.M. Gephyrin: A master regulator of neuronal function? *Nat. Rev. Neurosci.* **2014**, *15*, 141–156. [CrossRef] [PubMed]
24. Soykan, T.; Schneeberger, D.; Tria, G.; Buechner, C.; Bader, N.; Svergun, D.; Tessmer, I.; Poulopoulos, A.; Papadopoulos, T.; Varoqueaux, F.; et al. A conformational switch in collybistin determines the differentiation of inhibitory postsynapses. *EMBO J.* **2014**, *33*, 2113–2133. [CrossRef]
25. Essrich, C.; Lorez, M.; Benson, J.A.; Fritschy, J.M.; Lüscher, B. Postsynaptic clustering of major $GABA_A$ receptor subtypes requires the $\gamma 2$ subunit and gephyrin. *Nat. Neurosci.* **1998**, *1*, 563–571. [CrossRef] [PubMed]
26. Kneussel, M.; Brandstätter, J.H.; Laube, B.; Stahl, S.; Müller, U.; Betz, H. Loss of postsynaptic GABAA receptor clustering in gephyrin-deficient mice. *J. Neurosci.* **1999**, *19*, 9289–9297. [CrossRef]
27. Jacob, T.C.; Bogdanov, Y.D.; Magnus, C.; Saliba, R.S.; Kittler, J.T.; Haydon, P.G.; Moss, S.J. Gephyrin regulates the cell surface dynamics of synaptic GABAA receptors. *J. Neurosci.* **2005**, *25*, 10469–10478. [CrossRef]
28. Maas, C.; Tagnaouti, N.; Loebrich, S.; Behrend, B.; Lappe-Siefke, C.; Kneussel, M. Neuronal cotransport of glycine receptor and the scaffold protein gephyrin. *J. Cell Biol.* **2006**, *172*, 441–451. [CrossRef]
29. Bogdanov, Y.D. Dynamic regulation of GABAA receptor biosynthesis and transport. *Neurosci. Behav. Physiol.* **2019**, *49*, 838–846. [CrossRef]
30. Comenencia-Ortiz, E.; Moss, S.J.; Davies, P.A. Phosphorylation of GABAA receptors influences receptor trafficking and neurosteroid actions. *Psychopharmacology* **2014**, *231*, 3453–3465. [CrossRef]
31. Groeneweg, F.L.; Trattnig, C.; Kuhse, J.; Nawrotzki, R.A.; Kirsch, J. Gephyrin: A key regulatory protein of inhibitory synapses and beyond. *Histochem. Cell Biol.* **2018**, *150*, 489–508. [CrossRef] [PubMed]
32. Papadopoulos, T.; Soykan, T. The role of collybistin in gephyrin clustering at inhibitory synapses: Facts and open questions. *Front. Cell Neurosci.* **2011**, *5*. [CrossRef] [PubMed]
33. Papadopoulos, T.; Korte, M.; Eulenburg, V.; Kubota, H.; Retiounskaia, M.; Harvey, R.J.; Harvey, K.; O'Sullivan, G.A.; Laube, B.; Hülsmann, S.; et al. Impaired GABAergic transmission and altered hippocampal synaptic plasticity in collybistin-deficient mice. *EMBO J.* **2007**, *26*, 3888–3899. [CrossRef] [PubMed]
34. Papadopoulos, T.; Eulenburg, V.; Reddy-Alla, S.; Mansuy, I.M.; Li, Y.; Betz, H. Collybistin is required for both the formation and maintenance of GABAergic postsynapses in the hippocampus. *Mol. Cell Neurosci.* **2008**, *39*, 161–169. [CrossRef]
35. Poulopoulos, A.; Aramuni, G.; Meyer, G.; Soykan, T.; Hoon, M.; Papadopoulos, T.; Zhang, M.; Paarmann, I.; Fuchs, C.; Harvey, K.; et al. Neuroligin 2 drives postsynaptic assembly at perisomatic inhibitory synapses through gephyrin and collybistin. *Neuron* **2009**, *63*, 628–642. [CrossRef]
36. Mayer, S.; Kumar, R.; Jaiswal, M.; Soykan, T.; Ahmadian, M.R.; Brose, N.; Betz, H.; Rhee, J.S.; Papadopoulos, T. Collybistin activation by GTP-TC10 enhances postsynaptic gephyrin clustering and hippocampal GABAergic neurotransmission. *Proc. Natl. Acad. Sci. USA* **2013**, *110*, 20795–20800. [CrossRef]
37. Körber, C.; Richter, A.; Kaiser, M.; Schlicksupp, A.; Mükusch, S.; Kuner, T.; Kirsch, J.; Kuhse, J. Effects of distinct collybistin isoforms on the formation of GABAergic synapses in hippocampal neurons. *Mol. Cell Neurosci.* **2012**, *50*, 250–259. [CrossRef]
38. Fang, C.; Deng, L.; Keller, C.A.; Fukata, M.; Fukata, Y.; Chen, G.; Lüscher, B. GODZ-mediated palmitoylation of GABAA receptors is required for normal assembly and function of GABAergic inhibitory synapses. *J. Neurosci.* **2006**, *26*, 12758–12768. [CrossRef]
39. Barbiero, I.; De Rosa, R.; Kilstrup-Nielsen, C. Microtubules: A key to understand and correct neuronal defects in CDKL5 deficiency disorder? *Int. J. Mol. Sci.* **2019**, *20*, 4075. [CrossRef]
40. Baltussen, L.L.; Negraes, P.D.; Silvestre, M.; Claxton, S.; Moeskops, M.; Christodoulou, E.; Flynn, H.R.; Snijders, A.P.; Muotri, A.R.; Ultanir, S.K. Chemical genetic identification of CDKL5 substrates reveals its role in neuronal microtubule dynamics. *EMBO J.* **2018**, *37*, e99763. [CrossRef]

41. Muñoz, I.M.; Morgan, M.E.; Peltier, J.; Weiland, F.; Gregorczyk, M.; Brown, F.C.; Macartney, T.; Toth, R.; Trost, M.; Rouse, J. Phosphoproteomic screening identifies physiological substrates of the CDKL5 kinase. *EMBO J.* **2018**, *37*, e99559. [CrossRef] [PubMed]
42. Barbiero, I.; Bianchi, M.; Kilstrup-Nielsen, C. Therapeutic potential of pregnenolone and pregnenolone methyl ether on depressive and CDKL5 deficiency disorders: Focus on microtubule targeting. *J. Neuroendocrinol.* **2022**, *34*, e13033. [CrossRef] [PubMed]
43. Nirschl, J.J.; Magiera, M.M.; Lazarus, J.E.; Janke, C.; Holzbaur, E.L.F. α-Tubulin tyrosination and CLIP-170 phosphorylation regulate the initiation of dynein-driven transport in neurons. *Cell Rep.* **2016**, *14*, 2637–2652. [CrossRef]
44. Nakamura, T.; Arima-Yoshida, F.; Sakaue, F.; Nasu-Nishimura, Y.; Takeda, Y.; Matsuura, K.; Akshoomoff, N.; Mattson, S.N.; Grossfeld, P.D.; Manabe, T.; et al. PX-RICS-deficient mice mimic autism spectrum disorder in Jacobsen syndrome through impaired GABAA receptor trafficking. *Nat. Commun.* **2016**, *7*, 10861. [CrossRef] [PubMed]
45. Harvey, K.; Duguid, I.C.; Alldred, M.J.; Beatty, S.E.; Ward, H.; Keep, N.H.; Lingenfelter, S.E.; Pearce, B.R.; Lundgren, J.; Owen, M.J.; et al. The GDP-GTP exchange factor collybistin: An essential determinant of neuronal gephyrin clustering. *J. Neurosci.* **2004**, *24*, 5816–5826. [CrossRef]
46. Williamson, S.L.; Giudici, L.; Kilstrup-Nielsen, C.; Gold, W.; Pelka, G.J.; Tam, P.P.L.; Grimm, A.; Prodi, D.; Landsberger, N.; Christodoulou, J. A novel transcript of cyclin-dependent kinase-like 5 (CDKL5) has an alternative C-terminus and is the predominant transcript in brain. *Hum. Genet.* **2012**, *131*, 187–200. [CrossRef] [PubMed]
47. Rusconi, L.; Kilstrup-Nielsen, C.; Landsberger, N. Extrasynaptic N-Methyl-D-aspartate (NMDA) receptor stimulation induces cytoplasmic translocation of the CDKL5 kinase and its proteasomal degradation. *J. Biol. Chem.* **2011**, *286*, 36550–36558. [CrossRef]
48. Arancibia-Cárcamo, I.L.; Fairfax, B.P.; Moss, S.J.; Kittler, J.T. Studying the localization, surface stability and endocytosis of neurotransmitter receptors by antibody labeling and biotinylation approaches. In *The Dynamic Synapse: Molecular Methods in Ionotropic Receptor Biology*, Chapter 6, 1st ed.; Kittler, J.T., Moss, S.J., Eds.; CRC Press/Taylor & Francis: Boca Raton, FL, USA, 2006; ISBN 9780849318917.
49. Ferreira, J.S.; Schmidt, J.; Rio, P.; Águas, R.; Rooyakkers, A.; Li, K.W.; Smit, A.B.; Craig, A.M.; Carvalho, A.L. GluN2B-containing NMDA receptors regulate AMPA receptor traffic through anchoring of the synaptic proteasome. *J. Neurosci.* **2015**, *35*, 8462–8479. [CrossRef]

International Journal of *Molecular Sciences*

MDPI

Communication

Pattern of Mitochondrial Respiration in Peripheral Blood Cells of Patients with Parkinson's Disease

Tommaso Schirinzi [1,*,†], Illari Salvatori [2,†], Henri Zenuni [1], Piergiorgio Grillo [1], Cristiana Valle [2,3], Giuseppina Martella [2,*], Nicola Biagio Mercuri [1,2,‡] and Alberto Ferri [2,3,‡]

1 Unit of Neurology, Department of Systems Medicine, Tor Vergata University of Rome, 00133 Rome, Italy
2 IRCCS Fondazione Santa Lucia, 00179 Rome, Italy
3 National Research Council, Institute of Translational Pharmacology (IFT), 00179 Rome, Italy
* Correspondence: t.schirinzi@yahoo.com or tommaso.schirinzi@uniroma2.it (T.S.); g.martella@hsantalucia.it (G.M.)
† These authors equally contributed as first.
‡ These authors equally contributed as last.

Abstract: Mitochondria are central in the pathogenesis of Parkinson's disease (PD), as they are involved in oxidative stress, synaptopathy, and other immunometabolic pathways. Accordingly, they are emerging as a potential neuroprotection target, although further human-based evidence is needed for therapeutic advancements. This study aims to shape the pattern of mitochondrial respiration in the blood leukocytes of PD patients in relation to both clinical features and the profile of cerebrospinal fluid (CSF) biomarkers of neurodegeneration. Mitochondrial respirometry on the peripheral blood mononucleate cells (PBMCs) of 16 PD patients and 14 controls was conducted using Seahorse Bioscience technology. Bioenergetic parameters were correlated either with standard clinical scores for motor and non-motor disturbances or with CSF levels of α-synuclein, amyloid-β peptides, and tau proteins. In PD, PBMC mitochondrial basal respiration was normal; maximal and spare respiratory capacities were both increased; and ATP production was higher, although not significantly. Maximal and spare respiratory capacity was directly correlated with disease duration, MDS-UPDRS part III and Hoehn and Yahr motor scores; spare respiratory capacity was correlated with the CSF amyloid-β-42 to amyloid-β-42/40 ratio. We provided preliminary evidence showing that mitochondrial respiratory activity increases in the PBMCs of PD patients, probably following the compensatory adaptations to disease progression, in contrast to the bases of the neuropathological substrate.

Keywords: Parkinson's disease; PBMCs; mitochondria; Seahorse; immunometabolic pathway; neuroinflammation; biomarkers; synaptopathy

Citation: Schirinzi, T.; Salvatori, I.; Zenuni, H.; Grillo, P.; Valle, C.; Martella, G.; Mercuri, N.B.; Ferri, A. Pattern of Mitochondrial Respiration in Peripheral Blood Cells of Patients with Parkinson's Disease. *Int. J. Mol. Sci.* **2022**, *23*, 10863. https://doi.org/10.3390/ijms231810863

Academic Editor: Hari Shanker Sharma

Received: 18 August 2022
Accepted: 13 September 2022
Published: 17 September 2022

1. Introduction

Parkinson's disease (PD) is a common neurodegenerative disorder, responsible for both motor and non-motor disturbances, whose neuropathological hallmarks are the loss of dopaminergic nigral cells and the brain accumulation of α-synuclein (α-syn)-positive Lewy bodies [1,2].

Mitochondria are highly plastic and dynamic organelles, and are critical for cellular bioenergetics and neuronal homeostasis [3]. Mitochondrial dysfunction and PD pathogenesis have been historically connected. Solid evidence from genetics and experimental models [4,5] shows how dysfunctional mitochondria lead to oxidative stress and synaptopathy, namely the earliest neurodegeneration steps in PD [6,7]. Mitochondrial impairment also precipitates lysosomes activity, protein turn-over, and α-syn metabolism, triggering or fostering neurodegeneration at every stage of PD [8]. Moreover, mitochondria are central in neuroinflammation and immune response [9], other critical determinants of PD pathology [10,11]. Accordingly, mitochondria-related immunometabolic pathways could emerge as alternative targets for disease-modifying treatments in PD [12], useful for the substitution or integration of anti-α-syn drugs, which failed to halt disease progression alone [8,13,14].

From this perspective, the role of mitochondria in PD must be further elucidated. In particular, we have to overtake current knowledge, mostly relying on preclinical studies [15], and acquire evidence directly from patients.

Peripheral blood mononucleate cells (PBMCs) exhibit typical PD-neuropathology signatures [1,16]. In addition, immune cells directly participate in the pathogenic cascade of PD, either at the central or peripheral level [11], thus representing the circulating leukocytes as an ideal tissue to analyze in vivo molecular events underlying PD.

In this study, we shaped the pattern of mitochondrial bioenergetics in vivo in PD. We thus assessed mitochondrial respiration in the PBMCs of PD patients and examined the respective correlations with clinical features and levels of cerebrospinal fluid (CSF) neurodegeneration biomarkers (α-synuclein, amyloid-β peptides, tau proteins and lactate), which are considered as an indication of pathological changes in the brain [17].

Indeed, mitochondrial respiration is a key function in cellular homeostasis, an epicenter of many metabolic pathways that are crucial for the clinical–pathological progression of PD [18], whose deeper knowledge is fundamental to design novel therapeutic intervention strategies in PD.

2. Results

2.1. Study Population

Table 1 summarizes the clinical and demographic data of the study population. Groups were homogeneous regarding age and sex distribution.

Table 1. Demographics, clinical parameters and CSF biomarkers of the study population. Age and disease duration are expressed in years; biomarkers in pg/mL. F = female; M = male; ns = not significant; other abbreviations are spelled out in the text.

	PD		**Controls**		**Significance**
Sex F/M (%)	**6/10 (38/62%)**		**6/8 (42/58%)**		**ns**
	mean	*st.dev.*	*mean*	*st.dev.*	
Age	66.7	7.5	64.1	11.3	ns
Disease Duration	3.25	2.38	-		-
H & Y	2.1	0.68	-		-
MDS-UPDRS part III	34.1	12.39	-		-
MMSE	27.5	3.01	-		-
MOCA	25.9	4.7	-		-
NMSS	49.9	38.6	-		-
LEDD	350.25	384.6	-		-
α-syn	761.66	287.9	-		-
Aβ42	996.1	340.16	-		-
Aβ40	6169.33	2052.29	-		-
Aβ42/Aβ40	0.16	0.046	-		-
t-tau	251.89	156.29	-		-
p-tau	42.01	29.8	-		-
Aβ42/p-tau	34.7	19.7	-		-
Lactate	1.4	0.25	-		-
Albumin Ratio	7.16	2.15	-		-

2.2. Bioenergetic Parameters

PBMCs obtained from PD patients and controls had a similar baseline oxygen consumption rate (OCR) (mean ± st.dev. in pmol/min) (PD = 22.4 ± 10.6; controls = 20.8 ± 8.8). ATP-linked respiration was higher in PD (40.2 ± 23.4) than in the controls (26.3 ± 9.4), although it was not statistically significant ($p = 0.07$). Maximal respiration was significantly higher in PD (155.6 ± 115.0) than in the controls (79.3 ± 29.9, $p = 0.038$). The spare (or reserve) respiratory capacity was significantly higher in PD (134.8 ± 108.2) than in the controls (58.5 ± 28.04, $p = 0.02$) (Figure 1a–e). There were no significant gender differences in either of the groups. Sample normalization was optimized through Western Blot analysis (Supplementary Figure S1).

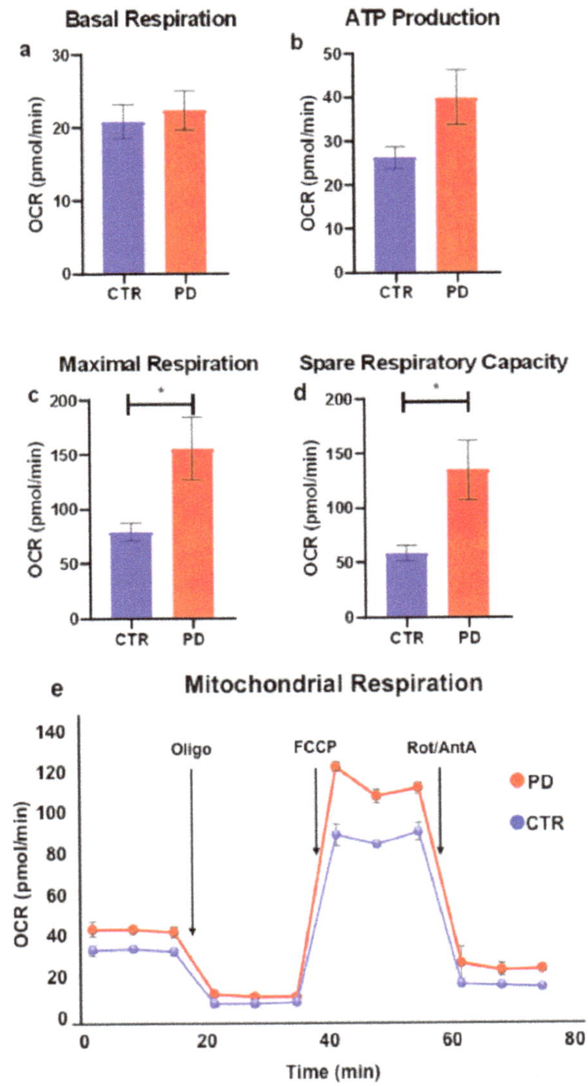

Figure 1. PBMC respirometry by Seahorse Bioscience. (**a–d**) Bar graphs showing differences in bioenergetic parameters between PD and controls (values are expressed as means ± S.E.M; * = $p < 0.05$). (**e**) Representative time course of OCR during an experimental session (one patient and one control).

2.3. Correlation Analysis

Maximal respiration was directly correlated with disease duration ($R = 0.5$, $p = 0.04$), Hoehn and Yahr scale (H & Y) stage ($R = 0.7$, $p = 0.003$), and MDS-UPDRS part III ($R = 0.62$,

$p = 0.01$). Spare respiratory capacity was directly correlated with disease duration (R = 0.5, $p = 0.04$), H & Y stage (R = 0.7, $p = 0.004$), and MDS-UPDRS part III (R = 0.63, $p = 0.009$). No further correlations emerged between bioenergetic parameters and clinical features (age, Mini-Mental State Examination (MMSE), Montreal Cognitive Assessment (MOCA) and Non-Motor Symptoms Scale (NMSS) scores, and levodopa equivalent daily dose (LEDD)).

Spare respiratory capacity was directly correlated with both amyloid-β-42 (Aβ42) and amyloid-β-42/amyloid-β-40 (Aβ42/Aβ40) ratio (R = 0.68, $p = 0.02$ and R = 0.66, $p = 0.04$, respectively). No further correlations between bioenergetic parameters and other CSF biomarkers (α-syn, total-tau (t-tau) and phosphorylated-tau (p-tau), lactate, and the CSF/blood albumin ratio) were found.

3. Discussion

PBMCs offer the opportunity to track the in vivo molecular events underlying PD, serving as a reliable model for central neuropathology [1,16]. Here, we shaped the pattern of mitochondrial respiration in PD by applying Seahorse Bioscience technology to the PBMCs of both PD patients and healthy controls, and examining the correlations of bioenergetic parameters with clinical features and levels of CSF biomarkers of neurodegeneration.

We found that, in PD, PBMCs had a peculiar pattern of mitochondrial respiration, with normal basal respiration, a significant increase in both maximal respiratory capacity and spare respiratory capacity, and a higher (but not statistically significant) ATP production. The increase in maximal respiratory capacity and spare respiratory capacity was directly associated with disease duration and severity of motor impairment, suggesting that mitochondrial respiration capability could rise in parallel with the clinical–pathological progression of PD. In addition, the spare respiratory capacity was greater in patients with higher CSF levels of Aβ42 and Aβ42/Aβ40 ratio, which identifies individuals with a lower burden of cerebral amyloidopathy [17].

In contrast with theoretical expectances, which would have predicted a reduction in bioenergetic activity in the mitochondria of PD patients [8,15,19], we discovered an increase in respiratory capacity. Albeit surprising, these data are consistent with previous findings from other experimental settings. In fact, lymphoblasts of patients with PD, namely immortalized cells derived from peripheral blood lymphocytes, also showed a dramatic increase in mitochondrial respiration, ATP synthesis and maximum OCR. The greater energy production found by Annesley et al. was assumed to satisfy an increased requirement due to the higher mitochondrial turn-over induced by pathological α-syn accumulation, or, in general, to an abnormal metabolic state with high energy consumption [20]. However, the changes in respiratory activity could follow modifications in the lipid composition of different cell compartments or in lipid trafficking overall, which markedly affect the mitochondrial architecture, the transport of proteins into mitochondria, and the functioning of respiratory proteins [21]. Indeed, mitochondrial membrane potential could be altered in the PBMCs of PD patients, suggesting a certain degree of stress in those organelles [22]. Alternatively, the growing neuroinflammatory state or the greater immune activation against progressive neurodegeneration might account for the increased bioenergetics of leukocytes in PD [20]. In addition to PBMCs, fibroblasts from PD patients also displayed higher mitochondrial respiratory rates compared to healthy controls. This has been observed either in Parkin mutant carriers [23] or in subjects with the idiopathic form [24], and interpreted as a compensatory bioenergetics activation or as a consequence of abnormal mitochondrial functioning.

In line with the hypothesis that the implementation of PBMC bioenergetic activity could help cellular compensations to the progression of neuropathology and neuroinflammation, we noticed that maximal and spare respiratory capacities tended to increase with disease evolution, either in terms of disease duration or clinical impairment.

Nuclear factor erythroid 2-related factor 2 (Nrf2) is a transcription factor regulating the cellular defense against oxidative stress, inflammation and neurodegeneration, by promoting overall mitochondrial respiration and metabolism [25]. We recently demonstrated that Nrf2 was highly expressed and its pathway activated in the PBMCs of PD patients, especially in those with a longer disease duration [1]. It is thus reasonable that the progressive increase in respiratory capacity over the clinical–pathological course of PD may follow an Nrf2-mediated mitochondrial activation aimed at sustaining a systemic defensive response. Of interest, similar findings were observed in the lymphoblasts of patients with Amyotrophic Lateral Sclerosis, where the respiratory activity was increased together with the upregulation of the Nrf2 pathway, suggesting this axis as a common compensation mechanism among different sporadic neurodegenerative diseases [26].

Then, we found that, in PD patients, the respiratory capacity was directly correlated with CSF levels of the Aβ42 to Aβ42/Aβ40 ratio. Experimental models of Alzheimer's disease showed that neurons exhibit mitochondrial dysfunction and respiration impairment in the presence of Aβ42 peptides [27]. A reduction in CSF Aβ42 corresponds to the greater brain accumulation of amyloid-β plaques even in PD [17]. Accordingly, we could hypothesize that the poorer respiratory activity in patients with lower CSF Aβ42 (and worse amyloidopathy in the CNS) reflects a detrimental association between mitochondrial performances and pathological amyloid peptides, as well as at a peripheral level. In fact, PD patients with lower CSF Aβ42 present a more malignant and frailer phenotype [28], which is consistent with the weakening of defensive or compensatory mechanisms.

In contrast, other CSF biomarkers were not correlated with bioenergetic parameters, preventing further suppositions on the interactions between central neuropathology and systemic reactions.

Although limited by the sample size, this study demonstrated that PBMC mitochondria in PD patients had a peculiar pattern of respiration, with increased maximal and spare respiratory capacities. Respiratory changes probably reflect the increased energetic requirement due to the clinical–pathological progression of the disease and the subsequent compensatory adaptations. Such changes vary depending on the disease stage and the neuropathological substrate; they are more contained in patients with biochemical signatures of frailty.

We could interpret our results in two ways. Accepting PBMCs as a model of central neuropathology, they allow for a better understanding of the role of mitochondria in PD, profiling those dynamics that may underlie PD pathology throughout its course. On the other hand, they provide additional evidence on the presence of metabolic abnormalities in systemic immune cells, which could be useful in developing novel therapeutic strategies. Indeed, mitochondria-related metabolic pathways could be specifically targeted in immune cells to modulate their activity either at the central or systemic level, counteracting neuroinflammation and neurodegeneration. Further investigations on larger replication cohorts are now necessary to confirm and extend our findings.

4. Methods

4.1. Study Population and Biosampling

This study involved sixteen PD patients and fourteen sex/age-matched healthy controls enrolled at Tor Vergata University Hospital (Rome, Italy) in 2021–2022. PD was diagnosed according to MDS 2015 Postuma's criteria. Controls were healthy volunteers without a history or clinical signs of neurological diseases. Subjects with main acute/chronic infectious/inflammatory/internal diseases and/or abnormal blood cell count were excluded [1].

For each subject, demographics, anthropometrics, and medical history were collected. PD patients were assessed using the Hoehn and Yahr scale (H & Y), MDS-UPDRS part III, Non-Motor Symptoms Scale (NMSS), Mini-Mental State Examination (MMSE) adjusted for age and educational level, Montreal Cognitive Assessment (MOCA), and the levodopa equivalent daily dose (LEDD) calculation.

All participants underwent venous blood sampling (20 mL), in the morning, after overnight fasting (morning drugs allowed). Blood was immediately processed to separate PBMCs through density gradient centrifugation with Ficoll-Hypaque (GE Healthcare Life

Sciences), according to standard procedures. PBMCs were carefully frozen in cryoSFM medium, stored in liquid nitrogen and subsequently thawed for bioenergetics analyses [29].

Ten PD patients also underwent cerebrospinal fluid (CSF) analysis for neurodegeneration related biomarker measurement. CSF was obtained through lumbar puncture, which was performed following standard procedures. The CSF levels of total α-synuclein (α-syn), amyloid-β-42 (Aβ42) and amyloid-β-40 (Aβ40), total-tau (t-tau) and phosphorylated-tau (p-tau), lactate, and the CSF/blood albumin ratio were quantified as previously described [2,30]. The Aβ42/Aβ40 and the Aβ42/p-tau ratios were then calculated.

The study was approved by the local EC (protocol n° 16.21), following the principles of the declaration of Helsinki. All participants signed informed consent.

4.2. Bioenergetics Analysis

Mitochondrial function was determined using a Seahorse XF96e Analyzer (Seahorse Bioscience—Agilent, Santa Clara, CA, USA) [31]. The PBMCs were plated at the density of 15×10^4 cells/well. An equal number of cells, from each sample, was processed for a quantitative Western Blot assay (as described below). A mitochondrial stress test was performed according to Agilent's recommendations. In brief, growth medium was replaced with XF test medium (Eagle's modified Dulbecco's medium, 0 mM glucose, pH = 7.4; Agilent Seahorse) supplemented with 1 mM pyruvate, 10 mM glucose and 2 mM L-glutamine. Before the assay, the PBMCs were incubated in a 37 °C incubator without CO_2 for 45 min to allow the pre-equilibration of the assay medium. The test was performed by first measuring the baseline oxygen consumption rate (OCR), followed by sequential OCR measurements after the injection of oligomycin (1.5 μM), carbonyl cyanide 4-(trifluoromethoxy) phenylhydrazone (1 μM) (FCCP) and Rotenone (0.5 μM) + Antimycin A (0.5 μM). This allowed the measurement of the key parameters of the mitochondrial function, including the basal respiration, the ATP-linked respiration (obtained through the oligomycin-induced inhibition of ATP synthase with a subsequent decrease in electron flow through the electron transport chain, ETC), the maximal respiration (obtained through the uncoupling effect of FCCP, which induces ETC to operate at maximal capacity), the spare respiratory capacity (the difference between maximal respiration and basal respiration), and the non-mitochondrial ATP production (obtained through the rotenone/antimycin A-induced inhibition of both complex I and III).

4.3. Western Blot Analysis

Western Blot analysis was performed on protein extracts according to Scaricamazza et al. 2022 [32]. Specifically, 15×10^4 PBMCs from each sample were lysed in 100 μL RIPA buffer (50 mM Tris–HCl pH 7.4, 0.5% Triton X-100, 0.25% Na-deoxycholate, 0.1% SDS, 250 mM NaCl, 1 mM EDTA and 5 mM $MgCl_2$), 25 μL of which was loaded on SDS–polyacrylamide gel electrophoresis and transferred to nitrocellulose membranes (Perkin Elmer, Cat# NBA085B). Membranes were probed using β-Actin antibody (Santa Cruz Biotechnology Inc., Cat# sc-47778, WB 1:5000).

4.4. Statistical Analysis

The distribution of variables was preliminarily examined using the Shapiro–Wilk test, and the non-normally distributed variables were Log10+1 transformed for analysis when necessary. Categorical variables were compared using a chi-square test. Group analysis was conducted via parametric (one-way ANOVA or Student's *t*-test) or non-parametric tests, as appropriate; correlations were evaluated using Spearman's test. Statistical significance was set at $p < 0.05$. A blind analysis was run using IBM-SPSS-23. Data are available from the authors upon reasonable request.

Supplementary Materials: The following supporting information can be downloaded at: https://www.mdpi.com/article/10.3390/ijms231810863/s1

Author Contributions: T.S., I.S. and A.F. conceived the study, analyzed the data and wrote the manuscript. H.Z., P.G. and C.V. collected the data and performed the experimental session. N.B.M. and G.M. provided supervision and edited the final manuscript. All authors have read and agreed to the published version of the manuscript.

Funding: This research received no external funding.

Institutional Review Board Statement: The study was conducted in accordance with the Declaration of Helsinki, and approved by the Ethics Committee of Fondazione PTV—Policlinico Tor Vergata (Rome, Italy) (protocol n° 16.21).

Informed Consent Statement: Informed consent was obtained from all subjects involved in the study.

Data Availability Statement: Data are available from the authors upon reasonable request.

Conflicts of Interest: The authors declare no conflict of interest.

References

1. Petrillo, S.; Schirinzi, T.; Di Lazzaro, G.; D'Amico, J.; Colona, V.L.; Bertini, E.; Pierantozzi, M.; Mari, L.; Mercuri, N.B.; Piemonte, F.; et al. Systemic activation of Nrf2 pathway in Parkinson's disease. *Mov. Disord.* **2020**, *35*, 180–184. [CrossRef] [PubMed]
2. Schirinzi, T.; Sancesario, G.M.; Di Lazzaro, G.; Biticchi, B.; Colona, V.L.; Mercuri, N.B.; Bernardini, S.; Pisani, A. CSF α-synuclein inversely correlates with non-motor symptoms in a cohort of PD patients. *Parkinsonism Relat. Disord.* **2018**, *61*, 203–206. [CrossRef] [PubMed]
3. Ma, K.; Chen, G.; Li, W.; Kepp, O.; Zhu, Y.; Chen, Q. Mitophagy, Mitochondrial Homeostasis, and Cell Fate. *Front. Cell Dev. Biol.* **2020**, *8*, 467. [CrossRef] [PubMed]
4. Martella, G.; Madeo, G.; Maltese, M.; Vanni, V.; Puglisi, F.; Ferraro, E.; Schirinzi, T.; Valente, E.M.; Bonanni, L.; Shen, J.; et al. Neurobiology of Disease Exposure to low-dose rotenone precipitates synaptic plasticity alterations in PINK1 heterozygous knockout mice. *Neurobiol. Dis.* **2016**, *91*, 21–36. [CrossRef]
5. Madeo, G.; Schirinzi, T.; Martella, G.; Latagliata, E.C.; Puglisi, F.; Shen, J.; Valente, E.M.; Federici, M.; Mercuri, N.B.; Puglisi-Allegra, S.; et al. PINK1 heterozygous mutations induce subtle alterations in dopamine-dependent synaptic plasticity. *Mov. Disord.* **2014**, *29*, 41–53. [CrossRef]
6. Schirinzi, T.; Madeo, G.; Martella, G.; Maltese, M.; Picconi, B.; Calabresi, P.; Pisani, A. Early synaptic dysfunction in Parkinson's disease: Insights from animal models. *Mov. Disord.* **2016**, *31*, 802–813. [CrossRef]
7. Imbriani, P.; Schirinzi, T.; Meringolo, M.; Mercuri, N.B.; Pisani, A. Centrality of Early Synaptopathy in Parkinson's Disease. *Front. Neurol.* **2018**, *9*, 103. [CrossRef]
8. Trinh, D.; Israwi, A.R.; Arathoon, L.R.; Gleave, J.A.; Nash, J.E. The multi-faceted role of mitochondria in the pathology of Parkinson's disease. *J. Neurochem.* **2021**, *156*, 715–752. [CrossRef]
9. Angajala, A.; Lim, S.; Phillips, J.B.; Kim, J.H.; Yates, C.; You, Z.; Tan, M. Diverse roles of mitochondria in immune responses: Novel insights into immuno-metabolism. *Front. Immunol.* **2018**, *9*, 1605. [CrossRef]
10. Krashia, P.; Cordella, A.; Nobili, A.; La Barbera, L.; Federici, M.; Leuti, A.; Campanelli, F.; Natale, G.; Marino, G.; Calabrese, V.; et al. Blunting neuroinflammation with resolvin D1 prevents early pathology in a rat model of Parkinson's disease. *Nat. Commun.* **2019**, *10*, 3945. Available online: http://www.nature.com/articles/s41467-019-11928-w (accessed on 1 August 2022). [CrossRef]
11. Magistrelli, L.; Storelli, E.; Rasini, E.; Contaldi, E.; Comi, C.; Cosentino, M.; Marino, F. Relationship between circulating CD4+ T lymphocytes and cognitive impairment in patients with Parkinson's disease. *Brain Behav. Immun.* **2020**, *89*, 668–674. [CrossRef] [PubMed]
12. de Oliveira, L.G.; Angelo Y de, S.; Iglesias, A.H.; Peron, J.P.S. Unraveling the Link Between Mitochondrial Dynamics and Neuroinflammation. *Front. Immunol.* **2021**, *12*, 752. [CrossRef] [PubMed]
13. Lang, A.E.; Espay, A.J. Disease Modification in Parkinson's Disease: Current Approaches, Challenges, and Future Considerations. *Mov. Disord.* **2018**, *33*, 660–677. [CrossRef]
14. Espay, A.J. Movement disorders research in 2021: Cracking the paradigm. *Lancet Neurol.* **2022**, *21*, 10–11. Available online: http://www.thelancet.com/article/S1474442221004130/fulltext (accessed on 1 August 2022). [CrossRef]
15. Nicoletti, V.; Palermo, G.; Del Prete, E.; Mancuso, M.; Ceravolo, R. Understanding the Multiple Role of Mitochondria in Parkinson's Disease and Related Disorders: Lesson From Genetics and Protein–Interaction Network. *Front. Cell Dev. Biol.* **2021**, *9*, 493. [CrossRef] [PubMed]
16. Avenali, M.; Cerri, S.; Ongari, G.; Ghezzi, C.; Pacchetti, C.; Tassorelli, C.; Valente, E.M.; Blandini, F. Profiling the Biochemical Signature of GBA-Related Parkinson's Disease in Peripheral Blood Mononuclear Cells. *Mov. Disord.* **2021**, *36*, 1267–1272. Available online: https://onlinelibrary.wiley.com/doi/full/10.1002/mds.28496 (accessed on 1 August 2022). [CrossRef] [PubMed]
17. Schirinzi, T.; Di Lazzaro, G.; Sancesario, G.M.; Summa, S.; Petrucci, S.; Colona, V.L.; Bernardini, S.; Pierantozzi, M.; Stefani, A.; Mercuri, N.B.; et al. Young-onset and late-onset Parkinson's disease exhibit a different profile of fluid biomarkers and clinical features. *Neurobiol. Aging* **2020**, *90*, 119–124. [CrossRef]

18. Area-Gomez, E.; Guardia-Laguarta, C.; Schon, E.A.; Przedborski, S. Mitochondria, OxPhos, and neurodegeneration: Cells are not just running out of gas. *J. Clin. Investig.* **2019**, *129*, 34–45. [CrossRef]
19. Imbriani, P.; D'Angelo, V.; Platania, P.; Di Lazzaro, G.; Scalise, S.; Salimei, C.; El Atiallah, I.; Colona, V.L.; Mercuri, N.B.; Bonsi, P.; et al. Ischemic injury precipitates neuronal vulnerability in Parkinson's disease: Insights from PINK1 mouse model study and clinical retrospective data. *Parkinsonism Relat. Disord.* **2020**, *74*, 57–63. [CrossRef]
20. Annesley, S.J.; Lay, S.T.; De Piazza, S.W.; Sanislav, O.; Hammersley, E.; Allan, C.Y.; Francione, L.M.; Bui, M.Q.; Chen, Z.P.; Ngoei, K.R.; et al. Immortalized Parkinson's disease lymphocytes have enhanced mitochondrial respiratory activity. *Dis. Model. Mech.* **2016**, *9*, 1295–1305. [CrossRef]
21. Fais, M.; Dore, A.; Galioto, M.; Galleri, G.; Crosio, C.; Iaccarino, C. Parkinson's Disease-Related Genes and Lipid Alteration. *Int. J. Mol. Sci.* **2021**, *22*, 7630. [CrossRef]
22. Qadri, R.; Namdeo, M.; Behari, M.; Goyal, V.; Sharma, S.; Mukhopadhyay, A.K. Alterations in mitochondrial membrane potential in peripheral blood mononuclear cells in Parkinson's Disease: Potential for a novel biomarker. *Restor. Neurol. Neurosci.* **2018**, *36*, 719–727. [CrossRef] [PubMed]
23. Haylett, W.; Swart, C.; Van der Westhuizen, F.; Van Dyk, H.; Van der Merwe, L.; Van der Merwe, C.; Loos, B.; Carr, J.; Kinnear, C.; Bardien, S. Altered Mitochondrial Respiration and Other Features of Mitochondrial Function in Parkin-Mutant Fibroblasts from Parkinson's Disease Patients. *Parkinson's Dis.* **2016**, *2016*, 6230370. [CrossRef]
24. Antony, P.M.; Kondratyeva, O.; Mommaerts, K.; Ostaszewski, M.; Sokolowska, K.; Baumuratov, A.S.; Longhino, L.; Poulain, J.F.; Grossmann, D.; Balling, R.; et al. Fibroblast mitochondria in idiopathic Parkinson's disease display morphological changes and enhanced resistance to depolarization. *Sci. Rep.* **2020**, *10*, 1569. [CrossRef] [PubMed]
25. Holmström, K.M.; Kostov, R.V.; Dinkova-Kostova, A.T. The multifaceted role of Nrf2 in mitochondrial function [Internet]. *Curr. Opin. Toxicol.* **2017**, *2*, 80–91. [CrossRef]
26. Lastres-Becker, I.; Porras, G.; Arribas-Blázquez, M.; Maestro, I.; Borrego-Hernández, D.; Boya, P.; Cerdán, S.; García-Redondo, A.; Martínez, A.; Martin-Requero, Á. Molecular Alterations in Sporadic and SOD1-ALS Immortalized Lymphocytes: Towards a Personalized Therapy. *Int. J. Mol. Sci.* **2021**, *22*, 3007. [CrossRef]
27. Zhang, L.; Trushin, S.; Christensen, T.A.; Tripathi, U.; Hong, C.; Geroux, R.E.; Howell, K.G.; Poduslo, J.F.; Trushina, E. Differential effect of amyloid beta peptides on mitochondrial axonal trafficking depends on their state of aggregation and binding to the plasma membrane. *Neurobiol. Dis.* **2018**, *114*, 1. [CrossRef]
28. Fereshtehnejad, S.M.; Zeighami, Y.; Dagher, A.; Postuma, R.B. Clinical criteria for subtyping Parkinson's disease: Biomarkers and longitudinal progression. *Brain* **2017**, *140*, 1959–1976. [CrossRef]
29. Tiberi, M.; Evron, T.; Saracini, S.; Boffa, L.; Mercuri, N.B.; Chintalacharuvu, S.R.; Atamas, S.P.; Chiurchiù, V. Potent T cell-mediated anti-inflammatory role of the selective CB2 agonist lenabasum in multiple sclerosis. *Neuropathol. Appl. Neurobiol.* **2022**, *48*, e12768. Available online: https://onlinelibrary.wiley.com/doi/full/10.1111/nan.12768 (accessed on 1 August 2022). [CrossRef]
30. Schirinzi, T.; Zenuni, H.; Grillo, P.; Bovenzi, R.; Guerrera, G.; Gargano, F.; Pieri, M.; Bernardini, S.; Mercuri, N.B.; Battisitini, L.; et al. Tau and Amyloid-β Peptides in Serum of Patients With Parkinson's Disease: Correlations With CSF Levels and Clinical Parameters. *Front. Neurol.* **2022**, *13*, 155. Available online: https://www.frontiersin.org/articles/10.3389/fneur.2022.748599/full (accessed on 1 August 2022). [CrossRef]
31. Salvatori, I.; Ferri, A.; Scaricamazza, S.; Giovannelli, I.; Serrano, A.; Rossi, S.; D'Ambrosi, N.; Cozzolino, M.; Giulio, A.D.; Moreno, S.; et al. Differential toxicity of TAR DNA-binding protein 43 isoforms depends on their submitochondrial localization in neuronal cells. *J. Neurochem.* **2018**, *146*, 585–597. [CrossRef] [PubMed]
32. Scaricamazza, S.; Salvatori, I.; Amadio, S.; Nesci, V.; Torcinaro, A.; Giacovazzo, G.; Primiano, A.; Gloriani, M.; Candelise, N.; Pieroni, L.; et al. Repurposing of Trimetazidine for amyotrophic lateral sclerosis: A study in SOD1G93A mice. *Br. J. Pharmacol.* **2022**, *179*, 1732–1752. [CrossRef] [PubMed]

International Journal of
Molecular Sciences

MDPI

Review

Synaptic Plasticity Dysfunctions in the Pathophysiology of 22q11 Deletion Syndrome: Is There a Role for Astrocytes?

Eva Cristina de Oliveira Figueiredo [1,†], Bianca Maria Bondiolotti [1,†], Anthony Laugeray [1] and Paola Bezzi [1,2,*]

1 Department of Fundamental Neurosciences, University of Lausanne, 1005 Lausanne, Switzerland; evacristina.deoliveirafigueiredo@unil.ch (E.C.d.O.F.); biancambondiolotti@gmail.com (B.M.B.); anthony.laugeray@unil.ch (A.L.)
2 Department of Pharmacology and Physiology, University of Rome Sapienza, 00185 Rome, Italy
* Correspondence: paola.bezzi@unil.ch or paola.bezzi@uniroma1.it
† These authors contributed equally to this work.

Abstract: The 22q11 deletion syndrome (DS) is the most common microdeletion syndrome in humans and gives a high probability of developing psychiatric disorders. Synaptic and neuronal malfunctions appear to be at the core of the symptoms presented by patients. In fact, it has long been suggested that the behavioural and cognitive impairments observed in 22q11DS are probably due to alterations in the mechanisms regulating synaptic function and plasticity. Often, synaptic changes are related to structural and functional changes observed in patients with cognitive dysfunctions, therefore suggesting that synaptic plasticity has a crucial role in the pathophysiology of the syndrome. Most interestingly, among the genes deleted in 22q11DS, six encode for mitochondrial proteins that, in mouse models, are highly expressed just after birth, when active synaptogenesis occurs, therefore indicating that mitochondrial processes are strictly related to synapse formation and maintenance of a correct synaptic signalling. Because correct synaptic functioning, not only requires correct neuronal function and metabolism, but also needs the active contribution of astrocytes, we summarize in this review recent studies showing the involvement of synaptic plasticity in the pathophysiology of 22q11DS and we discuss the relevance of mitochondria in these processes and the possible involvement of astrocytes.

Keywords: 22q11 deletion syndrome; synaptic plasticity; synapses; mitochondria; astrocytes

check for updates

Citation: de Oliveira Figueiredo, E.C.; Bondiolotti, B.M.; Laugeray, A.; Bezzi, P. Synaptic Plasticity Dysfunctions in the Pathophysiology of 22q11 Deletion Syndrome: Is There a Role for Astrocytes? *Int. J. Mol. Sci.* **2022**, *23*, 4412. https://doi.org/10.3390/ijms23084412

Academic Editor: Giuseppina Martella

Received: 15 March 2022
Accepted: 15 April 2022
Published: 16 April 2022

1. Introduction

The 22q11 deletion syndrome (DS), also known as DiGeorge or velocardiofacial syndrome, which occurs in one out of 4000 live births is caused by a hemizygous microdeletion in the long arm of chromosome 22, the most frequent of which are a three megabase (Mb) deletion that affects ~60 genes and a 1.5 Mb deletion affecting 35 genes [1,2]. It is thought that the deletion is caused by a non-allelic homologous recombination due to the presence of a low copy number of repeats flanking the region that predispose to ectopic recombination during meiosis [3]. Most affected subjects (85%) have a 3 Mb deletion encompassing approximately 30 contiguous genes, but ~15% have smaller atypical deletions of 1.5 Mb [4] that seem to contain all the genes necessary for the development of the syndrome [5] and an increased risk of psychosis [6,7].

The syndromic nature of the disorder gives rise to various phenotypes that can involve different organs and tissues, but most patients (~76%) have congenital heart defects, palatal and renal abnormalities, characteristic craniofacial dysmorphisms and hypocalcaemia, and are affected by immune deficiency and learning difficulties that lead to major developmental delays (see [3] for the detailed pathophysiological mechanisms and a comprehensive review). The 22q11DS is also associated with a strikingly high risk of developing neuropsychiatric disorders: about 25–30% of patients develop affective psychosis or schizophrenia, thus making the syndrome one of the greatest risk factors for psychotic disorders identified so far [8,9]. The 22q11.2 microdeletion accounts for up to 2% of all cases of schizophrenia and is the only known recurrent copy number variation (CNV) responsible for new cases of schizophrenia [4,7].

Furthermore, the prevalence of other neuropsychiatric disorders (Figure 1) such as autism spectrum disorder (ASD), anxiety, and attention deficit hyperactivity disorder (ADHD) [10,11] is considerably higher in 22q11DS patients than in the general population [12–14]. ADHD and anxiety disorders are the most frequent diagnoses during childhood, whereas the rates of psychosis and mood disorders increase dramatically during adolescence and young adulthood. Although the average age of 22q11DS patients at the time of the onset of overt psychotic disorders is 19–26 years [15], earlier manifestations of psychotic-like symptoms characterise almost one-third of adolescents and ~17% of pre-adolescent children [12,16], thus suggesting that the severity of the psychotic symptoms progressively worsens.

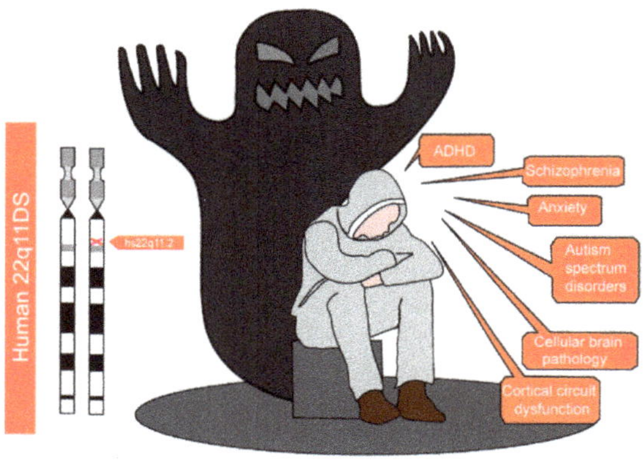

Figure 1. Phenotypic features of 22q11.2 Deletion Syndrome (22q11DS). The chromosomic location of the deletion (**left**) and the most common manifestations of the disease (**right**) are shown. Starting from childhood and early adolescence 22q11.2 subjects face a wide range of neuropsychiatric illnesses which include attention deficit disorder (ADHD), schizophrenia spectrum disorder, anxiety, and autism spectrum disorders. A baseline cellular brain pathology and the consequent cortical circuit dysfunction are at the root of these alterations.

The 22q11.2DS is associated with a wide range of cognitive impairments [13,17]. Patients generally have poor non-verbal, numerical, and spatio-temporal skills [18] and their impaired attention system [19] prevents them from distinguishing relevant from irrelevant information and fails to inhibit the impulsive responses that interfere with the brain's ability to focus on goal-relevant thoughts. Furthermore, those with psychotic symptoms are associated with cognitive deficits and a decline in academic performance. It has been reported that measures of executive functions such as working memory and sustained attention reveal greater deficits in 22q11.2DS patients with schizophrenia than in those without [20–22], and longitudinal studies following adolescents with 22q11.2DS into adulthood have shown that the IQ of patients with psychotic disorders is lower than in those without [23].

Studies of 22q11DS patients and mouse models (see next paragraph) have indicated that the cognitive deficits may be due to deficits in synaptic plasticity in the cortical [24,25] and hippocampal circuits that are known to be involved in various aspects of planning, working memory, rule-based learning, attention, and emotional regulation [26,27]. Plasticity refers to the ability of neural activity to modify the neural circuit functions generated by an experience (such as subsequent thoughts, feelings, and behaviour). "Synaptic plasticity" specifically refers to activity-dependent modifications of the strength or efficacy of synaptic transmission at pre-existing synapses. For more than a century, it has been suggested that this mechanism plays a central role in the capacity of the brain to incorporate transient experiences into persistent memory traces [28,29] and it is therefore also thought to be critically involved in the early development of neural circuits [30]. Many forms and mechanisms of short- and long-term synaptic plasticity have been described (including paired-pulse facilitation and depression, facilitation and depression following trains of stimuli, and the modulation of transmission by pre-synaptic receptors), and the temporal domains of such changes range from milliseconds to hours, days, and presumably even longer [30]. Over the last ten years, several studies have indicated that deficits in synaptic plasticity may contribute to the cognitive deficits associated with many neurodevelopmental disorders, including 22q11DS [6,31–36], which suggests that elucidating the mechanisms underlying synaptic plasticity may be a crucial step in improving our understanding of the disorders themselves.

Furthermore, there is increasing evidence suggesting that astrocytes may be involved in some forms of short-term and long-term plasticity [37–42] by changing their synaptic coverage, controlling the clearance of neurotransmitters, or by releasing neuroactive substances (i.e., gliotransmitters, cytokines, and chemokines) that can directly affect synaptic efficacy [37,39,42].

2. Synaptic Plasticity and the Pathophysiology of 22q11DS

After having exposed how behavioural and cognitive deficits are an integral part of the pathogeny of the 22q11DS and highly evolve throughout a patients' life, we now focus on details regarding the extent to which synaptic plasticity alterations are crucially involved in the occurrence of cognitive defects during the disease. With this aim, we first emphasize how 22q11 deleted genes are likely to affect synaptic plasticity processes in animal models of the disease, and then, describe how mitochondrial defects and altered astrocytic development are closely entangled and may ultimately underlie cognitive deficits associated with synaptic plasticity changes.

Over the last ten years, it has been suggested that behavioural and cognitive impairments underlying most neuropsychiatric diseases are probably due to alterations in the mechanisms regulating synaptic function and plasticity [43]. In the case of 22q11DS, many of the transcribed genes in the human minimally critical 1.5 MB and the larger 3 MB deleted region encode proteins involved in synaptic processes, and studies of animal models have confirmed the role of many 22q11 genes in regulating synaptic transmission and plasticity [4,35]. Furthermore, in comparison with controls, histological studies of cell-derived neurons (hiPSC cells and human organoids) from 22q11DS patients have confirmed the synaptic deficit by revealing a reduction in dendritic arborisation and spines [6,32,36] as well as in synapse markers [32].

Neuronal cell dysmorphologies are common to many 22q11DS phenotypes (cardiovascular, craniofacial and limb malformations, and thymic dysplasia), which suggests that the heterozygous 1.5 or 3 MB deletion and the consequent reduction in 22q11 genes disrupt the development of various systems, including the brain. Similarly, the significant susceptibility of 22q11DS patients to the development of schizophrenia, ASD, ADHD, language delay, and other behavioural alterations [15,44], all of which are features of neurodevelopmental disorders [45–47], suggests that 22q11DS disrupts brain development as well as peripheral and visceral morphogenesis.

It has also been suggested that 22q11 genes may differentially operate at distinct times during development [48]: for example, a reduction in a subset of genes expressed at sites of mesenchymal/epithelial interaction may compromise early brain, face, heart, and limb morphogenesis; a reduction in the genes regulating the cell cycle may disrupt neurogenesis in the cerebral cortex; and a reduction in a distinct subset of mitochondrial genes may compromise post-natal astrocyte formation and maturation, dendrite and axon formation, and synaptogenesis. Such combined and/or recurrent dysfunctions in 22q11DS genes may disrupt the differentiation and maturation of brain cells from early embryonic to late post-natal development, and this may explain the variable behavioural pathology and dysmorphologies associated with 22q11DS. Neuronal cell dysmorphologies are also consistent with widespread reductions in the volume of, particularly, the hippocampus and pre-frontal and temporal cortices [49–53] and this suggests alterations in the functional connectivity of, particularly, fronto-limbic circuitry [54–56].

However, although these cellular and morphological observations suggesting the possible involvement of altered synaptic plasticity in 22q11DS do not explain the complex symptoms associated with 22q11DS pathology, circuit-level explanations may be a means of integrating the information coming from studies of patients and animal models. Animal studies indicate that altered synaptic function in hippocampal and cortical brain structures is crucially involved in behavioural and cognitive impairments observed in most neuropsychiatric diseases [43]. In line with this, many laboratories have developed complementary models that relate synaptic changes to the structural and functional changes observed in patients with behavioural and cognitive dysfunctions [4,57].

The orthologous murine region of the human 22q11 locus lies on mouse chromosome 16, and all but one of the human genes in this region are represented, although they are organised in a different order [4,58]. Several mouse models carrying chromosomal deficiencies that are in synteny with the human 22q11.2 microdeletion have been generated [4,59,60], including $Df1/^{+}$ and $LgDel^{+/-}$ mice carrying a hemizygous deletion of 18 and 24 genes respectively in the 22q11DS-related region of mouse chromosome 16 [60–62] (Figure 2). Mouse models of 22q11DS are among the few animal models that replicate the abnormalities associated with the syndrome in humans: for example, they reveal cognitive deficits in the conditioned contextual fear paradigm, an assay that partially depends on the hippocampus and pre-frontal cortex (PFC) [6,63,64], and have shown abnormalities in corticogenesis and development of dendrites and dendritic spines in hippocampal and PFC pyramidal neurons [4,6,27,57,65,66], highlighting the importance of correct neuronal development. As previously said, cognitive deficits in 22q11 patients are mainly associated with a diagnosis of schizophrenia [67,68] and are now considered better predictors of disease progression than any other symptoms [23,69]. As in the case of schizophrenia, it is thought that 22q11DS-related cognitive symptoms (particularly deficits in spatial working memory) originate from the hippocampal and cortical regions involved in learning and memory [70,71]. Of note, these alterations have been shown to occur both in 22q11DS patients [14,51] and 22q11DS mouse models [26,27]. Furthermore, mouse models of 22q11DS reveal abnormal short- and long-term hippocampal synaptic plasticity [36,57,62,72,73], which is consistent with the idea that synaptic plasticity is a cellular mechanism of learning and memory [74,75].

Many studies have highlighted alterations in synaptic plasticity using animal models in which the genes forming the genetic basis of the psychiatric and cognitive symptoms of 22q11DS [76] have been deleted or mutated. For example, it has been found that a genetic variant of *ZDHHC8* (a gene predisposing to schizophrenia) [76,77] is causally related to a reduction in the strength of synaptic connections and alterations in the terminal arborisation of both cortical and hippocampal neurons [65]. *ZDHHC8* is a palmitoyltransferase that belongs to a 23-member family of enzymes sharing a conserved cysteine-rich signature catalytic domain (the DHHC domain) [78] and it has been shown that palmitoylation is a key reversible post-translational protein modification involved in protein trafficking and the regulation of various membrane and cytosolic proteins, especially in neurons [79]. The alterations in synaptic strength were accompanied by impaired mPFC-hippocampus connectivity and spatial working memory, one of the main cognitive features of early-stage schizophrenia [80]. Most of the synaptic dysfunctions observed in these *ZDHHC8*-deficient mice are also present in the $Df(16)A^{+}/^{-}$ mice model of 22q11DS carrying the 1.3 MB microdeletion in the mouse locus that is in synteny with the human 22q11.1 locus encompassing 27 genes [62].

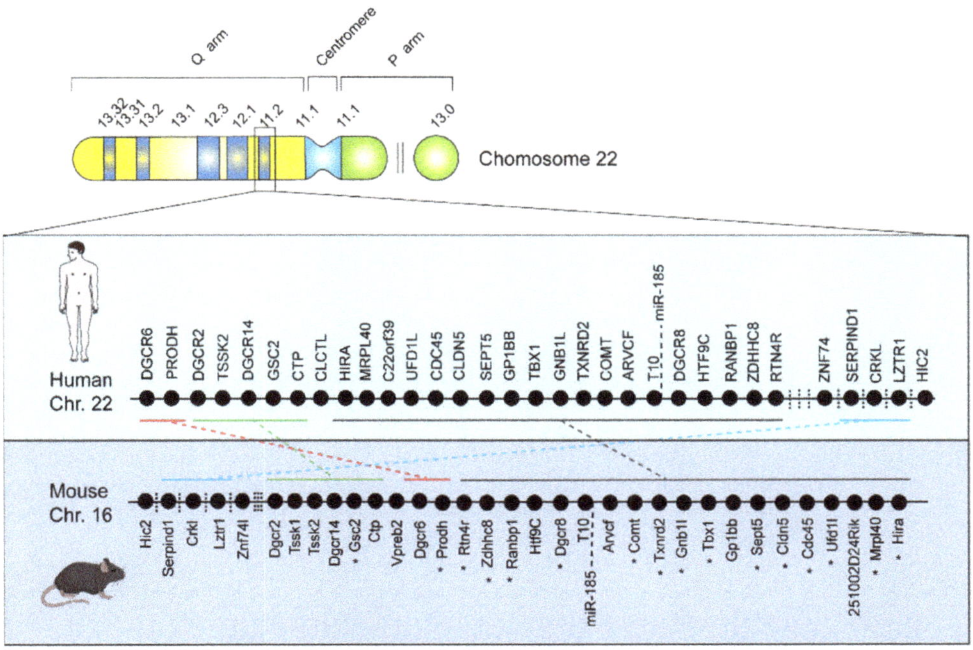

Figure 2. Genetic background of 22q11 DS. The ideogram of chromosome 22 is shown: there is a short (p) arm and long (q) arm, along with the centromere. The 22q11.2 deletion occurs in the q arm, as indicated by the box in the 22q11.2 band. The black dots represent the genes within the commonly deleted region of Chr. 22 in Humans (light blue box) and of the corresponding Chr. 16 in Mice (blue box). Dashed lines connecting groups of identical genes depict the important analogy between these genomic regions in Humans and in Mouse Models. Asterisks indicate mitochondrial genes. This characteristic allows the use of animal models to study molecular mechanisms that cannot be easily inferred working solely with human subjects.

It has also been found that the effects of ZDHHC8 are at least partially mediated by the Cdc42-dependent modulation of Akt/Gsk3b signalling, which is consistent with the increasingly supported association between dysregulated Akt/Cdc42 signalling dysregulation and schizophrenia in humans [81]. Furthermore, pre-clinical studies have demonstrated that impaired AKT signalling affects neuronal connectivity and neuromodulation within the PFC [82] and have identified AKT as a key signalling intermediary downstream of dopamine (DA) receptor 2 (DRD2), the most established target of the antipsychotic drugs used to treat schizophrenic patients. This strengthens the hypothesis of a relationship between 22q11 genes, impaired synaptic plasticity, and the cognitive and behavioural deficits observed in schizophrenia-expressing 22q11DS patients.

One of the genes deleted in most patients with 22q11DS is the DiGeorge critical region gene 8 (*DGCR8*), which encodes a crucial component of the micro-processing complex that contributes to the biogenesis of microRNA (miRNA). Approximately 22 nucleotides long, miRNAs regulate gene expression primarily by means of post-transcriptional gene silencing after binding to their target RNAs, and are involved in many biological processes including development, cell death, and cell metabolism [83,84]. Given the crucial role of 22q11 gene-mediated alterations in synaptic plasticity as the neural substrate underlying cognitive dysfunctions and increased risk of developing schizophrenia associated with 22q11 microdeletions, Fénelon et al. investigated the effect of *DGCR8* deficiency on the structure and function of cortical circuits by assessing their laminar organisation and the neuronal morphology and synaptic properties of layer 5 pyramidal neurons in the pre-frontal cortex of DGCR8$^{+}/^{-}$ mutant. They found that they had fewer cortical layer 2/4 neurons, and smaller spines in the basal dendrites of their layer 5 pyramidal neurons. In addition to these structural changes, field potential and whole-cell electrophysiological recordings of layer 5 of the PFC showed greater short-term synaptic depression in response to the stimulation of superficial cortical layers. As a key component of the micro-processing complex essential for miRNA production [85], it is likely that *DGCR8* significantly contributes to the miRNA dysregulation observed in the brain of Df(16)A$^{+}/^{-}$ mice [72].

It has been shown that a number of miRNAs regulate neuronal synaptic plasticity by means of the local synthesis of proteins at synaptic level [86]. In line with this, it has also been recently shown that the miRNA dysregulation due to *DGCR8* deficiency leads to deficits in various forms of prefrontal cortical synaptic plasticity, and is also likely to induce schizophrenia-related symptoms in Df(16)A$^{+}/^{-}$ mice [87]. Moreover, a reduction in the levels of *Mirta22* (an inhibitor of neuronal maturation whose expression is up-regulated in the brain of Df(16)A$^{+}/^{-}$ mice as a result of the hemizygosity of *DGCR8*) [88] rescued not only key schizophrenia-related cognitive and behavioural dysfunctions, but also abnormalities in PFC synaptic and structural plasticity.

Interestingly, Sivagnanasundram et al. found that the expression of 15 genes other than those involved in the initial deletion was significantly modified in the hippocampus of Df1/$^{+}$ mice, [89]. Five of these genes (*Calm1, Pcdh8, Ube2d2, Uble1b*, and *Ywhaz*) are known to be involved in learning and memory processes and/or schizophrenia-like phenotypes frequently observed in 22q11DS patients. Calmodulin 1, which is encoded by *Calm1* and plays a key role in synaptic plasticity by regulating a large number of enzymes and proteins [90], a process that shapes long-term neuronal function (particularly long-term hippocampal potentiation and spatial learning) by modifying the phosphorylation/dephosphorylation balance of downstream target proteins [91,92], was downregulated by 27%. Furthermore, a reduction in the expression of *Pcdh8* and *Ywhaz* may also contribute to impairing synaptic plasticity in Df1/$^{+}$ mice as the first encodes protocadherin 8, which modulate the synaptic plasticity that is essential to the process of learning and memory [93]. The second belongs to the 14-3-3 family and is thought to play a key role in neuronal differentiation, synaptic plasticity, and olfactory learning and memory [94]. Interestingly, it has been reported that disturbances in *Pcdh8* are closely associated with autistic-like symptoms [95], a clinical feature that is frequently observed in 22q11DS patients [96] and alterations in the expression of *Ywhaz* and other members of the 14-3-3 family have been identified in the PFC and cerebellum of schizophrenic patients [97]. These findings highlight the importance of impaired synaptic plasticity in the etiology of 22q11DS, particularly when it occurs within the PFC [98] or hippocampus [99] or along the PFC-hippocampus pathway [100].

3. Role of Mitochondrial Genes in the Dysfunctions of Synaptic Plasticity Associated with 22q11 DS

Intriguingly, nine of the many genes deleted in 22q11DS can induce mitochondrial homeostasis disorders (*COMT, UFD1L, DGCR8, MRPL40, PRODH, SLC25A1, TXNRD2, T10*, and *ZDHHC8*), but the first three may have no more than an indirect effect on mitochondrial function, and only the last six encode mitochondrial proteins [101]. These six genes are maximally expressed in mouse brain shortly after birth at a time of active synaptogenesis, which suggests that well-functioning mitochondrial processes may be intimately related to correct synapse formation. If this is so, alterations in mitochondrial metabolism during the early phases of brain development may pave the way for impaired neuronal metabolism or synaptic signalling, and thus partially explain the higher incidence of developmental and behavioural deficits in 22q11 patients. This review does not intend to cover precisely mitochondrial functions through brain development but, rather, to point out how deletion of the above-mentioned genes is causally related to cognitive deficits induced by synaptic plasticity disruption. In line with this, a recent study demonstrated a close correlation between the age-dependent decline in the working memory of monkeys and an increased prevalence of doughnut-shaped mitochondria in PFC presynaptic boutons [102] due to oxidative stress [103]. It is also worth noting that, of the six genes mentioned above, *TXNRD2* encodes the mitochondrial enzyme thioredoxin-reductase 2, which is of paramount importance for the mitochondrial scavenging of reactive oxygen species (ROS) [104]. Another recent study found that specific ROS accumulation in PFC layer 2-3 projecting neurons due to *TXNRD2* haploinsufficiency is not only causally related to alterations in local dendritic branching, but also to cognitive deficits measured during a touchscreen-mediated visual discrimination/reversal task [57]. It was also found that treating LgDel mouse models from birth to weaning with maternal drinking water containing the antioxidant N-acetylcysteine completely restored working memory in young adult mice, thus highlighting the central role of mitochondrial ROS-mediated stress in the occurrence of cognitive deficits in this 22q11DS mice model.

In the same year, Gokhale et al. published a study showing an association between proteome changes in the SLC25A1-SLC25A4 mitochondrial interactome and synapse function in the Df(16)$A^{+/-}$ mice model of 22q11DS [105]. *SLC25A1* encodes a mitochondrial citrate transport protein that allows exchanges of small metabolites between the mitochondrial matrix and the cytosol [106] and it was shown that its knockdown in zebrafish induced mitochondria depletion and proliferation defects, two phenotypes frequently described in 22q11DS patients [106]. More recently, another study demonstrated that the deletion of human *SLC25A1* compromises the integrity of the mitochondrial ribosome and down-regulates the expression of multiple ribosome subunits, including MRPL40 [107], which disrupts short-term synaptic plasticity and working memory by dysregulating mitochondrial calcium [73]. *MRPL40* encodes the mitochondrial ribosomal protein L40, a protein of the large subunit of the mitochondrial ribosome whose function is necessary for mitochondrial protein synthesis in eukaryotes [108]. As previously mentioned, *ZDHHC8* encodes a mitochondrial palmitoyl transferase enzyme that is involved in regulating the cell localisation of target proteins [109] and a reduction in its expression reduces the strength of synaptic connections and impairs terminal arborisation in the hippocampus-PFC circuit of a mice model of 22q11DS [65].

Lastly, it has been found that dysregulation of the *PRODH* gene, which encodes the mitochondrial enzyme proline dehydrogenase, is associated with the occurrence of schizophrenia-like phenotypes in humans [110] and animal models, probably because of alterations in glutamatergic and dopaminergic transmission, especially in the PFC and hippocampus [66]. Interestingly, *PRODH* and *ZDHHC8* both have single nucleotide polymorphisms (SNPs) that are closely associated with 22q11DS-related schizophrenia-like symptoms [111]. It has also been reported that the *T10* gene (also known as TANGO2), which encodes a member of the transport and Golgi organisation family that plays a role in endoplasmic reticulum (ER) secretory protein loading [112], has associated SNPs but at a less robust level [111].

Taken together, these findings not only underline the crucial role of mitochondrial alterations in the etiology of cognitive impairments observed in 22q11DS patients, but also indicate that these alterations are likely to be due to dysfunctions in synaptic processes, which is in line with the fact that two major mitochondrial functions (energy supply and calcium buffering) are proven regulators of synaptic plasticity [113]. In connection with this, Earls et al. showed that the dysregulation of calcium dynamics in the presynaptic terminals of CA3 neurons in Df1/$^{+}$ mice alters long-term potentiation at excitatory synapses [36]. They also showed that these changes are due to the up regulation of sarcoendoplasmic reticulum calcium (Ca^{2+}) ATPase (SERCA2), which is responsible for loading Ca^{2+} into the ER, and that the resulting increase in ER calcium load triggers enhanced neurotransmitter release and increases long term potentiation (LTP) levels [114]; furthermore, the depletion of calcium stores induced by a SERCA2 inhibitor fully rescues synaptic phenotypes observed in Df1/$^{+}$ mice [36]. Interestingly, it has been found that SERCA2 levels are increased in brains of patients with idiopathic schizophrenia, thus providing a direct link between the processes described above and 22q11Ds-related psychiatric deficits.

4. Astrocytes Can Modulate Synaptic Plasticity: Do They Play a Role in the Pathophysiology of 22q11DS?

As reviewed so far, altered synaptic plasticity through mitochondrial deficits seems to be strongly involved in cognitive symptoms reported in 22q11DS patients. Even though it has long been considered that synapses have the inherent property of plasticity, a property based on mechanistic changes occurring within neurons [115], neurons do not function in isolation but belong to elaborate glial networks in which they are intimately associated with astrocytes [116]. A growing body of evidence suggests that correct synaptic functioning may involve the active participation of astrocytes [116–118]. It is possible that astrocytes may also be at least partially responsible for the regulation of synaptic plasticity [119–121]. Therefore, in the next sections of this review, we focus on how developing astrocytes may be involved in 22q11DS related synaptic plasticity and cognitive deficits through changes in dopamine and mitochondrial homeostasis.

Astrocytes have functional and structural domains that are estimated to be in contact with 200–600 dendrites and about 10^5 synapses [122,123] and their extensive contacts with synaptic sites ensure strict control of local ions [124], pH homeostasis [125], the delivery of metabolic substrates to neurons [126–129], control of the microvasculature [130], and modulation of synaptic activity and plasticity by releasing neuroactive substances [42,131–138]. Perisynaptic astrocytes also express many transporters of amino acid neurotransmitters, including glutamate and GABA, and remove neurotransmitters to maintain transmitter homeostasis [139], thus assuring the rapid and efficient control of the speed and extent of neurotransmitter clearance, a mechanism involved in synaptic plasticity [140,141].

Recent studies showed that astrocytes also remove monoamine neurotransmitters, and various research groups demonstrated that cultured astrocytes and astrocytoma transport monoamines such as serotonin, dopamine, norepinephrine, and histamine [142–145]. Transcriptome and immunohistochemistry analyses confirmed the presence of monoamine transporters in in vivo astroglial cells, including organic cation transporter 3 (OCT3) [145–150] and plasma membrane monoamine transporter (PMAT) [150]. OCT3 and PMAT form part of the uptake system of low-affinity (higher Km values) and high-capacity transporters (higher Vmax values) of monoamines such as serotonin and dopamine regardless of extracellular Na^+/Cl^-, and recent studies indicated the critical role of low-affinity transporters in in vivo serotonin and dopamine clearance [151,152], although the relative importance of the two uptake systems is still unknown.

It was recently suggested that the taking up of dopamine by the astroglial OCT3 and PMAT transporters in the developing PFC indicates their involvement in controlling dopamine homeostasis [150]. The PFC is different from other dopaminergic brain areas such as the striatum insofar as it expresses much lower levels of the high-affinity dopamine transporter (DAT) and dopamine uptake by DAT plays a marginal role in clearing extra-cellular dopamine levels [153]. Studies of the mechanisms regulating dopamine homeostasis in the PFC showed that dopamine clearance in the presence of low DAT levels depends on secondary mechanisms such as the metabolic activity of monoamine oxidase B (MAOB) and cathecol-o-methyltransferase (COMT), and uptake of the norepinephrine transporter [154–157].

Astrocytes in the developing PFC are equipped to control dopamine homeostasis as those expressing OCT3/PMAT and MAOB/COMT contain vesicular monoamine transporter 2 (VMAT2), which acts in concert with OCT3 and MAOB to provide effective control of the metabolic capacity of dopamine [150]. The dopamine entering via OCT3/PMAT transporters accumulates in VMAT2-positive intra-cellular organelles (i.e., endosomes or lysosomes), which are highly dynamic in terms of fusions and fissions and passively leak dopamine into the cytoplasm, where it is degraded by metabolic enzymes. This leakage seems to be a crucial step in the metabolism of catecholamines (including dopamine) in various cells including neurons [158]. In the absence of the VMAT2-dependent control of dopamine storage in astrocytes, the dopamine taken up by plasma membrane transporters is immediately metabolised, which triggers the aberrant uptake of the transporters and a consequently significant decrease in the extra-cellular dopamine levels [150] whereas, by expressing the determinants of dopamine control, astrocytes maintain the correct equilibrium of extra-cellular dopamine levels. In line with the particular nature of the PFC, this mechanism seems to be quite PFC-specific as VMAT2 is highly expressed by astrocytes located in the frontal and pre-frontal cortical areas, but absent from those in other cortical areas such as the visual cortex [150]. Interestingly, when this astroglial mechanism is perturbed (for example, in the absence of VMAT2), the concomitant decrease in extra-cellular dopamine levels may have significant effects on synaptic plasticity [150] as it triggers the accelerated spine maturation of pyramidal neurons in layer 5 of the PFC, increases the frequency of miniature excitatory postsynaptic currents (mEPSCs), and compromises LTP. Furthermore, the effect on synaptic strength is consistent with accelerated synaptic development, and this suggests that appropriate dopamine levels act as a developmental repressor via the activation of pre-synaptic D2 receptors [150]. In the absence of VMAT2, the alteration in synaptic plasticity is also accompanied by defects in executive cognitive functions, one of the hallmarks of schizophrenia and 22q11DS. The fact that the deletion of VMAT2 and subsequent decrease in dopamine causes excessive neuronal excitation, strongly suggests a neural network that is resistant to experience-dependent refinement and therefore prone to cognitive and behavioural deficits.

Is it possible that these mechanisms are also involved in the pathophysiology of the cognitive deficits associated with 22q11? The *COMT* gene encoding the cytoplasmic COMT enzyme, which (in addition to mitochondrially located MAOB) acts as a key degrader of dopamine, is contained within the critical 1.5 Mb 22q11 DS microdeletion. The dopamine hypothesis of schizophrenia, which has greatly influenced research into the mechanisms underlying the onset of psychosis, suggests that *COMT* is an attractive candidate gene underlying susceptibility to schizophrenia [159–162], and studies of $Comt^{-}/^{-}$ mice have demonstrated the ability of the COMT enzyme to control the extra-cellular clearance of PFC dopamine and influence cognitive functions such as working memory [6,163,164]. It can therefore be argued that reduced COMT activity may slow extra-cellular clearance and lead to the aberrant tonic stimulation of dopamine receptors in the PFC, a situation that has been associated with perseverative behaviour and inflexible cognitive performances [165,166] which is a hallmark of the cognitive deficits associated with schizophrenia [166].

When considering the role of *COMT* in the pathophysiology of 22q11DS-related cognitive dysfunctions, it is necessary to see *COMT* in the context of an epitastic interaction with the *PRODH* gene, which is also within the 22q11 microdeletion and encodes mitochondrially expressed proline dehydrogenase (PRODH). As previously mentioned, the PRODH enzyme is the rate-limiting enzyme in proline degradation [167] and the homozygous deletion of *PRODH* is associated with hyperprolinemia, which significantly increases proline levels in the brain and leads to significant neuropsychiatric dysfunctions [6,167]. Studies of mouse models and 22q11DS patients have provided evidence supporting the idea of an epistatic interaction between *COMT* and *PRODH* as they both functionally converge on the dopaminergic system [66,168] and give rise to a hyper-dopaminergic state that may predispose to psychosis and schizophrenia [66]. Initial evidence of such an interaction came from a transcriptome analysis of a study of PRODH-deficient mice [66,169] showing the up-regulation of *COMT* and thus suggesting the possible up-regulation of the enzyme in order to compensate for the *PRODH* deficiency.

The mechanism by which *PRODH* deficiency leads to an over-activated dopaminergic system is still an important unknown, but one intriguing hypothesis is that it is due to dysregulated astroglial control of dopamine homeostasis. According to a recent transcriptome analysis of isolated astrocytes [146,148,170] contain all the genes deleted in 22q11DS to some extent or another, and *PRODH* seems to be one among the 30 genes that are highly expressed by astrocytes [146,148,170], thus suggesting that the PRODH enzyme plays an important role in regulating some astroglial functions. Although this role has not yet been investigated in detail, several studies have shown that it is involved in many cell functions other than proline catabolism, including providing energy, shuttling redox potentials between cell compartments, and producing reactive oxygen species [171]. Proline oxidation is an excellent source of energy and the oxidation of one molecule of proline can yield 30 ATP equivalents [171], thus suggesting that PRODH deficiency may cause an equivalent deficit in cell energy.

In line with the hypothesis that PRODH contributes to cell energy demand, it is highly expressed during post-natal development [172], when the brain grows to approximately 70% of its adult size and brain tissues undergo extensive remodelling. Interestingly, this post-natal growth is due to a rapid increase in the number of neuropil and glial cells, particularly astrocytes. During their post-natal maturation, astrocytes undergo extensive morphological and functional changes [132,159], begin to express the determinants of dopamine uptake, storage, and metabolism, while becoming competent in regulating dopamine homeostasis [150]. Over the last ten years, many studies have shown that defective astrocytic maturation has profound effects on developmental synaptogenesis and circuit formation and function [173,174] and any disruption in astrocyte maturation can be expected to confound the construction and functional architecture of neural networks significantly [175]. The mechanisms regulating post-natal astrocyte maturation have been extensively investigated [176], and the findings of a very recent study suggest that mitochondria biogenesis may play a role in the cellular processes regulating morphogenesis [177]. Developing astrocytes contain a highly interconnected network of mitochondria, but the conditional deletion of the metabolic regulator PPARγ co-activator 1α (PGC-1α) and the consequent interference with mitochondrial biogenesis impairs their morphological maturation and decreases the formation and function of excitatory synapses in the PFC [177], thus underlining the importance of mitochondria to post-natal astrocyte development. The PGC-1α in developing astrocytes can be modulated by the expression of metabotropic glutamate receptor 5 (mGluR5), one of the well characterized signalling ways in astrocytes that has been shown to be highly expressed in developing astrocytes [178,179]. Therefore, the temporal relationship between post-natal maturation of astrocytes and synapses suggests the existence of bidirectional interactions based on the release of glutamate that orchestrate the maturation of functional circuits [132,180]. The mechanism by which altered mitochondrial biogenesis in developing astrocytes affects the formation and maturation of synapses is not known but, as it has been suggested that mitochondrial impairments are also involved in 22q11DS, these findings offer an astrocyte-based mechanism of neural pathology. It is possible that the PRODH and/or mitochondrial deficiency caused by the microdeletion impairs the post-natal maturation of astrocytes and consequently causes the decrease in synaptogenesis and the neuronal cell dysmorphogenesis associated with the syndrome. It is also tempting to speculate that the lack of post-natal maturation somehow affects the way in which astrocytes acquire their dopaminergic competence and therefore gives rise to the hyperdopaminergic state and dysregulated synaptic plasticity associated with 22q11DS.

5. Conclusions and Future Directions

Studies in 22q11DS mouse models have revealed neural circuit dysfunctions and a variety of synaptic plasticity alterations that may be responsible for the cognitive impairments observed in the syndrome. Given the critical role of synaptic plasticity in sculpting and in passaging information within neural circuits, network alterations arise because neuronal structure and synapse formation are affected. To investigate these features of the 22q11DS, different studies have focused on hippocampal and cortical neurons, both in vivo and in vitro.

Changes in synaptic transmission, strength, and number of synapses and spines, as well as changes in neuronal excitability have been observed extensively in 22q11DS mouse models. Taken together, all recent findings confirm that synaptic plasticity is involved in the pathophysiological mechanisms of 22q11DS.

Because active synaptogenesis occurs straight after birth and coincides with the increased expression of specific mitochondrial proteins, which are deleted in the syndrome, the possibility of a direct link between mitochondrial metabolism and neuronal function and synapse formation is almost certain. The relationship between mitochondria and neuronal maturation is incredibly tight; indeed, not only the behaviour of these organelles is shaped by development, but they themselves appear to regulate different stages of neurogenesis. Indeed, their importance has been pointed out in different neuronal studies and also recently in glial cells. In fact, we propose a possible role of astrocytes in the pathophysiology of 22q11DS, because of their active participation in synapse formation and function. Mitochondrial impairment in astrocytes lead to important dysfunctions in neurons, therefore their importance is further supported.

Indeed, as a result of the complexity of the syndrome and because of the need for new therapeutical strategies and targets, mitochondria seem to be a good potential target.

Individuals suffering from 22q11DS present an important decrease in their quality of life, mostly because of their psychiatric manifestations. From a therapeutic point of view, not much has been explored, not only because of the complexity of the developing brain and the hard task of reaching specific brain structures solely with a pharmacological product, but also mainly because of the phenotypic variability of the syndrome. Considerable research is still needed in order to learn about the molecular aspects of this condition and find a potential molecular mechanism to target the treatment of the disease.

6. Strengths and limitations and Search Strategy

The main asset of the present review is to provide new insights on how synaptic plasticity disturbances described in 22q11DS may be related to abnormal maturation of astrocytes because of mitochondrial defects, making such disturbances a new potential target for treating psychiatric and cognitive problems associated with this syndrome. To avoid complexifying this assumption, we voluntary restricted our analysis to studies not dealing with in-depth mechanistic information, but simply gave an overview of the most recent findings regarding 22q11DS and the involvement of synaptic plasticity deficits in the occurrence of cognitive abnormalities, and then to link it to developing astrocytes and mitochondria disturbances. In addition, even though the findings reported in the present review originate from animal studies, they have the merit of pointing out the importance of implementing translational studies by using, for instance, hiPSC [181] or the cutting-edge technology of human cerebral organoids that has recently been shown to be suitable to study synaptic processes [182].

7. Search Strategy and Selection Criteria

Data for this review were identified by searches of PubMed mainly focused over the last 20 years. References from relevant articles were obtained using the search terms (i) mitochondria or mitochondrial biogenesis; (ii) 22q11 DS or 22q11 deletion syndrome; (iii) schizophrenia or psychosis; (iv) developmental disorders or neuropsychiatric disorders; (v) cognition deficits; (vi) astrocytes, and (vii) synaptic plasticity. In addition, we paid particular attention to use both historical and the most up-to-date references to respectively address well-known assertions and the more recent and pioneering works.

Author Contributions: E.C.d.O.F., A.L., B.M.B. and P.B. contributed to drafting the manuscript and revising the manuscript. All authors have read and agreed to the published version of the manuscript.

Funding: This study was supported by grants from the Swiss National Foundation NCCR Switzerland "Synapsy" (51NF40-158776), "Transcure" (51NF6240-16) (to P.B.), Swiss National Science Foundation SNSF (310030_185363) (to P.B.) and Telethon Italy (GGP20037) (to P.B.).

Acknowledgments: We would like to thank Vigna Graphic Design for assistance in designing, developing, and reviewing graphics and artwork.

Conflicts of Interest: The authors declare no conflict of interest.

References

1. Gur, R.E.; Bassett, A.S.; McDonald-Mcginn, D.M.; Bearden, C.E.; Chow, E.; Emanuel, B.S.; Owen, M.; Swillen, A.; Van den Bree, M.; Vermeesch, J. A neurogenetic model for the study of schizophrenia spectrum disorders: The International 22q11.2 Deletion Syndrome Brain Behavior Consortium. *Mol. Psychiatry* **2017**, *22*, 1664–1672. [CrossRef] [PubMed]
2. Maynard, T.M.; Haskell, G.T.; Lieberman, J.A.; Lamantia, A.-S. 22q11 DS: Genomic mechanisms and gene function in DiGeorge/velocardiofacial syndrome. *Int. J. Dev. Neurosci.* **2002**, *20*, 407–419. [CrossRef]
3. McDonald-McGinn, D.M.; Sullivan, K.E.; Marino, B.; Philip, N.; Swillen, A.; Vorstman, J.A.S.; Zackai, E.H.; Emanuel, B.S.; Vermeesch, J.; Morrow, B.E. 22Q11.2 Deletion Syndrome. *Nat. Rev. Dis. Primers* **2015**, *1*, 1–19. [CrossRef] [PubMed]
4. Karayiorgou, M.; Simon, T.J.; Gogos, J.A. 22q11.2 microdeletions: Linking DNA structural variation to brain dysfunction and schizophrenia. *Nat. Rev. Neurosci.* **2010**, *11*, 402–416. [CrossRef]
5. Carlson, C.; Sirotkin, H.; Pandita, R.; Goldberg, R.; McKie, J.; Wadey, R.; Patanjali, S.R.; Weissman, S.M.; Anyane-Yeboa, K.; Warburton, D. Molecular definition of 22q11 deletions in 151 velo-cardio-facial syndrome patients. *Am. J. Hum. Genet.* **1997**, *61*, 620–629. [CrossRef]
6. Drew, L.J.; Crabtree, G.W.; Markx, S.; Stark, K.L.; Chaverneff, F.; Xu, B.; Mukai, J.; Fenelon, F.; Hsu, P.K.; Gogos, J.A.; et al. The 22q11.2 microdeletion: Fifteen years of insights into the genetic and neural complexity of psychiatric disorders. *Int. J. Dev. Neurosci.* **2011**, *29*, 259–281. [CrossRef]
7. Karayiorgou, M.; Morris, M.A.; Morrow, B.; Shprintzen, R.J.; Goldberg, R.; Borrow, J.; Gos, A.; Nestadt, G.; Wolyniec, P.S.; Lasseter, V.K. Schizophrenia susceptibility associated with interstitial deletions of chromosome 22q11. *Proc. Natl. Acad. Sci. USA* **1995**, *92*, 7612–7616. [CrossRef]
8. Murphy, K.C. Schizophrenia and velo-cardio-facial syndrome. *Lancet* **2002**, *359*, 426–430. [CrossRef]
9. Bassett, A.S.; Chow, E.W.C. Chromosomal Abnormalities and Schizophrenia and a senior psychiatric genetics researcher. *Am. J. Med. Genet.* **2000**, *97*, 45–51. [CrossRef]
10. Schneider, M.; Debbané, M.; Bassett, A.S.; Chow, E.W.C.; Fung, W.L.A.; Van Den Bree, M.B.M.; Owen, M.; Murphy, K.C.; Niarchou, M.; Kates, W.R.; et al. Psychiatric disorders from childhood to adulthood in 22q11.2 deletion syndrome: Results from the international consortium on brain and behavior in 22q11.2 deletion syndrome. *Am. J. Psychiatry* **2014**, *171*, 627–639. [CrossRef]
11. Jolin, E.M.; Weller, R.A.; Jessani, N.R.; Zackai, E.H.; McDonald-McGinn, D.M.; Weller, E.B. Affective disorders and other psychiatric diagnoses in children and adolescents with 22q11.2 Deletion Syndrome. *J. Affect. Disord.* **2009**, *119*, 177–180. [CrossRef] [PubMed]
12. Vorstman, J.A.S.; Morcus, M.E.J.; Duijff, S.N.; Klaassen, P.W.J.; Heineman-De Boer, J.A.; Beemer, F.A.; Swaab, H.; Kahn, R.S.; Van Engeland, H. The 22q11.2 deletion in children: High rate of autistic disorders and early onset of psychotic symptoms. *J. Am. Acad. Child Adolesc. Psychiatry* **2006**, *45*, 1104–1113. [CrossRef] [PubMed]
13. Gothelf, D.; Schneider, M.; Green, T.; Debbané, M.; Frisch, A.; Glaser, B.; Zilkha, H.; Schaer, M.; Weizman, A.; Eliez, S. Risk factors and the evolution of psychosis in 22q11.2 deletion syndrome: A longitudinal 2-site study. *J. Am. Acad. Child Adolesc. Psychiatry* **2013**, *52*, 112–1203. [CrossRef] [PubMed]
14. McDonald-McGinn, D.M.; Sullivan, K.E. Chromosome 22q11.2 deletion syndrome (DiGeorge syndrome/velocardiofacial syndrome). *Medicine* **2011**, *90*, 1–18. [CrossRef]
15. Murphy, K.C.; Jones, L.A.; Owen, M.J. High rates of schizophrenia in adults with velo-cardio-facial syndrome. *Arch. Gen. Psychiatry* **1999**, *56*, 940–945. [CrossRef]
16. Debbané, M.; Glaser, B.; David, M.K.; Feinstein, C.; Eliez, S. Psychotic symptoms in children and adolescents with 22q11.2 deletion syndrome: Neuropsychological and behavioral implications. *Schizophr. Res.* **2006**, *84*, 187–193. [CrossRef]
17. Hooper, S.R.; Curtiss, K.; Schoch, K.; Keshavan, M.S.; Allen, A.; Shashi, V. A Longitudinal Examination of the Psychoeducational, Neurocognitive, and Psychiatric Functioning in Children with 22q11.2 Deletion Syndrome. *Res. Dev. Disabil.* **2013**, *34*, 1758. [CrossRef]
18. Simon, T.; Takarae, Y.; DeBoer, T.; McDonald-McGinn, D.; Zackai, E.; Ross, J. Overlapping numerical cognition impairments in children with chromosome 22q11.2 deletion or Turner syndromes. *Neuropsychologia* **2008**, *46*, 82–94. [CrossRef]
19. Simon, T.J.; Bish, J.P.; Bearden, C.E.; Ding, L.; Ferrante, S.; Nguyen, V.; Gee, J.G.; McDonald-McGinn, D.M.; Zackai, E.H.; Emanuel, S.E. A multilevel analysis of cognitive dysfunction and psychopathology associated with chromosome 22q11.2 deletion syndrome in children. *Dev. Psychopathol.* **2005**, *17*, 753–784. [CrossRef]
20. Amelsvoort, T.; Van Henry, J.; Morris, R.; Owen, M.; Linszen, D.; Murphy, K.; Murphy, D. Cognitive deficits associated with schizophrenia in velo-cardio-facial syndrome. *Schizophr. Res.* **2004**, *70*, 223–232. [CrossRef]
21. Weinberger, R.; Yi, J.; Calkins, M.; Guri, Y.; McDonald-McGinn, D.M.; Emanuel, B.S.; Zackai, E.H.; Ruparel, K.; Carmel, M.; Michaelovsky, E. Neurocognitive profile in psychotic versus nonpsychotic individuals with 22q11.2 deletion syndrome. *Eur. Neuropsychopharmacol. J. Eur. Coll. Neuropsychopharmacol.* **2016**, *26*, 1610–1618. [CrossRef] [PubMed]
22. Fiksinski, A.M.; Breetvelt, E.J.; Lee, Y.J.; Boot, E.; Butcher, N.; Palmer, L.; Chow, E.W.C.; Kahn, R.S.; Vorstman, J.A.S.; Bassett, A.S. Neurocognition and adaptive functioning in a genetic high risk model of schizophrenia. *Psychol. Med.* **2019**, *49*, 1047–1054. [CrossRef] [PubMed]
23. Vorstman, J.A.S.; Breetvelt, E.J.; Duijff, S.N.; Eliez, S.; Schneider, M.; Jalbrzikowski, M.; Armando, M.; Vicari, S.; Shashi, V.; Hooper, S.R. Cognitive decline preceding the onset of psychosis in patients with 22q11.2 deletion syndrome. *JAMA Psychiatry* **2015**, *72*, 377–385. [CrossRef] [PubMed]

24. Fénelon, K.; Mukai, J.; Xu, B.; Hsu, P.K.; Drew, L.J.; Karayiorgou, M.; Fischbach, G.D.; Macdermott, A.B.; Gogos, J.A. Deficiency of Dgcr8, a gene disrupted by the 22q11.2 microdeletion, results in altered short-term plasticity in the prefrontal cortex. *Proc. Natl. Acad. Sci. USA* **2011**, *108*, 4447–4452. [CrossRef]
25. Tripathi, A.; Spedding, M.; Schenker, E.; Didriksen, M.; Cressant, A.; Jay, T.M. Cognition- and circuit-based dysfunction in a mouse model of 22q11.2 microdeletion syndrome: Effects of stress. *Transl. Psychiatry* **2020**, *10*, 1–15. [CrossRef]
26. Kimber, W.L.; Hsieh, P.; Hirotsune, S.; Yuva-Paylor, L.; Sutherland, H.F.; Chen, A.; Ruiz-Lozano, P.; Hoogstraten-Miller, S.L.; Chien, K.R.; Paylor, R.; et al. Deletion of 150 kb in the Minimal Digeorge/Velocardiofacial Syndrome Critical Region in Mouse. *Hum. Mol. Genet.* **1999**, *8*, 2229–2237. [CrossRef]
27. Meechan, D.W.; Maynard, T.M.; Fernandez, A.; Karpinski, B.A.; Rothblat, L.A.; LaMantia, A.S. Modeling a model: Mouse genetics, 22q11.2 Deletion Syndrome, and disorders of cortical circuit development. *Prog. Neurobiol.* **2015**, *130*, 1–28. [CrossRef]
28. Morris, R.G.M. Elements of a neurobiological theory of hippocampal function: The role of synaptic plasticity, synaptic tagging and schemas. *Eur. J. Neurosci.* **2006**, *23*, 2829–2846. [CrossRef]
29. Martin, S.J.; Grimwood, P.D.; Morris, R.G.M. Synaptic Plasticity and Memory: An Evaluation of the Hypothesis. *Annu. Rev. Neurosci.* **2000**, *23*, 649–711. [CrossRef]
30. Citri, A.; Malenka, R.C. Synaptic Plasticity: Multiple Forms, Functions, and Mechanisms. *Neuropsychopharmacology* **2008**, *33*, 18–41. [CrossRef]
31. Marissal, T.; Salazar, R.F.; Bertollini, C.; Mutel, S.; De Roo, M.; Rodriguez, I.; Müller, D.; Carleton, A. Restoring wild-type-like CA1 network dynamics and behavior during adulthood in a mouse model of schizophrenia. *Nat. Neurosci.* **2018**, *21*, 1412–1420. [CrossRef] [PubMed]
32. Mukherjee, A.; Carvalho, F.; Eliez, S.; Caroni, P. Long-Lasting Rescue of Network and Cognitive Dysfunction in a Genetic Schizophrenia Model. *Cell* **2019**, *178*, 1387–1402.e14. [PubMed]
33. Kimura, H.; Fujita, Y.; Kawabata, T.; Ishizuka, K.; Wang, C.; Iwayama, Y.; Okahisa, Y.; Kushima, I.; Morikawa, M.; Uno, Y. A novel rare variant R292H in RTN4R affects growth cone formation and possibly contributes to schizophrenia susceptibility. *Transl. Psychiatry* **2017**, *7*, e1214. [PubMed]
34. Budel, S.; Padukkavidana, T.; Liu, B.P.; Feng, Z.; Hu, F.; Johnson, S.; Lauren, J.; Park, J.H.; McGee, A.W.; Liao, J. Genetic variants of Nogo-66 receptor with possible association to schizophrenia block myelin inhibition of axon growth. *J. Neurosci.* **2008**, *28*, 13161–13172. [PubMed]
35. Crabtree, G.; Gogos, J.A. Synaptic plasticity, neural circuits and the emerging role of altered short-term information processing in schizophrenia. *Front. Synaptic Neurosci.* **2014**, *6*, 28. [CrossRef] [PubMed]
36. Earls, L.R.; Bayazitov, I.T.; Fricke, R.G.; Berry, R.B.; Illingworth, E.; Mittleman, G.; Zakharenko, S.S. Dysregulation of presynaptic calcium and synaptic plasticity in a mouse model of 22q11 deletion syndrome. *J. Neurosci.* **2010**, *30*, 15843–15855. [CrossRef]
37. Haydon, P.G. GLIA: Listening and talking to the synapse Septal cholinergic neuromodulation tunes the astrocyte-dependent gating of hippocampal NMDA receptors to wakefulness View project. *Nat. Rev. Neurosci.* **2001**, *2*, 185–193. [CrossRef]
38. Barker, A.J.; Ullian, E.M. Astrocytes and synaptic plasticity. The Neuroscientist: A Review Journal Bringing Neurobiology. *Neurol. Psychiatry* **2010**, *16*, 40–50.
39. Ben Achour, S.; Pascual, O. Glia: The many ways to modulate synaptic plasticity. *Neurochem. Int.* **2010**, *57*, 440–445. [CrossRef]
40. Sancho, L.; Contreras, M.; Allen, N.J. Glia as sculptors of synaptic plasticity. *Neurosci. Res.* **2021**, *167*, 17–29. [CrossRef]
41. De Pittà, M.; Brunel, N.; Volterra, A. Astrocytes: Orchestrating synaptic plasticity? *Neuroscience* **2016**, *323*, 43–61. [CrossRef] [PubMed]
42. Araque, A.; Carmignoto, G.; Haydon, P.G.; Oliet, S.H.R.; Robitaille, R.; Volterra, A. Gliotransmitters Travel in Time and Space. *Neuron* **2014**, *81*, 728–739. [CrossRef] [PubMed]
43. Zoghbi, H.Y.; Bear, M.F. Synaptic dysfunction in neurodevelopmental disorders associated with autism and intellectual disabilities. *Cold Spring Harb. Perspect. Biol.* **2012**, *4*, a009886. [CrossRef] [PubMed]
44. Niklasson, L.; Rasmussen, P.; Óskarsdóttir, S.; Gillberg, C. Autism, ADHD, mental retardation and behavior problems in 100 individuals with 22q11 deletion syndrome. *Res. Dev. Disabil.* **2009**, *30*, 763–773. [CrossRef]
45. Geschwind, D.H.; Levitt, P. Autism spectrum disorders: Developmental disconnection syndromes. *Curr. Opin. Neurobiol.* **2007**, *17*, 103–111. [CrossRef]
46. Weinberger, D.R. Implications of Normal Brain Development for the Pathogenesis of Schizophrenia. *Arch. Gen. Psychiatry* **1987**, *44*, 660–669. [CrossRef]
47. Gourovitch, M.L.; Goldberg, T.E.; Weinberger, D.R. Verbal fluency deficits in patients with schizophrenia: Semantic fluency is differentially impaired as compared with phonologic fluency. *Neuropsychology* **1996**, *10*, 573–577. [CrossRef]
48. Meechan, D.W.; Maynard, T.M.; Tucker, E.S.; LaMantia, A.S. Three phases of DiGeorge/22q11 deletion syndrome pathogenesis during brain development: Patterning, proliferation, and mitochondrial functions of 22q11 genes. *Int. J. Dev. Neurosci.* **2011**, *29*, 283–294. [CrossRef]
49. Eliez, S.; Schmitt, J.E.; White, C.D.; Reiss, A.L. Children and adolescents with velocardiofacial syndrome: A volumetric MRI study. *Am. J. Psychiatry* **2000**, *157*, 409–415. [CrossRef]
50. Bassett, A.S.; Chow, E.W.C.; AbdelMalik, P.; Gheorghiu, M.; Husted, J.; Weksberg, R. The schizophrenia phenotype in 22q11 deletion syndrome. *Am. J. Psychiatry* **2003**, *160*, 1580–1586. [CrossRef]

51. Kates, W.R.; Burnette, C.P.; Bessette, B.A.; Folley, B.S.; Strunge, L.; Jabs, E.W.; Pearlson, G.D. Frontal and caudate alterations in velocardiofacial syndrome (deletion at chromosome 22q11.2). *J. Child Neurol.* **2004**, *19*, 337–342. [CrossRef] [PubMed]
52. Ching, C.R.K.; Gutman, B.A.; Sun, D.; Reina, J.V.; Ragothaman, A.; Isaev, D.; Zavaliangos-Petropulu, A.; Lin, A.; Jonas, R.K.; Kushan, L. Mapping subcortical brain alterations in 22q11.2 deletion syndrome: Effects of deletion size and convergence with idiopathic neuropsychiatric illness. *Am. J. Psychiatry* **2020**, *177*, 589. [CrossRef] [PubMed]
53. Rogdaki, M.; Gudbrandsen, M.; McCutcheon, R.A.; Blackmore, C.E.; Brugger, S.; Ecker, C.; Craig, M.C.; Daly, E.; Murphy, D.G.M.; Howes, O. Magnitude and heterogeneity of brain structural abnormalities in 22q11.2 deletion syndrome: A meta-analysis. *Mol. Psychiatry* **2020**, *25*, 1704–1717. [CrossRef] [PubMed]
54. Kebets, V.; Favre, P.; Houenou, J.; Polosan, M.; Perroud, N.; Aubry, J.-M.; Van De Ville, D.; Piguet, C. Fronto-limbic neural variability as a transdiagnostic correlate of emotion dysregulation. *Transl. Psychiatry* **2021**, *11*, 1–8. [CrossRef]
55. Ottet, M.C.; Schaer, M.; Cammoun, L.; Schneider, M.; Debbané, M.; Thiran, J.P.; Eliez, S. Reduced Fronto-Temporal and Limbic Connectivity in the 22q11.2 Deletion Syndrome: Vulnerability Markers for Developing Schizophrenia? *PLoS ONE* **2013**, *8*, e58429. [CrossRef]
56. Pelgrim, T.A.D.; Bossong, M.G.; Cuiza, A.; Alliende, L.M.; Mena, C.; Tepper, A.; Ramirez-Mahaluf, J.P.; Iruretagoyena, B.; Ornstein, C.; Fritsch, R. Abnormal nodal and global network organization in resting state functional MRI from subjects with the 22q11 deletion syndrome. *Sci. Rep.* **2021**, *11*, 1–9. [CrossRef]
57. Fernandez, A.; Meechan, D.W.; Karpinski, B.A.; Paronett, E.M.; Bryan, C.A.; Rutz, H.L.; Radin, E.A.; Lubin, N.; Bonner, E.R.; Popratiloff, A. Mitochondrial Dysfunction Leads to Cortical Under-Connectivity and Cognitive Impairment. *Neuron* **2019**, *102*, 1127–1142.e3. [CrossRef]
58. Puech, A.; Saint-Jore, B.; Funke, B.; Gilbert, D.J.; Sirotkin, H.; Copeland, N.G.; Jenkins, N.A.; Kucherlapati, R.; Morrow, B.; Skoultchi, A.I. Comparative mapping of the human 22q11 chromosomal region and the orthologous region in mice reveals complex changes in gene organization. *Proc. Natl. Acad. Sci. USA* **1997**, *94*, 14608. [CrossRef]
59. Lindsay, E.A.; Botta, A.; Jurecic, V.; Carattini-Rivera, S.; Cheah, Y.-C.; Rosenblatt, H.M.; Bradley, A.; Baldini, A. Congenital heart disease in mice deficient for the DiGeorge syndrome region. *Nature* **1999**, *401*, 379–383. [CrossRef]
60. Merscher, S.; Funke, B.; Epstein, J.A.; Heyer, J.; Puech, A.; Lu, M.M.; Xavier, R.J.; Demay, M.B.; Russel, R.G.; Factor, S. TBX1 is responsible for cardiovascular defects in velo-cardio-facial/DiGeorge syndrome. *Cell* **2001**, *104*, 619–629. [CrossRef]
61. Paylor, R.; McIlwain, K.L.; McAninch, R.; Nellis, A.; Yuva-Paylor, L.A.; Baldini, A.; Lindsay, E.A. Mice deleted for the DiGeorge/velocardiofacial syndrome region show abnormal sensorimotor gating and learning and memory impairments. *Hum. Mol. Genet.* **2001**, *10*, 2645–2650. [CrossRef] [PubMed]
62. Stark, K.L.; Xu, B.; Bagchi, A.; Lai, W.S.; Liu, H.; Hsu, R.; Wan, X.; Pavlidis, P.; Mills, A.A.; Karayiorgou, M. Altered brain microRNA biogenesis contributes to phenotypic deficits in a 22q11-deletion mouse model. *Nat. Genet.* **2008**, *40*, 751–760. [CrossRef] [PubMed]
63. Kim, J.J.; Fanselow, M.S. Modality-specific retrograde amnesia of fear. *Science* **1992**, *256*, 675–677. [CrossRef] [PubMed]
64. Phillips, R.G.; LeDoux, J.E. Differential contribution of amygdala and hippocampus to cued and contextual fear conditioning. *Behav. Neurosci.* **1992**, *106*, 274–285. [CrossRef]
65. Mukai, J.; Tamura, M.; Fénelon, K.; Rosen, A.M.; Spellman, T.J.; Kang, R.; MacDermott, A.B.; Karayiorgou, M.; Gordon, J.A.; Gogos, J.A. Molecular Substrates of Altered Axonal Growth and Brain Connectivity in a Mouse Model of Schizophrenia. *Neuron* **2015**, *86*, 680–695. [CrossRef]
66. Paterlini, M.; Zakharenko, S.S.; Lai, W.S.; Qin, J.; Zhang, H.; Mukai, J.; Westphal, K.G.C.; Olivier, B.; Sulzer, D.; Pavlidis, P. Transcriptional and behavioral interaction between 22q11.2 orthologs modulates schizophrenia-related phenotypes in mice. *Nat. Neurosci.* **2005**, *8*, 1586–1594. [CrossRef]
67. Pulver, A.E.; Nestadt, G.; Goldberg, R.; Shprintzen, R.J.; Lamacz, M.; Wolyniec, P.S.; Morrow, B.; Karayiorgou, M.; Antonarakis, S.E.; Housman, D. Psychotic illness in patients diagnosed with velo-cardio-facial syndrome and their relatives. *J. Nerv. Ment. Dis.* **1994**, *182*, 476–478. [CrossRef]
68. Chow, E.W.C.; Watson, M.; Young, D.A.; Bassett, A.S. Neurocognitive profile in 22q11 deletion syndrome and schizophrenia. *Schizophr. Res.* **2006**, *87*, 270–278. [CrossRef]
69. Lewis, D.A. Cortical circuit dysfunction and cognitive deficits in schizophrenia—Implications for preemptive interventions. *Eur. J. Neurosci.* **2012**, *35*, 1871–1878. [CrossRef]
70. Heckers, S.; Rauch, S.L.; Goff, D.; Savage, C.R.; Schacter, D.L.; Fischman, A.J.; Alpert, N.M. Impaired recruitment of the hippocampus during conscious recollection in schizophrenia. *Nat. Neurosci.* **1998**, *1*, 318–323. [CrossRef]
71. Tamminga, C.A.; Stan, A.D.; Wagner, A.D. The hippocampal formation in schizophrenia. *Am. J. Psychiatry* **2010**, *167*, 1178–1193. [CrossRef] [PubMed]
72. Fénelon, K.; Xu, B.; Lai, C.S.; Mukai, J.; Markx, S.; Stark, K.L.; Hsu, P.K.; Gan, W.B.; Fischbach, G.D.; MacDermott, A.B.; et al. The Pattern of Cortical Dysfunction in a Mouse Model of a Schizophrenia-Related Microdeletion. *J. Neurosci.* **2013**, *33*, 14825. [CrossRef] [PubMed]
73. Devaraju, P.; Yu, J.; Eddins, D.; Mellado-Lagarde, M.M.; Earls, L.R.; Westmoreland, J.J.; Quarato, G.; Green, D.R.; Zakharenko, S.S. Haploinsufficiency of the 22q11.2 microdeletion gene Mrpl40 disrupts short-term synaptic plasticity and working memory through dysregulation of mitochondrial calcium. *Mol. Psychiatry* **2017**, *22*, 1313–1326. [CrossRef] [PubMed]

74. Howland, J.G.; Wang, Y.T. Synaptic plasticity in learning and memory: Stress effects in the hippocampus. *Prog. Brain Res.* **2008**, *169*, 145–158. [PubMed]
75. Neves, G.; Cooke, S.F.; Bliss, T.V.P. Synaptic plasticity, memory and the hippocampus: A neural network approach to causality. *Nat. Rev. Neurosci.* **2008**, *9*, 65–75. [CrossRef] [PubMed]
76. Mukai, J.; Liu, H.; Burt, R.A.; Swor, D.E.; Lai, W.S.; Karayiorgou, M.; Gogos, J.A. Evidence that the gene encoding ZDHHC8 contributes to the risk of schizophrenia. *Nat. Genet.* **2004**, *36*, 725–731. [CrossRef] [PubMed]
77. Chen, W.Y.; Shi, Y.Y.; Zheng, Y.L.; Zhao, X.Z.; Zhang, G.J.; Chen, S.Q.; Yang, P.D.; He, L. Case-control study and transmission disequilibrium test provide consistent evidence for association between schizophrenia and genetic variation in the 22q11 gene ZDHHC8. *Hum. Mol. Genet.* **2004**, *13*, 2991–2995. [CrossRef]
78. Fukata, Y.; Fukata, M. Protein palmitoylation in neuronal development and synaptic plasticity. *Nat. Rev. Neurosci.* **2010**, *11*, 161–175. [CrossRef]
79. El-Husseini, A.E.D.; Bredt, D.S. Protein palmitoylation: A regulator of neuronal development and function. *Nat. Rev. Neurosci.* **2002**, *3*, 791–802. [CrossRef]
80. Eryilmaz, H.; Tanner, A.S.; Ho, N.F.; Nitenson, A.Z.; Silverstein, N.J.; Petruzzi, L.J.; Goff, D.C.; Manoach, D.S.; Roffman, J.L. Disrupted Working Memory Circuitry in Schizophrenia: Disentangling fMRI Markers of Core Pathology vs. Other Aspects of Impaired Performance. *Neuropsychopharmacol. Off. Publ. Am. Coll. Neuropsychopharmacol.* **2016**, *41*, 2411–2420. [CrossRef]
81. Gilks, W.P.; Hill, M.; Gill, M.; Donohoe, G.; Corvin, A.P.; Morris, D.W. Functional investigation of a schizophrenia GWAS signal at the CDC42 gene. *World J. Biol. Psychiatry Off. J. World Fed. Soc. Biol. Psychiatry* **2012**, *13*, 550–554. [CrossRef] [PubMed]
82. Lai, W.S.; Xu, B.; Westphal, K.G.C.; Paterlini, M.; Olivier, B.; Pavlidis, P.; Karayiorgou, M.; Gogos, J.A. Akt1 deficiency affects neuronal morphology and predisposes to abnormalities in prefrontal cortex functioning. *Proc. Natl. Acad. Sci. USA* **2006**, *103*, 16906–16911. [CrossRef] [PubMed]
83. Lewis, B.P.; Shih, I.H.; Jones-Rhoades, M.W.; Bartel, D.P.; Burge, C.B. Prediction of mammalian microRNA targets. *Cell* **2003**, *115*, 787–798. [CrossRef]
84. Kolfschoten, I.G.M.; Regazzi, R. Technology Insight: Small, noncoding RNA molecules as tools to study and treat endocrine diseases. *Nat. Clin. Pract. Endocrinol. Metab.* **2007**, *3*, 827–834. [CrossRef]
85. Han, J.; Lee, Y.; Yeom, K.H.; Kim, Y.K.; Jin, H.; Kim, V.N. The Drosha-DGCR8 complex in primary microRNA processing. *Genes Dev.* **2004**, *18*, 3016–3027. [CrossRef]
86. Schratt, G. microRNAs at the synapse. *Nat. Rev. Neurosci.* **2009**, *10*, 842–849. [CrossRef]
87. Diamantopoulou, A.; Sun, Z.; Mukai, J.; Xu, B.; Fenelon, K.; Karayiorgou, M.; Gogos, J.A. Loss-of-function mutation in Mirta22/Emc10 rescues specific schizophrenia-related phenotypes in a mouse model of the 22q11.2 deletion. *Proc. Natl. Acad. Sci. USA* **2017**, *114*, E6127–E6136. [CrossRef]
88. Xu, B.; Hsu, P.K.; Stark, K.L.; Karayiorgou, M.; Gogos, J.A. Derepression of a neuronal inhibitor due to miRNA dysregulation in a schizophrenia-related microdeletion. *Cell* **2013**, *152*, 262–275. [CrossRef]
89. Sivagnanasundaram, S.; Fletcher, D.; Hubank, M.; Illingworth, E.; Skuse, D.; Scambler, P. Differential gene expression in the hippocampus of the Df1/+ mice: A model for 22q11.2 deletion syndrome and schizophrenia. *Brain Res.* **2007**, *1139*, 48–59. [CrossRef]
90. Elgersma, Y.; Sweatt, J.D.; Giese, K.P. Mouse Genetic Approaches to Investigating Calcium/Calmodulin-Dependent Protein Kinase II Function in Plasticity and Cognition. *J. Neurosci.* **2004**, *24*, 8410. [CrossRef]
91. Cammarota, M.; Bevilaqua, L.R.M.; Viola, H.; Kerr, D.S.; Reichmann, B.; Teixeira, V.; Bulla, M.; Izquierdo, I.; Medina, J.H. Participation of CaMKII in neuronal plasticity and memory formation. *Cell. Mol. Neurobiol.* **2002**, *22*, 259–267. [CrossRef] [PubMed]
92. Colbran, R.J.; Brown, A.M. Calcium/calmodulin-dependent protein kinase II and synaptic plasticity. *Curr. Opin. Neurobiol.* **2004**, *14*, 318–327. [PubMed]
93. Yamagata, K.; Andreasson, K.I.; Sugiura, H.; Maru, E.; Dominique, M.; Irie, Y.; Miki, N.; Hayashi, Y.; Yoshioka, M.; Kaneko, K. Arcadlin Is a Neural Activity-regulated Cadherin Involved in Long Term Potentiation. *J. Biol. Chem.* **1999**, *274*, 19473–19479. [CrossRef] [PubMed]
94. Philip, N.; Acevedo, S.F.; Skoulakis, E.M.C. Conditional rescue of olfactory learning and memory defects in mutants of the 14-3-3zeta gene leonardo. *J. Neurosci. Off. J. Soc. Neurosci.* **2001**, *21*, 8417–8425. [CrossRef]
95. Butler, M.G.; Rafi, S.K.; Hossain, W.; Stephan, D.A.; Manzardo, A.M. Whole exome sequencing in females with autism implicates novel and candidate genes. *Int. J. Mol. Sci.* **2015**, *16*, 1312–1335. [CrossRef]
96. Clements, C.C.; Wenger, T.L.; Zoltowski, A.R.; Bertollo, J.R.; Miller, J.S.; De Marchena, A.B.; Mitteer, L.A.; Carey, J.C.; Yerys, B.E.; Zackai, E.H. Critical region within 22q11.2 linked to higher rate of autism spectrum disorder. *Mol. Autism* **2017**, *8*, 58. [CrossRef]
97. Middleton, F.A.; Peng, L.; Lewis, D.A.; Levitt, P.; Mirnics, K. Altered Expression of 14-3-3 Genes in the Prefrontal Cortex of Subjects with Schizophrenia. *Neuropsychopharmacology* **2005**, *30*, 974–983. [CrossRef]
98. Meechan, D.W.; Rutz, H.L.H.; Fralish, M.S.; Maynard, T.M.; Rothblat, L.A.; Lamantia, A.S. Cognitive ability is associated with altered medial frontal cortical circuits in the LgDel mouse model of 22q11.2DS. *Cereb. Cortex* **2015**, *25*, 1143–1151. [CrossRef]
99. Piskorowski, R.A.; Nasrallah, K.; Diamantopoulou, A.; Mukai, J.; Hassan, S.I.; Siegelbaum, S.A.; Siegelbaum, S.A.; Gogos, J.A.; Chevaleyre, V. Age-Dependent Specific Changes in Area CA2 of the Hippocampus and Social Memory Deficit in a Mouse Model of the 22q11.2 Deletion Syndrome. *Neuron* **2016**, *89*, 163–176. [CrossRef]

100. Sigurdsson, T.; Stark, K.L.; Karayiorgou, M.; Gogos, J.A.; Gordon, J.A. Impaired hippocampal–prefrontal synchrony in a genetic mouse model of schizophrenia. *Nature* **2010**, *464*, 763–767.
101. Maynard, T.M.; Meechan, D.W.; Dudevoir, M.L.; Gopalakrishna, D.; Peters, A.Z.; Heindel, C.C.; Sugimoto, T.J.; Wu, Y.; Lieberman, J.A.; LaMantia, A.S. Mitochondrial localization and function of a subset of 22q11 deletion syndrome candidate genes. *Mol. Cell. Neurosci.* **2008**, *39*, 439–451. [CrossRef] [PubMed]
102. Hara, Y.; Yuk, F.; Puri, R.; Janssen, W.G.M.; Rapp, P.R.; Morrison, J.H. Presynaptic mitochondrial morphology in monkey prefrontal cortex correlates with working memory and is improved with estrogen treatment. *Proc. Natl. Acad. Sci. USA* **2014**, *111*, 486–491. [CrossRef] [PubMed]
103. Ahmad, T.; Aggarwal, K.; Pattnaik, B.; Mukherjee, S.; Sethi, T.; Tiwari, B.K.; Kumar, M.; Micheal, A.; Mabalirajan, U.; Ghosh, B. Computational classification of mitochondrial shapes reflects stress and redox state. *Cell Death Dis.* **2013**, *4*, e461. [PubMed]
104. Arnér, E.S.J.; Holmgren, A. Physiological functions of thioredoxin and thioredoxin reductase. *Eur. J. Biochem.* **2000**, *267*, 6102–6109. [CrossRef] [PubMed]
105. Gokhale, A.; Hartwig, C.; Freeman, A.A.H.; Bassell, J.L.; Zlatic, S.A.; Savas, C.S.; Vadlamudi, T.; Abudulai, F.; Pham, T.T.; Crocker, A. Systems analysis of the 22q11.2 microdeletion syndrome converges on a mitochondrial interactome necessary for synapse function and behavior. *J. Neurosci.* **2019**, *39*, 3561–3581. [CrossRef]
106. Catalina-Rodriguez, O.; Kolukula, V.K.; Tomita, Y.; Preet, A.; Palmieri, F.; Wellstein, A.; Byers, S.; Giaccia, A.J.; Glasgow, E.; Albanese, C. The mitochondrial citrate transporter, CIC, is essential for mitochondrial homeostasis. *Oncotarget* **2012**, *3*, 1220–1235. [CrossRef]
107. Gokhale, A.; Lee, C.E.; Zlatic, S.A.; Freeman, A.A.H.; Shearing, N.; Hartwig, C.; Ogunbona, O.; Bassell, J.L.; Wynne, M.E.; Werner, E. Mitochondrial Proteostasis Requires Genes Encoded in a Neurodevelopmental Syndrome Locus. *J. Neurosci. Off. J. Soc. Neurosci.* **2021**, *41*, 6596–6616.
108. Amunts, A.; Brown, A.; Bai, X.C.; Llácer, J.L.; Hussain, T.; Emsley, P.; Long, F.; Murshudov, G.; Scheres, S.H.W.; Ramakrishnan, V. Structure of the yeast mitochondrial large ribosomal subunit. *Science* **2014**, *343*, 1485–1489. [CrossRef]
109. Tang, M.; Lu, L.; Huang, Z.; Chen, L. Palmitoylation signaling: A novel mechanism of mitochondria dynamics and diverse pathologies. *Acta Biochim. Biophys. Sin.* **2018**, *50*, 831–833. [CrossRef]
110. Liu, H.; Heath, S.C.; Sobin, C.; Roos, J.L.; Galke, B.L.; Blundell, M.L.; Lenane, M.; Robertson, B.; Wijsman, E.M.; Rapoport, J.L. Genetic variation at the 22q11 PRODH2/DGCR6 locus presents an unusual pattern and increases susceptibility to schizophrenia. *Proc. Natl. Acad. Sci. USA* **2002**, *99*, 3717–3722. [CrossRef]
111. Liu, H.; Abecasis, G.R.; Heath, S.C.; Knowles, A.; Demars, S.; Chen, Y.J.; Roos, J.L.; Rapoport, J.L.; Gogos, J.A.; Karayiorgou, M. Genetic variation in the 22q11 locus and susceptibility to schizophrenia. *Proc. Natl. Acad. Sci. USA* **2002**, *99*, 16859–16864. [CrossRef] [PubMed]
112. Bard, F.; Casano, L.; Mallabiabarrena, A.; Wallace, E.; Saito, K.; Kitayama, H.; Guizzunti, G.; Hu, Y.; Wendler, F.; Dasgupta, R. Functional genomics reveals genes involved in protein secretion and Golgi organization. *Nature* **2006**, *439*, 604–607. [CrossRef] [PubMed]
113. Mattson, M.P.; Gleichmann, M.; Cheng, A. Mitochondria in neuroplasticity and neurological disorders. *Neuron* **2008**, *60*, 748–766. [CrossRef] [PubMed]
114. Earls, L.R.; Gaines Fricke, R.; Yu, J.; Berry, R.B.; Baldwin, L.T.; Zakharenko, S.S. Age-Dependent MicroRNA Control of Synaptic Plasticity in 22q11 Deletion Syndrome and Schizophrenia. *J. Neurosci.* **2012**, *32*, 14132. [CrossRef]
115. Cooke, S.F.; Bliss, T.V.P. Plasticity in the human central nervous system. *Brain J. Neurol.* **2006**, *129*, 1659–1673. [CrossRef]
116. Auld, D.S.; Robitaille, R. Glial Cells and Neurotransmission: An Inclusive View of Synaptic Function. *Neuron* **2003**, *40*, 389–400. [CrossRef]
117. Pannasch, U.; Rouach, N. Emerging role for astroglial networks in information processing: From synapse to behavior. *Trends Neurosci.* **2013**, *36*, 405–417. [CrossRef]
118. Volterra, A.; Meldolesi, J. Astrocytes, from brain glue to communication elements: The revolution continues. *Nat. Rev. Neurosci.* **2005**, *6*, 626–640. [CrossRef]
119. Allen, N.J.; Barres, B.A. Signaling between glia and neurons: Focus on synaptic plasticity. *Curr. Opin. Neurobiol.* **2005**, *15*, 542–548. [CrossRef]
120. Perea, G.; Navarrete, M.; Araque, A. Tripartite synapses: Astrocytes process and control synaptic information. *Trends Neurosci.* **2009**, *32*, 421–431. [CrossRef]
121. López-Hidalgo, M.; Schummers, J. Cortical maps: A role for astrocytes? *Curr. Opin. Neurobiol.* **2014**, *24*, 176–189. [CrossRef] [PubMed]
122. Bushong, E.A.; Martone, M.E.; Jones, Y.Z.; Ellisman, M.H. Protoplasmic astrocytes in CA1 stratum radiatum occupy separate anatomical domains. *J. Neurosci.* **2002**, *22*, 183–192. [CrossRef] [PubMed]
123. Halassa, M.M.; Fellin, T.; Takano, H.; Dong, J.-H.; Haydon, P.G. Synaptic Islands Defined by the Territory of a Single Astrocyte. *J. Neurosci.* **2007**, *27*, 6473–6477. [CrossRef] [PubMed]
124. Simard, M.; Nedergaard, M. The neurobiology of glia in the context of water and ion homeostasis. *Neuroscience* **2004**, *129*, 877–896. [CrossRef] [PubMed]
125. Deitmer, J.W.; Theparambil, S.M.; Ruminot, I.; Noor, S.I.; Becker, H.M. Energy Dynamics in the Brain: Contributions of Astrocytes to Metabolism and pH Homeostasis. *Front. Neurosci.* **2019**, *13*, 1301. [CrossRef] [PubMed]

126. Magistretti, P.J. Role of glutamate in neuron-glia metabolic coupling. *Am. J. Clin. Nutr.* **2009**, *90*, 875S–880S. [CrossRef] [PubMed]
127. Magistretti Pj Sorg, O.; Martin, J.-L. *Astrocytes* **1993**, 243–265. Available online: https://books.google.ca/books?hl=en&lr=&id=4tpGN-_X6m4C&oi=fnd&pg=PA243&dq=Regulation+of+glycogen+metabolism+in+astrocytes:+physiological,+pharmacological+and+pathological+aspects+&ots=fc7SacoArw&sig=yxUNMa-kDtT1taLkbH2iJf6EOyk#v=onepage&q=Regulation%20of%20glycogen%20metabolism%20in%20astrocytes%3A%20physiological%2C%20pharmacological%20and%20pathological%20aspects&f=false (accessed on 1 March 2022).
128. Magistretti, P.J.; Allaman, I. Brain energy metabolism. In *Neuroscience in the 21st Century: From Basic to Clinical*; Springer: New York, NY, USA, 2013; pp. 1591–1620.
129. Bélanger, M.; Allaman, I.; Magistretti, P.J. Brain Energy Metabolism: Focus on Astrocyte-Neuron Metabolic Cooperation. *Cell Metab.* **2011**, *14*, 724–738. [CrossRef]
130. Attwell, D.; Buchan, A.M.; Charpak, S.; Lauritzen, M.; MacVicar, B.A.; Newman, E.A. Glial and neuronal control of brain blood flow. *Nature* **2010**, *468*, 232. [CrossRef]
131. Bezzi, P.; Volterra, A. A neuron–glia signalling network in the active brain. *Curr. Opin. Neurobiol.* **2001**, *11*, 387–394. [CrossRef]
132. Petrelli, F.; Bezzi, P. Novel insights into gliotransmitters. *Curr. Opin. Pharmacol.* **2016**, *26*, 138–145. [CrossRef] [PubMed]
133. Santello, M.; Calì, C.; Bezzi, P. Gliotransmission and the Tripartite Synapse2012. *Synaptic Plast.* **2012**, *970*, 307–331.
134. Bezzi, P.; Gundersen, V.; Galbete, J.L.; Seifert, G.; Steinhäuser, C.; Pilati, E.; Volterra, A. Astrocytes contain a vesicular compartment that is competent for regulated exocytosis of glutamate. *Nat. Neurosci.* **2004**, *7*, 613–620. [CrossRef]
135. Jourdain, P.; Bergersen, L.H.; Bhaukaurally, K.; Bezzi, P.; Santello, M.; Domercq, M.; Matute, C.; Tonello, F.; Gundersen, V.; Volterra, A. Glutamate exocytosis from astrocytes controls synaptic strength. *Nat. Neurosci.* **2007**, *10*, 331–339. [CrossRef] [PubMed]
136. Vesce, S.; Bezzi, P.; Volterra, A. The active role of astrocytes in synaptic transmission. *Cell. Mol. Life Sci.* **1999**, *56*, 991–1000. [CrossRef] [PubMed]
137. Calì, C.; Marchaland, J.; Spagnuolo, P.; Gremion, J.; Bezzi, P. Chapter 20 Regulated Exocytosis from Astrocytes2009. *Int. Rev. Neurobiol.* **2009**, *85*, 261–293. [PubMed]
138. Gómez-Gonzalo, M.; Zehnder, T.; Requie, L.M.; Bezzi, P.; Carmignoto, G. Insights into the release mechanism of astrocytic glutamate evoking in neurons NMDA receptor-mediated slow depolarizing inward currents. *GLIA* **2018**, *66*, 2188–2199. [CrossRef]
139. Schousboe, A.; Sarup, A.; Bak, L.K.; Waagepetersen, H.S.; Larsson, O.M. Role of astrocytic transport processes in glutamatergic and GABAergic neurotransmission. *Neurochem. Int.* **2004**, *45*, 521–527. [CrossRef]
140. Huang, Y.H.; Sinha, S.R.; Tanaka, K.; Rothstein, J.D.; Bergles, D.E. Astrocyte Glutamate Transporters Regulate Metabotropic Glutamate Receptor-Mediated Excitation of Hippocampal Interneurons. *J. Neurosci.* **2004**, *24*, 4551. [CrossRef]
141. Tzingounis, A.V.; Wadiche, J.I. Glutamate transporters: Confining runaway excitation by shaping synaptic transmission. *Nat. Rev. Neurosci.* **2007**, *8*, 935–947. [CrossRef]
142. Hirst, W.D.; Price, G.W.; Rattray, M.; Wilkin, G.P. Serotonin transporters in adult rat brain astrocytes revealed by [3H]5-HT uptake into glial plasmalemmal vesicles. *Neurochem. Int.* **1998**, *33*, 11–22. [CrossRef]
143. Inazu, M.; Takeda, H.; Matsumiya, T. The role of glial monoamine transporters in the central nervous system. *Nihon Shinkei Seishin Yakurigaku Zasshi* **2003**, *23*, 171–178. [PubMed]
144. Perdan-Pirkmajer, K.; Mavri, J.; Kržan, M. Histamine (re)uptake by astrocytes: An experimental and computational study. *J. Mol. Modeling* **2010**, *16*, 1151–1158. [CrossRef] [PubMed]
145. Naganuma, F.; Yoshikawa, T.; Nakamura, T.; Iida, T.; Harada, R.; Mohsen, A.S.; Miura, Y.; Yanai, K. Predominant role of plasma membrane monoamine transporters in monoamine transport in 1321N1, a human astrocytoma-derived cell line. *J. Neurochem.* **2014**, *129*, 591–601. [CrossRef]
146. Cahoy, J.D.; Emery, B.; Kaushal, A.; Foo, L.C.; Zamanian, J.L.; Christopherson, K.S.; Xing, Y.; Lubischer, J.L.; Krieg, P.A.; Krupenko, S.A. A transcriptome database for astrocytes, neurons, and oligodendrocytes: A new resource for understanding brain development and function. *J. Neurosci.* **2008**, *28*, 264–278. [CrossRef]
147. Cui, M.; Aras, R.; Christian, W.V.; Rappold, P.M.; Hatwar, M.; Panza, J.; Jackson-Lewis, V.; Javitch, J.A.; Ballatori, N.; Przedborski, S. The organic cation transporter-3 is a pivotal modulator of neurodegeneration in the nigrostriatal dopaminergic pathway. *Proc. Natl. Acad. Sci. USA* **2009**, *106*, 8043–8048. [CrossRef]
148. Zhang, Y.; Sloan, S.A.; Clarke, L.E.; Caneda, C.; Plaza, C.A.; Blumenthal, P.D.; Vogel, H.; Steinberg, G.K.; Edwards, M.S.B.; Li, G. Purification and Characterization of Progenitor and Mature Human Astrocytes Reveals Transcriptional and Functional Differences with Mouse. *Neuron* **2016**, *89*, 37–53. [CrossRef]
149. Yoshikawa, T.; Naganuma, F.; Iida, T.; Nakamura, T.; Harada, R.; Mohsen, A.S.; Kasajima, A.; Sasano, H.; Yanai, K. Molecular mechanism of histamine clearance by primary human astrocytes. *Glia* **2013**, *61*, 905–916. [CrossRef]
150. Petrelli, F.; Dallérac, G.; Pucci, L.; Calì, C.; Zehnder, T.; Sultan, S.; Lecca, S.; Chicca, A.; Ivanov, A.; Asensio, C.S. Dysfunction of homeostatic control of dopamine by astrocytes in the developing prefrontal cortex leads to cognitive impairments. *Mol. Psychiatry* **2020**, *25*, 732–749. [CrossRef]
151. Baganz, N.L.; Horton, R.E.; Calderon, A.S.; Owens, W.A.; Munn, J.L.; Watts, L.T.; Koldzic-Zivanovic, N.; Jeske, N.A.; Koek, W.; Toney, G.M. Organic cation transporter 3, Keeping the brake on extracellular serotonin in serotonin-transporter-deficient mice. *Proc. Natl. Acad. Sci. USA* **2008**, *105*, 18976. [CrossRef]

152. Bacq, A.; Balasse, L.; Biala, G.; Guiard, B.; Gardier, A.M.; Schinkel, A.; Louis, F.; Vialou, V.; Martres, M.P.; Chevarin, C. Organic cation transporter 2 controls brain norepinephrine and serotonin clearance and antidepressant response. *Mol. Psychiatry* **2012**, *17*, 926–939. [CrossRef] [PubMed]
153. Giros, B.; Jaber, M.; Jones, S.R.; Wightman, R.M.; Caron, M.G. Hyperlocomotion and indifference to cocaine and amphetamine in mice lacking the dopamine transporter. *Nature* **1996**, *379*, 606–612. [CrossRef] [PubMed]
154. Käenmäki, M.; Tammimäki, A.; Myöhänen, T.; Pakarinen, K.; Amberg, C.; Karayiorgou, M.; Gogos, J.A.; Männistö, P.T. Quantitative role of COMT in dopamine clearance in the prefrontal cortex of freely moving mice. *J. Neurochem.* **2010**, *114*, 1745–1755. [CrossRef] [PubMed]
155. Carboni, E.; Tanda, G.L.; Frau, R.; Chiara, G.D. Blockade of the noradrenaline carrier increases extracellular dopamine concentrations in the prefrontal cortex: Evidence that dopamine is taken up in vivo by noradrenergic terminals. *J. Neurochem.* **1990**, *55*, 1067–1070. [CrossRef] [PubMed]
156. Mundorf, M.L.; Joseph, J.D.; Austin, C.M.; Caron, M.G.; Wightman, R.M. Catecholamine release and uptake in the mouse prefrontal cortex. *J. Neurochem.* **2001**, *79*, 130–142. [CrossRef] [PubMed]
157. Mazei, M.S.; Pluto, C.P.; Kirkbride, B.; Pehek, E.A. Effects of catecholamine uptake blockers in the caudate-putamen and subregions of the medial prefrontal cortex of the rat. *Brain Res.* **2002**, *936*, 58–67. [CrossRef]
158. Eisenhofer, G.; Kopin, I.J.; Goldstein, D.S. Catecholamine Metabolism: A Contemporary View with Implications for Physiology and Medicine. *Pharmacol. Rev.* **2004**, *56*, 331–349. [CrossRef]
159. Howes, O.D.; Kapur, S. The dopamine hypothesis of schizophrenia: Version III—The final common pathway. *Schizophr. Bull.* **2009**, *35*, 549–562. [CrossRef]
160. Moncrieff, J. A critique of the dopamine hypothesis of schizophrenia and psychosis. *Harv. Rev. Psychiatry* **2009**, *17*, 214–225. [CrossRef]
161. Glatt, S.J.; Faraone Ming TTsuang, S.V. Association between a Functional Catechol O-Methyltransferase Gene Polymorphism and Schizophrenia: Meta-Analysis of Case-Control and Family-Based Studies. *Am. J. Psychiatry* **2003**, *160*, 3. [CrossRef]
162. Weinberger, D.R.; Egan, M.F.; Bertolino, A.; Callicott, J.H.; Mattay, V.S.; Lipska, B.K.; Berman, K.F.; Goldberg, T.E. Prefrontal neurons and the genetics of schizophrenia. *Biol. Psychiatry* **2001**, *50*, 825–844. [CrossRef]
163. Huotari, M.; Gogos, J.A.; Karayiorgou, M.; Koponen, O.; Forsberg, M.; Raasmaja, A.; Hyttinen, J.; Männistö, P.T. Brain catecholamine metabolism in catechol-O-methyltransferase (COMT)-deficient mice. *Eur. J. Neurosci.* **2002**, *15*, 246–256. [CrossRef] [PubMed]
164. Yavich, L.; Forsberg, M.M.; Karayiorgou, M.; Gogos, J.A.; Männistö, P.T. Site-Specific Role of Catechol-O-Methyltransferase in Dopamine Overflow within Prefrontal Cortex and Dorsal Striatum. *J. Neurosci.* **2007**, *27*, 10196. [CrossRef] [PubMed]
165. Vijayraghavan, S.; Wang, M.; Birnbaum, S.G.; Williams, G.V.; Arnsten, A.F.T. Inverted-U dopamine D1 receptor actions on prefrontal neurons engaged in working memory. *Nat. Neurosci.* **2007**, *10*, 376–384. [CrossRef]
166. Dickinson, D.; Elvevåg, B. Genes, cognition and brain through a COMT lens. *Neuroscience* **2009**, *164*, 72–87. [CrossRef]
167. Phang, J.M.; Hu, C.A.; Valle, D. Disorders of proline and hydroxyproline metabolism. *Metab. Mol. Basis Inherit. Dis.* **2001**, 1821–1838. Available online: https://www.researchgate.net/publication/284684119_Disorders_of_proline_and_hydroxyproline_metabolism (accessed on 1 March 2022).
168. Raux, G.; Bumsel, E.; Hecketsweiler, B.; van Amelsvoort, T.; Zinkstok, J.; Manouvrier-Hanu, S.; Fantini, C.; Brévière, G.M.; Di Rosa, G.; Pustorino, G.; et al. Involvement of hyperprolinemia in cognitive and psychiatric features of the 22q11 deletion syndrome. *Hum. Mol. Genet.* **2007**, *16*, 83–91. [CrossRef]
169. Gogos, J.A.; Santha, M.; Takacs, Z.; Beck, K.D.; Luine, V.; Lucas, L.R.; Nadler, J.V.; Karayiorgou, M. The gene encoding proline dehydrogenase modulates sensorimotor gating in mice. *Nat. Genet.* **1999**, *21*, 434–439. [CrossRef]
170. Boisvert, M.M.; Erikson, G.A.; Shokhirev, M.N.; Allen, N.J. The Aging Astrocyte Transcriptome from Multiple Regions of the Mouse Brain. *Cell Rep.* **2018**, *22*, 269–285. [CrossRef]
171. Servet, C.; Ghelis, T.; Richard, L.; Zilberstein, A.; Savoure, A. Proline dehydrogenase: A key enzyme in controlling cellular homeostasis. *Front. Biosci.* **2012**, *17*, 607. [CrossRef]
172. Maynard, T.M.; Haskell, G.T.; Peters, A.Z.; Sikich, L.; Lieberman, J.A.; Lamantia, A.-S. A comprehensive analysis of 22q11 gene expression in the developing and adult brain. *Proc. Natl. Acad. Sci. USA* **2003**, *100*, 14433–14438. [CrossRef] [PubMed]
173. Clarke, L.E.; Barres, B.A. Emerging roles of astrocytes in neural circuit development. *Nat. Rev. Neurosci.* **2013**, *14*, 311–321. [CrossRef] [PubMed]
174. Ullian, E.M.; Sapperstein, S.K.; Christopherson, K.S.; Barres, B.A. Control of synapse number by glia. *Science* **2001**, *291*, 657–661. [CrossRef] [PubMed]
175. Dietz, A.G.; Goldman, S.A.; Nedergaard, M. Glial cells in schizophrenia: A unified hypothesis. *Lancet Psychiatry* **2020**, *7*, 272–281. [CrossRef]
176. Allen, N.J.; Eroglu, C. Cell Biology of Astrocyte-Synapse Interactions. *Neuron* **2017**, *96*, 697–708. [CrossRef]
177. Zehnder, T.; Petrelli, F.; Romanos, J.; de Oliveira Figueiredo, E.C.; Lewis, T.L.; Polleux, F.; Santello, M.; Bezzi, P. Mitochondrial biogenesis in developing astrocytes regulates astrocyte maturation and synapse formation. *Cell Rep.* **2021**, *35*, 108952. [CrossRef]
178. Buscemi, L.; Ginet, V.; Lopatar, J.; Montana, V.; Pucci, L.; Spagnuolo, P.; Zehnder, T.; Grubišic, V.; Truttman, A.; Sala, C. Homer1 Scaffold Proteins Govern Ca^{2+} Dynamics in Normal and Reactive Astrocytes. *Cereb. Cortex* **2017**, *27*, 2365–2384. [CrossRef]

179. Sun, W.; McConnell, E.; Pare, J.F.; Xu, Q.; Chen, M.; Peng, W.; Lovatt, D.; Han, X.; Smith, Y.; Nedergaard, M. Glutamate-Dependent Neuroglial Calcium Signaling Differs between Young and Adult Brain. *Science* **2013**, *339*, 197–200. [CrossRef]
180. Petrelli, F.; Bezzi, P. mGlu5-mediated signalling in developing astrocyte and the pathogenesis of autism spectrum disorders. *Curr. Opin. Neurobiol.* **2018**, *48*, 139–145. [CrossRef]
181. Yakoub, A.M.; Sadek, M. Analysis of Synapses in Cerebral Organoids. *Cell Transplant.* **2019**, *28*, 1173–1182. [CrossRef]
182. Crockett, A.M.; Ryan, S.K.; Vasquez, A.H.; Canning, C.; Kanyuch, N.; Kebir, H.; Ceja, G.; Gesualdi, J.; Zackai, E.; McDon-ald-McGinn, D.; et al. Disruption of the blood-brain barrier in 22q11.2 deletion syndrome. *Brain* **2021**, *144*, 1351–1360. [CrossRef] [PubMed]

International Journal of *Molecular Sciences*

MDPI

Article

Comparison of Cerebellar Grey Matter Alterations in Bipolar and Cerebellar Patients: Evidence from Voxel-Based Analysis

Michela Lupo [1,*,†], Giusy Olivito [1,2,†], Andrea Gragnani [3,4], Marco Saettoni [3,5], Libera Siciliano [6], Corinna Pancheri [7], Matteo Panfili [7], Marco Bozzali [8], Roberto Delle Chiaie [7] and Maria Leggio [1,2]

1 Ataxia Laboratory, Fondazione Santa Lucia IRCCS, 00179 Rome, Italy; g.olivito@hsantalucia.it (G.O.); maria.leggio@uniroma1.it (M.L.)
2 Department of Psychology, Sapienza University of Rome, 00185 Rome, Italy
3 Scuola di Psicoterapia Cognitiva SPC, 58100 Grosseto, Italy; gragnani@apc.it (A.G.); marcosaettoni@gmail.com (M.S.)
4 Associazione Psicologia Cognitiva (APC)/Scuola di Psicoterapia Cognitiva (SPC), 00185 Rome, Italy
5 Unità Funzionale Salute Mentale Adulti ASL Toscana Nord-Ovest Valle del Serchio, 56121 Pisa, Italy
6 PhD Program in Behavioral Neuroscience, Sapienza University of Rome, 00185 Rome, Italy; libera.siciliano@uniroma1.it
7 Departement of Neuroscience and Mental Health–Policlinico Umberto I Hospital, Sapienza University of Rome, 00161 Rome, Italy; corinnapancheri@hotmail.it (C.P.); matteopanfili93@gmail.com (M.P.); r.dellechiaie@centrokahlbaum.it (R.D.C.)
8 Clinical Imaging Science Center, Brighton and Sussex Medical School, Brighton BN1 9RR, UK; m.bozzali@hsantalucia.it
* Correspondence: m.lupo@hsantalucia.it; Tel.: +39-065-150-1115
† Michela Lupo and Giusy Olivito are co-first author.

Citation: Lupo, M.; Olivito, G.; Gragnani, A.; Saettoni, M.; Siciliano, L.; Pancheri, C.; Panfili, M.; Bozzali, M.; Delle Chiaie, R.; Leggio, M. Comparison of Cerebellar Grey Matter Alterations in Bipolar and Cerebellar Patients: Evidence from Voxel-Based Analysis. *Int. J. Mol. Sci.* **2021**, *22*, 3511. https://doi.org/10.3390/ijms22073511

Academic Editor: Giuseppina Martella

Received: 8 March 2021
Accepted: 26 March 2021
Published: 29 March 2021

Abstract: The aim of this study was to compare the patterns of cerebellar alterations associated with bipolar disease with those induced by the presence of cerebellar neurodegenerative pathologies to clarify the potential cerebellar contribution to bipolar affective disturbance. Twenty-nine patients affected by bipolar disorder, 32 subjects affected by cerebellar neurodegenerative pathologies, and 37 age-matched healthy subjects underwent a 3T MRI protocol. A voxel-based morphometry analysis was used to show similarities and differences in cerebellar grey matter (GM) loss between the groups. We found a pattern of GM cerebellar alterations in both bipolar and cerebellar groups that involved the anterior and posterior cerebellar regions ($p = 0.05$). The direct comparison between bipolar and cerebellar patients demonstrated a significant difference in GM loss in cerebellar neurodegenerative patients in the bilateral anterior and posterior motor cerebellar regions, such as lobules I–IV, V, VI, VIIIa, VIIIb, IX, VIIb and vermis VI, while a pattern of overlapping GM loss was evident in right lobule V, right crus I and bilateral crus II. Our findings showed, for the first time, common and different alteration patterns of specific cerebellar lobules in bipolar and neurodegenerative cerebellar patients, which allowed us to hypothesize a cerebellar role in the cognitive and mood dysregulation symptoms that characterize bipolar disorder.

Keywords: cerebellar atrophy; bipolar disorder; voxel-based morphometry; cerebellar grey matter volume

1. Introduction

Bipolar disorder (BD) is a severe and chronic psychiatric disease that is often associated with several medical conditions, including cardiovascular problems, diabetes mellitus, and neurovascular disease [1,2]. This disorder is characterized by episodes of mania (or hypomania) and depression that can lead to cognitive impairments with a considerable impact on one's quality of life [1,3–7]. Cognitive impairments, mainly involving executive function, attention, verbal and episodic memory, persist during euthymic phases [8–10].

Furthermore, neuroimaging studies have shown that bipolar disorder is characterized by several structural modifications in the cortical and subcortical areas, such as grey matter volume abnormalities in the frontal and temporal regions and in limbic regions, such as the cingulate cortex [11–13]. Among subcortical structures, specific structural alterations have also been reported in the cerebellum.

Indeed, recent literature has shown cerebellar involvement in psychiatric diseases and neurodevelopmental disorders, specifically in schizophrenia, autism and obsessive-compulsive disorder [14–21]. Furthermore, in recent years, some evidence has highlighted the potential role of the cerebellum in other psychiatric conditions characterized by mood swings, such as bipolar disorder [22,23]. It has been shown that alterations of the cortico-cerebellar network in patients with bipolar disorder (type 1 and type 2) are present during earlier stages of the disease and remain stable over time, thus suggesting a possible neurodevelopmental involvement of this network in the mechanism of bipolar disorder, with no differences between bipolar disorder subtypes [23]. Despite this evidence, the specific cerebellar contribution to the neuropathophysiological mechanisms underlying bipolar disorder still needs to be clarified.

Indeed, neuroimaging studies have described cerebellar alterations in patients with bipolar disorder [24–27] over the last twenty years and researchers have focused on this structure [28–32] in light of the cerebellar connections with cortical areas involved in the pathophysiology of bipolar disorder [13,28,31,33,34] and the cerebellar role in emotion [35–39], social cognition [16,40,41] and cognitive functions [42–45]. Moreover, it is worth noting that very recently the onset of mood disorders (manic and depressive symptoms) has been demonstrated in the presence of isolated cerebellar lesions and neurodegenerative cerebellar pathologies [46–49]. These data confirm that cerebellar alterations are associated with mood symptoms, as reported in cerebellar-cognitive affective syndrome (CCAS) [50].

In light of these observations, it is reasonable to hypothesize that the cerebellar alteration in BD may contribute to specific features of bipolar symptoms (i.e., mania, hypomania, depression) and further insight can be gained from the comparison with neurodegenerative disorders that selectively affect the cerebellar cortex, such as spinocerebellar ataxia (SCA). To this aim, in the present study, cerebellar structural patterns were compared between BD patients and patients affected by cerebellar degenerative disorders (CD) of different etiology. Indeed, according to the growing evidence of a cerebellar involvement in the manic symptoms of bipolar disorder and the presence of mood disturbance in cerebellar diseases, the comparison between BD and CD patients will provide a crucial insight into understanding the cerebellum role in the pathophysiology of mood disturbances and in maintaining symptoms related to bipolar disorder.

2. Results

All CD patients enrolled in the present study obtained a normal Intelligence Quotient (IQ > 70). There were no differences in terms of age ($F = 0.995$; $p = 0.374$) or sex distribution ($F = 0.042$; $p = 0.959$) between BD, CD and healthy subjects (HS), as assessed by one-way ANOVA (Table 1).

Table 1. Main demographic and clinical characteristics of the BD, CD and HS groups.

Characteristic	BD (n = 29)	CD (n = 32)	HS (n = 37)
Age, years, mean ± SD	42.69 ± 10.53	46.81 ± 11.48	45.75 ± 14.26
Males/females	13/16	18/14	15/22
ICARS mean ± SD	1.10 ± 2.06	25.78 ± 12.94	–

SD = standard deviation; BD = bipolar disorder group; CD = cerebellar neurodegenerative disorders group; HS = healthy subjects group; ICARS = international cooperative ataxia rating scale.

As assessed by the international cooperative ataxia rating scale (ICARS), neurological examination revealed the presence of a pure cerebellar motor syndrome in CD. No extracerebellar symptoms were present at the time of the study. Moreover, 10 out of 29 BD patients showed cerebellar symptoms as evidenced by the ICARS scale (see Table 1 for clinical details). Specifically, three patients were affected by standing balance and gait ataxia, three patients were affected by upper limb ataxia, and one patient was affected by dysmetric eye movements. In addition, 7 out of these 10 patients were also affected by fine postural hand tremor.

Voxel-wise analysis of cerebellar grey matter (GM) maps showed a significant pattern of GM loss in the cerebellar cortex of both BD and CD patients compared to the HS group ($p = 0.05$ family-wise error-FEW-corrected). More specifically, BD patients showed several clusters of significantly decreased GM volume that included right lobules I—IV and V, crus I, VIIB, IX and vermis crus II as well as left lobule VI and bilateral crus II (Figure 1a). When compared to HSs, CD patients also showed different clusters of cerebellar GM loss that diffusively affected both anterior and posterior cerebellar regions, including bilateral lobule VI, bilateral lobules I—IV, right V, bilateral crus I and Crus II (Figure 1b). Finally, the direct comparison between the two groups demonstrated a significant difference in GM loss in CD patients compared to BD patients, mainly involving anterior and posterior motor cerebellar regions such as bilateral lobules I—IV, V, VI, VIIIa and VIIIb, with extension in bilateral lobule IX and vermis VI (Figure 1c). Detailed statistics and peak voxels of voxel-based morphometry (VBM) analysis are reported in Table 2. Interestingly, the pattern of overlapping GM reduction in CD and BD patients was evident in the right lobule V, right crus I and bilateral crus II (Figure 2a,b).

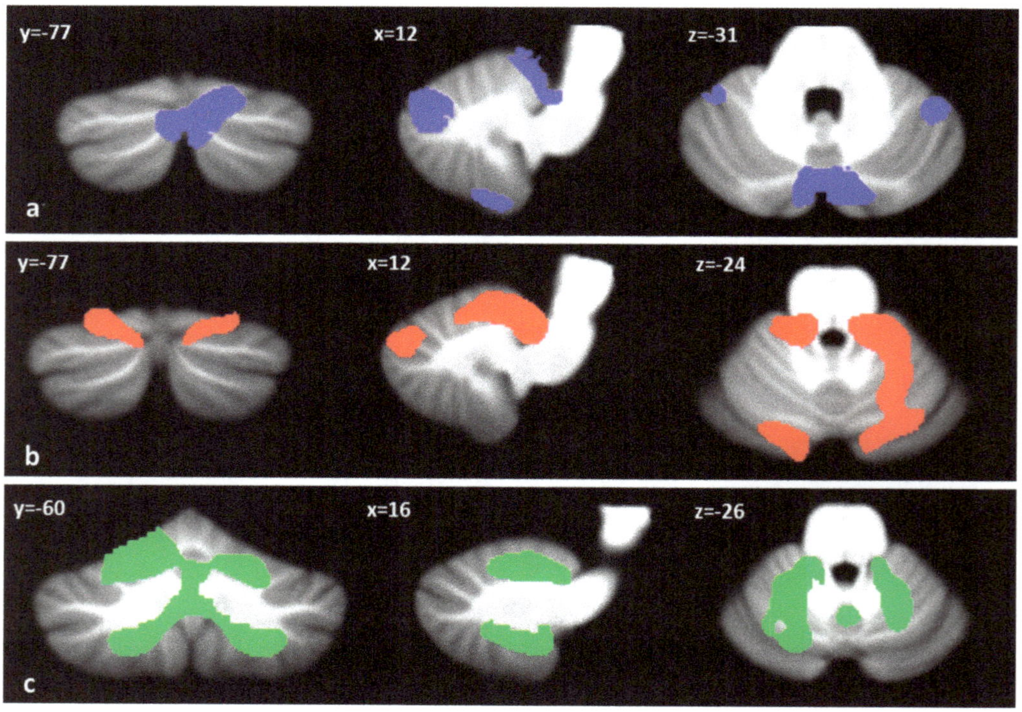

Figure 1. Cerebellar regions showing patterns of significantly reduced cerebellar grey matter (GM) are reported and superimposed on the spatially unbiased infratentorial template (SUIT) [51] in coronal (y), axial (z) and sagittal (x) sections. (**a**) BD < HS (in blue); (**b**) CD < HS (in red); (**c**) CD < BD (in green). The results are considered significant at p-values < 0.05 FWE corrected at the cluster level. Images are shown in the radiological convention.

Table 2. Cerebellar voxel-wise analyses between BD patients, CD patients and HSs. Detailed statistics and z-scores of peak voxels showing greatest statistical significance in the cluster. Cluster-forming threshold $p < 0.05$ FWE.

	Cluster Size (NoV)	Coordinates			Peak z-Score	Cerebellar Region
		x	z	y		
BD < HS	4667	7	−78	−33	5.35	R-Crus II
	3358	17	−42	−12	6.86	R-Lobule V
		20	−31	−19	5.95	R-Lobule V
	1024	45	−46	−27	4.36	R-Crus I
	892	−44	−47	−45	4.38	L-Crus II
	223	−41	−43	−28	4.37	L-lobule VI
CD < HS	17246	−8	−37	−19	5.99	L-Lobule I–IV
		12	−37	−22	5.75	R-Lobule I–IV
		13	−48	−16	5.21	R-Lobule V
	2047	−12	−87	−29	4.77	L-Crus II
		−18	−79	−21	4.65	L-Crus I
CD < BD	28567	22	−55	−45	6.78	R-Lobule VIIIb
		−20	−61	−48	5.53	L-Lobule VIIIa
		−14	−57	−19	5.52	L-Lobule VI

NoV = number of voxels; R = right; L = left.

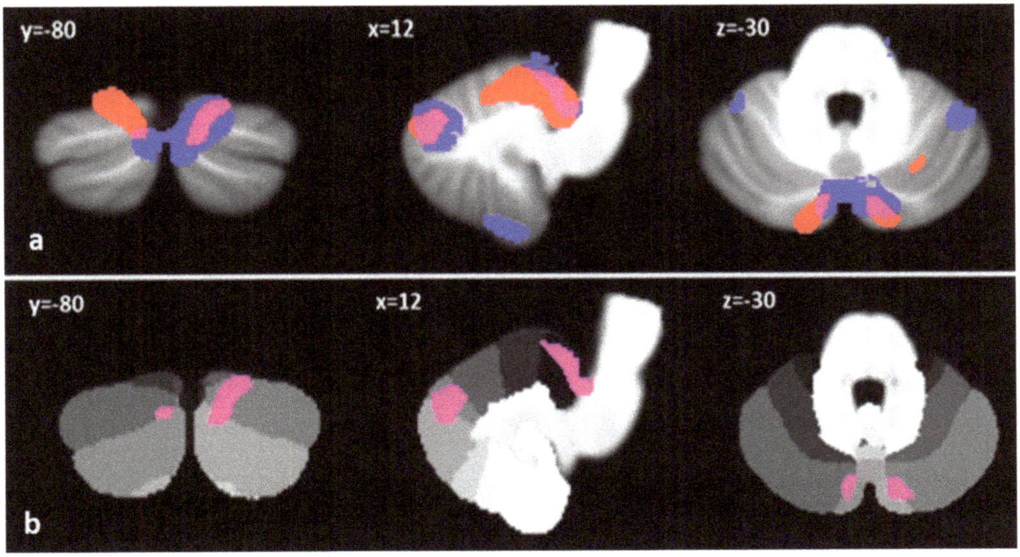

Figure 2. (**a**) Clusters of reduced cerebellar GM in BD and CD compared to HSs are reported in blue and red, respectively. Regions of overlapped cerebellar GM loss between BD and CD are shown in violet on the spatially unbiased infratentorial template (SUIT) [51]. (**b**) Only cerebellar regions showing patterns of significantly reduced GM in both BD and CD patients are shown in violet and superimposed on the probabilistic SUIT cerebellar atlas [51], which provides anatomical subdivision of cerebellar lobules (in greyscale). Images are shown according to radiological convention in coronal (y), axial (z) and sagittal (x) sections.

3. Discussion

The aim of the present study was to clarify the potential role of the cerebellum in bipolar disorder by analysing the patterns of cerebellar alterations in BD patients compared to patients with neurodegenerative disorders selectively affecting the cerebellum. In this framework, the CD group was used as a model of cerebellar degeneration in order to highlight the common and distinct patterns of grey matter loss in comparison to BD in which cerebellar alterations have been detected [23].

The comparison between cerebellar-related neurodegenerative syndromes (such as SCA) and BD mainly arises from the growing evidence of a cerebellar involvement in manic symptoms of BD [49] and the presence of mood disturbance in cerebellar diseases [46]. As largely demonstrated, the posterior cerebellum is involved in the processing of cognitive and emotional information and takes part in the network involved in mentalizing and social interactions [16,40,41,49,52–56]. Indeed, the cerebellar cortex receives direct and indirect inputs from cortical associative areas and the midbrain, and in turn, the cerebellar nuclei send signals back to the limbic lobe and hypothalamic and thalamic nuclei, important relay stations that connect cortical and subcortical structures [19]. However, although cognitive and emotional cerebellar functions are widely described, the cerebellar role in bipolar disorder has been scarcely explored [22,23,30]. In light of these observations, our findings provide great insight into understanding the specific cerebellar contribution to the underlying neuropathophysiological mechanism of bipolar disorder.

Compared to controls, BD patients showed significant GM reduction in anterior cerebellar regions, including the right I–IV and V lobules; in posterior cerebellar areas, including right crus I and lobule IX; and in cerebellar vermis, including crus II, left lobule VI, and bilateral crus II.

According to the motor function of anterior cerebellar regions, the patterns of GM loss may be more related to the presence of clinical motor manifestations, i.e., psychomotor agitation [23,57], that, by definition, is influenced by the presence of the mania, and hypomania and mixed states of bipolar disorder throughout the progression of the disease (American Psychiatric Association. Diagnostic and Statistical Manual of Medical Disorders. 5th ed Washington, DC: American Psychiatric Association [58]). In light of these findings, it is reasonable to assume that the repeated mood relapses that characterize BD could have affected these cerebellar areas leading to structural modifications.

On the other hand, the extended pattern of GM reductions in hemispheric and vermal posterior cerebellar regions may be linked to the specific cognitive alterations described in BD studies in light of the cerebellar role in several emotional and cognitive domains [16,19,40,48,49,54,56,59], that characterize bipolar disorder [6,7,38,60–63].

It is worth noting that previous research reporting cerebellar alterations in bipolar disorder mainly used a whole-brain approach and did not focus on the cerebellum. In the present study, we implemented a procedure (see Methods section) that allowed us to restrict the analysis to the cerebellum and achieve anatomical localization of the specific cerebellar subregions involved [51]. Indeed, according to the cerebellar functional topography [19], we were able to identify the pattern of GM reduction in cerebellar portions, which have been linked to the cognitive, emotional and mood symptoms by several studies on bipolar patients [11,27,64–68].

When compared to controls, the CD group showed a diffuse pattern of GM loss throughout the cerebellar cortex. In line with the presence of typical cerebellar motor syndrome [19,55,69], an extensive pattern of GM loss involved motor anterior (i.e., I–IV, V) and posterior cerebellar regions (i.e., VIIIA and VIIIB). On the other hand, a pattern of GM loss was also found to extensively affect cognitive posterior cerebellar lobules, specifically crus I, crus II and lobe VI, in line with the presence of cognitive and emotional alterations as reported in CCAS [46,54,55,69–72]. Finally, when directly comparing BD and CD patients, significantly reduced cerebellar GM was found in the CD compared to the BD patients, only involving motor anterior cerebellar regions. This is in line with the presence of the cerebellar motor syndrome that is specific of CD patients and was not detected in our BD patients [19,55,69]. As previously stated, structural alterations in anterior cerebellar regions are also found in BD but they may be more related to the psychomotor agitation that typically accompanies the affective episodes. In spite of the presence of anterior cerebellar atrophy, CD patients did not show psychomotor agitation. This might mean that the involvement of the motor anterior cerebellar regions per se is not enough to evoke psychomotor agitation, which, by definition, is related to affective episodes that characterize bipolar patients and are subtended by a more complex dysfunctional network in which the cerebellum acts.

Interestingly, as depicted in Figure 2, BD and CD patients showed a common atrophy pattern specifically affecting right lobule V, right crus I and bilateral crus II, known to be typically involved in higher-order cognitive functions, whose alterations widely characterize both disorders [6,7,9,16,19,35,37–42,44,45,60–63].

Altogether, these results show for the first time common and different patterns of cerebellar alterations in BD and CD patients that may shed light on the potential cerebellar contribution to the neuropathophysiology of bipolar disorder. Specifically, according to these findings, it is possible to hypothesize that in BD patients, the pattern of cerebellar atrophy is specifically related to the cognitive dysfunctions and mood dysregulation typically reported in bipolar disorder. Indeed, the atrophy of posterior cerebellar lobules, in particular crus I and crus II, may be related the cognitive dysfunctions in line with the extensive connections with cortical association areas and, in particular, prefrontal cortical regions [19], while posterior vermal regions in the limbic cerebellum contribute to emotional lability and a flattening effect consistent with its extensive connections with the limbic system [73].

Interestingly, the specific involvement of the limbic cerebellum in mood regulation and behavior has been previously reported in a single-case study of a patient who showed behavioral abnormalities after the rupture of a cerebellar arteriovenous malformation [49]. Specifically, the behavioral symptoms were characterized by disinhibition or inappropriate behavior, emotional lability, irritability, aggressiveness, affective instability and impulsiveness together with the onset of a manic state. Data from MRI scans demonstrated an involvement of the posterior area of the cerebellar vermis, and cerebello-cerebral functional connectivity analysis revealed a pattern of altered connectivity in specific areas of the prefrontal-striatal-thalamic circuits that are typically altered in bipolar subjects during the manic state, suggesting a cerebellar role in mood regulation [49].

To conclude, there are some potential limitations that need to be discussed. With regard to the CD group and specifically to SCA2 sample, it has to be considered that extracerebellar signs may occur by definition on a subclinical level in SCA2 patients. However, as evidenced by the neurological examination and MRI analysis, the absence of extracerebellar signs and cortical atrophy has been ensured. Moreover, it must be taken into account that the choice of grouping cerebellar patients with different etiology depended on the rarity of this neurodegenerative condition that clearly affects the inclusion rate and makes it difficult to find large numbers of cerebellar patients with the same diagnosis. Furthermore, due to the retrospective nature of the study on CD patients, data on the presence of manic/mood symptoms were not available. Future studies may further address these issues and overcome these shortcomings. Another potential limitation is that all our BD patients were on medication, often involving polypharmacy, which is typical in patients with severe illness [74]. However, while different medication approaches can differently affect the central nervous system, it has to be considered that enrolling drug-free patients presents important ethical and clinical concerns.

Another issue that needs to be mentioned concerns the different profiles described in the literature (in terms of cognitive and mood symptoms) between the subtypes of BD patients. In the present study, the number of BD patients enrolled did not allow us to divide our sample into BD type 1 and type 2. Future studies that consider this different classification will aim to fill this gap, and to investigate the relationship between cerebellar structural changes and the severity of the BD symptoms providing further support for the present conclusions and opening new avenues for the therapeutic treatment of bipolar disorder specifically targeting the cerebellum.

4. Materials and Methods

4.1. Participants

Two groups of adult patients were enrolled in the study: subjects affected by BD and subjects affected by CDs.

BD patients were recruited from the bipolar disorder outpatient ward of the Department of Psychiatry, Policlinico Umberto I Hospital. All of the patients met the Diagnostic and Statistical Manual of Mental Disorders, Fifth Edition criteria [58] for BD, according to a diagnostic assessment performed with the Italian version of the Structured Clinical Interview for DSM-5—Clinician Version (SCID-5-CV) [75]. All patients had been euthymic for at least three months. The euthymic phase was established by using the Hamilton Depression Rating Scale (HDRS score < 10) [76] and Young Mania Rating Scale administration (YMRS score < 13) [77].

The inclusion criteria for BD patients were as follows: (i) aged between 18 and 60 years, (ii) euthymic mood for at least 3 months, (iii) first examination by a psychiatrist performed before age 40, and (iv) suitability for magnetic resonance imaging (MRI).

The exclusion criteria for BD patients were (i) having other Axis-I psychiatric disorders; (ii) exhibiting lifetime alcohol/substance abuse; (iii) having a history of an organic brain disorder or neurological disorder; (iv) having mental retardation; and (v) having a medical condition such as pregnancy, cardiovascular disease or diabetes. All BD patients were recruited by an expert clinical psychiatrist of the Department of Psychiatry, Policlinico Umberto I Hospital. Moreover, for each patient, the clinical diagnosis of BD in the euthymic phase was confirmed by another senior psychotherapist by means of a clinical interview, the HDRS [76] and YMRS [77].

Thirty-six patients with BD were initially included in the present study. Of the original group of patients, three refused to undergo the MRI exam, and four were excluded for the presence of moderate to severe brain vascular lesions (see exclusion criteria). Thus, the final sample was of 29 BD patients, and all participants underwent medical treatment (Table 3).

Table 3. Clinical details regarding the BD group.

Medical Treatment	Mean ± SD	N°
HDRS	1.82 ± 2.68	29
YMRS	1.61 ± 3.01	29
	Current pharmacotherapy	
Antipsychotics		13
Lithium		14
Antiepileptics		22
Antidepressants		2
Anxiolytic		3
Polypharmacy		17

HDRS = Hamilton Depression Rating Scale; YMRS = Young Mania Rating Scale; SD = standard deviation.

Thirty-two patients affected by CDs of different etiologies were also recruited (Table 4).

The inclusion criteria for CD patients were (i) more than 6 months of illness and (ii) evidence of diffuse cerebellar atrophy. The exclusion criteria for CD patients were (i) the presence of other pathological conditions (ii) the presence of any cortical lesion on conventional MRI scans and (iii) the presence of mental retardation.

Additionally, 37 sex- and age-matched HSs with no history of neurological or psychiatric illness were recruited as the control group.

All BD and CD patients underwent neurological evaluations. To quantify cerebellar motor deficits, the international cooperative ataxia rating scale [78], which ranges from 0 (absence of a motor deficit) to 100 (presence of motor deficits at the highest degree), was used.

Table 4. Clinical details regarding the CD group.

Diagnosis	N°
SCA type1	1
SCA type 2	12
SCA type 6	1
SCA type 15	1
SCA type 28	1
SPG7	6
FRDA	2
ICA	8

SCA = spinocerebellar ataxia; SPG7 = spastic paraplegia type 7; FRDA = Friedreich's ataxia; ICA = idiopathic cerebellar atrophy.

In CD patients, the Wechsler adult intelligence scale—IV edition [79] was performed to exclude the presence of mental retardation.

Since these patients had already taken part in another study from our group, see Table 2 in Clausi and colleagues [40] for the cognitive profile.

This research study was approved by the Ethics Committee of Fondazione Santa Lucia IRCCS according to the principles expressed in the Declaration of Helsinki. Written informed consent was obtained from each subject. The main demographic and clinical characteristics are summarized in Table 1.

4.2. MRI Acquisition Protocol

Patients and HSs underwent MRI examination at 3T (Magnetom Allegra, Siemens, Erlangen, Germany) that included the following acquisitions: (1) dual-echo turbo spin echo [TSE] (TR = 6190 ms, TE = 12/109 ms); (2) fast-FLAIR (TR = 8170 ms, 204TE = 96 ms, TI = 2100 ms); and (3) 3D modified driven equilibrium Fourier transform (MDEFT) scans (TR = 1338 ms, TE = 2.4 ms, matrix = 256 × 224 × 176, in-plane FOV = 250 × 250 mm^2, slice thickness = 1 mm) to perform voxel-based morphometry on cerebellar grey matter (GM) maps. To characterize the brain anatomy and determine the presence of macroscopic structural abnormalities, the TSE scans of patients were visually inspected by an expert neuro-radiologist. For HSs, conventional MRI scans were reviewed to ensure the absence of any macroscopic brain abnormality.

4.3. Image Processing and Analysis

The cerebellum was preprocessed individually using the SUIT toolbox [51] implemented in Statistical Parametric Mapping version 8 (Wellcome Department of Imaging Neuroscience; SPM-8 (http://www.fl.ion.ucl.ac.uk/spm/, accessed on 2 April 2009). The procedure was performed on individual T1 anatomical images and included isolating the cerebellum and then normalizing each cropped image into SUIT space and reslicing the probabilistic cerebellar atlas into individual subject spaces using the deformation parameters obtained by normalization. Finally, each segmented cerebellar GM map was smoothed using an 8-mm FWHM Gaussian kernel. Voxel-based morphometry was used to characterize the patterns of regional cerebellar GM atrophy in BD and CD patients compared to those of HSs. Additionally, a direct comparison between BD and CD patients was also carried out. Voxel-wise two-sample t-tests as implemented in SPM-8 were used. Sex was entered as a variable of no interest, and the analysis was restricted only to the voxels of the cerebellum by using an explicit exclusion mask. The results were considered significant at p values < 0.05 FWE corrected at the cluster level.

5. Conclusions

In conclusion, the present study demonstrates that BD and CD patients show a common pattern of altered cerebellar regions in lobules that are involved in cognitive and emotional abilities, such as crus I and crus II. Consistent with previous evidence, the alteration of these specific cerebellar portions may be the anatomical substrate that contributes to manic symptoms of bipolar disorders and the mood disturbances often observed in cerebellar diseases.

Altogether, these results provide useful insights for understanding and clarifying the cerebellar contribution to the pathophysiology of bipolar disorder and its potential role as a target for future treatments.

Author Contributions: Conceptualization, M.L. (Michela Lupo), G.O. and M.L. (Maria Leggio); methodology, M.L. (Michela Lupo) and G.O.; formal analysis, M.L. (Michela Lupo) and G.O.; investigation, M.L. (Michela Lupo), G.O., A.G., M.S., L.S., C.P., M.P.; data curation, M.L. (Michela Lupo) and G.O.; writing—original draft preparation, M.L. (Michela Lupo), G.O. and M.L. (Maria Leggio); writing—review and editing, M.L. (Michela Lupo), G.O. and M.L. (Maria Leggio); supervision, M.B., R.D.C. and M.L. (Maria Leggio); All authors have read and agreed to the published version of the manuscript.

Funding: This research received no external funding.

Institutional Review Board Statement: The study was conducted according to the guidelines of the Declaration of Helsinki, and approved by the Ethics Committee of Fondazione Santa Lucia IRCCS (Prot. CE/PROG.579 and 22 November 2016).

Informed Consent Statement: Informed consent was obtained from all subjects involved in the study.

Data Availability Statement: Study data will be shared upon request.

Conflicts of Interest: The authors declare no conflict of interest.

References

1. Merikangas, K.R.; Akiskal, H.S.; Angst, J.; Greenberg, P.E.; Hirschfeld, R.M.; Petukhova, M.; Kessler, R.C. Lifetime and 12-month prevalence of bipolar spectrum disorder in the National Comorbidity Survey replication. *Arch. Gen. Psychiatry* **2007**, *64*, 543–552. [CrossRef]
2. Prieto, M.L.; Cuéllar-Barboza, A.B.; Bobo, W.V.; Roger, V.L.; Bellivier, F.; Leboyer, M.; West, C.P.; Frye, M.A. Risk of myocardial infarction and stroke in bipolar disorder: A systematic review and exploratory meta-analysis. *Acta Psychiatr. Scand* **2014**, *130*, 342–353. [CrossRef] [PubMed]
3. Dean, B.B.; Gerner, D.; Gerner, R.H. A systematic review evaluating health-related quality of life, work impairment, and healthcare costs and utilization in bipolar disorder. *Curr. Med. Res. Opin.* **2004**, *20*, 139–154. [CrossRef]
4. Grande, I.; Goikolea, J.M.; de Dios, C.; González-Pinto, A.; Montes, J.M.; Saiz-Ruiz, J.; Prieto, E.; Vieta, E.; PREBIS group. Occupational disability in bipolar disorder: Analysis of predictors of being on severe disablement benefit (PREBIS study data). *Acta Psychiatr. Scand* **2013**, *127*, 403–411. [CrossRef] [PubMed]
5. Martinez-Aran, A.; Vieta, E.; Torrent, C.; Sanchez-Moreno, J.; Goikolea, J.M.; Salamero, M.; Malhi, G.S.; Gonzalez-Pinto, A.; Daban, C.; Alvarez-Grandi, S.; et al. Functional outcome in bipolar disorder: The role of clinical and cognitive factors. *Bipolar Disord.* **2007**, *9*, 103–113. [CrossRef] [PubMed]
6. Palazzo, M.C.; Arici, C.; Cremaschi, L.; Cristoffanini, M.; Dobrea, C.; Dell'Osso, B.; Altamura, A.C. Cognitive Performance in Euthymic Patients with Bipolar Disorder vs Healthy Controls: A Neuropsychological Investigation. *Clin. Pract. Epidemiol. Ment Health* **2017**, *13*, 71–81. [CrossRef] [PubMed]
7. Sparding, T.; Silander, K.; Pålsson, E.; Östlind, J.; Sellgren, C.; Ekman, C.J.; Joas, E.; Hansen, S.; Landén, M. Cognitive Functioning in Clinically Stable Patients With Bipolar Disorder I and II. *PLoS ONE* **2015**, *10*, e0115562. [CrossRef]
8. Mann-Wrobel, M.C.; Carreno, J.T.; Dickinson, D. Meta-analysis of neuropsychological functioning in euthymic bipolar disorder: An update and investigation of moderate variables. *Bipolar Disord.* **2011**, *13*, 334–342. [CrossRef]
9. Robinson, L.J.; Thompson, J.M.; Gallagher, P.; Goswami, U.; Young, A.H.; Ferrier, I.N.; Moore, P.B. A meta-analysis of cognitive deficits in euthymic patients with bipolar disorder. *J. Affect. Disord.* **2006**, *93*, 105–115. [CrossRef]
10. Torres, I.J.; Boudreau, V.G.; Yatham, L.N. Neuropsychological functioning in euthymic bipolar disorder: A meta-analysis. *Acta Psychiatr. Scand Suppl.* **2007**, *434*, 17–26. [CrossRef]
11. Adler, C.M.; Del Bello, M.P.; Jarvis, K.; Levine, A.; Adams, J.; Strakowski, S.M. Voxel-based study of structural changes in first-episode patients with bipolar disorder. *Biol. Psychiatry* **2007**, *61*, 776–781. [CrossRef]
12. Lyoo, I.K.; Kim, M.J.; Stoll, A.L.; Demopulos, C.M.; Parow, A.M.; Dager, S.R.; Friedman, S.D.; Dunner, D.L.; Renshaw, P.F. Frontal lobe gray matter density decreases in bipolar I disorder. *Biol. Psychiatry* **2004**, *55*, 648–651. [CrossRef]
13. Shaffer, J.J., Jr.; Johnson, C.P.; Fiedorowicz, J.G.; Christensen, G.E.; Wemmie, J.A.; Magnotta, V.A. Impaired Sensory Processing Measured by Functional MRI in Bipolar Disorder Manic and Depressed Mood States. *Brain Imaging Behav.* **2018**, *12*, 837–847. [CrossRef]
14. Baumann, O.; Mattingley, J.B. Functional topography of primary emotion processing in the human cerebellum. *Neuroimage* **2012**, *61*, 805–811. [CrossRef] [PubMed]

15. Dong, D.; Luo, C.; Guell, X.; Wang, Y.; He, H.; Duan, M.; Eickhoff, S.B.; Yao, D. Compression of Cerebellar Functional Gradients in Schizophrenia. *Schizophr. Bull.* **2020**. [CrossRef] [PubMed]
16. Leggio, M.; Olivito, G. Topography of the cerebellum in relation to social brain regions and emotions. *Handb. Clin. Neurol.* **2018**, *154*, 71–84. [CrossRef]
17. Sha, Z.; Edmiston, E.K.; Versace, A.; Fournier, J.C.; Graur, S.; Greenberg, T.; Lima Santos, J.P.; Chase, H.W.; Stiffler, R.S.; Bonar, L.; et al. Functional Disruption of Cerebello-thalamo-cortical Networks in Obsessive-Compulsive Disorder. *Biol. Psychiatry Cogn. Neurosci. Neuroimaging* **2019**, *5*, 438–447. [CrossRef]
18. Siciliano, L.; Clausi, S. Implicit vs. Explicit Emotion Processing in Autism Spectrum Disorders: An Opinion on the Role of the Cerebellum. *Front. Psychol.* **2020**, *11*, 96. [CrossRef] [PubMed]
19. Stoodley, C.J.; Schmahmann, J.D. Functional topography in the human cerebellum: A meta-analysis of neuroimaging studies. *Neuroimage* **2009**, *44*, 489–501. [CrossRef]
20. Zhang, H.; Wang, B.; Li, K.; Wang, X.; Li, X.; Zhu, J.; Zhao, Q.; Yang, Y.; Lv, L.; Zhang, M.; et al. Altered Functional Connectivity Between the Cerebellum and the Cortico-Striato-Thalamo-Cortical Circuit in Obsessive-Compulsive Disorder. *Front. Psychiatry* **2019**, *10*, 522. [CrossRef] [PubMed]
21. Zhuo, C.; Wang, C.; Wang, L.; Guo, X.; Xu, Q.; Liu, Y.; Zhu, J. Altered resting-state functional connectivity of the cerebellum in schizophrenia. *Brain Imaging Behav.* **2018**, *12*, 383–389. [CrossRef]
22. Lupo, M.; Siciliano, L.; Leggio, M. From cerebellar alterations to mood disorders: A systematic review. *Neurosci. Biobehav. Rev.* **2019**, *103*, 21–28. [CrossRef]
23. Sani, G.; Chiapponi, C.; Piras, F.; Ambrosi, E.; Simonetti, A.; Danese, E.; Janiri, D.; Brugnoli, R.; De Filippis, S.; Caltagirone, C.; et al. Gray and white matter trajectories in patients with bipolar disorder. *Bipolar Disord.* **2016**, *18*, 52–62. [CrossRef] [PubMed]
24. Baldaçara, L.; Borgio, J.G.; Lacerda, A.L.; Jackowski, A.P. Cerebellum and psychiatric disorders. *Braz. J. Psychiatry* **2008**, *30*, 281–289. [CrossRef] [PubMed]
25. Del Bello, M.P.; Strakowski, S.M.; Zimmermann, M.E.; Hawkins, J.M.; Sax, K.W. MRI analysis of the cerebellum in bipolar disorder: A pilot study. *Neuropsychopharmacology* **1999**, *21*, 63–68. [CrossRef]
26. Kim, D.; Cho, H.B.; Dager, S.R.; Yurgelun-Todd, D.A.; Yoon, S.; Lee, J.H.; Lee, S.H.; Lee, S.; Renshaw, P.F.; Lyoo, I.K. Posterior Cerebellar Vermal Deficits in Bipolar Disorder. *J. Affect. Disord.* **2013**, *150*, 499–506. [CrossRef]
27. Mills, N.P.; Del Bello, M.P.; Adler, C.M.; Strakowski, S.M. MRI analysis of cerebellar vermal abnormalities in bipolar disorder. *Am. J. Psychiatry* **2005**, *162*, 1530–1532. [CrossRef]
28. Chen, G.; Zhao, L.; Jia, Y.; Zhong, S.; Chen, F.; Luo, X.; Qiu, S.; Lai, S.; Qi, Z.; Huang, L.; et al. Abnormal cerebellum-DMN Regions Connectivity in Unmedicated Bipolar II Disorder. *J. Affect. Disord.* **2019**, *243*, 441–447. [CrossRef]
29. Johnson, C.P.; Christensen, G.E.; Fiedorowicz, J.G.; Mani, M.; Shaffer, J.J., Jr.; Magnotta, V.A.; Wemmie, J.A. Alterations of the cerebellum and basal ganglia in bipolar disorder mood states detected by quantitative T1ρ mapping. *Bipolar Disord.* **2018**, *20*, 381–390. [CrossRef] [PubMed]
30. Shinn, A.K.; Roh, Y.S.; Ravichandran, C.T.; Baker, J.T.; Öngür, D.; Cohen, B.M. Aberrant cerebellar connectivity in bipolar disorder with psychosis. *Biol. Psychiatry Cogn. Neurosci. Neuroimaging* **2017**, *2*, 438–448. [CrossRef] [PubMed]
31. Zhao, L.; Wang, Y.; Jia, Y.; Zhong, S.; Sun, Y.; Zhou, Z.; Zhang, Z.; Huang, L. Cerebellar microstructural abnormalities in bipolar depression and unipolar depression: A diffusion kurtosis and perfusion imaging study. *J. Affect. Disord.* **2016**, *195*, 21–31. [CrossRef]
32. Zhao, L.; Luo, Z.; Qiu, S.; Jia, Y.; Zhong, S.; Chen, G.; Lai, S.; Qi, Z.; Luo, X.; Huang, G.; et al. Abnormalities of aquaporin-4 in the cerebellum in bipolar II disorder: An ultra-high b values diffusion weighted imaging study. *J. Affect. Disord.* **2020**, *274*, 136–143. [CrossRef]
33. Deng, W.; Zhang, B.; Zou, W.; Zhang, X.; Cheng, X.; Guan, L.; Lin, Y.; Lao, G.; Ye, B.; Li, X.; et al. Abnormal Degree Centrality Associated With Cognitive Dysfunctions in Early Bipolar Disorder. *Front. Psychiatry* **2019**, *10*, 140. [CrossRef]
34. Luo, X.; Chen, G.; Jia, Y.; Gong, J.; Qiu, S.; Zhong, S.; Zhao, L.; Chen, F.; Lai, S.; Qi, Z.; et al. Disrupted Cerebellar Connectivity With the Central Executive Network and the Default-Mode Network in Unmedicated Bipolar II Disorder. *Front. Psychiatry* **2018**, *9*, 705. [CrossRef]
35. Clausi, S.; Iacobacci, C.; Lupo, M.; Olivito, G.; Molinari, M.; Leggio, M. The Role of the Cerebellum in Unconscious and Conscious Processing of Emotions: A Review. *Appl. Sci.* **2017**, *7*, 521. [CrossRef]
36. Heath, R.G.; Dempesy, C.W.; Fontana, C.J.; Myers, W.A. Cerebellar stimulation: Effects on septal region, hippocampus, and amygdala of cats and rats. *Biol. Psychiatry* **1978**, *13*, 501–529.
37. Lupo, M.; Troisi, E.; Chiricozzi, F.R.; Clausi, S.; Molinari, M.; Leggio, M. Inability to Process Negative Emotions in Cerebellar Damage: A Functional Transcranial Doppler Sonographic Study. *Cerebellum* **2015**, *14*, 663–669. [CrossRef] [PubMed]
38. Strata, P. The emotional cerebellum. *Cerebellum* **2015**, *14*, 570–577. [CrossRef] [PubMed]
39. Van den Berg, N.S.; Huitema, R.B.; Spikman, J.M.; Luijckx, G.J.; de Haan, E.H.F. Impairments in Emotion Recognition and Risk-Taking Behavior After Isolated, Cerebellar Stroke. *Cerebellum* **2020**, *19*, 419–425. [CrossRef] [PubMed]
40. Clausi, S.; Olivito, G.; Lupo, M.; Siciliano, L.; Bozzali, M.; Leggio, M. The Cerebellar Predictions for Social Interactions: Theory of Mind Abilities in Patients With Degenerative Cerebellar Atrophy. *Front. Cell Neurosci.* **2019**, *12*, 510. [CrossRef] [PubMed]

41. Olivito, G.; Siciliano, L.; Clausi, S.; Lupo, M.; Romano, S.; Masciullo, M.; Molinari, M.; Cercignani, M.; Bozzali, M.; Leggio, M. Functional Changes of Mentalizing Network in SCA2 Patients: Novel Insights into Understanding the Social Cerebellum. *Cerebellum* **2020**, *19*, 235–242. [CrossRef]
42. Koziol, L.F.; Budding, D.; Andreasen, N.; D'Arrigo, S.; Bulgheroni, S.; Imamizu, H.; Ito, M.; Manto, M.; Marvel, C.; Parker, K.; et al. Consensus paper: The cerebellum's role in movement and cognition. *Cerebellum* **2014**, *13*, 151–177. [CrossRef]
43. Leggio, M.G.; Chiricozzi, F.R.; Clausi, S.; Tedesco, A.M.; Molinari, M. The neuropsychological profile of cerebellar damage: The sequencing hypothesis. *Cortex* **2011**, *47*, 137–144. [CrossRef]
44. Schmahmann, J.D. The cerebellum and cognition. *Neurosci. Lett.* **2019**, *688*, 62–75. [CrossRef]
45. Tedesco, A.M.; Chiricozzi, F.R.; Clausi, S.; Lupo, M.; Molinari, M.; Leggio, M.G. The cerebellar cognitive profile. *Brain* **2011**, *134*, 3672–3686. [CrossRef] [PubMed]
46. Clausi, S.; Lupo, M.; Olivito, G.; Siciliano, L.; Contento, M.P.; Aloise, F.; Pizzamiglio, L.; Molinari, M.; Leggio, M. Depression disorder in patients with cerebellar damage: Awareness of the mood state. *J. Affect. Disord.* **2019**, *245*, 386–393. [CrossRef] [PubMed]
47. Delle Chiaie, R.; Minichino, A.; Salviati, M.; Fiorentini, S.; Tonini, A.; Bersani, F.S.; De Michele, F.; Caredda, M.; Biondi, M. Bipolar spectrum disorders in patients with cerebellar lesions. A comparison with Parkinson's Disease. *J. Nerv. Ment. Dis.* **2015**, *203*, 725–729. [CrossRef] [PubMed]
48. Lupo, M.; Olivito, G.; Siciliano, L.; Masciullo, M.; Bozzali, M.; Molinari, M.; Leggio, M. Development of a Psychiatric Disorder Linked to Cerebellar Lesions. *Cerebellum* **2018**, *17*, 438–446. [CrossRef] [PubMed]
49. Lupo, M.; Olivito, G.; Siciliano, L.; Masciullo, M.; Molinari, M.; Cercignani, M.; Bozzali, M.; Leggio, M. Evidence of Cerebellar Involvement in the Onset of a Manic State. *Front. Neurol.* **2018**, *9*, 774. [CrossRef]
50. Schmahmann, J.D.; Sherman, J.C. The cerebellar cognitive affective syndrome. *Brain* **1998**, *121*, 561–579. [CrossRef]
51. Diedrichsen, J.; Balsters, J.H.; Flavell, J.; Cussans, E.; Ramnani, N. A probabilistic MR atlas of the human cerebellum. *Neuroimage* **2009**, *46*, 39–46. [CrossRef]
52. Buckner, R.L.; Krienen, F.M.; Castellanos, A.; Diaz, J.C.; Yeo, B.T. The organization of the human cerebellum estimated by intrinsic functional connectivity. *J. Neurophysiol.* **2011**, *106*, 2322–2345. [CrossRef]
53. Habas, C.; Kamdar, N.; Nguyen, D.; Prater, K.; Beckmann, C.F.; Menon, V.; Greicius, M.D. Version 2. Distinct cerebellar contributions to intrinsic connectivity networks. *J. Neurosci.* **2009**, *29*, 8586–8594. [CrossRef]
54. Lupo, M.; Olivito, G.; Clausi, S.; Siciliano, L.; Riso, V.; Bozzali, M.; Santorelli, F.M.; Silvestri, G.; Leggio, M. Cerebello-Cortical Alterations Linked to Cognitive and Social Problems in Patients With Spastic Paraplegia Type 7: A Preliminary Study. *Front. Neurol.* **2020**, *11*, 82. [CrossRef] [PubMed]
55. Olivito, G.; Lupo, M.; Iacobacci, C.; Clausi, S.; Romano, S.; Masciullo, M.; Molinari, M.; Cercignani, M.; Bozzali, M.; Leggio, M. Structural cerebellar correlates of cognitive functions in spinocerebellar ataxia type 2. *J. Neurol.* **2018**, *265*, 597–606. [CrossRef] [PubMed]
56. Van Overwalle, F.; D'aes, T.; Mariën, P. Social cognition and the cerebellum: A meta-analytic connectivity analysis. *Hum. Brain Mapp.* **2015**, *36*, 5137–5154. [CrossRef]
57. Kraepein, E. *Manic-Depressive Insanity and Paranoia*; Livingston: Edinburgh, UK, 1921.
58. *Diagnostic and Statistical Manual of Mental Disorders*, 5th ed.; American Psychiatric Association: Washington, DC, USA, 2013.
59. Toniolo, S.; Serra, L.; Olivito, G.; Marra, C.; Bozzali, M.; Cercignani, M. Patterns of cerebellar gray matter atrophy across Alzheimer's disease progression. *Front. Cell. Neurosci.* **2018**, *12*, 430. [CrossRef] [PubMed]
60. Bourne, C.; Aydemir, Ö.; Balanzá-Martínez, V.; Bora, E.; Brissos, S.; Cavanagh, J.T.; Clark, L.; Cubukcuoglu, Z.; Dias, V.V.; Dittmann, S.; et al. Neuropsychological testing of cognitive impairment in euthymic bipolar disorder: An individual patient data meta-analysis. *Acta Psychiatr. Scand.* **2013**, *128*, 149–162. [CrossRef] [PubMed]
61. Ekinci, O.; Albayrak, Y.; Ekinci, A.E.; Caykoylu, A. Relationship of trait impulsivity with clinical presentation in euthymic bipolar disorder patients. *Psychiatry Res.* **2011**, *190*, 259–264. [CrossRef] [PubMed]
62. Gallagher, P.; Gray, J.M.; Watson, S.; Young, A.H.; Ferrier, I.N. Neurocognitive functioning in bipolar depression: A component structure analysis. *Psychol. Med.* **2014**, *44*, 961–974. [CrossRef]
63. O'Shea, R.; Poz, R.; Michael, A.; Berrios, G.E.; Evans, J.J.; Rubinsztein, J.S. Ecologically valid cognitive tests and everyday functioning in euthymic bipolar disorder patients. *J. Affect. Disord.* **2010**, *125*, 336–340. [CrossRef]
64. Ambrosi, E.; Chiapponi, C.; Sani, G.; Manfredi, G.; Piras, F.; Caltagirone, C.; Spalletta, G. White matter microstructural characteristics in Bipolar I and Bipolar II Disorder: A diffusion tensor imaging study. *J. Affect. Disord.* **2016**, *189*, 176–183. [CrossRef] [PubMed]
65. Mahon, K.; Burdick, K.E.; Szeszko, P.R. A role for white matter abnormalities in the pathophysiology of bipolar disorder. *Neurosci. Biobehav. Rev.* **2010**, *34*, 533–554. [CrossRef]
66. Narita, K.; Suda, M.; Takei, Y.; Aoyama, Y.; Majima, T.; Kameyama, M.; Kosaka, H.; Amanuma, M.; Fukuda, M.; Mikuni, M. Volume reduction of ventromedial prefrontal cortex in bipolar II patients with rapid cycling: A voxel-based morphometric study. *Prog. Neuropsychopharmacol. Biol. Psychiatry* **2011**, *35*, 439–445. [CrossRef] [PubMed]
67. Strakowski, S.M.; Eliassen, J.C.; Lamy, M.; Cerullo, M.A.; Allendorfer, J.B.; Madore, M.; Lee, J.H.; Welge, J.A.; Del Bello, M.P.; Fleck, D.E.; et al. Functional magnetic resonance imaging brain activation in bipolar mania: Evidence for disruption of the ventrolateral prefrontal-amygdala emotional pathway. *Biol. Psychiatry* **2011**, *69*, 381–388. [CrossRef] [PubMed]

68. Wang, Y.; Zhong, S.; Jia, Y.; Zhou, Z.; Wang, B.; Pan, J.; Huang, L. Interhemispheric Resting State Functional Connectivity Abnormalities in Unipolar Depression and Bipolar Depression. *Bipolar Disord.* **2015**, *17*, 486–495. [CrossRef] [PubMed]
69. Olivito, G.; Cercignani, M.; Lupo, M.; Iacobacci, C.; Clausi, S.; Romano, S.; Masciullo, M.; Molinari, M.; Bozzali, M.; Leggio, M. Neural substrates of motor and cognitive dysfunctions in SCA2 patients: A network based statistics analysis. *Neuroimage Clin.* **2017**, *14*, 719–725. [CrossRef]
70. Dayan, M.; Olivito, G.; Molinari, M.; Cercignani, M.; Bozzali, M.; Leggio, M. Impact of Cerebellar Atrophy on Cortical Gray Matter and Cerebellar Peduncles as Assessed by Voxel-Based Morphometry and High Angular Resolution Diffusion Imaging. *Funct. Neurol.* **2016**, *31*, 239–248. [CrossRef]
71. Ginestroni, A.; Della Nave, R.; Tessa, C.; Giannelli, M.; De Grandis, D.; Plasmati, R.; Salvi, F.; Piacentini, S.; Mascalchi, M. Brain structural damage in spinocerebellar ataxia type 1: A VBM study. *J. Neurol.* **2008**, *255*, 1153–1158. [CrossRef]
72. Stoodley, C.J.; MacMore, J.P.; Makris, N.; Sherman, J.C.; Schmahmann, J.D. Location of lesion determines motor vs. cognitive consequences in patients with cerebellar stroke. *Neuroimage Clin.* **2016**, *12*, 765–775. [CrossRef]
73. Stoodley, C.J.; Limperopoulos, C. Structure-function Relationships in the Developing Cerebellum: Evidence From Early-Life Cerebellar Injury and Neurodevelopmental Disorders. *Semin. Fetal. Neonatal Med.* **2016**, *21*, 356–364. [CrossRef]
74. Centorrino, F.; Fogarty, K.V.; Sani, G.; Salvatore, P.; Cincotta, S.L.; Hennen, J.; Guzzetta, F.; Talamo, A.; Saadeh, M.G.; Baldessarini, R.J. Use of combinations of antipsychotics: McLean Hospital inpatients, 2002. *Hum. Psychopharmacol.* **2005**, *20*, 485–492. [CrossRef] [PubMed]
75. First, M.B.; Williams, J.B.W.; Karg, R.S.; Spitzer, R.L. *SCID-5-CV. Intervista Clinica Strutturata per i Disturbi del DSM-5, Versione per il Clinico;* Italiana a cura di Andrea Fossati e Serena Borroni; Raffaello Cortina Editore: Milano, Italy, 2017.
76. Hamilton, M. Development of a rating scale for primary depressive illness. *Br. J. Soc. Clin. Psychol.* **1967**, *6*, 278–296. [CrossRef]
77. Young, R.C.; Biggs, J.T.; Ziegler, V.E.; Meyer, D.A. A rating scale for mania: Reliability, validity and sensitivity. *Br. J. Psychiatry* **1978**, *133*, 429–435. [CrossRef] [PubMed]
78. Trouillas, P.; Takayanagi, T.; Hallett, M.; Currier, R.D.; Subramony, S.H.; Wessel, K.; Bryer, A.; Diener, H.C.; Massaquoi, S.; Gomez, C.M.; et al. International Cooperative Ataxia Rating Scale for pharmacological assessment of the cerebellar syndrome. The Ataxia Neuropharmacology Committee of the World Federation of Neurology. *J. Neurol. Sci.* **1977**, *145*, 205–211. [CrossRef]
79. Wechsler, D. *Wechsler Adult Intelligence Scale*, 4th ed.; Psychological Corporation: San Antonio, TX, USA, 2008. [CrossRef]

International Journal of *Molecular Sciences*

MDPI

Article

A2A Receptor Dysregulation in Dystonia DYT1 Knock-Out Mice

Vincenza D'Angelo [1,†], Mauro Giorgi [2,†], Emanuela Paldino [3], Silvia Cardarelli [2], Francesca R. Fusco [3], Ilaria Saverioni [2], Roberto Sorge [1], Giuseppina Martella [1,3], Stefano Biagioni [2], Nicola B. Mercuri [1], Antonio Pisani [4,5] and Giuseppe Sancesario [1,*]

1 Department of Systems Medicine, Tor Vergata University of Rome, 00133 Rome, Italy; dangelo@med.uniroma2.it (V.D.); sorge@uniroma2.it (R.S.); martella@med.uniroma2.it (G.M.); mercurin@med.uniroma2.it (N.B.M.)
2 Department of Biology and Biotechnology "Charles Darwin", Sapienza University of Rome, 00185 Rome, Italy; mauro.giorgi@uniroma1.it (M.G.); silvia.cardarelli@uniroma1.it (S.C.); ilaria.saverioni@gmail.com (I.S.); stefano.biagioni@uniroma1.it (S.B.)
3 IRCCS Santa Lucia Foundation, 00179 Rome, Italy; e.paldino@hsantalucia.it (E.P.); f.fusco@hsantalucia.it (F.R.F.)
4 IRCCS Mondino Foundation, 27100 Pavia, Italy; pisani@uniroma2.it
5 Department of Brain and Behavioral Sciences, University of Pavia, 27100 Pavia, Italy
* Correspondence: sancesario@med.uniroma2.it; Tel.: +39-33-8146-6058
† These authors to be considered equally first author.

Abstract: We aimed to investigate A2A receptors in the basal ganglia of a DYT1 mouse model of dystonia. A2A was studied in control Tor1a+/+ and Tor1a+/− knock-out mice. A2A expression was assessed by anti-A2A antibody immunofluorescence and Western blotting. The co-localization of A2A was studied in striatal cholinergic interneurons identified by anti-choline-acetyltransferase (ChAT) antibody. A2A mRNA and cyclic adenosine monophosphate (cAMP) contents were also assessed. In Tor1a+/+, Western blotting detected an A2A 45 kDa band, which was stronger in the striatum and the globus pallidus than in the entopeduncular nucleus. Moreover, in Tor1a+/+, immunofluorescence showed A2A roundish aggregates, 0.3–0.4 μm in diameter, denser in the neuropil of the striatum and the globus pallidus than in the entopeduncular nucleus. In Tor1a+/−, A2A Western blotting expression and immunofluorescence aggregates appeared either increased in the striatum and the globus pallidus, or reduced in the entopeduncular nucleus. Moreover, in Tor1a+/−, A2A aggregates appeared increased in number on ChAT positive interneurons compared to Tor1a+/+. Finally, in Tor1a+/−, an increased content of cAMP signal was detected in the striatum, while significant levels of A2A mRNA were neo-expressed in the globus pallidus. In Tor1a+/−, opposite changes of A2A receptors' expression in the striatal-pallidal complex and the entopeduncular nucleus suggest that the pathophysiology of dystonia is critically dependent on a composite functional imbalance of the indirect over the direct pathway in basal ganglia.

Keywords: A2A; cAMP; A2A mRNA; dystonia; DYT1; basal ganglia; D2

Citation: D'Angelo, V.; Giorgi, M.; Paldino, E.; Cardarelli, S.; Fusco, F.R.; Saverioni, I.; Sorge, R.; Martella, G.; Biagioni, S.; Mercuri, N.B.; et al. A2A Receptor Dysregulation in Dystonia DYT1 Knock-Out Mice. *Int. J. Mol. Sci.* **2021**, *22*, 2691. https://doi.org/10.3390/ijms22052691

Academic Editor: José L. Lanciego

Received: 14 February 2021
Accepted: 1 March 2021
Published: 7 March 2021

1. Introduction

The clinical features of dystonia are relatively well-defined, produced by the ill-timed activation of agonist-antagonist muscles, and presenting either as focal or generalized. Conversely, the pathological mechanisms underlying such a heterogeneous group of movement disorders remain elusive [1–3]. Unlike symptomatic dystonia secondary to histological damages affecting basal ganglia, no neuropathologic correlates are detectable at microscopic levels for primary dystonia [4,5].

A common form of primary early onset generalized dystonia is caused by 3 bp deletion (GAG) in the coding region of the TOR1A (DYT1) gene, which results in a defective protein called torsinA, whose role in dystonia pathology is unclear [6]. In animal models for

DYT1 dystonia, multiple lines of evidence revealed the impairment of dopamine receptor type 2 (D2 receptor), with D2 downregulation, sparse D2 synapses, reduced coupling between the D2 receptor and its cognate G proteins, as well as the loss of D2 dependent electrophysiological inhibition and severely altered synaptic plasticity in medium spiny neurons and cholinergic interneurons in the striatum [7–18]. Therefore, the striatum and D2 receptors have been considered central in the cellular pathomechanism underlying DYT1 dystonia [15].

Noteworthy, the deficit in D2 receptor-mediated transmission can be counteracted by A2A receptor pharmacological antagonism in the striatum of a genetic mouse model of DYT1, and also of DYT11 and DYT25 dystonia, suggesting that an imbalance in D2/A2A functions may represent a convergent pathogenetic mechanism [11,19,20]. Moreover, a peculiarity of D2 receptors is their selective co-localization with A2A receptors in striatal cholinergic interneurons [21,22], and mainly in the sub-population of striatal medium spiny GABAergic neurons containing enkephalin, which are known to lead the indirect striatal-pallidal pathway [23,24]. D2 and A2A receptors in enkephalin neurons form complex heteromers and exert reciprocal antagonistic interaction, inhibiting or stimulating the second messenger cyclic adenosine monophosphate (cAMP) synthesis, respectively [25,26].

Therefore, besides the dysfunction of D2 receptors, A2A receptors may be of interest in dystonia pathophysiology for several reasons. In this paper, we aimed to investigate whether the D2 receptor dysfunction is coupled with any change in the A2A receptor expression in the basal ganglia of a DYT1 mouse model. We report that opposite to the downregulation of D2 receptors in the striatum, an up-regulation of the A2A receptors occurs in the striatal-pallidal complex, whereas A2A down-regulation occurs in the entopeduncular nucleus of a DYT1 dystonia mouse model.

2. Results

2.1. Expression of A2A Receptors

We first quantified the presence of A2A receptors by Western blot on proteins extracted from the striatum, globus pallidus and entopeduncular nucleus of Tor1a+/+ and Tor1a+/−. A single specific band was detected in all areas of the basal ganglia approximately at 45 kDa, which corresponds to the migration level of A2A receptor monomers (Supplementary Materials Figure S1). In Tor1a+/+, a higher A2A expression was detected in the striatum that was not significantly different from the globus pallidus, whereas the lowest A2A expression was detected in the entopeduncular nucleus ($p < 0.001$). Moreover, Western blotting analysis revealed a significant increase in A2A receptor levels in the striatum ($n = 11$, df = 1, F = 5.189, $p = 0.034$) and globus pallidus ($n = 9$, df = 1, F = 6.791, $p = 0.020$), but a significant decrease in A2A expression in the entopeduncular nucleus ($n = 8$, df = 1, F = 7.629, $p = 0.016$) of mutant Tor1a+/− compared to control Tor1a+/+ mice (Figure 1).

2.2. Cyclic AMP Levels

We then investigated whether the changes in A2A expression in the basal ganglia of Tor1a+/− mice consistently affected second messenger cAMP content in tissue extracts. Actually, the cAMP level was significantly increased in the striatum (df = 1, F = 4.828, $p < 0.05$), but was unchanged in the globus pallidus (df = 1, F = 0.151, $p > 0.5$) and in the entopeduncular nucleus (df = 1, F = 0.403, $p > 0.5$) of Tor1a+/− mice ($n = 7$) compared with Tor1a+/+ ($n = 7$) (Figure 2).

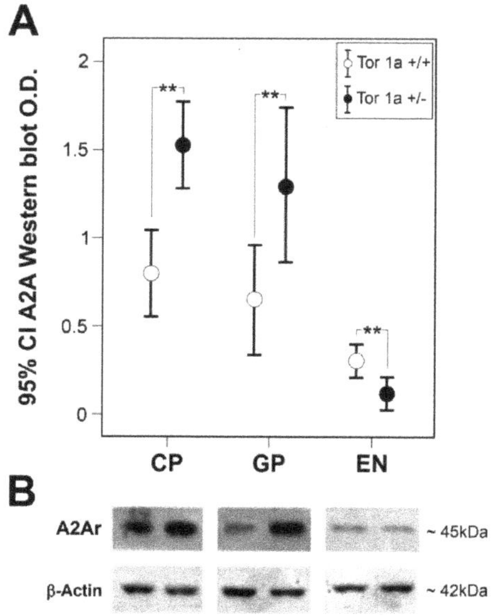

Figure 1. Expression of A2A in basal ganglia of Tor1a+/+ and Tor1a+/− mice. (**A**) Densitometric analysis optical density (O.D.) of A2A-immunostained bands in correspondent areas of caudate–putamen (CP), globus pallidus (GP), and entopeduncular nucleus (EN) of Tor1a+/+ and Tor1a+/−. Results were expressed as the mean ± SD of the values obtained in each group. One-way ANOVA, ** $p < 0.020$. (**B**) Representative immunoblots of A2A content in correspondent area of CP, GP and EN of Tor1a+/+ and Tor1a+/− mice. As an internal reference standard, the β-actin content was detected in each lane of the same blots to correct for protein loading.

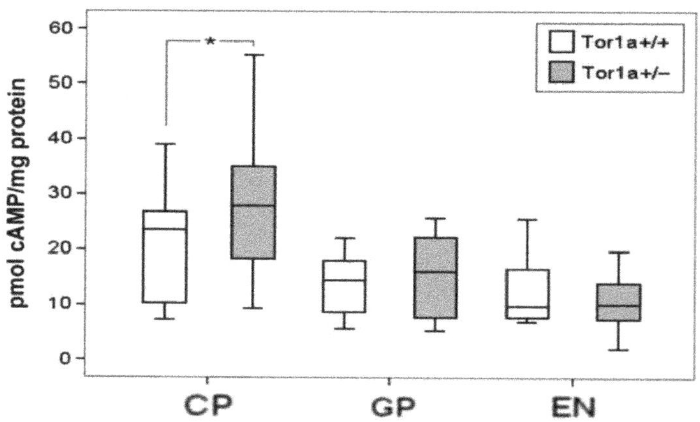

Figure 2. Cyclic adenosine monophosphate (cAMP) level in tissue homogenates of caudate–putamen (CP), globus pallidus (GP), and entopeduncular nucleus (EN) in control Tor1a+/+ and in mutant Tor1a+/− mice. Values, expressed as pmol of cAMP/ mg protein, are represented as mean ± SD. Significant increase in cAMP content was detected in the CP of Tor1a+/− versus Tor1a+/+. One-way ANOVA, * $p < 0.05$.

2.3. Expression of A2A mRNA

Real-time PCR amplification of A2A mRNA was performed in the two groups of animals to investigate A2A changes at the transcription level in basal ganglia. The primers, which amplified a specific mRNA fragment (140 bp), have shown that the A2A mRNA expression level in the caudate–putamen of Torla+/+ (n = 6) is much higher than in the globus pallidus (n = 8) (Figure 3A), whereas in the entopeduncular nucleus its expression is under the detection level (data not shown). The pattern and relative intensity of A2A mRNA expression detected in the caudate–putamen of Torla+/− mice (n = 8) were similar compared to the caudate–putamen of Torla+/+ mice (Figure 3B). Surprisingly, a significant increase in A2A mRNA expression was detected in the globus pallidus of Torla+/−mice (n = 8) compared to Torla+/+ mice (Figure 3C).

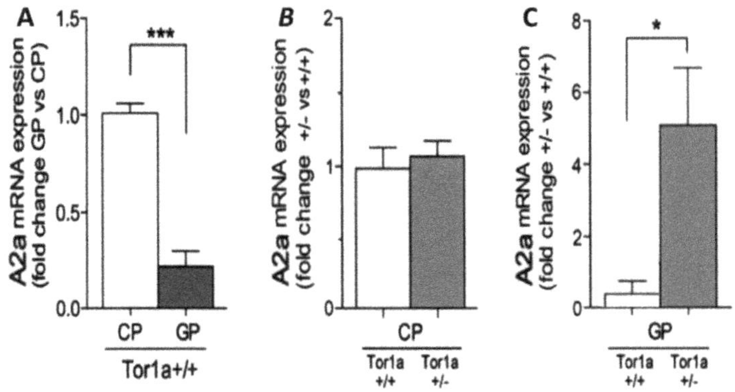

Figure 3. (**A**)—Comparative real-time PCR of A2A mRNA in caudate–putamen (CP) and globus pallidus (GP) of Torla+/+ control mice (n = 6, df = 1, F = 71.172, p = 0.001). (**B**)—The expression of A2A mRNA in the CP of Torla+/+ mice is compared to the levels in the CP of Torla+/− mice (n = 8, df = 1, F = 0.311, p = 0.586). (**C**)—The expression of A2A mRNA in the GP of Torla+/+ mice is compared to the levels in the GP of Torla+/− mice (n = 7, df = 1, F = 4.743, p = 0.048). Results were expressed as the mean ± SD of values obtained from each group. Data in (**A–C**) were analyzed with one-way ANOVA followed by Bonferroni's correction for multiple data. * $p < 0.05$, *** $p < 0.001$

2.4. A2A Immunofluorescence in Confocal Microscopy

To allow a precise morphological definition of the A2A receptor aggregates on the striatal-pallidal complex in control and mutant mice, we performed a fluorescence staining, followed by detection with confocal microscopy. A better understanding of A2A receptors' subcellular distribution came out in images acquired using 63× oil immersion objective (1.4 numerical aperture) with an additional digital zoom factor (1×–1.5×–2×). The immuno-fluorescent signal appeared extremely specific without background staining, showing A2A positive small grains with elliptical or roundish shape about 0.3–0.4 μm in diameter, isolated or contiguous, composing irregular clusters of size variable in the different areas of the basal ganglia of Torla+/+ mice (Figure 4A,C,E).

Noteworthy, in Torla+/+ mice, A2A small grains appeared to be diffusely covering the neuronal compartments of the striatum, globus pallidus and entopeduncular nucleus, uniformly distributed in the neuropil, whereas grains were rare and almost absent on the cell nuclei and in striatal axonal bundles (Figure 4A,C,E). Moreover, among the basal ganglia regions, the number of A2A receptors per microscopic field was much higher in the striatum and globus pallidus than in the entopeduncular nucleus.

In Torla+/− mice, A2A receptor grains appeared either increased in number in the striatum and globus pallidus, or reduced in number in the entopeduncular nucleus (Figure 4B,D,F), compared with control Torla+/+ (Figure 4A,C,E). Moreover, the distribution pattern of A2A positive fluorescent grains was clearly different in the basal ganglia of Torla+/− mice, wherein A2A receptor positive grains appeared as tiny spots uniformly distributed in the neuropil, but also clustered on neuronal bodies of the striatum and globus pallidus (Figure 4B,D), unlike in Torla+/+ mice (Figure 4A,C).

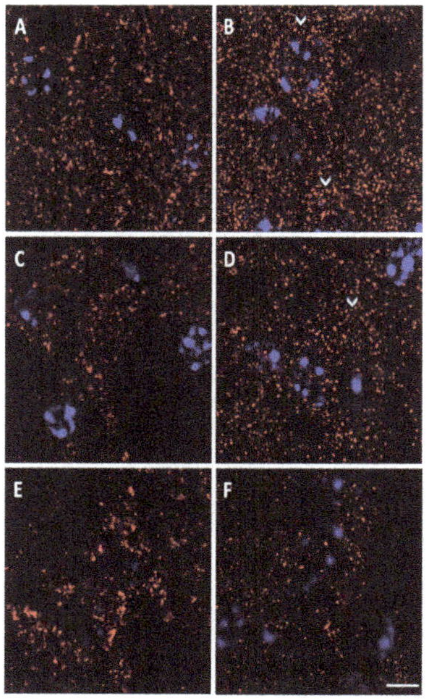

Figure 4. Representative immuno-fluorescence microphotographs of high magnification (63× oil immersion objective plus zoom factor 1×–1.5×–2.5×) confocal laser scanning microscopy, showing the roundish morphology of A2A positive tiny spots and their distributions in the striatum (**A**,**B**), globus pallidus (**C**,**D**), and entopeduncular nucleus (**E**,**F**) of control Tor1a+/+ (**A**,**C**,**E**), and of mutant Tor1a+/− (**B**,**D**,**F**) mice. A2A receptor labeling is visualized in red-Cy3 fluorescence, while cell nuclei are visualized by 4′,6–diamidino–2–phenylindole (DAPI) fluorescence in blue. White arrows in B and D point to clustering of A2A grains around neuronal bodies. Scale Bar in F = 5 μm.

Densitometric analysis confirmed a higher density of the number of A2A receptors per microscopic field in the striatum (df = 1, F = 14.039, p = 0.013) and in the globus pallidus (df = 1, F = 22.859, p = 0.003), and their lower density in the entopeduncular nucleus (df = 1, F = 9.031, p = 0.024) of Tor1a+/− (n = 4), compared with correspondent basal ganglia areas of the Tor1a+/+ mice (n = 4) (Figure 5).

While the numbers of A2A receptors were either increased in the striatal-pallidal complex or decreased in the entopeduncular nucleus, their size appeared smaller in all the basal ganglia structures of TOR1a+/−. A representative number of A2A positive grains were randomly selected from the basal ganglia of control and mutant mice: the mean perimeters per A2A positive grain in TOR1a+/+ versus TOR1a+/− were, respectively, (two tailed T test): in the striatum, 2.06 ± 0.19 > 1.62 ± 0.08 SEM μm (n = 13, t = 2.077 df = 24, p = 0.04); in the globus pallidus, 2.342 ± 0.14 > 1.378 ± 0.06 SEM μm (n = 15, t = 6.088 df = 28, p = 0.0001); and in the entopeduncular nucleus, 4.938 ± 0.39 > 1.61 ± 0.08 SEM μm (n = 14, t = 8.238, df = 26, p = 0.0001). However, the fluorescence intensity of the A2A positive grains was comparable in the correspondent areas of TOR1a+/+ versus TOR1a+/− (data not shown).

2.5. Colocalization of A2A in Cholinergic Neurons

Finally, we evaluated whether any change also occurs in the expression of A2A receptors in the striatal choline-acetyltransferase (ChAT) positive cholinergic interneurons of mutant Tor1a+/− mice. Double-labeling fluorescence was obtained, with ChAT reactive large neuronal bodies and thick primary dendrites marked in red (Figure 6B,E), and A2A receptor positive grains marked in green in the neuropil (Figure 6A,D). In the merged ChAT/A2A images, A2A receptors appeared as yellow grains uniformly diffused on

ChAT neuronal bodies and thick dendrites (Figure 6C,F), but absent in the white matter bundles appearing as black holes in the microscopic field (Figure 6A,C,D,F), confirming the specificity of A2A receptor staining on neurons. Compared to Tor1a+/+ mice (Figure 6C), in Tor1a+/− the A2A receptor grains (Figure 6F) appeared markedly more numerous on the surface of ChAT neurons (yellow grains) and in the surrounding neuropil (green grains).

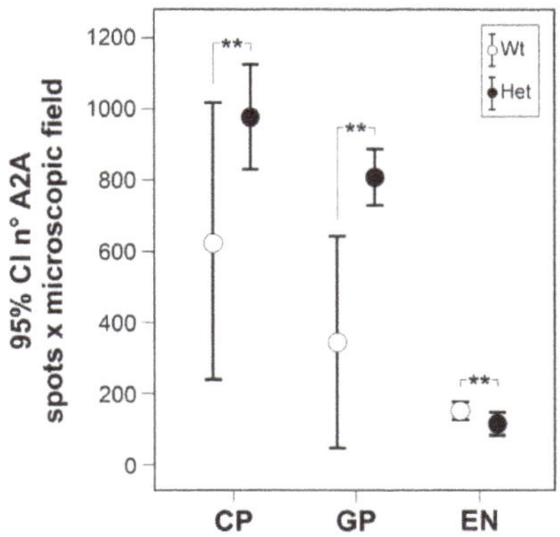

Figure 5. Comparative densitometric analysis of the number (n°) of A2A positive grains per microscopic field in caudate–putamen (CP), globus pallidus (GP), and entopeduncular nucleus (EN) of control Tor1a+/+ and of mutant Tor1a+/− mice. One-way ANOVA, ** $p < 0.02$.

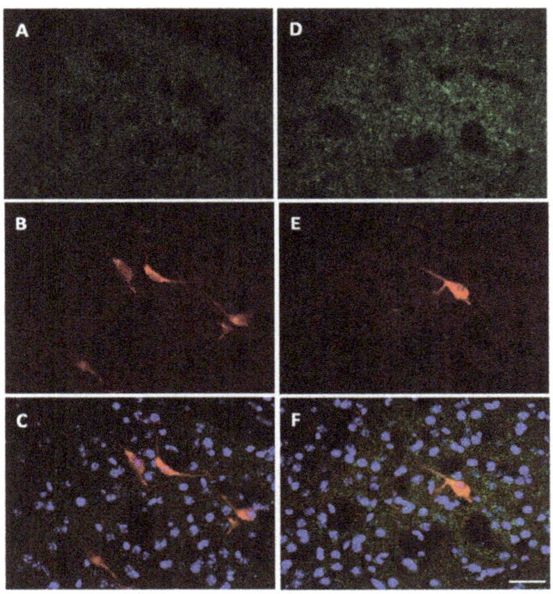

Figure 6. Localization of A2A receptors in cholinergic striatal interneurons. Representative confocal laser scanning microscopy images double-labeled with antibody for anti A2A receptors and with antibody anti-ChAT for cholinergic interneurons. A2A receptors are visualized in green-cy2 fluorescence (**A**,**D**); ChAT positive neurons are visualized in red-cy3 fluorescence (**B**,**E**); merged images in C, F with A2A positive grains visualized in yellow on ChAT-positive interneurons, and in green in the neuropil. Cell nuclei are visualized by DAPI fluorescence in blue. (**A**–**C**) Tor1a+/+ mice; (**D**–**F**) Tor1a+/− mice. Scale bar in F = 50 μm.

The overlapping areas of A2A receptors on striatal ChAT neuronal bodies were 36.37 + 2.92 SEM μm^2 (neuronal bodies n = 12 from 3 mice) in Tor1a+/+ mice, and 47.44 + 2.98 SEM μm^2 (neuronal bodies n = 13 from 3 mice) in Tor1a+/− mice (unpaired two-tailed T test, t = 2.645, df = 23, p = 0.014).

3. Discussion

In our study, different techniques demonstrate specific changes in A2A receptors in the basal ganglia circuits of Tor1a+/− mice, with increased or decreased expression, respectively, in the striatal-pallidal complex and entopeduncular nucleus. We briefly discuss the limits of our morpho-chemical study, and the possible relevance of A2A changes to the pathophysiology of dystonia in this DYT1 animal model.

The high specificity of the polyclonal antibody was demonstrated in the Western blot analysis, with a selective band corresponding to the 45 kDa migration level of A2A receptor monomers [27], and by the well-defined immunofluorescent reaction product, configuring the sharp morphology of A2A receptor aggregates without surrounding diffuse background staining. Moreover, our Western blot analysis and immunohistochemical method were extremely sensitive, also detecting the low level of A2A receptors in the entopeduncular nucleus of control Tor1a+/+ mice, and even their decrease in this nucleus, contemporary to the opposite changes of A2A expression in the contiguous striatal-pallidal complex on the same tissue section of Tor1a+/− mice.

3.1. Morphological Characteristics of A2A Receptor Aggregates

In previous studies with electron microscopy, the A2A labeling in the striatum was most commonly localized in dendrites and dendritic spines, in fewer amounts in axon terminals and in very low levels in neuronal soma and glia, whereas in the external globus pallidus (GP), the A2A labeling was mainly presynaptic [28,29]. In our study, the immunofluorescence A2A aggregates appeared arranged in a homogenous spot, with a diameter between 0.3 and 0.4 μm, which approaches the dimensions and morphology of synaptic complexes [30]. However, we cannot define the subcellular localization of the A2A aggregates at synaptic and extra synaptic levels since we cannot detect anything but the fluorescent A2A signals, losing the surrounding subcellular structures configuring the synaptic complex as detected in electron microscopy. Instead, the advantage of the immuno-fluorescence technique and confocal microscopy with high power objective is that they can allow at a time detection of A2A aggregates, approaching synaptic size at cellular level, as well as their distribution in the tissue in a relatively large microscopic field.

3.2. Distribution of A2A Aggregates

The high distribution of adenosine A2A receptors in the striatum and globus pallidus of control Tor1a+/+ mice and their selective increase in the same areas of Tor1a+/− mice suggest a major role of A2A within the indirect pathway for basal ganglia physiology and dystonia pathophysiology. However, it is worth noting that besides the selective postsynaptic localization of A2A receptors on striatal medium spiny enkephalin neurons in the indirect pathway [25], a morphological segregation of A2A receptors has been demonstrated at presynaptic levels in glutamatergic terminals that make synapses on striatal medium spiny neurons of the direct pathway, where they exert a selective facilitator modulation of cortico-striatal neurotransmission [31]. Further studies should investigate the localization of A2A receptors at synaptic and extra-synaptic levels in dystonic mice, clarifying the distinct changes of A2A receptors in basal ganglia sub-regions.

Our immuno-fluorescence technique and confocal microscopy with high power objective demonstrate the increased distribution density of A2A in Tor1a+/− mice in the neuropil, likely targeting dendrites and spines in the striatum and globus pallidus, as well as on the soma of the striatal projecting neurons, striatal cholinergic interneurons, and neurons in the globus pallidus. The widespread increase in A2A receptors in the striatal-pallidal complex may have functional relevance, affecting acetylcholine release in the striatum and regulating neuronal excitability and plasticity on cortico-striatal glutamatergic terminals, on GABAergic projecting neurons, and on GABAergic terminals in the globus pallidus [21,25,31,32].

Moreover, the light A2A expression generally observed in the entopeduncular nucleus of Tor1a+/+ mice [33] and its significant decrease in Tor1a+/− mice, detected with different techniques, cannot be considered functionally irrelevant, and this suggests that A2A receptors in the direct pathway could also be involved in basal ganglia function and in dystonia pathophysiology. We hypothesize that the increased expression of A2A in the striatum and in the globus pallidus is likely involved in increased inhibitory GABA input at striatal-pallidal synapses demonstrated in globus pallidus in DYT1 dystonia [32], disinhibiting the subthalamic nucleus and its excitatory glutamatergic input to the entopeduncular nucleus [34]. In this scenario, the reduced A2A expression in the entopeduncular nucleus may be viewed, at least in part, as an adaptive mechanism to regulate increased glutamatergic input from the subthalamic nucleus to basal ganglia output. Future studies should investigate the elaborated role of pre- and postsynaptic A2A receptors in the glutamatergic synapses of the direct- and indirect pathway neurons, respectively, on GABAergic medium spiny neurons' soma and dendrites in the striatum, and of their terminals in the entopeduncular nucleus and globus pallidus in the pathophysiology of dystonia. While the numbers of A2A positive dots were either increased or decreased, respectively, in the striatal-pallidal complex and entopeduncular nucleus of Tor1a+/− mice, at the same time, the size of A2A positive dots was significantly decreased in all three structures. Such a decrease in the size of A2A receptor aggregates could be associated with the contemporary decrease in the D2 synapses [18], and the consequent deficiency in A2A/D2 colocalization and the formation of heteromers, at least in the striatum. The functional relevance of such A2A microstructural change should be checked in the direct and indirect pathway since, in general, the efficiency of neuronal connectivity is directly related to the synapses' size [35]. Further studies should investigate whether the opposite changes in the numbers of D2 and A2A receptors in dystonic mice could also compromise the formation of D2/A2A etheromers, changing their interaction not only quantitatively but also qualitatively.

3.3. Changes in Second Messenger cAMP

To verify whether the tissue levels of cAMP are influenced by the changes in A2A expression, we evaluated the basal content of cAMP in tissue homogenates of the different basal ganglia nuclei from Tor1a+/+ and Tor1a+/− knock-out mice. Surprisingly, we found significantly increased cAMP levels only in the striatum of Tor1a+/−, whereas in the globus pallidus and in the entopeduncular nucleus, cAMP levels were similar in Tor1a+/+ and in Tor1a+/− mice. The cAMP synthesis in striatal neurons is stimulated by dopamine D1 receptors, which are significantly decreased in DYT1 transgenic mice [36], and by A2A receptors, which are significantly increased in Tor1a+/− mice (present work), and so likely determining the increased striatal cAMP levels in the Tor1a+/− striatum.

To explain the discrepancy between the higher levels of cAMP in the Tor1a+/− striatum and the unchanged cAMP levels in the globus pallidus and entopeduncular nucleus, we should consider the interference of other related factors in addition to the increased or decreased stimulating action of the A2A receptors. We can hypothesize that the A2A receptor dependent cAMP synthesis is out of the inhibitory control of the deficient D2 receptors, but the D2 receptors appear to be functionally impaired both in the striatum as well as in the globus pallidus [8,32].

In a previous work, we reported that PDE10A-dependent cAMP hydrolyzing activity as well as total PDE activity are unaffected in the striatum of a mutant mouse model of DYT1 dystonia, wherein at the same time the PDE10A-dependent cAMP hydrolyzing activity was increased in the globus pallidus and decreased in the entopeduncular nucleus [37]. Therefore, we could hypothesize that in the mutant DYT1 striatum, an insufficient total PDE activity cannot timely catabolize the A2A dependent increased cAMP synthesis, whereas the equilibrium between cAMP synthesis and catabolism is preserved in the globus pallidus and in the entopeduncular nucleus by a contemporary increase or decrease in PDE activity, respectively. The reasons for the discrepant compensatory PDE activity in front of A2A-dependent cAMP synthesis in the striatum of mutant DYT1 mice are unknown. We can just speculate on the different cellular distributions of the PDE10A isoform in different striatal neurons.

Indeed, PDE10A is selectively expressed in medium spiny neurons, but not in striatal interneurons, and in particular it is absent in cholinergic interneurons [38,39]. A2A receptors on cholinergic interneurons play a crucial role in striatal network function, activating the release of acetylcholine in the striatum, and striatal cholinergic dysfunction has been shown to have a widespread role in the pathophysiology of dystonia [21,40]. Therefore, we can hypothesize that the increased levels of cAMP in the striatum may reflect, at least in part, the A2A-dependent cAMP synthesis in cholinergic interneurons, wherein cAMP is not hydrolysable by the absent PDE10A, nor adequately hydrolyzed by another unknown phosphodiesterase isoform.

Anyway, the relevant impact of cAMP on the membrane excitability of striatal medium-sized spiny neurons and cholinergic interneurons has been widely described [41–43]. Our observation of a chronic elevation in neuronal cAMP in the striatum of Tor1a+/− can open a new window to investigate the role of cAMP changes in the severely altered synaptic plasticity of medium spiny neurons and cholinergic interneurons in dystonia mouse models [8,11–16,40,44].

3.4. Changes in A2A mRNA

Finally, to evaluate whether the changes in A2A expression in Tor1a+/− mice are regulated at translational or post-translational levels, we used the sensitive real-time PCR technique analysis of A2A mRNA in basal ganglia. According to previous studies, the A2A mRNA can be found only on the striatum in cell bodies of medium spiny striatal neurons and occasionally in dendrites, coding for the synthesis of A2A receptors, which would be transported to the terminals of striatal-pallidal efferent neurons till the globus pallidus [45]. Within the basal ganglia, we detected significant levels of A2A mRNA only in the striatum of Tor1a+/+ that were similar to A2A mRNA levels in the striatum of Tor1a+/−, suggesting that the increased A2A expression in the Tor1a+/− striatum can be regulated at post-translational levels. However, the striatal results of A2A mRNA should be kept with caution since in tissue homogenates we cannot distinguish eventual differentiated A2A mRNA levels in different classes of striatal neurons either to the direct or indirect pathway, obtaining on the whole apparently unchanged striatal A2A mRNA levels.

Instead, a surprising A2A mRNA significant increase was observed in the globus pallidus of Tor1a+/−, whereas A2A mRNA was detected just in a very low amount in the globus pallidus of Tor1a+/+. Therefore, in Tor1a+/− it is likely that the globus pallidus is coding for its own A2A receptors, and such A2A neo-expression may change, at least in part, its dependent role from the striatal A2A inputs. The A2A mRNA changes in the basal ganglia of Tor1a+/− appear to be specific, since in Parkinson's disease A2A mRNA appears either decreased in the striatum or unchanged in the globus pallidus [46]. Future studies should clarify the cell types expressing A2A mRNA in the globus pallidus of Tor1a+/− mice, and the function of their intrinsic A2A receptors.

4. Materials and Methods

4.1. Animals

C57BL/6 Tor1a+/− knock-out mice, which mimic the loss of function of the DYT1 dystonia *Tor1a* mutation [47], were bred at Santa Lucia Foundation Animal Facility. Control Tor1a+/+, and Tor1a+/− knock-out mice were kept under an artificial day-night cycle, with free access to food and water. Male control and mutant mice were sacrificed at the age of 5–6 months. DNA was isolated and amplified from 1 to 2 mm tail fragments with the Extract-N-Amp Tissue polymerase chain reaction (PCR) kit (XNAT2 kit; Sigma-Aldrich Merck Life Science, Milan, Italy), and genotyping was performed as previously reported [17]. All the efforts were made to minimize the number of animals utilized and their suffering. Treatment and handling of mice were carried out in compliance with both the European Council and Italian guidelines (2010/63EU, 20 October 2010; D.L. 26/2014, 4 March 2014; 86/609/EEC, 24 November 1986; D.L. 116/1992—27 January 1992), according to experimental protocols approved by the Animal Ethics Committee of the University of Rome Tor Vergata (D.M. 153/2001-A, 9 April 2001, and 43/2002-A, 4 May 2002) and by the IRCCS Santa Lucia Foundation Animal Care and Use Committee (D.M.9/2006-A, 2006), and authorized by the Italian Ministry of Health (authorization 223/2017-PR, May 2017).

For biochemical studies, the animals were killed by cervical dislocation, and the brains were removed rapidly and placed on an ice-cold plate. Thick brain sections were cut with an Oxford vibratome, and the caudate–putamen, globus pallidus and entopeduncular nucleus were dissected out rapidly from both hemispheres under a stereomicroscope, and promptly frozen in liquid nitrogen and stored at −80 °C [48].

For morphological studies, the animals were deeply anesthetized with tiletamine/zolazepam (80 mg/kg) and xylazine (10 mg/kg) and perfused trans-cardially with 1% heparin in 50 mL 0.1 M sodium phosphate buffer (PBS), and with 250 mL 4% para-formaldehyde in 0.1 M in PBS (pH 7.4). The brains were removed immediately and post-fixed in the same fixative solution overnight at 4 °C; then they were equilibrated with 30% sucrose overnight, and finally frozen in liquid nitrogen and stored at −80 °C [48].

4.2. Quantitative Analysis of A2A Protein

The quantitative analysis of A2A expressions in basal ganglia was assessed by Western blotting. Tissues were lysated in 20 mM Tris-HCl buffer pH 7.2 containing 0.2 mM EGTA, 5 mM MgCl2, 0.1% v/v Triton X-100, 5 mM β–mercaptoethanol, 1 mM PMSF and 2% v/v antiprotease cocktail (Sigma–Aldrich Merck Life Science, Milan, Italy). Thirty micrograms of proteins were loaded on a 9% SDS polyacrylamide gel and subjected to electrophoresis under a reducing condition. The proteins were then transferred to a nitrocellulose membrane (Bio–Rad, Milan, Italy). The blots were incubated overnight at 4 °C with a rabbit polyclonal anti–A2A receptor antibody (1:1000; BML–SA654, Enzo Life Sciences, Farmingdale, NY, USA), or mouse anti-β-actin (1:10,000; Sigma-Aldrich Merck Life Science, Milan, Italy) as a reference standard. A2A receptors-reactive bands were revealed by horseradish peroxidase-conjugated secondary antibodies (1:10,000, Jackson Immunoresearch, Cambridgeshire, UK), incubated in a lumi-light-enhanced chemiluminescence substrate (Bio-Rad, Milan, Italy), and exposed to Chemidoc (Bio–Rad, Milan, Italy). Densitometric analysis of scanned blots was performed using the NIH ImageJ version 1.29 program (NIH, Bethesda, MD, USA).

4.3. cAMP Measurement

The frozen samples were immediately submerged in 0.1 M HCl, and rapidly homogenized. The homogenates were centrifuged at 13,000× g for 30 min at room temperature and the supernatants were acetylated, according to the instructions of the commercial kit manufacturer (Direct EIA kit, Enzo Life Sciences, Farmingdale, NY, USA). The kit uses a polyclonal antibody to cAMP, which in a competitive manner binds the cyclic nucleotides in the sample, or cyclic nucleotides conjugated with alkaline phosphatase molecules added to the incubation medium. The samples were incubated for 2 h at room temperature, and thereafter the substrate p-nitrophenyl phosphate (pNPP) was added for 1 h to reveal the residual alkaline phosphatase activity in the medium, which generates a yellow product, readable on a microplate reader (Multiskan™ FC Microplate Photometer, Thermo Fisher Scientific, Monza, Italy) at 405 nm. The intensity of the yellow color in the medium is inversely proportional to the concentration of cyclic nucleotides in the samples and in comparative standards.

4.4. RNA Extraction and Real–Time PCR

Total RNA was prepared using the TRI Reagent (Sigma-Aldrich Merck Life Science, Milan, Italy), according to the manufacturer's instructions, and quantified with a NanoDrop 1000 spectrophotometer (Thermo Fisher Scientific, Monza, Italy). One microgram of RNA was reverse-transcribed using the QuantiTect reverse transcription kit (Qiagen, Milan, Italy). Real-time PCR was performed on reverse-transcription products with the PowerUp™ SYBR™ Green Master Mix in the QuantStudio 3 Real-Time PCR System (Thermo Fisher Scientific, Monza, Italy). Thermal cycling conditions consisted of an initial denaturation step at 95 °C for 2 min, followed by 40 cycles at 95 °C for 15 s, 60 °C for 15 s, and 72 °C for 1 min. The threshold cycle (Ct) (defined as the fractional PCR cycle number at which fluorescence reaches 10 times the baseline SD) was used for comparison analysis. The $2^{-\Delta\Delta Ct}$ method was used to evaluate the relative expression ratio for A2A receptors compared with the internal control gene β–actin. The following sequences of primers were used: A2A receptors (a) forward 5′–TCTTCTTCGCCTGCTTTGTCC–3′, (b) reverse 5′–GCCCTCATACCCGTCACCA–3′; β-actin (a) forward 5′–GCGCAAGTACTCTGTGTGGA–3′, (b) reverse 5′–AAGGGTGTAAAACGCAGCT–3′.

4.5. A2A Immunofluorescence in Confocal Microscopy

Coronal brain sections (40 μm thick) were cut with a freezing microtome. Tissue sections were then incubated for 1 h at room temperature in 10% donkey serum solution in PBS 0.25%–Triton X-100 (PBS-Tx). The sections were incubated with the primary rabbit anti-A2A antibody (BML–SA654, Enzo Life Sciences, Farmingdale, NY, USA) 1:200, 3 days at 4 °C; thereafter, the sections were incubated with cyanine 3 (cy3) conjugated secondary antibody (Jackson Immuno Research, Cambridgeshire, UK) 1:200, 2 h at room temperature. For negative controls, representative sections were processed with the omission of the primary or of the secondary antibody. Cell nuclei were detected with a blue-fluorescent DNA stain by 4′,6–diamidino–2–phenylindole (DAPI) (D9542, Sigma-Aldrich Merck Life Science, Milan, Italy). After washout, tissue sections were mounted on plus polarized glass slides with Vectashield mounting medium (Super Frost Plus, Thermo Fisher Scientific, Monza, Italy) and cover-slipped.

Fluorescent images were acquired with a LSM700 Zeiss confocal laser scanning microscope (Zeiss, Oberkochen, Germany), with a 5×, a 20× objective, or a 63× oil immersion lens (1.4 numerical aperture) with an additional digital zoom factor (1×–1.5×–2×) under no saturation conditions. Single-section images (1024 × 1024) or z-stack projections in the z-dimension (z-spacing, 1 μm) were collected. Z-stack images were acquired to analyze the whole neuronal soma, which spans multiple confocal planes. The confocal pinhole was kept at 1, the gain and the offset were adjusted to prevent saturation of the brightest signal and sequential scanning for each channel was performed. The confocal settings, including laser power, photomultiplier gain, and offset, were kept constant for each marker.

A negative staining was obtained omitting the primary or the secondary antibody. The 5× and 20× objectives were used to define areas of interest in the dorsolateral striatum, globus pallidus, and entopeduncular nucleus; the distribution of A2A receptors was first acquired using 20× objective with an additional digital zoom factor (1×–1.5×–2×), and thereafter 63× oil immersion objective (1.4 numerical aperture).

For quantitative analysis of the number of A2A positive spots per microscopic field and of their perimeters, images were collected from at least 3–4 slices per animal, processed simultaneously from each basal ganglia nucleus ($n \geq 4$ mice/genotype), and exported for analysis with ImageJ software (NIH, Bethesda, MD, USA). Software background subtraction was utilized to reduce noise.

4.6. A2A Localization in Striatal Cholinergic Interneurons

A double immunofluorescent staining for choline-acetyltransferase (ChAT) and A2A was used to evaluate the colocalization of A2A receptors in striatal cholinergic interneurons. Brain sections were incubated with goat anti-ChAT (Nova biological, CA, USA) and rabbit anti-A2A receptors (BML–SA654, Enzo Life Sciences, Farmingdale, NY, USA). Both primary antibodies were used at a 1:200 dilution, in 0.1 M PBSS containing 0.3% Triton X–100 for 72 h at 4 °C. Sections were then rinsed three times for 5 min at room temperature and subsequently incubated with cyanine 2 (Cy2) and cyanine 3 (cy3) conjugated secondary antibodies (Jackson Immuno Research, Cambridgeshire, UK) for 2 h at room temperature at 1:200 dilution in a 0.1 M PB solution containing 0.3% Triton X–100. For negative controls, representative sections were processed with the omission of the primary or of the secondary antibodies. Cell nuclei were detected with a blue-fluorescent DNA stain DAPI (D9542, Sigma-Aldrich Merck Life Science, Milan, Italy). Subsequently, sections were rinsed in PBS, mounted on plus polarized glass slides with Vectashield mounting medium (Super Frost Plus, Thermo Fisher Scientific, Monza, Italy) and cover-slipped.

The sections were preliminary examined under an epi-illumination fluorescence microscope (Zeiss Axioskop 2, Oberkochen, Germany). Confocal laser scanner microscopy (Zeiss LSM800, Oberkochen, Germany) was used to acquire immunofluorescent images as reported in the previous subsection.

Immunofluorescence intensity and colocalization analysis were evaluated by using the Java image processing and plugin analysis program included in Fiji ImageJ (NIH, Bethesda, MD, USA). To quantify the density of the A2A specific marker on a defined area, the "overlapping signal" of the A2A signal on ChAT neuronal bodies was calculated.

4.7. *Experimental Design and Statistical Analysis*

Biological samples and microscopic fields were randomly selected from the basal ganglia areas of the two experimental groups. All data were initially entered into an Excel database (Microsoft, Redmond, WA, USA) and the analysis was performed using the Statistical Package for the Social Sciences Windows, version 15.0 (SPSS, Chicago, IL, USA). Descriptive statistics consisted of the mean ± standard deviation (SD) for parameters with gaussian distributions (after confirmation with histograms and the Kolgomorov–Smirnov test), or of the mean ± standard error mean (SEM). Comparisons were performed in compliance with data characteristics either with the Paired Samples Test, or ANOVA one-way or ANOVA two factors and multiple comparisons by Bonferroni test. A p value of < 0.05 was considered statistically significant.

5. Conclusions

The increased expression of A2A receptors in the striatal-pallidal complex of Tor1a+/− knock-out mice may suggest a gaining of function of such receptors, overwhelming the D2 loss of function in the indirect pathway; moreover, the associated decrease in A2A expression in the entopeduncular nucleus may be functionally significative. Future studies should clarify the relevance of A2A changes in the direct and indirect pathway in dysregulating the basal ganglia network in dystonia pathophysiology.

Supplementary Materials: The following are available online at https://www.mdpi.com/1422-0067/22/5/2691/s1. Figure S1—Image of Western blot analysis of A2A receptors in brain tissue.

Author Contributions: V.D. and M.G. designed research, performed research, and were involved in drafting the manuscript; E.P., S.C. and I.S. performed research; R.S. statistically analyzed data; F.R.F. and A.P. interpreted data and critically revised the manuscript; S.B. resources; S.B., G.M., N.B.M. critically revised the manuscript; G.S. conceived research, interpreted data and wrote the paper. All authors have read and approved the final version of the manuscript.

Funding: This research received no external funding.

Institutional Review Board Statement: The study was conducted according to the guidelines of the Declaration of Helsinki, and of the European Council (2010/63EU, 20 October 2010), approved by the Institutional Ethics Committee of University of Rome Tor Vergata (43/2002-A, 4 May 2002), and of the IRCCS Santa Lucia Foundation Animal Care and Use Committee (D.M.9/2006-A, 2006), and authorized by the Italian Ministry of Health (authorization 223/2017-PR, May 2017).

Informed Consent Statement: Not applicable.

Data Availability Statement: Data and methods used in the research have been presented in sufficient detail in the paper. Additional data can be requested to the corresponding author.

Acknowledgments: We kindly acknowledge Graziano Bonelli for excellent technical assistance, and James Lynch for style reviewing.

Conflicts of Interest: The authors declare no competing financial interests.

References

1. Marsden, C.D.; Sheehy, M.P. Writer's cramp. *Trends Neurosci.* **1990**, *13*, 148–153. [CrossRef]
2. Berardelli, A.; Rothwell, J.C.; Hallett, M.; Thompson, P.D.; Manfredi, M.; Marsden, C.D. The pathophysiology of primary dystonia. *Brain* **1998**, *121*, 1195–1212. [CrossRef] [PubMed]
3. Albanese, A.; Bhatia, K.; Bressman, S.B.; Delong, M.R.; Fahn, S.; Fung, V.S.; Hallett, M.; Jankovic, J.; Jinnah, H.A.; Klein, C.; et al. Phenomenology and classification of dystonia: A consensus update. *Mov. Disord.* **2013**, *28*, 863–873. [CrossRef] [PubMed]
4. Lanska, D.J. Chapter 33: The History of Movement Disorders. In *The History of Neurology. Handbook of Clinical Neurology*; Aminoff, M., Boller, F., Swaab, D.F., Eds.; Elsevier: Amsterdam, The Netherlands, 2009; Volume 95, pp. 501–546.
5. Standaert, D.G. Update on the pathology of dystonia. *Neurobiol. Dis.* **2011**, *42*, 148–151. [CrossRef] [PubMed]
6. Ozelius, L.J.; Hewett, J.W.; Page, C.E.; Bressman, S.B.; Kramer, P.L.; Shalish, C.; de Leon, D.; Brin, M.F.; Raymond, D.; Corey, D.P.; et al. The early-onset torsion dystonia gene (DYT1) encodes an ATP-binding protein. *Nat. Genet.* **1997**, *17*, 40–48. [CrossRef] [PubMed]
7. Asanuma, K.; Ma, Y.; Okulski, J.; Dhawan, V.; Chaly, T.; Carbon, M.; Bressman, S.B.; Eidelberg, D. Decreased striatal D2 receptor binding in non-manifesting carriers of the DYT1 dystonia mutation. *Neurology* **2005**, *64*, 347–349. [CrossRef] [PubMed]
8. Pisani, A.; Martella, G.; Tscherter, A.; Bonsi, P.; Sharma, N.; Bernardi, G.; Standaert, D.G. Altered responses to dopaminergic D2 receptor activation and N-type calcium currents in striatal cholinergic interneurons in a mouse model of DYT1 dystonia. *Neurobiol. Dis.* **2006**, *24*, 318–325. [CrossRef] [PubMed]
9. Yao, W.D.; Spealman, R.D.; Zhang, J. Dopaminergic signaling in dendritic spines. *Biochem. Pharmacol.* **2008**, *75*, 2055–2069. [CrossRef] [PubMed]
10. Carbon, M.; Niethammer, M.; Peng, S.; Raymond, D.; Dhawan, V.; Chaly, T.; Ma, Y.; Bressman, S.; Eidelberg, D. Abnormal striatal and thalamic dopamine neurotransmission: Genotype-related features of dystonia. *Neurology* **2009**, *72*, 2097–2103. [CrossRef]
11. Napolitano, F.; Pasqualetti, M.; Usiello, A.; Santini, E.; Pacini, G.; Sciamanna, G.; Errico, F.; Tassone, A.; Di Dato, V.; Martella, G.; et al. Dopamine D2 receptor dysfunction is rescued by adenosine A2A receptor antagonism in a model of DYT1 dystonia. *Neurobiol. Dis.* **2010**, *38*, 434–445. [CrossRef]
12. Sciamanna, G.; Tassone, A.; Martella, G.; Mandolesi, G.; Puglisi, F.; Cuomo, D.; Madeo, G.; Ponterio, G.; Standaert, D.G.; Bonsi, P.; et al. Developmental profile of the aberrant dopamine D2 receptors response in striatal cholinergic interneurons in DYT1 dystonia. *PLoS ONE* **2011**, *6*, e24261. [CrossRef] [PubMed]
13. Sciamanna, G.; Hollis, R.; Ball, C.; Martella, G.; Tassone, A.; Marshall, A.; Parsons, D.; Li, X.; Yokoi, F.; Zhang, L.; et al. Cholinergic dysregulation produced by selective inactivation of the dystonia-associated protein torsinA. *Neurobiol. Dis.* **2012**, *47*, 416–427. [CrossRef] [PubMed]
14. Yokoi, F.; Dang, M.T.; Li, J.; Standaert, D.G.; Li, Y. Motor deficits and decreased striatal dopamine receptor 2 binding activity in the striatum-specific Dyt1 conditional knock- out mice. *PLoS ONE* **2011**, *6*, e24539. [CrossRef] [PubMed]
15. Martella, G.; Maltese, M.; Nistico, R.; Schirinzi, T.; Madeo, G.; Sciamanna, G.; Ponterio, G.; Tassone, A.; Mandolesi, G.; Vanni, V.; et al. Regional specificity of synaptic plasticity deficits in a knock-in mouse model of DYT1 dystonia. *Neurobiol. Dis.* **2014**, *65*, 124–132. [CrossRef] [PubMed]
16. Scarduzio, M.; Zimmerman, C.N.; Jaunarais, K.L.; Wang, Q.; Standaert, D.G.; McMahon, L.L. Strength of cholinergic tone dictates the polarity of dopamine D2 receptor modulation of striatal cholinergic interneuron excitability in DYT1 dystonia. *Exp. Neurol.* **2017**, *295*, 162–175. [CrossRef]
17. Bonsi, P.; Ponterio, G.; Vanni, V.; Tassone, A.; Sciamanna, G.; Migliarini, S.; Martella, G.; Meringolo, M.; Dehay, B.; Doudnikoff, E.; et al. RGS 9-2 rescues dopamine D2 receptor levels and signaling in DYT1 dystonia mouse models. *EMBO Mol. Med.* **2019**, *11*, e9283. [CrossRef]
18. D'Angelo, V.; Paldino, E.; Cardarelli, S.; Sorge, R.; Fusco, F.R.; Biagioni, S.; Mercuri, N.B.; Giorgi, M.; Sancesario, G. Dystonia: Sparse synapses for D2 receptors in the striatum of a DYT1 knock-out mouse model. *Int J. Mol. Sci.* **2020**, *21*, 1073. [CrossRef] [PubMed]
19. Maltese, M.; Martella, G.; Imbriani, P.; Schuermans, J.; Billion, K.; Sciamanna, G.; Farook, F.; Ponterio, G.; Tassone, A.; Santoro, M.; et al. Abnormal striatal plasticity in a DYT11/SGCE myoclonus dystonia mouse model is reversed by adenosine A2A receptor inhibition. *Neurobiol. Dis.* **2017**, *108*, 128–139. [CrossRef] [PubMed]
20. Yu-Taeger, L.; Ott, T.; Bonsi, P.; Tomczak, C.; Wassouf, Z.; Martella, G.; Sciamanna, G.; Imbriani, P.; Ponterio, G.; Tassone, A.; et al. Impaired dopamine- and adenosine-mediated signaling and plasticity in a novel rodent model for DYT25 dystonia. *Neurobiol. Dis.* **2020**, *134*, 104634. [CrossRef] [PubMed]

21. Preston, Z.; Lee, K.; Widdowson, L.; Freeman, T.; Dixon, A.K.; Richardson, P.J. Adenosine receptor expression and function in rat striatal cholinergic interneurons. *Brit. J. Pharm.* **2000**, *130*, 886–890. [CrossRef]
22. Tozzi, A.; de Iure, A.; Di Filippo, M.; Tantucci, M.; Costa, C.; Borsini, F.; Ghiglieri, V.; Giampà, C.; Fusco, F.R.; Picconi, B.; et al. The Distinct Role of Medium Spiny Neurons and Cholinergic Interneurons in the D_2/A_{2A} Receptor Interaction in the Striatum: Implications for Parkinson's Disease. *J. Neurosci.* **2011**, *31*, 1850–1862. [CrossRef] [PubMed]
23. Schiffmann, S.; Jacobs, O.; Vanderhaeghen, J.J. Striatal restricted adenosine A2 receptor (RDC8) is expressed by enkephalin but not by substance P neurons: An in-situ hybridization histochemistry study. *J. Neurochem.* **1991**, *57*, 1062–1067. [CrossRef]
24. Fink, J.S.; Weaver, D.R.; Rivkees, S.A.; Peterfreund, R.A.; Pollack, A.E.; Adler, E.M.; Reppert, S.M. Molecular cloning of the rat A2 adenosine receptor: Selective co-expression with D2 dopamine receptors in rat striatum. *Mol. Brain Res.* **1992**, *14*, 186–195. [CrossRef]
25. Schiffmann, S.N.; Fisone, G.; Moresco, R.; Cunha, R.A.; Ferré, S. Adenosine A2A receptors and basal ganglia physiology. *Prog. Neurobiol.* **2007**, *83*, 277–292. [CrossRef] [PubMed]
26. Fuxe, K.; Marcellino, D.; Borroto-Escuela, D.O.; Guescini, M.; Fernández-Dueñas, V.; Tanganelli, S.; Rivera, A.; Ciruela, F.; Agnati, L.F. Adenosine-dopamine interactions in the pathophysiology and treatment of CNS disorders. *CNS Neurosci. Ther.* **2010**, *16*, e18–e42. [CrossRef] [PubMed]
27. Rosin, D.L.; Robeva, A.; Woodard, R.L.; Guyenet, P.G.; Linden, J. Immunohistochemical localization of adenosine A2A receptors in the rat central nervous system. *J. Comp. Neurol.* **1998**, *401*, 163–186. [CrossRef]
28. Hettinger, B.D.; Lee, A.; Linden, J.; Rosin, D.L. Ultrastructural Localization of Adenosine A2A Receptors Suggests Multiple Cellular Sites for Modulation of GABAergic Neurons in Rat Striatum. *J. Comp. Neurol.* **2001**, *431*, 331–346. [CrossRef]
29. Bogenpohl, J.W.; Ritter, S.L.; Hall, R.A.; Smith, Y. Adenosine A2A Receptor in the Monkey Basal Ganglia: Ultrastructural Localization and Colocalization with the Metabotropic Glutamate Receptor 5 in the Striatum. *J. Comp. Neurol.* **2012**, *520*, 570–589. [CrossRef]
30. Santuy, A.; Rodríguez, J.R.; De Felipe, J.; Merchán-Pérez, A. Study of the size and shape of synapses in the juvenile rat somatosensory cortex with 3D electron microscopy. *eNeuro* **2018**, *5*. [CrossRef] [PubMed]
31. Quiroz, C.; Luján, R.; Uchigashima, M.; Simoes, A.P.; Lerner, T.N. Key Modulatory Role of Presynaptic Adenosine A2A Receptors in Cortical Neurotransmission to the Striatal Direct Pathway. *Sci. World J.* **2009**, *9*, 1321–1344. [CrossRef] [PubMed]
32. Sciamanna, G.; Ponterio, G.; Vanni, V.; Laricchiuta, D.; Martella, G.; Bonsi, P.; Meringolo, M.; Tassone, A.; Mercuri, N.B.; Pisani, A. Optogenetic Activation of Striatopallidal Neurons Reveals Altered HCN Gating in DYT1 Dystonia. *Cell Rep.* **2020**, *31*, 107644. [CrossRef] [PubMed]
33. Ferraro, L.; Beggiato, S.; Tomasini, M.C.; Fuxe, K.; Antonelli, T.; Tanganelli, S. A2A/D2 receptor heteromerization in a model of Parkinson's disease. Focus on striatal aminoacidergic signaling. *Brain Res.* **2012**, *1476*, 96–107. [CrossRef] [PubMed]
34. Mori, A.; Shindou, T. Modulation of GABAergic transmission in the striatopallidal system by adenosine A2A receptors: A potential mechanism for the antiparkinsonian effects of A2A antagonists. *Neurology* **2003**, *61*, S44–S448. [CrossRef]
35. Montes, J.; Pena, J.M.; De Felipe, J.; Herreras, O.; Merchan-Perez, A. The Influence of Synaptic Size on AMPA Receptor Activation: A Monte Carlo Model. *PLoS ONE* **2015**, *10*, e0130924. [CrossRef] [PubMed]
36. Yokoi, F.; Dang, M.T.; Liu, J.; Gandre, J.R.; Kwon, K.; Yuen, R.; Li, Y. Decreased dopamine receptor 1 activity and impaired motor-skill transfer in Dyt1ΔGAG heterozygous knock-in mice. *Behav. Brain Res.* **2015**, *279*, 202–210. [CrossRef] [PubMed]
37. D'Angelo, V.; Castelli, V.; Giorgi, M.; Cardarelli, S.; Saverioni, I.; Palumbo, F.; Bonsi, P.; Pisani, A.; Giampà, C.; Sorge, R.; et al. Phosphodiesterase-10A Inverse Changes in Striatopallidal and Striatoentopeduncular Pathways of a Transgenic Mouse Model of *DYT1* Dystonia. *J. Neurosci.* **2017**, *37*, 2112–2124. [CrossRef] [PubMed]
38. Xie, Z.; Adamowicz, W.; Eldren, W.D.; Jacowski, A.B.; Kleiman, R.J.; Morton, D.G.; Stephenson, D.T.; Strick, C.A.; Williams, R.D.; Menniti, F.S. Cellular and Subcellular localization of PDE10A, a striatum-enriched phosphodiesterase. *Neuroscience* **2006**, *139*, 597–607. [CrossRef] [PubMed]
39. Cardinale, A.; Fusco, F.R. Inhibition of phosphodiesterases as a strategy to achieve neuroprotection in Huntington's disease. *CNS Neurosci. Ther.* **2018**, *24*, 319–328. [CrossRef] [PubMed]
40. Eskow Jaunarais, K.L.; Bonsi, P.; Chesselet, M.F.; Standaert, D.G.; Pisani, A. Striatal cholinergic dysfunction as a unifying theme in the pathophysiology of dystonia. *Prog. Neurobiol.* **2015**, *127–128*, 91–107. [CrossRef] [PubMed]
41. Lee, D.D.; O'Dowd, D.K. cAMP-Dependent Plasticity at Excitatory Cholinergic Synapses in Drosophila Neurons: Alterations in the Memory Mutant Dunce. *J. Neurosci.* **2000**, *20*, 2104–2111. [CrossRef] [PubMed]
42. Threlfell, S.; West, A.R. Review: Modulation of striatal neuron activity by cyclic nucleotide signaling and phosphodiesterase inhibition. *Basal Ganglia* **2013**, *3*, 137–146. [CrossRef] [PubMed]
43. Lee, D.D. Global and local missions of cAMP signaling in neural plasticity, learning, and memory. *Front Pharmacol.* **2015**, *6*, 161. [CrossRef] [PubMed]
44. Bonsi, P.; Cuomo, D.; Martella, G.; Madeo, G.; Schirinzi, T.; Puglisi, F.; Ponterio, G.; Pisani, A. Centrality of Striatal Cholinergic Transmission in Basal Ganglia Function. *Front. Neuroanat.* **2011**, *5*, 6. [CrossRef] [PubMed]
45. Dixon, A.K.; Gubiz, A.K.; Sirinathsinghji, D.J.S.; Richardson, P.J.; Freeman, T.C. Tissue distribution of adenosine receptor mRNAs in the rat. *Br. J. Pharmacol.* **1996**, *18*, 1461–1468. [CrossRef] [PubMed]
46. Hurley, M.J.; Mash, D.C.; Jenner, P. Adenosine A2A receptor mRNA expression in Parkinson's disease. *Neurosci. Lett.* **2000**, *291*, 54–58. [CrossRef]

47. Goodchild, R.E.; Kim, C.E.; Dauer, W.T. Loss of the dystonia-associated protein torsinA selectively disrupts the neuronal nuclear envelope. *Neuron* **2005**, *48*, 923–932. [CrossRef]
48. Giorgi, M.; D'Angelo, V.; Esposito, Z.; Nuccetelli, V.; Sorge, R.; Martorana, A.; Stefani, A.; Bernardi, G.; Sancesario, G. Lowered cAMP and cGMP signalling in the brain during levodopa-induced dyskinesias in hemiparkinsonian rats: New aspects in the pathogenetic mechanisms. *Eur. J. Neurosci.* **2008**, *28*, 941–950. [CrossRef] [PubMed]

International Journal of *Molecular Sciences*

MDPI

Article

Long-Term Shaping of Corticostriatal Synaptic Activity by Acute Fasting

Federica Campanelli [1,2,†], Daniela Laricchiuta [3,†], Giuseppina Natale [1,2], Gioia Marino [1,2], Valeria Calabrese [1,4], Barbara Picconi [4,5], Laura Petrosini [3], Paolo Calabresi [2,6] and Veronica Ghiglieri [3,5,*]

1 Dipartmento di Medicina, Università di Perugia, 06129 Perugia, Italy; federica.campanelli@unicatt.it (F.C.); giuseppina.natale@studenti.unipg.it (G.N.); gioia.marino@gmail.com (G.M.); valeria.calabrese@sanraffaele.it (V.C.)

2 Dipartimento di Neuroscienze, Facoltà di Medicina e Chirurgia, Università Cattolica del Sacro Cuore, 00168 Rome, Italy; paolo.calabresi@policlinicogemelli.it

3 Laboratorio di Neurofisiologia Sperimentale e del Comportamento, IRCCS Fondazione Santa Lucia c/o CERC, 00143 Rome, Italy; daniela.laricchiuta@gmail.com (D.L.); laura.petrosini@uniroma1.it (L.P.)

4 IRCCS San Raffaele Pisana, Rome 00176, Italy; barbara.picconi@uniroma5.it

5 Università Telematica San Raffaele, 00166 Rome, Italy

6 Clinica Neurologica, Fondazione Policlinico Universitario Agostino Gemelli IRCCS, 00168 Rome, Italy

* Correspondence: veronica.ghiglieri@uniroma5.it

† These authors contributed equally to this work.

Abstract: Food restriction is a robust nongenic, nonsurgical and nonpharmacologic intervention known to improve health and extend lifespan in various species. Food is considered the most essential and frequently consumed natural reward, and current observations have demonstrated homeostatic responses and neuroadaptations to sustained intermittent or chronic deprivation. Results obtained to date indicate that food deprivation affects glutamatergic synapses, favoring the insertion of GluA2-lacking α-Ammino-3-idrossi-5-Metil-4-idrossazol-Propionic Acid receptors (AMPARs) in postsynaptic membranes. Despite an increasing number of studies pointing towards specific changes in response to dietary restrictions in brain regions, such as the nucleus accumbens and hippocampus, none have investigated the long-term effects of such practice in the dorsal striatum. This basal ganglia nucleus is involved in habit formation and in eating behavior, especially that based on dopaminergic control of motivation for food in both humans and animals. Here, we explored whether we could retrieve long-term signs of changes in AMPARs subunit composition in dorsal striatal neurons of mice acutely deprived for 12 h/day for two consecutive days by analyzing glutamatergic neurotransmission and the principal forms of dopamine and glutamate-dependent synaptic plasticity. Overall, our data show that a moderate food deprivation in experimental animals is a salient event mirrored by a series of neuroadaptations and suggest that dietary restriction may be determinant in shaping striatal synaptic plasticity in the physiological state.

Keywords: food deprivation; dietary restriction; dorsolateral striatum; GluA1; calcium-permeable AMPA; naphthyl-acetyl spermine

Citation: Campanelli, F.; Laricchiuta, D.; Natale, G.; Marino, G.; Calabrese, V.; Picconi, B.; Petrosini, L.; Calabresi, P.; Ghiglieri, V. Long-Term Shaping of Corticostriatal Synaptic Activity by Acute Fasting. *Int. J. Mol. Sci.* **2021**, *22*, 1916. https://doi.org/10.3390/ijms22041916

Academic Editor: Volkmar Lessmann

Received: 15 January 2021

Accepted: 11 February 2021

Published: 15 February 2021

1. Introduction

Dietary restriction and acute food deprivation (fasting) are voluntary practices currently used in many cultures for religious and health reasons [1–3]. The popularity of abstaining habits relies on their durable beneficial effects at a systemic level. In humans, fasting has effects on the activity of brain areas involved in working memory tasks and on performance in different cognitive domains [4,5]. It also improves executive functions, such as mental flexibility and set-shifting [6], to produce long-lasting beneficial effects when combined with conditioning tests [7].

Despite much evidence supporting the view of a general enhancement of cognition, the effects of food restriction on the central nervous system are still debated due to experimental differences that include statistical issues, diet composition, protocols and settings [4]. Potential applications of fasting or dietary restriction to the most common neurological diseases remain poorly investigated [8].

Animal studies, which provide mechanistic insights into the effects of food deprivation or restriction, mostly focused on the hippocampus for its involvement in cognitive functions and on the nucleus accumbens for its role in reward and motivation. Food restriction has mainly been tested as a tool to enhance hippocampal-dependent memory performance [9,10], leading to improvement in cognition and synaptic efficacy. In the hippocampal CA (Cornu Ammonis)1 area, short-term dietary restriction increases the number of glutamate receptors at the synapses responsible for an enhanced long-term potentiation (LTP) and α-Ammino-3-idrossi-5-Metil-4-idrossazol-Propionic Acid receptor (AMPAR) membrane incorporation [11,12]. Dietary restrictions are also associated with a reorganization of glutamatergic synapses, increasing surface expression and postsynaptic density abundance of GluA1 subunits of AMPARs, suggesting synaptic incorporation of these GluA2-lacking Ca^{2+}-permeable AMPARs [13]. Other studies reported a compensatory upregulation of D1 Dopamine (DA) receptors [14,15] with enhanced phosphorylation of the glutamatergic AMPAR GluA1 at Ser845. This downstream effect increases peak current, facilitates the trafficking to the cell surface [16–18] and stabilizes the membrane Ca^{2+}-permeable AMPARs [19]. In cells expressing these receptors, a combination of fast decay kinetics and large conductances that enhance synaptic transmission create the conditions for metaplasticity, in which the synapses become primed for a preferential direction of plasticity [20,21].

The dorsolateral part of the nucleus striatum hosts two forms of D1 DA-dependent synaptic plasticity that encode specific action-outcome associations in goal-directed behaviors. An intact function of striatal spiny projection neurons (SPNs) is needed for action selection and initiation through the integration of sensorimotor, cognitive, and motivational/emotional information [22]. Relevant to eating behaviors, the dorsal striatum is a site of action of DA control of motivation for food in both humans and animals [23,24].

Based on previous findings showing that food deprivation may impact glutamatergic synapses through enhancement of GluA2-lacking AMPARs-mediated activity in inhibitory neurons [25], we explored whether in SPNs of mice acutely deprived for 12 hours/day, for two consecutive days, we could retrieve signs of the synaptic insertion of GluA2-lacking AMPARs. Towards this aim, we analyzed the glutamatergic neurotransmission and the principal forms of DA-and glutamate-dependent synaptic plasticity: the corticostriatal long-term depression (LTD) and LTP.

2. Results

2.1. Acute Food Restriction Protocol Induced Long-term Changes in Spontaneous Glutamatergic Synaptic Currents in the Corticostriatal Synapses

To explore the long-term effects of acute food restriction on the activity of striatal SPNs, we first analyzed the basal membrane properties through ex vivo patch-clamp and intracellular recordings from SPNs in corticostriatal slices obtained from 40-days-old Food Restricted (FR) and aged-matched C57BL/6J male mice (Naïve) (Figure 1A). The current-voltage relationship, obtained by applying hyperpolarizing and depolarizing current steps to SPNs, showed no significant differences between the two experimental groups. No differences were observed in the resting membrane potential (RPMs) and in the firing patterns (Figure 1B,C, RPM -85.89 $\pm$ 0.95 mV for Naïve and -87.54 ± 0.66 mV for FR; Student's t-test $p > 0.05$, Naïve $n = 18$, FR $n = 28$). The firing rate was also unchanged as the mean number of spikes was similar between FR and Naïve mice (Figure 1D, 14.65 $\pm$ 0.51 $n = 17$ for Naïve and 16.00 $\pm$ 0.71 $n = 15$ for FR).

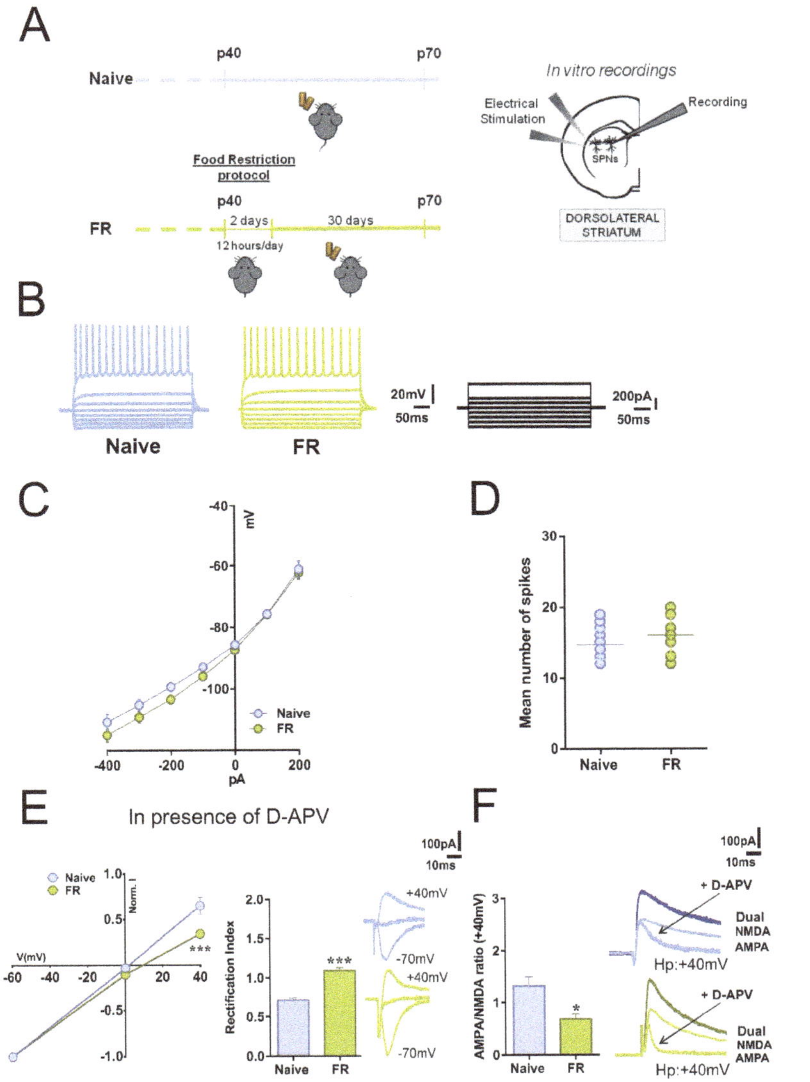

Figure 1. Acute food restriction is associated with changes in glutamatergic transmission in spiny projection neurons (SPNs). (**A**) Experimental plan of Naïve and food-restricted condition. (**B**) Representative firing traces and (**C**) current-voltage (I/V) graphs of Naïve (n = 17) and FR mice (n = 15), obtained after applying hyperpolarizing and depolarizing steps of current to SPNs recorded in dorsolateral striatum. (**D**) The aligned dot plot shows the mean number of spikes triggered by a step that generates a maximum response. (**E**) Current/voltage curve and bar graph show the rectification pattern and the rectification index in SPNs of Naïve and FR mice. The α-Ammino-3-idrossi-5-Metil-4-idrossazol-Propionic Acid receptor (AMPAR)-excitatory postsynaptic currents (EPSCs) were pharmacologically isolated by application of the N-Methyl-d-aspartate (NMDAR) antagonist D-(-)-2-Amino-5-phosphonopentanoic acid (D-APV, 50 μM) (current/voltage curve, two-way ANOVA time x group interaction, Naïve n = 19 *vs*. FR n = 11, $F_{(2,56)}$ = 5.36, *** $p < 0.001$; bar graph, unpaired t-test, Naïve n = 13 *vs*. FR n = 15, t = 7.942, df = 28, *** $p < 0.001$). Example traces of evoked AMPAR-EPSCs recorded at -70, 0, and+40 mV. Scale bar: 100 ms, 100 pA. (**F**) Group mean AMPA:NMDA ratio calculated in Naïve and FR SPNs in the presence of D-APV (unpaired t-test, Naïve n = 10, *vs*. FR n = 8, t = 2.911, df = 16, * $p < 0.05$); example traces of evoked AMPA- and NMDA-EPSCs at +40 mV (Dual: AMPA + NMDA EPSCs; NMDA EPSCs: obtained by subtraction of the EPSCs measured before and after the application of 50 μM D-APV; AMPA EPSCs: isolated by application of 50 μM D-APV). Scale bar: 100 ms, 100 pA.

2.2. SPNs of FR Mice Showed Increased Inwardly Rectifying AMPARs Currents and Unbalanced AMPA:NMDA Ratio

In a separate set of experiments, in corticostriatal slices, we recorded combined AMPAR-N-methyl-D-aspartate receptor (NMDAR)–mediated excitatory postsynaptic currents (EPSCs) at +40 mV, and then to determine the rectification index (RI), we recorded AMPAR EPSCs at different holding potentials (−70 mV, 0mV, +40 mV) in the presence of D-(-)-2-Amino-5-phosphonopentanoic acid (D-APV, 50 µM), a selective NMDAR antagonist. In contrast with Naïve mice, we found that FR mice showed an increase in RI whose value differed significantly from the control mice (Figure 1E, current/voltage curve two-way ANOVA time x group interaction, Naïve $n = 19$ *vs.* FR $n = 11$, $F_{(2,56)} = 5.36$, *** $p < 0.001$; bar graph unpaired t-test, Naïve $n = 13$ *vs.* FR $n = 15$, $t = 7.942$, df = 28, *** $p < 0.001$). We then examined the AMPAR: NMDAR ratio at +40 mV and observed a significant difference between the two experimental groups (Figure 1F, unpaired t-test, Naïve $n = 10$ *vs.* FR $n = 8$, $t = 2.911$, df = 16, * $p < 0.05$). The decreased ratio in FR mice suggests a correlation between this parameter and the RI, indicating an increased contribution of the GluA2-lacking AMPAR to the EPSC. Notably, in SPNs of FR mice, AMPA-mediated currents showed a marked difference in kinetics, with a more rapid decay of EPSCs (Figure 1F, bottom right panel, green lines).

2.3. Enhanced GluA1-mediated Function in Striatal SPNs of FR Mice was associated with a Change in the Direction of Corticostriatal Synaptic Plasticity

To explore if changes in AMPAR subunit composition could affect the corticostriatal glutamatergic transmission, we examined spontaneous EPSCs (sEPSCs) in SPNs of mice of the two experimental groups. As reported in Figure 2B, sEPSCs frequency was significantly increased in FR mice compared with Naïve mice (Figure 2A; unpaired t-test, Naïve $n = 10$, vs. FR $n = 8$, $t = 3.088$, df = 19, ** $p < 0.01$). Conversely, the amplitude of sEPSCs was comparable in SPNs between the two groups (Figure 2A).

Subsequently, we tested the ability of SPNs to express the long-term depression (LTD). In control condition, in which the bathing solution contained a physiological concentration of magnesium ions, a high-frequency stimulation (HFS) protocol of the corticostriatal fibers induced a robust LTD of the excitatory postsynaptic potentials (EPSPs) in the SPNs of Naïve mice (Figure 2B, paired t-test pre- *vs.* 20 min post-HFS, Naïve $n = 7$, $t = 12.10$, df = 12, *** $p < 0.001$). In contrast, the induction of this form of synaptic plasticity was impaired in FR mice and, interestingly, we observed a long-term potentiation (LTP) (Figure 2B, paired t-test pre vs. 20 min post-HFS, FR $n = 9$, $t = 7.299$, df = 17, *** $p < 0.001$), resulting in a significant difference on the response to HFS between the two groups (Figure 2B, two-way ANOVA: time x group interaction $F_{(24,336)} = 18.85$, Bonferroni's post hoc $^{###}p < 0.001$).

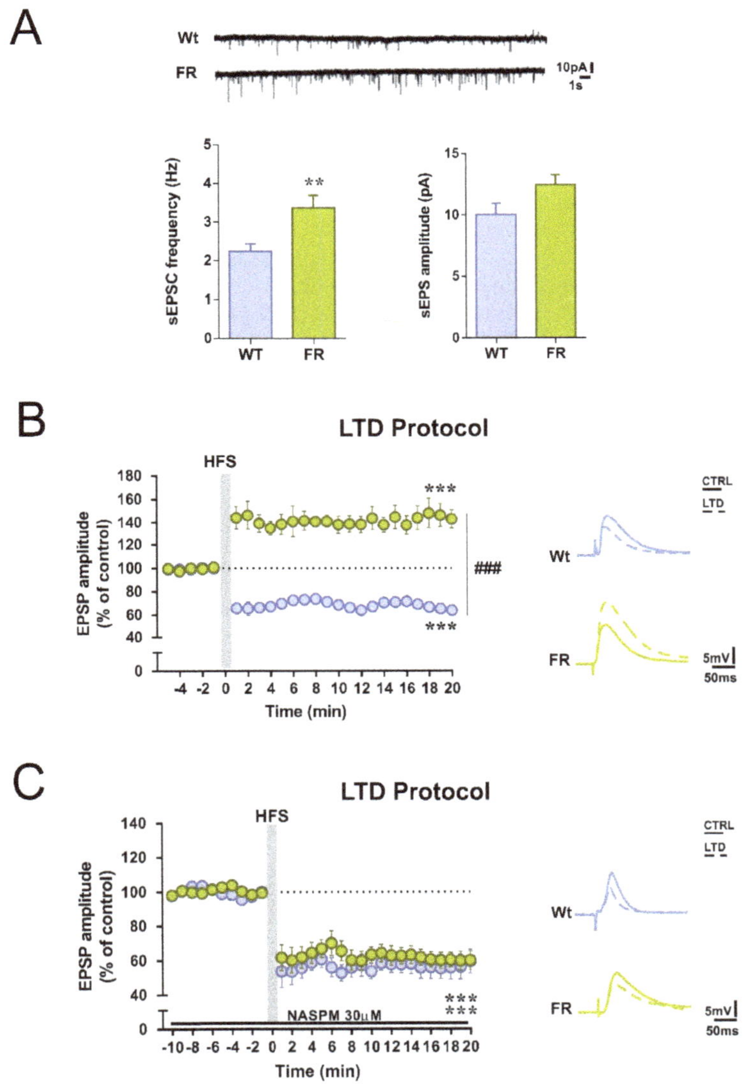

Figure 2. Enhanced activity of GluA2-lacking AMPARs in food-restricted (FR) mice is associated with changes in the direction of corticostriatal synaptic plasticity in SPNs. (**A**) Frequency and amplitude of sEPSCs glutamatergic transmission in Naïve and FR mice. In the upper part, comparison traces of spontaneous activity recorded from groups are shown. The frequency of sEPSC is increased in FR mice compared to Naïve mice (unpaired t-test, Naïve $n = 10$, *vs.* FR $n = 8$, $t = 3.088$, df = 19, ** $p < 0.01$). (**B**) Left panel: time course of excitatory postsynaptic potential (EPSP) amplitude of SPNs from Naïve and FR mice after induction of long-term depression (LTD) protocol (high-frequency stimulation, HFS) (paired t-test pre *vs.* 20 min post-HFS, Naïve $n = 7$, $t = 12.10$, df = 12, *** $p < 0.001$, FR $n = 9$, $t = 7.299$, df = 17, *** $p < 0.001$). Grouped analysis shows significant group effect (two-way ANOVA: time x group interaction $F_{(24,336)} = 18.85$, Bonferroni's post hoc ###$p < 0.001$). The scale factor is 50 ms/5 mV for all traces. Right panel, representative traces of single SPNs recorded from Naïve and FR mice before (solid lines) and after HFS (dotted lines). (**C**) Left panel: time course of SPNs EPSP amplitude, recorded from Naïve and FR mice in the presence of 30 μM 1-naphthylacetyl spermine (NASPM) bath application for the whole duration of the experiment (paired t-test pre *vs.* 20 min post-HFS, Naïve $n = 8$, $t = 8.017$, df = 15; FR $n = 9$, $t = 10.55$, df = 17, *** $p < 0.001$ for both groups). The scale factor is 50 ms/5 mV for all traces. Right panel, representative traces of single SPNs recorded from Naïve and FR mice before (solid lines) and after HFS (dotted lines).

2.4. Selective GluA1 Antagonism was Associated with the Reappearance of Corticostriatal LTD in FR Mice

To test if this unexpected form of plasticity could depend on the increase in GluA1-mediated activity, we analyzed the LTD expression in the presence of a selective blocker of GluA2-lacking AMPARs, 1-naphthylacetyl spermine (NASPM, 30 μM). EPSPs were recorded for 10 min to obtain a stable baseline and then for 20 min after applying the HFS protocol. We found that in the presence of NASPM, SPNs recorded in Naïve and FR mice showed LTDs of similar amplitudes (Figure 2C, paired t-test pre *vs.* 20 min post-HFS, Naïve $n = 8$, $t = 8.017$, df = 15; FR $n = 9$, $t = 10.55$, df = 17, *** $p < 0.001$ for both groups). Bath application of 30 μM NASPM in corticostriatal slices from FR mice efficiently reduced the EPSPs amplitude after HFS, contrasting the unphysiological potentiation observed in the untreated condition, demonstrating that blocking GluA2-lacking AMPARs prevents the shift in synaptic plasticity direction.

2.5. Enhanced GluA1-AMPARs Function was Associated with Changes in LTP Maintenance in Striatal SPNs of FR Mice

Since AMPARs play a variety of roles in shaping synaptic plasticity and are important for both LTD induction and LTP maintenance, we explored if the time course of LTP was also changed by using whole-cell patch-clamp recordings of SPNs in corticostriatal slices. To study this form of plasticity, Mg^{2+} ions were removed from the medium to promote the activation of glutamate NMDARs.

Under these experimental conditions, an initial post-tetanic potentiation was normally induced by the application of an HFS protocol in both Naïve and FR mice. Although comparisons between EPSP amplitudes before and 20 minutes after HFS indicates that a slight potentiation could still be observed in FR mice (paired t-test pre *vs.* 20 min post-HFS, Naïve $n = 5$, $t = 12.58$, df = 9, $p < 0.0001$, FR $n = 8$, $t = 4.075$, df = 15, $p < 0.001$), the LTP maintenance was different in the two groups. In FR mice, the amplitude of EPSPs decreased over time, bringing to a significant difference in the strength of LTP between the two groups (Figure 3, two-way ANOVA: time × group interaction $F_{(24,264)} = 4.23$, 11–20 min, Bonferroni's post hoc # $p < 0.05$).

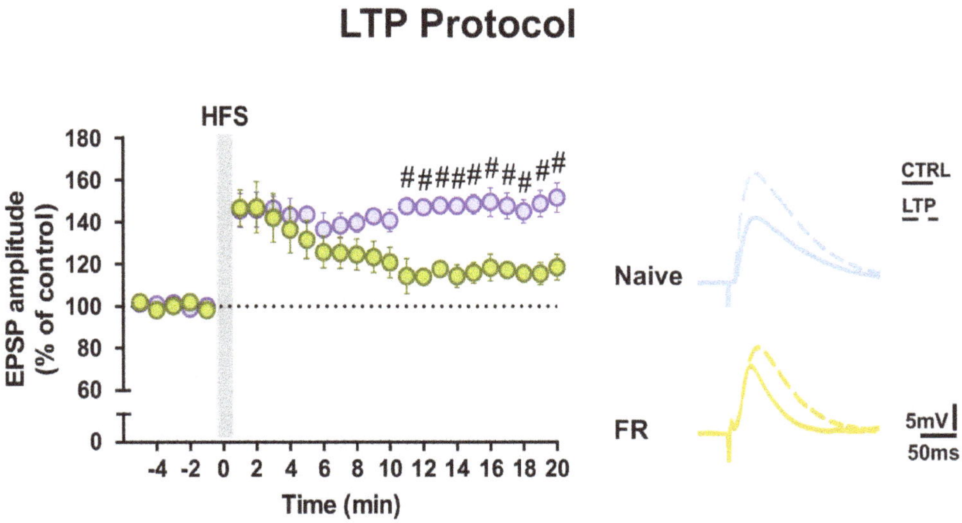

Figure 3. AMPAR subunit composition is critical for the maintenance of long-term potentiation (LTP). Left panel: time course of EPSP amplitude of SPNs from Naïve and FR mice after induction of LTP protocol. Grouped analysis shows significant group effect (two-way ANOVA: time x group interaction $F_{(24,264)} = 4.23$, 11–20 min, Bonferroni's post hoc $^{\#}$ $p < 0.05$). The scale factor is 50 ms/5 mV for all traces. Right panel: representative traces of single SPNs recorded from Naïve and FR mice before (solid lines) and after HFS (dotted lines).

3. Discussion

In this paper, we present evidence that acute but moderate (12 hours/day, for two consecutive days) food restriction can be related to persistent and subunit-specific enhancement in AMPARs function in SPNs of the dorsolateral striatum that emerges under specific stimulation of the corticostriatal pathway.

In our experiments, striatal SPNs of FR mice displayed intrinsic membrane properties and amplitude of spontaneous glutamatergic-mediated activity comparable with ad libitum fed controls. However, we observed significant changes in other aspects of glutamatergic transmission with substantial modifications in the glutamate-dependent synaptic plasticity linked to an enhanced function of GluA2-lacking AMPARs.

According to findings showing that food deprivation is associated with an enhanced abundance of Ca^{2+}-permeable AMPARs at glutamatergic synapses [13], here we describe a significant increase of the AMPAR rectification index. In our paradigm, this measure provided an indication of the changed proportion of the GluA1 over the GluA2 AMPAR subunits. These latter are the most expressed subunits of AMPARs in the striatum of adult rodents [26–28], characterized by a unique editing at the mRNA level where a glutamine codon is switched to arginine that confers channel resistance to Ca^{2+}. As a result, the neurons with increased insertion of homomeric GluA1 AMPARs show inward rectification that becomes linear when the GluA1 subunits are coexpressed with the GluA2 subunits [29,30]. Therefore, AMPAR complexes that lack GluA2 are permeable to sodium and Ca^{2+} ions, resulting in a higher single channel conductance and fast decay kinetics.

Since striatal AMPARs and NMDARs act in concert with dopaminergic neurotransmission to shape the direction of corticostriatal synaptic plasticity, we explored a possible imbalance between AMPA- and NMDA-mediated glutamatergic transmission to find a link between the different AMPAR subunit composition and the changes in the AMPA:NMDA ratio. Our data show a reduction of such a ratio in SPNs of Naïve and FR mice and, as confirmation of the increased GluA1-mediated activity, AMPAR-mediated currents were markedly different in their decay slope, exhibiting faster channel deactivation kinetics [31] and more rapidly decaying EPSPs [32].

We then tested the hypothesis that in deprived mice, the corticostriatal LTD, which depends on the activation of AMPARs [33], was also changed.

In SPNs recorded from FR mice, a high-frequency stimulation protocol of the corticostriatal fibers that, in physiological condition, elicited LTD, induced a shift toward LTP. This form of synaptic plasticity, whose induction is typically dependent on NMDAR activation, was not observed at physiological concentrations of magnesium ion, which acts as a natural blocker of the receptor pore. Such a change in synaptic plasticity direction might depend on increased Ca^{2+} entrance due to the enhancement of GluA2-lacking activity after a strong afferent stimulation. In such conditions, a massive corticostriatal stimulation would produce a greater depolarization of the SPNs membrane, relieving the NMDA receptors from the magnesium block and facilitating the induction of LTP.

This interpretation was substantiated by the present electrophysiological findings showing that NASPM, a selective blocker of GluA1-bearing AMPARs, restored a physiological LTD in FR mice at a dose that did not induce any effects in Naïve mice.

A possible explanation for the effect of such a moderate restriction protocol could be sought in a coincident impact of food deprivation on AMPARs changes during the development. In the newborn striatum, GluR1 immunoreactivity was observed in the presynaptic neurites, forming synapses with a more pronounced presence at the postsynaptic level in morphologically mature synapses as shown in seminal immunoelectron microscopy studies [34]. Moreover, the expression of GluA2-lacking AMPA receptors at excitatory synapses have been detected in many brain regions in the early postnatal development [35], and the switch from this subunit to the GluA2-containing AMPARs subunit occurs within the first two to three postnatal weeks [36–38]. However, we tested the animals far beyond these time points, when subunit composition had reached a steady-state with a net prevalence of GluA2-containing AMPARs. Thus, we excluded possible confusing developmental factor's effect on corticostriatal synaptic activity.

Given that AMPARs govern a variety of functions, including a fine regulation of both Ca^{2+} influx and LTP maintenance, and the link with Ca^{2+}-mediated kinase II [17,18,39], we explored the possibility that a change in their function could also affect the maintenance of LTP. While LTP was typically induced in SPNs from both groups, the amplitude of LTP in FR mice degraded over time.

Although SPNs show a peculiar pharmacological modulation of LTP, due to co-activation of DA and glutamate receptors and concurrent modulation by interneuronal activity, a possible explanation would be that the phase transition between induction and maintenance of LTP requires a change in membrane insertion of AMPARs subunits, as observed in the hippocampal CA1 area. In fact, in pyramidal neurons of CA1, during LTP induction, an initial incorporation of GluA2-lacking Ca^{2+}-permeable AMPARs is followed by a replacement with GluA2-containing Ca^{2+}-impermeable receptors [40]. However, given that this aspect has not been investigated in striatal SPNs, additional analyses are required to clarify these dynamics and the relevant contribution of AMPAR- and NMDAR-mediated components in the reduction of LTP.

These results support the view that a sudden, although moderate, food deprivation in experimental animals that were fed ad libitum since birth could be a salient event that may be encoded into a series of synaptic adaptations associated with long-term effects with adaptive changes in the AMPAR-mediated functions. These changes observed in isolated currents also emerged upon electrical stimulation of afferents without affecting the NMDAR-dependent phases of corticostriatal plasticity. This is in agreement with other studies reporting that increased Ca^{2+} influx via AMPARs lacking the GluA2 subunit does not have an impact on the NMDA component of LTP [41] and might be managed in homeostatic conditions.

Further investigations should clarify if the changes in AMPARs subunit composition are only limited to the postsynaptic level. In fact, presynaptic GluA1-AMPARs have been identified in corticostriatal and thalamostriatal axon terminals [42,43]. A possible increased insertion of these receptors [13] may explain the increased sEPSC frequency observed in our experimental setting.

A concept of AMPA-dependent changes in presynaptic activity has already been put forward in past studies using in vivo microdialysis and showing enhanced release of glutamate in the striatum upon perfusion of AMPA that was blocked by AMPA antagonists [43], an effect associated with presynaptic adenylyl cyclase-dependent processes [44]. These data are in agreement with more recent findings demonstrating that AMPARs localize at presynaptic sites on glutamatergic afferents [42], and that AMPA autoreceptors would assure a positive feedback control of glutamate release that modulates synaptic scaling with nonlinear characteristics [42]. This system would work in balance with presynaptic metabotropic glutamate receptors, which, on the contrary, suppress or inhibit the release from axon terminals [45].

Relevant to our observations, these findings suggest that a dynamic regulation of synaptic scaling might also occur in FR mice, where an LTP is observed instead of an LTD. Given the increase of sEPSC frequency, this shaping in the direction of plasticity would be not only associated with an adaptive enhancement of postsynaptic GluA2-lacking-mediated function (resulting in an increase of rectification index) but also extend to changes in the AMPAR activity at presynaptic levels. A point that still awaits to be clarified is the identification of the duration of possible adaptive changes in AMPAR subunits composition in cortico- and thalamostriatal afferents, and if a switch toward GluA2-lacking subunit-dependent function can also be detected at presynaptic sites.

In conclusion, with the present electrophysiological results, this study provides new insights into the importance of Ca^{2+}-permeable GluA1 AMPARs and their involvement in the two primary forms of striatal plasticity, LTD and LTP. Dissecting the role of GluA2-lacking AMPARs, although less expressed in the adult striatum, can be critical for a full understanding of the mechanisms of compensative synaptic regulations that may occur in physiological conditions when the limits of homeostasis are challenged.

4. Materials and Methods

4.1. Animals

Male C57BL/6JO1aHsd mice (n = 23; approximately 40 days old at study onset) (Harlan, Italy) were used. All animals were housed four per cage, under a controlled 12-h light/12-h dark cycle and temperature (22–23 °C), with food and water ad libitum. All efforts were made to minimize the number of animals used and their suffering, in accordance with the European Directive (2010/63/EU). All procedures were approved by the institutional review board and ethics committee (IRCCS Fondazione Santa Lucia) and by the Italian Ministry of Health (Project identification code: 534/2019-PR). The animals were randomly allotted into two groups, Food Restricted (FR) and aged-matched C57BL/6J male mice (Naïve). The FR mice were subjected to a moderate food deprivation protocol that limited their access to food for 12 hours, during the dark phase, for two consecutive days [46]. Such a regimen resulted in no significant body weight loss. Thirty days later, both Naïve and FR mice were sacrificed for the electrophysiological recordings (Figure 1A).

4.2. Slices Preparation

FR male mice and Naïve aged-matched C57BL/6J male mice were used in electrophysiological experiments. All mice were sacrificed by cervical dislocation and the brain was rapidly removed from the skull. Corticostriatal slices were cut from mice brains (thickness, 240–280 μm) using a vibratome in Krebs' solution (in mmol/L: 126 NaCl, 2.5 KCl, 1.2 $MgCl_2$, 1.2 NaH_2PO_4, 2.4 $CaCl_2$, 10 glucose, and 25 $NaHCO_3$) bubbled with a 95% O_2–5% CO_2 gas mixture. After at least 1 hr recovery, individual slices including the cortex and the striatum were transferred to a recording chamber and continuously superfused with oxygenated Krebs' medium, at 2.5–3 mL/min and maintained at 32–33 °C.

4.3. Whole-cell Patch-clamp Recordings

Current-clamp recordings from spiny projection neurons (SPNs) were performed using the whole-cell patch-clamp technique. Neurons of the dorsal striatum were visualized using infrared differential interference contrast microscopy (Eclipse FN1, Nikon, Tokyo, Japan) [47,48]. Recordings were made with a Multiclamp 700B amplifier (Molecular Devices, San José, CA, USA) and stored on PC using pClamp 9 (Molecular Devices, San José, CA, USA). Borosilicate glass pipettes (6–9 MΩ) were filled with the following internal solutions (in mM): 120 K-gluconate, 0.1 $CaCl_2$, 2 $MgCl_2$, 0.1 EGTA, 10 N-(2-hydroxyethyl)-piperazine-N-s-ethanesulfonic acid, 0.3 Na-guanosine triphosphate, and 2 Mg-adenosine triphosphate (Mg-ATP), adjusted to pH 7.3 with KOH. For recordings in voltage-clamp mode the internal solution of glass pipettes contained (in mM): 120 $CsMeSO_3$, 10 CsCl, 8 NaCl, 2 $MgCl_2$, 10 HEPES, 0.2 EGTA, 10 TEA, 5 QX314, 0.3 NaGTP and 2 Mg-ATP. Striatal neurons were clamped at the holding potential of -80 mV and identified by the absence of spontaneous action potential discharge.

Whole-cell access resistance was 15–30 MΩ. Picrotoxin (50 μM) was added to block $GABA_A$-currents. D-APV (50 μM), an N-methyl-D-aspartate receptor (NMDAR) antagonist, was used to pharmacologically isolate the Ammino-3-idrossi-5-Metil-4-idrossazol-Propionic Acid receptor (AMPAR)-excitatory postsynaptic currents (EPSCs). To obtain the NMDA current, an evoked current at +40 mV was subtracted before and after the application of this antagonist. To calculate the AMPAR:NMDAR ratio, the AMPAR EPSC amplitude was divided by the NMDAR EPSC amplitude, both measured at +40 mV. The rectification index (RI) of AMPARs was obtained by dividing the chord conductance calculated at negative potential (−70 mV) and positive potential (+40 mV). Injected currents and input resistances were checked throughout the experiments. Cells with variations in these parameters >20% were rejected.

4.4. Intracellular Recordings with Sharp Electrodes

Current-clamp recordings, with an intracellular technique, were performed blindly using sharp electrodes filled with 2 M KCl (30−60 MΩ). Signal acquisition was performed with an Axoclamp 2B amplifier (Molecular Devices, San José, CA, USA), displayed on a separate oscilloscope, stored. Online and offline analyses were performed using a digital system (pClamp 9, Molecular Devices, San José, CA, USA).

To evoke EPSCs and glutamatergic excitatory postsynaptic potentials (EPSPs), the stimulating bipolar electrode and the recording electrodes were located in the white matter and within the dorsolateral striatum, respectively.

In both patch-clamp and intracellular recordings, EPSPs were evoked by electrical stimulation every 10 s and EPSP peak amplitudes were used as a measure of SPNs activity, according to previous studies from our and other research groups, which consider this parameter an optimal index to evaluate the extent of evoked striatal responses when recording in current-clamp mode [33,49–52]. To induce long-term depression (LTD) and long-term potentiation (LTP), we used a high-frequency stimulation (HFS) protocol, consisting of three trains of 100 Hz, 3 s of duration, and 20 s of interval. For the LTP protocol, magnesium (Mg^{2+}) ions were omitted from the Kreb's solution to remove the Mg^{2+}-dependent block of NMDA receptors [49]. During tetanic stimulation, the intensity of stimulation was increased to suprathreshold levels. EPSP modifications induced by HFS protocol were expressed as a percentage of control, the latter representing the mean of responses recorded during a stable period (15–20 min) before the tetanic stimulation. Current-voltage relationships were obtained by applying steps of current of 200 pA in both hyperpolarizing and depolarizing direction (from −400 to +200 pA). Firing frequency was calculated as the mean number of spikes in response to a step of 600 pA and shown as averaged values in scatter dot plots.

4.5. Chemicals

D-(-)-2-Amino-5-phosphonopentanoic acid (D-APV) was from Tocris Bioscience (Bristol, U.K.); 1-naphthylacetyl spermine trihydrochloride (NASPM) and Picrotoxin were from Sigma-Aldrich (Milan, Italy). Drugs dissolved in the final concentration were bath applied by switching the control perfusion to drug-containing solution. During the cell's recording, an aliquot of the stock APV solution was diluted in Krebs' solution to 50 μM and kept in a syringe for the entire duration of the experiment. For the NASPM and the Picrotoxin an aliquot of the stock solution was diluted in Krebs' solution to 30 μM and 50 μM, respectively.

4.6. Statistical Analyses

Analyses were performed using Prism 6.0 (GraphPad software). Electrophysiological results are presented as mean ± SEM. Paired Student's *t*-test was used for analysis of the mean pre vs post-HFS in the same cell population. Analysis of variance (ANOVA) test with a post hoc Bonferroni test were performed among different neuronal populations. Sample size was calculated with G*Power software (5% type I error; 80% power).

Author Contributions: Conceptualization, F.C. and V.G.; Data curation, F.C., D.L., G.M. and V.C.; Formal analysis, F.C., D.L., G.N. and G.M.; Funding acquisition, D.L., L.P.; Investigation, F.C., G.M. and G.N.; Methodology, G.N. and G.M.; Project administration, V.G.; Resources, D.L., L.P. and V.G.; Supervision, B.P., P.C. and V.G.; Visualization, V.C. and B.P.; Writing—original draft, F.C., G.M., G.N.; Writing—review & editing, D.L., L.P., P.C. and V.G. All authors have read and agreed to the published version of the manuscript.

Funding: This work was supported by the Italian Ministry of Health, Ricerca Corrente (IRCCS Fondazione Santa Lucia).

Institutional Review Board Statement: The study was conducted according to the guidelines of the national legislation on the use of animals for research (D.Lgs 26/2014) and to the EU Directive 86/609/EEC on the Protection of Animals used for Experimental and other Scientific Purposes. The study was approved by the IRCCS Fondazione Santa Lucia review board and ethics committee and by the Italian Ministry of Health (Project identification code: 534/2019-PR).

Informed Consent Statement: Not applicable.

Data Availability Statement: The data presented in this study are available on request from the corresponding author.

Conflicts of Interest: The authors declare no conflict of interest.

References

1. Robertson, L.T.; Mitchell, J.R. Benefits of short-term dietary restriction in mammals. *Exp. Gerontol.* **2013**, *48*, 1043–1048. [CrossRef]
2. Trepanowski, J.F.; Bloomer, R.J. The impact of religious fasting on human health. *Nutr. J.* **2010**, *9*, 57. [CrossRef]
3. Trepanowski, J.F.; Canale, R.E.; Marshall, K.E.; Kabir, M.M.; Bloomer, R.J. Impact of caloric and dietary restriction regimens on markers of health and longevity in humans and animals: A summary of available findings. *Nutr. J.* **2011**, *10*, 107. [CrossRef] [PubMed]
4. Benau, E.M.; Orloff, N.C.; Janke, E.A.; Serpell, L.; Timko, C.A. A systematic review of the effects of experimental fasting on cognition. *Appetite* **2014**, *77*, 52–61. [CrossRef]
5. Chechko, N.; Vocke, S.; Habel, U.; Toygar, T.; Kuckartz, L.; Berthold-Losleben, M.; Laoutidis, Z.G.; Orfanos, S.; Wassenberg, A.; Karges, W.; et al. Effects of overnight fasting on working memory-related brain network: An fMRI study. *Hum. Brain Mapp.* **2015**, *36*, 839–851. [CrossRef] [PubMed]
6. Solianik, R.; Sujeta, A.; Terentjevienė, A.; Skurvydas, A. Effect of 48 h Fasting on Autonomic Function, Brain Activity, Cognition, and Mood in Amateur Weight Lifters. *Biomed. Res. Int.* **2016**, *2016*. [CrossRef] [PubMed]
7. Shi, L.; Deng, J.; Chen, S.; Que, J.; Sun, Y.; Wang, Z.; Guo, X.; Han, Y.; Zhou, Y.; Zhang, X.; et al. Fasting enhances extinction retention and prevents the return of fear in humans. *Transl. Psychiatry* **2018**, *8*, 214. [CrossRef] [PubMed]
8. Phillips, M.C.L. Fasting as a Therapy in Neurological Disease. *Nutrients* **2019**, *11*, 2501. [CrossRef]
9. Fontán-Lozano, A.; Sáez-Cassanelli, J.L.; Inda, M.C.; de los Santos-Arteaga, M.; Sierra-Domínguez, S.A.; López-Lluch, G.; Delgado-García, J.M.; Carrión, A.M. Caloric restriction increases learning consolidation and facilitates synaptic plasticity through mechanisms dependent on NR2B subunits of the NMDA receptor. *J. Neurosci* **2007**, *27*, 10185–10195. [CrossRef]

10. Li, L.; Wang, Z.; Zuo, Z. Chronic intermittent fasting improves cognitive functions and brain structures in mice. *PLoS ONE* **2013**, *8*, e66069. [CrossRef]
11. Babits, R.; Szőke, B.; Sótonyi, P.; Rácz, B. Food restriction modifies ultrastructure of hippocampal synapses. *Hippocampus* **2016**, *26*, 437–444. [CrossRef]
12. Ribeiro, L.F.; Catarino, T.; Santos, S.D.; Benoist, M.; van Leeuwen, J.F.; Esteban, J.A.; Carvalho, A.L. Ghrelin triggers the synaptic incorporation of AMPA receptors in the hippocampus. *Proc. Natl. Acad. Sci. USA* **2014**, *111*, E149–E158. [CrossRef] [PubMed]
13. Ouyang, J.; Carcea, I.; Schiavo, J.K.; Jones, K.T.; Rabinowitsch, A.; Kolaric, R.; Cabeza de Vaca, S.; Froemke, R.C.; Carr, K.D. Food restriction induces synaptic incorporation of calcium-permeable AMPA receptors in nucleus accumbens. *Eur. J. Neurosci.* **2017**, *45*, 826–836. [CrossRef] [PubMed]
14. Carr, K.D.; Tsimberg, Y.; Berman, Y.; Yamamoto, N. Evidence of increased dopamine receptor signaling in food-restricted rats. *Neuroscience* **2003**, *119*, 1157–1167. [CrossRef]
15. Haberny, S.L.; Carr, K.D. Comparison of basal and D-1 dopamine receptor agonist-stimulated neuropeptide gene expression in caudate-putamen and nucleus accumbens of ad libitum fed and food-restricted rats. *Brain Res. Mol. Brain Res.* **2005**, *141*, 121–127. [CrossRef]
16. Shi, S.; Hayashi, Y.; Esteban, J.A.; Malinow, R. Subunit-specific rules governing AMPA receptor trafficking to synapses in hippocampal pyramidal neurons. *Cell* **2001**, *105*, 331–343. [CrossRef]
17. Esteban, J.A.; Shi, S.H.; Wilson, C.; Nuriya, M.; Huganir, R.L.; Malinow, R. PKA phosphorylation of AMPA receptor subunits controls synaptic trafficking underlying plasticity. *Nat. Neurosci.* **2003**, *6*, 136–143. [CrossRef]
18. Man, H.Y.; Sekine-Aizawa, Y.; Huganir, R.L. Regulation of {alpha}-amino-3-hydroxy-5-methyl-4-isoxazolepropionic acid receptor trafficking through PKA phosphorylation of the Glu receptor 1 subunit. *Proc. Natl. Acad. Sci. USA* **2007**, *104*, 3579–3584. [CrossRef]
19. He, K.; Song, L.; Cummings, L.W.; Goldman, J.; Huganir, R.L.; Lee, H.K. Stabilization of Ca2+-permeable AMPA receptors at perisynaptic sites by GluR1-S845 phosphorylation. *Proc. Natl. Acad. Sci. USA* **2009**, *106*, 20033–20038. [CrossRef] [PubMed]
20. Clem, R.L.; Huganir, R.L. Calcium-permeable AMPA receptor dynamics mediate fear memory erasure. *Science* **2010**, *330*, 1108–1112. [CrossRef]
21. Sanderson, J.L.; Gorski, J.A.; Dell'Acqua, M.L. NMDA Receptor-Dependent LTD Requires Transient Synaptic Incorporation of Ca(2)(+)-Permeable AMPARs Mediated by AKAP150-Anchored PKA and Calcineurin. *Neuron* **2016**, *89*, 1000–1015. [CrossRef] [PubMed]
22. Balleine, B.W.; Delgado, M.R.; Hikosaka, O. The role of the dorsal striatum in reward and decision-making. *J. Neurosci.* **2007**, *27*, 8161–8165. [CrossRef] [PubMed]
23. Wallace, D.L.; Aarts, E.; Dang, L.C.; Greer, S.M.; Jagust, W.J.; D'Esposito, M. Dorsal striatal dopamine, food preference and health perception in humans. *PLoS ONE* **2014**, *9*, e96319. [CrossRef]
24. Volkow, N.D.; Wang, G.J.; Fowler, J.S.; Tomasi, D.; Telang, F. Addiction: Beyond dopamine reward circuitry. *Proc. Natl. Acad. Sci. USA* **2011**, *108*, 15037–15042. [CrossRef]
25. Carr, K.D.; Cabeza de Vaca, S.; Sun, Y.; Chau, L.S. Reward-potentiating effects of D-1 dopamine receptor agonist and AMPAR GluR1 antagonist in nucleus accumbens shell and their modulation by food restriction. *Psychopharmacology* **2009**, *202*, 731–743. [CrossRef]
26. Jeun, S.H.; Cho, H.S.; Kim, K.J.; Li, Q.Z.; Sung, K.W. Electrophysiological Characterization of AMPA and NMDA Receptors in Rat Dorsal Striatum. *Korean J. Physiol. Pharmacol.* **2009**, *13*, 209–214. [CrossRef] [PubMed]
27. Greger, I.H.; Khatri, L.; Ziff, E.B. RNA editing at arg607 controls AMPA receptor exit from the endoplasmic reticulum. *Neuron* **2002**, *34*, 759–772. [CrossRef]
28. Nansen, E.A.; Jokel, E.S.; Lobo, M.K.; Micevych, P.E.; Ariano, M.A.; Levine, M.S. Striatal ionotropic glutamate receptor ontogeny in the rat. *Dev. Neurosci.* **2000**, *22*, 329–340. [CrossRef]
29. Boulter, J.; Hollmann, M.; O'Shea-Greenfield, A.; Hartley, M.; Deneris, E.; Maron, C.; Heinemann, S. Molecular cloning and functional expression of glutamate receptor subunit genes. *Science* **1990**, *249*, 1033–1037. [CrossRef]
30. Verdoorn, T.A.; Burnashev, N.; Monyer, H.; Seeburg, P.H.; Sakmann, B. Structural determinants of ion flow through recombinant glutamate receptor channels. *Science* **1991**, *252*, 1715–1718. [CrossRef]
31. Geiger, J.R.; Melcher, T.; Koh, D.S.; Sakmann, B.; Seeburg, P.H.; Jonas, P.; Monyer, H. Relative abundance of subunit mRNAs determines gating and Ca2+ permeability of AMPA receptors in principal neurons and interneurons in rat CNS. *Neuron* **1995**, *15*, 193–204. [CrossRef]
32. Lawrence, J.J.; Brenowitz, S.; Trussell, L.O. The mechanism of action of aniracetam at synaptic alpha-amino-3-hydroxy-5-methyl-4-isoxazolepropionic acid (AMPA) receptors: Indirect and direct effects on desensitization. *Mol. Pharmacol.* **2003**, *64*, 269–278. [CrossRef]
33. Calabresi, P.; Maj, R.; Pisani, A.; Mercuri, N.B.; Bernardi, G. Long-term synaptic depression in the striatum: Physiological and pharmacological characterization. *J. Neurosci.* **1992**, *12*, 4224–4233. [CrossRef]
34. Martin, L.J.; Furuta, A.; Blackstone, C.D. AMPA receptor protein in developing rat brain: Glutamate receptor-1 expression and localization change at regional, cellular, and subcellular levels with maturation. *Neuroscience* **1998**, *83*, 917–928. [CrossRef]
35. Kumar, S.S.; Bacci, A.; Kharazia, V.; Huguenard, J.R. A developmental switch of AMPA receptor subunits in neocortical pyramidal neurons. *J. Neurosci.* **2002**, *22*, 3005–3015. [CrossRef]

36. Bellone, C.; Mameli, M.; Luscher, C. In utero exposure to cocaine delays postnatal synaptic maturation of glutamatergic transmission in the VTA. *Nat. Neurosci.* **2011**, *14*, 1439–1446. [CrossRef]
37. Eybalin, M.; Caicedo, A.; Renard, N.; Ruel, J.; Puel, J.L. Transient Ca2+-permeable AMPA receptors in postnatal rat primary auditory neurons. *Eur. J. Neurosci.* **2004**, *20*, 2981–2989. [CrossRef] [PubMed]
38. Ho, M.T.; Pelkey, K.A.; Topolnik, L.; Petralia, R.S.; Takamiya, K.; Xia, J.; Huganir, R.L.; Lacaille, J.C.; McBain, C.J. Developmental expression of Ca2+-permeable AMPA receptors underlies depolarization-induced long-term depression at mossy fiber CA3 pyramid synapses. *J. Neurosci.* **2007**, *27*, 11651–11662. [CrossRef] [PubMed]
39. Derkach, V.; Barria, A.; Soderling, T.R. Ca2+/calmodulin-kinase II enhances channel conductance of alpha-amino-3-hydroxy-5-methyl-4-isoxazolepropionate type glutamate receptors. *Proc. Natl. Acad. Sci. USA* **1999**, *96*, 3269–3274. [CrossRef] [PubMed]
40. Isaac, J.T.; Ashby, M.C.; McBain, C.J. The role of the GluR2 subunit in AMPA receptor function and synaptic plasticity. *Neuron* **2007**, *54*, 859–871. [CrossRef] [PubMed]
41. Jia, Z.; Agopyan, N.; Miu, P.; Xiong, Z.; Henderson, J.; Gerlai, R.; Taverna, F.A.; Velumian, A.; MacDonald, J.; Carlen, P.; et al. Enhanced LTP in mice deficient in the AMPA receptor GluR2. *Neuron* **1996**, *17*, 945–956. [CrossRef]
42. Fujiyama, F.; Kuramoto, E.; Okamoto, K.; Hioki, H.; Furuta, T.; Zhou, L.; Nomura, S.; Kaneko, T. Presynaptic localization of an AMPA-type glutamate receptor in corticostriatal and thalamostriatal axon terminals. *Eur. J. Neurosci.* **2004**, *20*, 3322–3330. [CrossRef]
43. Patel, D.R.; Young, A.M.; Croucher, M.J. Presynaptic alpha-amino-3-hydroxy-5-methyl-4-isoxazole propionate receptor-mediated stimulation of glutamate and GABA release in the rat striatum in vivo: A dual-label microdialysis study. *Neuroscience* **2001**, *102*, 101–111. [CrossRef]
44. Dohovics, R.; Janaky, R.; Varga, V.; Hermann, A.; Saransaari, P.; Oja, S.S. Regulation of glutamatergic neurotransmission in the striatum by presynaptic adenylyl cyclase-dependent processes. *Neurochem. Int.* **2003**, *42*, 1–7. [CrossRef]
45. Rouse, S.T.; Marino, M.J.; Bradley, S.R.; Awad, H.; Wittmann, M.; Conn, P.J. Distribution and roles of metabotropic glutamate receptors in the basal ganglia motor circuit: Implications for treatment of Parkinson's disease and related disorders. *Pharmacol. Ther.* **2000**, *88*, 427–435. [CrossRef]
46. Laricchiuta, D.; Rossi, S.; Musella, A.; De Chiara, V.; Cutuli, D.; Centonze, D.; Petrosini, L. Differences in spontaneously avoiding or approaching mice reflect differences in CB1-mediated signaling of dorsal striatal transmission. *PLoS ONE* **2012**, *7*, e33260. [CrossRef] [PubMed]
47. Bagetta, V.; Picconi, B.; Marinucci, S.; Sgobio, C.; Pendolino, V.; Ghiglieri, V.; Fusco, F.R.; Giampà, C.; Calabresi, P. Dopamine-dependent long-term depression is expressed in striatal spiny neurons of both direct and indirect pathways: Implications for Parkinson's disease. *J. Neurosci.* **2011**, *31*, 12513–12522. [CrossRef]
48. Bagetta, V.; Sgobio, C.; Pendolino, V.; Del Papa, G.; Tozzi, A.; Ghiglieri, V.; Giampà, C.; Zianni, E.; Gardoni, F.; Calabresi, P.; et al. Rebalance of striatal NMDA/AMPA receptor ratio underlies the reduced emergence of dyskinesia during D2-like dopamine agonist treatment in experimental Parkinson's disease. *J. Neurosci.* **2012**, *32*, 17921–17931. [CrossRef]
49. Calabresi, P.; Pisani, A.; Mercuri, N.B.; Bernardi, G. Long-term Potentiation in the Striatum is Unmasked by Removing the Voltage-dependent Magnesium Block of NMDA Receptor Channels. *Eur. J. Neurosci.* **1992**, *4*, 929–935. [CrossRef] [PubMed]
50. Fino, E.; Vandecasteele, M.; Perez, S.; Saudou, F.; Venance, L. Region-specific and state-dependent action of striatal GABAergic interneurons. *Nat. Commun.* **2018**, *9*, 3339. [CrossRef]
51. Hawes, S.L.; Salinas, A.G.; Lovinger, D.M.; Blackwell, K.T. Long-term plasticity of corticostriatal synapses is modulated by pathway-specific co-release of opioids through kappa-opioid receptors. *J. Physiol.* **2017**, *595*, 5637–5652. [CrossRef] [PubMed]
52. Pawlak, V.; Kerr, J.N. Dopamine receptor activation is required for corticostriatal spike-timing-dependent plasticity. *J. Neurosci.* **2008**, *28*, 2435–2446. [CrossRef] [PubMed]

International Journal of *Molecular Sciences*

MDPI

Article

Optogenetic Stimulation of Prelimbic Pyramidal Neurons Maintains Fear Memories and Modulates Amygdala Pyramidal Neuron Transcriptome

Daniela Laricchiuta [1,*], Giuseppe Sciamanna [1], Juliette Gimenez [1], Andrea Termine [1,2], Carlo Fabrizio [1,2], Silvia Caioli [3], Francesca Balsamo [1], Anna Panuccio [1,4], Marco De Bardi [1], Luana Saba [1], Noemi Passarello [1], Debora Cutuli [1,4], Anna Mattioni [1], Cristina Zona [2], Valerio Orlando [1,5] and Laura Petrosini [1]

1 Department of Experimental Neuroscience, IRCCS Fondazione Santa Lucia, 00143 Rome, Italy; g.sciamanna@hsantalucia.it (G.S.); j.gimenez@hsantalucia.it (J.G.); andreatermine544@gmail.com (A.T.); carlo.fabrizio217@gmail.com (C.F.); francesca.balsamo93@gmail.com (F.B.); anna.panuccio@gmail.com (A.P.); m.debardi@hsantalucia.it (M.D.B.); luana_saba@libero.it (L.S.); noemi.passarello@gmail.com (N.P.); debora_cutuli@yahoo.it (D.C.); Anna.Mattioni@gmail.com (A.M.); valerio.orlando@kaust.edu.sa (V.O.); laura.petrosini@uniroma1.it (L.P.)

2 Department of Systems Medicine, Tor Vergata University of Rome, 00133 Rome, Italy; zona@uniroma2.it

3 Unit of Neurology, IRCCS Neuromed, 86077 Pozzilli, Italy; silviacaioli@yahoo.it

4 Department of Psychology, University "Sapienza" of Rome, 00185 Rome, Italy

5 Biological Environmental Science and Engineering Division, KAUST Environmental Epigenetics Program, Thuwal 23955-6900, Saudi Arabia

* Correspondence: daniela.laricchiuta@gmail.com

Abstract: Fear extinction requires coordinated neural activity within the amygdala and medial prefrontal cortex (mPFC). Any behavior has a transcriptomic signature that is modified by environmental experiences, and specific genes are involved in functional plasticity and synaptic wiring during fear extinction. Here, we investigated the effects of optogenetic manipulations of prelimbic (PrL) pyramidal neurons and amygdala gene expression to analyze the specific transcriptional pathways associated to adaptive and maladaptive fear extinction. To this aim, transgenic mice were (or not) fear-conditioned and during the extinction phase they received optogenetic (or sham) stimulations over photo-activable PrL pyramidal neurons. At the end of behavioral testing, electrophysiological (neural cellular excitability and Excitatory Post-Synaptic Currents) and morphological (spinogenesis) correlates were evaluated in the PrL pyramidal neurons. Furthermore, transcriptomic cell-specific RNA-analyses (differential gene expression profiling and functional enrichment analyses) were performed in amygdala pyramidal neurons. Our results show that the optogenetic activation of PrL pyramidal neurons in fear-conditioned mice induces fear extinction deficits, reflected in an increase of cellular excitability, excitatory neurotransmission, and spinogenesis of PrL pyramidal neurons, and associated to strong modifications of the transcriptome of amygdala pyramidal neurons. Understanding the electrophysiological, morphological, and transcriptomic architecture of fear extinction may facilitate the comprehension of fear-related disorders.

Keywords: fear extinction; fear conditioning; medial prefrontal cortex; RNA sequencing; differential gene expression; electrophysiological recordings; excitatory post-synaptic currents; spinogenesis; fear-related disorders

check for updates

Citation: Laricchiuta, D.; Sciamanna, G.; Gimenez, J.; Termine, A.; Fabrizio, C.; Caioli, S.; Balsamo, F.; Panuccio, A.; De Bardi, M.; Saba, L.; et al. Optogenetic Stimulation of Prelimbic Pyramidal Neurons Maintains Fear Memories and Modulates Amygdala Pyramidal Neuron Transcriptome. *Int. J. Mol. Sci.* **2021**, *22*, 810. https://doi.org/10.3390/ijms22020810

Received: 4 December 2020
Accepted: 12 January 2021
Published: 15 January 2021

1. Introduction

In variable and challenging environments with various contextual situations, the individuals face a number of approaching dangers. The knowledge of potential threats allows developing fear of threatening situations, choosing among the various behaviors the safest ones, and detecting future dangers. To develop adaptive fear responses, the brain has to discriminate different sensory cues and associate relevant stimuli with aversive events [1]. Thus, when a relevant stimulus (or a context) is associated with an aversive event, fear associations are learned and form a memory.

Learned fear has been widely studied using Contextual Fear Conditioning (CFC), a very useful paradigm to analyze the neuronal and molecular bases of fear associative learning and memory [2,3]. In experimental models in which the CFC paradigm is implemented, the conditioned stimulus (CS), such as a specific cue or context, is associated with the unconditioned stimulus (US), such as foot-shock [3,4]. After the association has taken place, CS alone is able to induce the conditioned response (CR) of fear, such as freezing behavior. The repeated exposure to unreinforced (presented without US) CS gradually weakens the fear CR. In fact, when in the same context the danger is no more present, fear memory goes extinct and other survival functions may be implemented. Interestingly, extinguishing a fear response does not simply involve the fading away of the previous learning. Rather, during extinction process the subject learns something new - the cue no longer predicts that the fearful event will occur [5]. Notably, the impairment of the coping mechanisms dampening fear memory results in maladaptive behaviors. Compromised fear extinction is the key clinical feature of several fear-related disorders, such as post-traumatic stress disorders (PTSD), generalized anxiety, major depression, and phobias [5]. The acquisition of CS-US associative memory as well as the acquisition and maintenance of extinction memory require coordinated neural activity within the fear matrix, encompassing amygdala, medial prefrontal cortex (mPFC), and hippocampus [6–11]. Specifically, the activation of pyramidal neurons of the basolateral amygdala (BLA) is necessary to associate sensory input with US [12,13]. Interacting with the amygdala, the mPFC exerts top-down control allowing for appropriate fear responses. In this framework, it has been demonstrated that the balance between expression and extinction of fear CR is modulated by bidirectional inputs from/to the amygdala to/from two sub-regions of the mPFC: the prelimbic (PrL) cortex, which promotes fear responses, and the infralimbic (IL) cortex, which promotes fear extinction [14,15]. In literature are reported in vivo optogenetic manipulations performed by stimulating or inhibiting, with advanced spatial and temporal precision, specific neurons in the fear matrix. These studies demonstrate that the balanced firing activity between amygdala and mPFC is causally involved in fear processing [14,16–20]. Fear extinction decreases the efficacy of excitatory synaptic transmission in the projections from mPFC to amygdala, and promotes the inhibition in mPFC-amygdala pathway [21]. In parallel, projections from the amygdala to PrL cortex exhibit cell-type-specific plasticity during states of high fear, whereas projections from the amygdala to IL cortex are activated during states of low fear [19].

It has to be noted that any behavior has a specific genomic, transcriptomic, and epigenetic signature. Namely, transcriptomic changes are a crucial component of the neuronal modifications that underlie learning and memory [22–24], also after the fear exposure [25–27]. To date, specific genes involved in functional plasticity and synaptic wiring during fear memory are retained possible disease contributors and potential therapeutic targets for fear-related disorders [28–32]. The environmental experiences may modify the transcriptome [33–36], and these modifications, in turn, may explain the variability in resilience or predisposition to fear-related disorders as well as the severity of their symptomatology [37]. Understanding the transcriptomic architecture of compromised fear inhibition may thus facilitate the comprehension of fear-related disorders and the identification of potential therapeutic targets.

In the present study, adult transgenic mice, endowed with light-activated ion channel-expressing pyramidal neurons, were used. These transgenic mice were submitted to CFC receiving (or not) the US, to test the synergy between the activity of the amygdala and mPFC in fear learning. In vivo optogenetic manipulations (photo-activations) of the PrL pyramidal neurons were delivered during fear extinction. At the end of behavioral testing, electrophysiological, morphological, and transcriptomic correlates were evaluated. Namely, in PrL pyramidal neurons neural cellular excitability, excitatory neurotransmission, and spinogenesis were evaluated. In parallel, amygdala pyramidal neurons were sorted to perform cell-specific RNA-analyses (differential gene expression profiling and functional enrichment analysis). This methodology permitted conducting a genome-wide investigation of pyramidal neuron expression patterns without a priori selection of specific gene factors, and preventing bias in gene expression brought by bulk tissue analysis and biological subject pooling. Notably, investigating the gene expression of amygdala pyramidal neurons revealed the specific transcriptional pathways associated to impaired fear extinction.

2. Results

2.1. *Behavioral Results: In Vivo Optogenetics of the PrL Pyramidal Neurons during CFC*

The animals were (or not) fear-conditioned by using the CFC with repetitive sessions on day 1 (Conditioning phase), and 2, 3, 4, 7, and 14 (Extinction phase) (Figure 1A). During the extinction phase, the mice received optogenetic (OPTO FEAR and OPTO NOT FEAR groups) or sham (SHAM FEAR and SHAM NOT FEAR groups) stimulations. A three-way ANOVA (stimulation × fear × day) on freezing behavior (measured during 0–3 min of CFC) (Figure 1B, data expressed in percentage of freezing times are showed in Supplementary Figure S1A) revealed significant stimulation ($F_{1,36} = 6.13$; $p = 0.018$), fear ($F_{1,36} = 63.29$; $p < 0.0001$), and day ($F_{5,36} = 50.27$; $p < 0.0001$) effects. The first-order interactions (at least $p = 0.05$), as well as the second-order interaction (stimulation × fear × day) ($F_{5,180} = 3.00$; $p = 0.013$) were significant. As revealed by Newman-Keuls post-hoc comparisons, while all animals showed similar responses in the 0–3 min of Conditioning phase, evidence of consolidation was seen on the first extinction day only in the fear-conditioned animals (OPTO FEAR and SHAM FEAR groups), as revealed by their significantly increased freezing times between the Conditioning phase and day 2 ($p = 0.00004$ for both OPTO FEAR and SHAM FEAR groups), and by the similar freezing times of the two groups ($p = 0.90$). As expected, SHAM FEAR animals progressively extinguished fear memories over time and on day 14 their freezing behavior returned to a level similar ($p = 0.17$) to that showed during the Conditioning phase. Interestingly, OPTO FEAR group showed impaired extinction of fear memories. On day 14, OPTO FEAR mice still displayed freezing times significantly longer than those of the Conditioning phase ($p = 0.00002$). One-way ANOVA on freezing behavior (measured during 0–6 min of day 14) of all groups revealed a significant group effect ($F_{3,36} = 11.05$; $p = 0.00003$) (Figure 1C, data expressed in percentage of freezing times are showed in Supplementary Figure S1B). As revealed by Newman-Keuls post-hoc comparisons, OPTO FEAR group showed the highest freezing times in comparison to the remaining groups (at least $p = 0.0005$).

To control for the effects of learned but not yet extinguished fear, other mice (No-EX group) were fear-conditioned by using the CFC paradigm without the extinction protocol. Freezing times in the 0–3 min of Conditioning phase and day 2 shown by No-EX group were similar to those shown by OPTO FEAR and SHAM FEAR groups, in the same time-points. A two-way ANOVA (group × day) on freezing times (measured during 0–3 min of CFC) revealed significant group ($F_{4,41} = 24.57$; $p < 0.0001$) and day ($F_{1,41} = 268.50$; $p < 0.00001$) effects. The interaction was significant ($F_{4,41} = 35.87$; $p < 0.0001$). As revealed by Newman-Keuls post-hoc comparisons, only the fear-conditioned animals (No-EX, OPTO FEAR, and SHAM FEAR groups) increased their freezing times between day 1 and day 2, showing similar consolidation of fear memory.

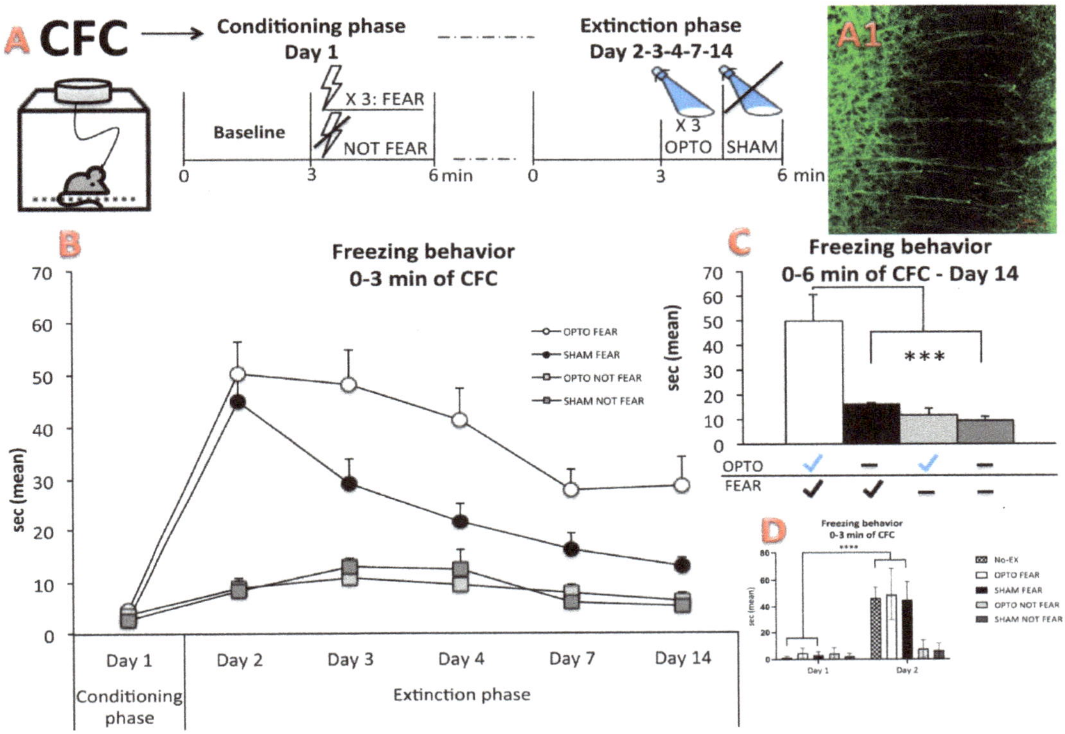

Figure 1. Experimental procedures and behavioral results of in vivo optogenetics of the Prelimbic (PrL) pyramidal neurons during Contextual Fear Conditioning (CFC). (**A**) On day 1 (Conditioning phase) of CFC, each Thy1-COP4 mouse was allowed to explore the conditioning chamber for 3 min (Baseline). Afterward, only a part of the entire sample received three foot-shocks. On days 2, 3, 4, 7, and 14 (Extinction phase), the fear-conditioned ($n = 20$) and not fear-conditioned ($n = 20$) mice were placed again in the conditioning chamber for 6 min. During the Extinction phase, no shock was delivered and the mice received three optogenetic (OPTO FEAR and OPTO NOT FEAR groups, $n = 10$/group) or sham (SHAM FEAR and SHAM NOT FEAR groups, $n = 10$/group) stimulations of PrL pyramidal neurons. To control for the effects of learned but not yet extinguished fear, on day 1 (Conditioning phase) of CFC other Thy1-COP4 mice (No-EX group, $n = 6$) were allowed to explore the conditioning chamber for 3 min (Baseline) and then they received three foot-shocks. The day after the Conditioning phase (day 2) these animals were placed again in the conditioning chamber for 6 min (without receiving any optogenetic stimulation) and then sacrificed for electrophysiological and morphological analyses. A1) Representative image of the expression of the transgenic ChR2-YFP fusion protein detected in pyramidal cortical layer 5 neurons of PrL cortex. Scale bar 50 μm. (**B**) Freezing behavior measured during 0–3 min of CFC. The animals belonging to OPTO FEAR, OPTO NOT FEAR, SHAM FEAR, and SHAM NOT FEAR groups showed similar responses in the Conditioning phase and only the fear-conditioned animals (OPTO FEAR and SHAM FEAR groups) showed increased freezing times on day 2. While SHAM FEAR group progressively extinguished fear memories over time, impaired extinction of fear memories was observed in OPTO FEAR group. (**C**) Freezing times measured during 0–6 min of day 14. OPTO FEAR group showed the highest freezing times in comparison to the remaining groups (*** $p = 0.0005$). (**D**) Freezing behavior measured during 0–3 min of day 1 and 2 of CFC. Only the fear-conditioned animals (No-EX, OPTO FEAR, and SHAM FEAR groups) increased their freezing times between day 1 and day 2 (**** $p < 0.0001$), showing similar consolidation of fear memory. Data are reported as mean ± SEM.

2.2. Electrophysiological Results: Cellular Excitability of PrL Pyramidal Neurons

To test the ability of the optogenetics to modulate the activity of PrL cortex in fear-conditioned or not fear-conditioned animals, an extensive characterization of the cellular excitability of pyramidal layer 5 neurons after sham or optogenetic stimulation was performed.

In OPTO FEAR group, the optogenetic stimulation induced a robust increase in the evoked firing activities triggered by growing pulses of depolarizing current (0 to 400 pA). Neurons of OPTO FEAR animals clearly showed a higher number of action potentials in comparison to neurons of SHAM FEAR animals, at all current levels considered as indicated by cumulative curve (Correlation test, SHAM FEAR group: Pearson $r = 0.99$, $R^2 = 0.99$, $n = 8$ neurons from 5 mice; OPTO FEAR group: Pearson $r = 0.98$, $R^2 = 0.97$, $n = 8$ neurons from 5 mice; $p < 0.0001$, Figure 2A). Rheobase that represents the lowest current amplitude able to generate an action potential is a further useful parameter to investigate changes in neural excitability. PrL pyramidal neurons of OPTO FEAR group recorded after optogenetic stimulation showed a clear reduction in the rheobase value in comparison to neurons of SHAM FEAR group (SHAM FEAR group: 69.88 ± 5.5 pA, $n = 8$ neurons from 5 mice, OPTO FEAR group: 50.50 ± 3.7 pA, $n = 8$ neurons from 5 mice, Mann-Whitney U Test $p = 0.002$; Figure 2B). To test the effect of optogenetic stimulation upon glutamatergic input to pyramidal neurons, the Excitatory Post-Synaptic Currents (EPSC) have been recorded in OPTO FEAR and SHAM FEAR groups. To avoid any inhibitory current contamination, the recordings have been made under pharmacological isolation by bath application of the GABAA blocker, Picrotoxin (10 min, 50 μM). Pyramidal PrL neurons of OPTO FEAR group showed a significantly higher frequency with respect to neurons of SHAM FEAR group (SHAM FEAR group: 4.11 ± 0.26 Hz, $n = 8$ neurons from 5 mice, OPTO FEAR group: 6.01 ± 0.25 Hz, $n = 8$ neurons from 5 mice, Mann-Whitney U Test $p = 0.002$, Figure 2C). Cumulative plot clearly corroborated such an effect (Correlation test, SHAM FEAR group: Pearson $r = 0.85$, $R^2 = 0.72$, $n = 8$ neurons from 5 mice; OPTO FEAR group: Pearson $r = 0.75$, $R^2 = 0.56$, $n = 8$ neurons from 5 mice, $p = 0.0004$, Figure 2C). No significant differences in EPSC amplitudes were found between SHAM FEAR and OPTO FEAR groups (SHAM FEAR group: 17.24 ± 0.81 pA, $n = 8$ neurons from 3 mice; OPTO FEAR group: 18.31 ± 0.69 pA, $n = 8$ neurons from 4 mice, Mann-Whitney U Test $p = 0.37$, data not shown).

To control for the effects of learned but not yet extinguished fear on cellular excitability, electrophysiological data recorded from PrL pyramidal neurons of No-EX group were compared with those of OPTO FEAR and SHAM FEAR groups. The excitability of PrL neurons in No-EX mice appeared significantly higher when compared with that showed by SHAM FEAR mice and quite similar to that showed by OPTO FEAR mice. In particular when compared with SHAM FEAR mice, No-EX mice showed an increased evoked firing activities (No-EX group Correlation test: Pearson $r = 0.96$, $R^2 = 0.92$, $n = 6$ neurons from 3 mice; $p < 0.0001$, Figure 2A), a lower rheobase value (No-EX group: 48.60 ± 4.1 pA, $n = 5$ neurons from 3 mice; Mann-Whitney U Test $p = 0.01$; Figure 2B) and an higher EPSC frequency (No-EX group: 5.77 ± 0.42 Hz, $n = 5$ neurons from 3 mice, Mann-Whitney U Test $p = 0.01$, Figure 2C).

Conversely, in both groups of not fear-conditioned animals (SHAM NOT FEAR and OPTO NOT FEAR groups) the optogenetic stimulation did not produce any significant difference in cellular excitability. Neither the evoked firing nor the rheobase nor EPSC frequency showed any significant difference between SHAM NOT FEAR and OPTO NOT FEAR groups (Figure 2D).

Taken together these data demonstrate that optogenetic stimulation was able to modify the intrinsic cellular excitability of PrL pyramidal neurons supporting the assumption that this area is involved in modulation of the fear responses during extinction phase.

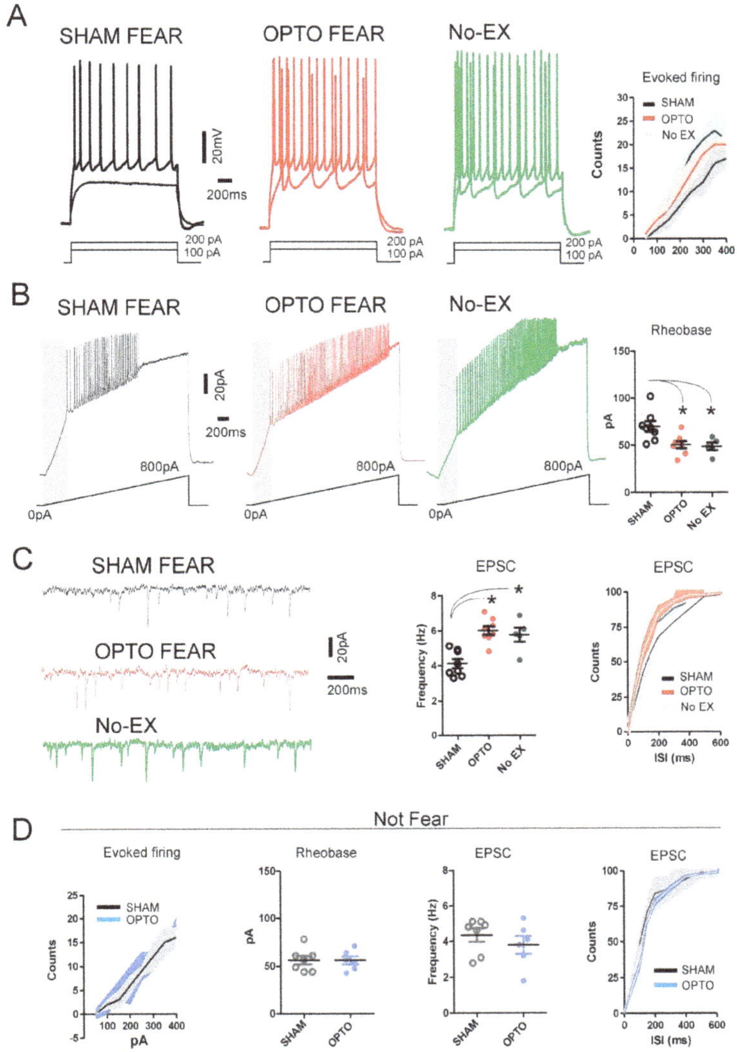

Figure 2. Modulation of cellular excitability of Prelimbic (PrL) pyramidal neurons by optogenetic stimulation. (**A**) Representative traces in current-clamp configuration reporting evoked firing activity triggered by a series of depolarizing current steps (0 to 400 pA) applied to PrL pyramidal neurons of SHAM FEAR (black, $n = 8$ neurons from 5 mice), OPTO FEAR (red, $n = 8$ neurons from 5 mice), and No-EX (green, $n = 5$ neurons from 3 mice) groups. The cumulative plot shows the changes in firing activity. (**B**) Representative traces of PrL pyramidal neurons of SHAM FEAR (black, $n = 8$ neurons from 5 mice), OPTO FEAR (red, $n = 8$ neurons from 5 mice), and No-EX (green, $n = 5$ neurons from 3 mice) groups showing the firing activity triggered by linear depolarization from 0 to 800 pA. Graph (on the right) reports the effects of optogenetic stimulation on rheobase value. Namely, PrL pyramidal neurons of OPTO FEAR and No-EX groups recorded after optogenetic stimulation showed a clear reduction in the rheobase value in comparison to neurons of SHAM FEAR group (* at least $p = 0.01$). (**C**) Representative traces of Excitatory Post-Synaptic Currents (EPSC) of PrL pyramidal neurons of SHAM FEAR (black, $n = 8$ neurons from 5 mice), OPTO FEAR (red, $n = 8$ neurons from 5 mice), and No-EX (green, $n = 5$ neurons from 3 mice) groups. Graph plot (in the middle) and cumulative curve (on the right) depict the clear increase in firing frequency in PrL pyramidal neurons of OPTO FEAR and No-EX groups (* at least $p = 0.01$). (**D**) Graphs and cumulative curves report no significant differences in cellular excitability in PrL pyramidal neurons of SHAM NOT FEAR (black) and OPTO NOT FEAR (blue) groups. Data are reported as median with interquartile range.

2.3. Morphological Results: Spine Counting of PrL Pyramidal Neurons

The five experimental groups exhibited different spine number and density in apical arborizations of PrL pyramidal neurons (Figure 3), as demonstrated by one-way ANOVAs on number ($F_{4,18} = 34.47$; $p < 0.0001$) and density ($F_{4,18} = 37.15$; $p < 0.0001$) of dendritic spines. Newman-Keuls post-hoc comparisons indicated that No-EX group had the highest number and density of dendritic spines in comparison to the other groups (at least $p = 0.00001$). Furthermore, OPTO FEAR group showed higher spine number and density in comparison to the other groups (at least $p = 0.0001$) that, in turn, exhibited similar spine number and density.

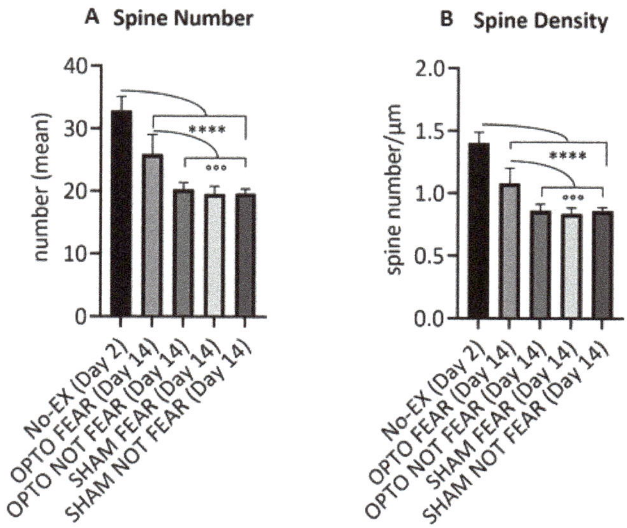

Figure 3. Spine counting of the apical arborizations of Prelimbic pyramidal neurons. No-EX group, encompassing animals submitted to fear learning but not to fear extinction (sacrificed at Day 2), had the highest number (**A**) and density (**B**) of dendritic spines in comparison to OPTO FEAR, OPTO NOT FEAR, SHAM FEAR, and SHAM NOT FEAR groups (**** at least $p = 0.0001$). The OPTO FEAR group showed higher spine number (**A**) and density (**B**) in comparison to the other groups (°°° at least $p = 0.0001$) that, in turn, exhibited similar spinogenesis. Data are reported as mean ± SEM.

2.4. Transcriptomic Results: Differential Gene Expression Profiling and Functional Enrichment Analysis of RNA Extracted by Sorted Amygdala Pyramidal Neurons

Gene expression profiling resulted in 4550 protein-coding genes with a reliable expression and they underwent the downstream analysis. A Principal Component Analysis (PCA) was performed to assess sample clustering based on gene expression profiles (Figure 4). Notably, individual samples belonging to the same group clustered well together. Further, while gene expression profiles of mice belonging to SHAM FEAR and SHAM NOT FEAR groups appeared clustered, those of individuals belonging to OPTO FEAR and OPTO NOT FEAR groups were markedly segregated. Differential expression analysis was performed on 2 × 2 design (Group × Condition) and Differentially Expressed Genes (DEGs) were identified. Significant differentially expressed genes were identified for a $q > 0.95$, equivalent to an FDR-corrected $p < 0.05$. Subsequently, Gene Ontology (GO) and Kyoto Encyclopedia of Genes and Genomes (KEGG) annotations and over-representation analyses (ORA) were performed using clusterProfiler (extended results are reported as Supplementary Results 1 and in Supplementary Figure S2). Significant pathways were shown by means of enrichment map method.

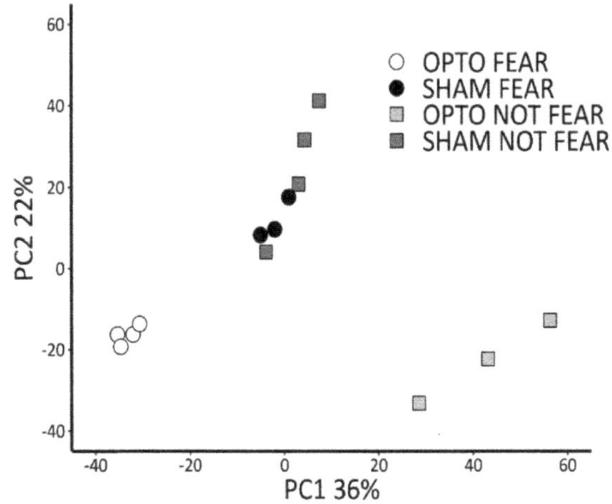

Figure 4. Principal Component Analysis revealing sample clustering based on gene expression profiles. While gene expression profile of mice belonging to SHAM FEAR and SHAM NOT FEAR groups appeared clustered, those of individuals belonging to OPTO FEAR and OPTO NOT FEAR groups were markedly segregated.

2.4.1. Comparison between OPTO FEAR vs. SHAM FEAR Groups

Differential expression analysis showed 2417 significant DEGs (1043 up; 1374 down, with reference level set on the SHAM FEAR condition) (Figure 5A,B, Table 1). To explore further the biological significance of the transcriptomic modulations caused by optogenetic stimulation in the presence of fear memory, ORA was performed on the obtained DEGs and resulted in 137 significantly enriched GO (Biological Processes, BP: 57; Cellular Component, CC: 61; Molecular Function, MF: 19) terms and 132 significantly enriched KEGG terms. The top twenty significant GO terms belonged to BP and CC and highlighted a differential involvement of the pathways associated with neuronal plasticity and synaptic signaling, resulting in an overall modulation of synaptic organization (Figure 5C) (namely: GO:0098590 Plasma membrane region, GO:0031226 Intrinsic component of plasma membrane, GO:009897 Glutamatergic synapse, GO:0005887 Integral component of plasma membrane, GO:0044456 Synapse part, GO:0030424 Axon, GO:0042734 Presynaptic membrane, GO:0031253 Cell projection membrane, GO:0098794 Postsynapse, GO:0099537 Trans-synaptic signaling, GO:0098609 Cell–cell adhesion, GO:0007155 Cell adhesion, GO:0099536 Synaptic signaling, GO:0022610 Biological adhesion, GO:0007268 Chemical synaptic transmission, GO:0098916 Anterograde trans-synaptic signaling, GO:0030425 Dendrite, GO:0097447 Dendritic tree, GO:0099572 Postsynaptic specialization, and GO:0097060 Synaptic membrane).

The top twenty significant KEGG terms resulted associated to signaling processes related to several neurodegenerative diseases and metabolic pathways (Figure 5D) (namely: mmu05022 Pathways of neurodegeneration—multiple diseases, mmu04144 Endocytosis, mmu05010 Alzheimer Disease, mmu05012 Parkinson Disease, mmu04140 Autophagy—animal, mmu05014 Amyotrophic Lateral Sclerosis, mmu04137 Mitophagy—animal, mmu05132 Salmonella infection, mmu04714 Thermogenesis, mmu04360 Axon guidance, mmu04919 Thyroid hormone signaling pathway, mmu04141 Protein processing in endoplasmic reticulum, mmu05231 Choline metabolism in cancer, mmu04722 Neurotrophin signaling pathway, mmu03010 Ribosome, mmu04072 Phospholipase D signaling pathway, mmu01212 Fatty acid metabolism, mmu01200 Carbon metabolism, mmu05020 Prion disease, and mmu00190 Oxidative phosphorylation).

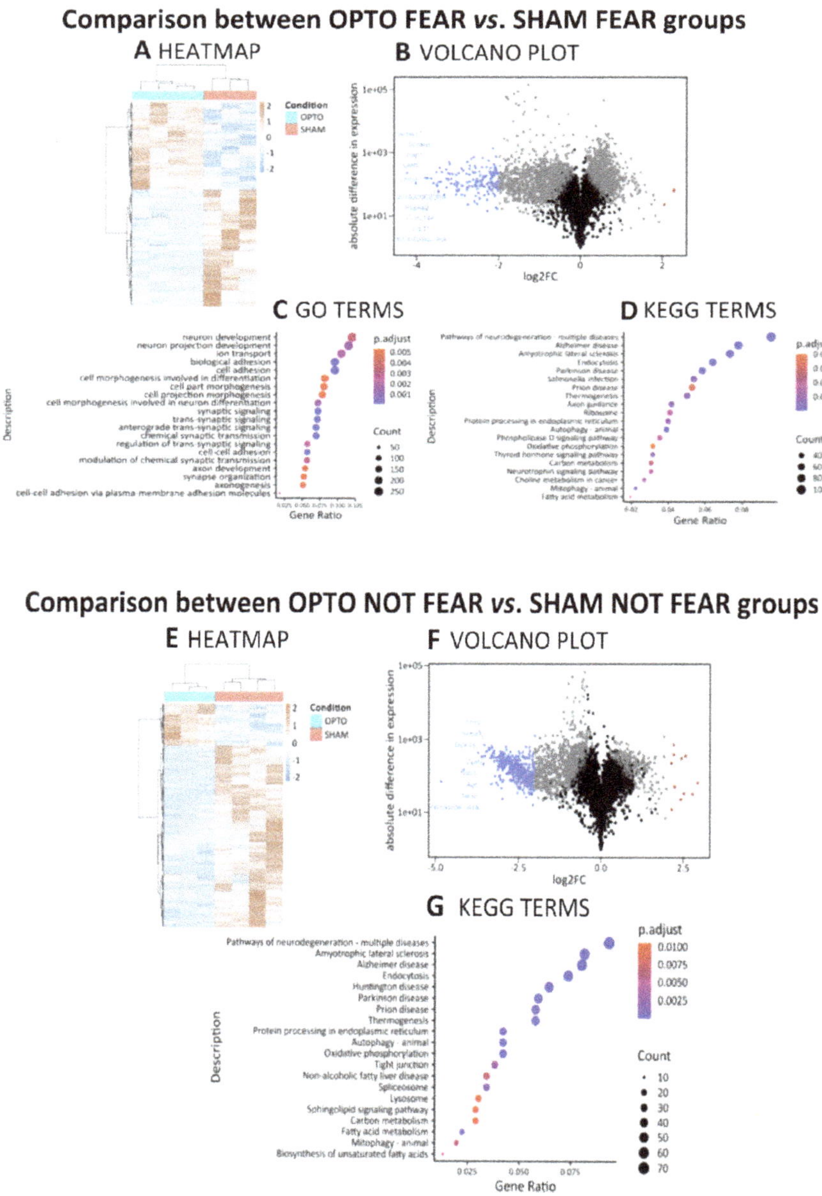

Figure 5. Differential Gene Expression profiling and functional enrichment analysis of RNA extracted by amygdala pyramidal neurons. Comparisons between OPTO FEAR vs. SHAM FEAR groups (upper part) and OPTO NOT FEAR vs. SHAM NOT FEAR groups (lower part). (**A**,**E**) Heatmaps showing gene expression values for the Differentially Expressed Genes (DEGs). (**B**,**F**) Volcano plots highlighting DEGs. The x-axis is the log2 fold change (log2FC) in normalized gene expression and the y-axis is for the $\log_{10}$ absolute value of the difference in expression between conditions. Each dot represents a gene. Grey dots are for DEGs, blue and red dots are for <-2 and $>+2$ log2FC genes, respectively. The top ten genes with the highest absolute log2FC values are labeled. (**C**,**D**,**G**) Dot plots representing the top twenty enriched terms from over-representation analyses (ORA) in Gene Ontology (GO) and Kyoto Encyclopedia of Genes and Genomes (KEGG) databases. GO terms from different domains (Biological Processes, Cellular Component, and Molecular Function) were sorted by q-value before plotting them together.

Table 1. The top 20 DEGs identified by differential expression profiling from RNA extracted by amygdala pyramidal neurons. Comparison between OPTO FEAR vs. SHAM FEAR groups.

Gene Symbol	Gene Name	Fold Change (log2 Scale)	OPTO FEAR adj.mean	SHAM FEAR adj.mean
Slc6a17	solute carrier family 6 (neurotransmitter transporter), member 17	−4.24	12.86	242.95
2010300C02Rik	RIKEN cDNA 2010300C02 gene	−4.02	6.95	113.03
Lmo3	LIM domain only 3	−3.80	12.43	172.75
Celf5	CUGBP, Elav-like family member 5	−3.74	12.58	168.18
Hap1	huntingtin-associated protein 1	−3.65	13.22	166.13
Papss2	3′-phosphoadenosine 5′-phosphosulfate synthase 2	−3.60	9.79	118.38
Sphkap	SPHK1 interactor, AKAP domain containing	−3.52	23.12	265.56
Ccdc148	coiled-coil domain containing 148	−3.51	9.18	104.53
Lrp11	low density lipoprotein receptor-related protein 11	−3.50	8.59	97.11
6330403K07Rik	RIKEN cDNA 6330403K07 gene	−3.48	6.08	67.76
Ube2ql1	ubiquitin-conjugating enzyme E2Q family-like 1	−3.47	5.54	61.52
Brinp1	bone morphogenic protein/retinoic acid inducible neural specific 1	−3.47	11.69	129.49
Arpp21	cyclic AMP-regulated phosphoprotein, 21	−3.44	33.80	366.38
Gabra4	gamma-aminobutyric acid (GABA) A receptor, subunit alpha 4	−3.41	12.24	130.15
Csmd1	CUB and Sushi multiple domains 1	−3.39	25.79	269.68
Kcnh3	potassium voltage-gated channel, subfamily H (eag-related), member 3	−3.34	5.02	50.67
Agt	angiotensinogen (serpin peptidase inhibitor, clade A, member 8)	−3.28	33.63	327.00
St6gal2	beta galactoside alpha 2,6 sialyltransferase 2	−3.28	9.03	87.47
Bcl2	B cell leukemia/lymphoma 2	−3.27	17.48	169.00
Cbarp	calcium channel, voltage-dependent, beta subunit associated regulatory protein	−3.23	16.69	156.53

2.4.2. Comparison between OPTO NOT FEAR vs. SHAM NOT FEAR Groups

Differential expression analysis showed 1661 significant DEGs (308 up; 1353 down, with reference level set on the SHAM NOT FEAR condition) (Figure 5E,F; Table 2). The ORA performed on the obtained DEGs resulted only in 36 significantly enriched KEGG terms and 0 significantly enriched GO terms. Here, the genes universe did not allow enriching many GO terms because of sampling bias correction [38]. Again, the top twenty over-represented KEGG terms were related to neurodegenerative diseases and metabolic pathways (Figure 5G) (namely: mmu04144 Endocytosis, mmu00190 Oxidative phosphorylation, mmu04140 Autophagy—animal, mmu05014 Amyotrophic Lateral Sclerosis, mmu04714 Thermogenesis, mmu05010 Alzheimer Disease, mmu05012 Parkinson Disease, mmu05022 Pathways of neurodegeneration—multiple diseases, mmu05016 Huntington Disease, mmu01212 Fatty acid metabolism, mmu05020 Prion Disease, mmu04141 Protein processing in endoplasmic reticulum, mmu03040 Spliceosome, mmu04530 Tight junction, mmu04137 Mitophagy—animal, mmu01040 Biosynthesis of unsaturated fatty acids, mmu04932 Non-alcoholic fatty liver disease, mmu04142 Lysosome, mmu01200 Carbon metabolism, and mmu04071 Sphingolipid signaling pathway).

Table 2. The top 20 DEGs identified by differential expression profiling from RNA extracted by amygdala pyramidal neurons. Comparison between OPTO NOT FEAR vs. SHAM NOT FEAR groups.

Gene Symbol	Gene Name	Fold Change (log2 Scale)	OPTO NOT FEAR adj.mean	SHAM NOT FEAR adj.mean
0610010K14Rik	RIKEN cDNA 0610010K14 gene	−4.83	2.44	69.50
Gm10163	predicted pseudogene 10163	−4.15	8.60	153.17
Dipk2a	divergent protein kinase domain 2A	−3.72	34.80	457.61
Nab1	Ngfi-A binding protein 1	−3.61	30.00	366.69
Nkap	NFKB activating protein	−3.56	19.30	227.98
Slc6a9	solute carrier family 6 (neurotransmitter transporter, glycine), member 9	−3.53	67.13	778.02
Agt	angiotensinogen (serpin peptidase inhibitor, clade A, member 8)	−3.51	33.15	378.72
Cpm	carboxypeptidase M	−3.47	45.96	509.16
Ermp1	endoplasmic reticulum metallopeptidase 1	−3.40	61.64	652.31
Gm15500	predicted pseudogene 15500	−3.40	54.43	573.54
Dazap2	DAZ associated protein 2	−3.36	59.16	608.47
Rtl8b	retrotransposon Gag like 8B	−3.35	33.48	340.80
Polm	polymerase (DNA directed), mu	−3.32	6.50	64.73
Gpx4	glutathione peroxidase 4	−3.31	39.34	391.42
Sos1	SOS Ras/Rac guanine nucleotide exchange factor 1	−3.31	73.41	727.62
Ctr9	CTR9 homolog, Paf1/RNA polymerase II complex component	−3.30	67.94	667.74
Tmem184c	transmembrane protein 184C	−3.29	47.58	464.32
Dzank1	double zinc ribbon and ankyrin repeat domains 1	−3.27	36.48	352.91
D430019H16Rik	RIKEN cDNA D430019H16 gene	−3.27	35.99	346.81
Klhl9	kelch-like 9	−3.24	57.35	543.23

2.4.3. Comparison between OPTO FEAR vs. OPTO NOT FEAR Groups

Differential expression analysis showed 3506 significant DEGs (2104 up; 1402 down, with reference level set on the OPTO NOT FEAR condition) (Figure 6A–C, Table 3). Despite the high number of DEGs, the ORA performed on the obtained DEGs resulted only in 3 significantly enriched GO terms (BP: 0; MF: 0; CC: 3, i.e.,: GO:0005887 integral component of plasma membrane, GO:0031226 intrinsic component of plasma membrane, and GO:0098590 plasma membrane region), indicating a broad modulation of genes without a significant enrichment for specific pathways. These large differences in transcriptome include crucial pathways linked to neuron morphogenesis and differentiation (e.g., GO:0031175 Neuron projection development, GO:0048667 Cell morphogenesis involved in neuron differentiation, GO:0007409 Axonogenesis, GO:0061564 Axon development, GO:0000904 Cell morphogenesis involved in differentiation) (Supplementary Results 1).

In parallel, 169 KEGG terms resulted significantly enriched. The top twenty over-represented KEGG terms were associated to neurodegeneration, metabolic regulation, regulation of actin cytoskeleton, and cancer (namely: mmu04144 Endocytosis, mmu05132 Salmonella infection, mmu05022 Pathways of neurodegeneration—multiple diseases, mmu04140 Autophagy—animal, mmu05014 Amyotrophic Lateral Sclerosis, mmu05012 Parkinson Disease, mmu05010 Alzheimer Disease, mmu04530 Tight junction, mmu04137 Mitophagy—animal, mmu04810 Regulation of actin cytoskeleton, mmu05020 Prion disease, mmu04360 Axon guidance, mmu04071 Sphingolipid signaling pathway, mmu05016 Huntington Disease, mmu04141 Protein processing in endoplasmic reticulum, mmu05017 Spinocerebellar ataxia, mmu05211 Renal cell carcinoma, mmu04928 Parathyroid hormone synthesis, secretion and action, mmu01212 Fatty acid metabolism, and mmu04714 Thermogenesis).

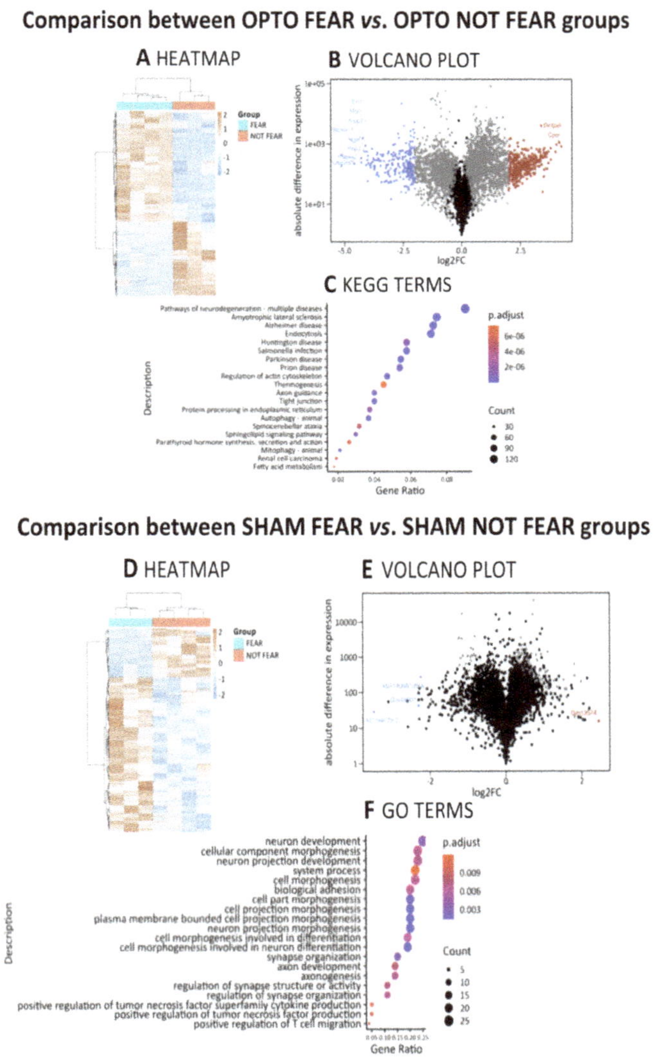

Figure 6. Differential Gene Expression profiling and functional enrichment analysis of RNA extracted by amygdala pyramidal neurons. Comparisons between OPTO FEAR vs. OPTO NOT FEAR groups (upper part) and SHAM FEAR vs. SHAM NOT FEAR groups (lower part). (**A**,**D**) Heatmaps showing gene expression values for the Differentially Expressed Genes (DEGs). (**B**,**E**) Volcano plots highlighting DEGs. The *x*-axis is the log2 fold change (log2FC) in normalized gene expression and the *y*-axis is for the $\log_{10}$ absolute value of the difference in expression between conditions. Each dot represents a gene. Grey dots are for DEGs, blue and red dots are for <−2 and > + 2 log2FC genes, respectively. The top ten genes with the highest absolute log2FC values are labeled. (**C**,**F**) Dot plots representing the top twenty enriched terms from over-representation analyses (ORA) in Gene Ontology (GO) and Kyoto Encyclopedia of Genes and Genomes (KEGG) databases. GO terms from different domains (Biological Processes, Cellular Component, and Molecular Function) were sorted by *q*-value before plotting them together.

Table 3. The top 20 DEGs identified by differential expression profiling from RNA extracted by amygdala pyramidal neurons Comparison between OPTO FEAR vs. OPTO NOT FEAR groups.

Gene Symbol	Gene Name	Fold Change (log2 Scale)	OPTO FEAR adj.mean	OPTO NOT FEAR adj.mean
Ube2ql1	ubiquitin-conjugating enzyme E2Q family-like 1	−5.13	5.54	194.27
Slc6a17	solute carrier family 6 (neurotransmitter transporter), member 17	−5.12	12.86	448.21
Papss2	3′-phosphoadenosine 5′-phosphosulfate synthase 2	−4.57	9.79	232.51
Hap1	huntingtin-associated protein 1	−4.29	13.22	258.18
Ngef	neuronal guanine nucleotide exchange factor	−4.28	18.41	358.22
Cpm	carboxypeptidase M	4.24	871.36	45.96
Thy1	thymus cell antigen 1, theta	−4.22	88.57	1649.90
Slc6a9	solute carrier family 6 (neurotransmitter transporter, glycine), member 9	4.16	1202.20	67.13
Arpp21	cyclic AMP-regulated phosphoprotein, 21	−4.15	33.80	599.42
Enc1	ectodermal-neural cortex 1	−4.10	54.26	928.57
Celf5	CUGBP, Elav-like family member 5	−4.08	12.58	212.32
Ermp1	endoplasmic reticulum metallopeptidase 1	4.04	1016.54	61.64
Ptprg	protein tyrosine phosphatase, receptor type, G	−4.04	36.77	603.55
Cxcl14	chemokine (C-X-C motif) ligand 14	−4.02	19.49	315.59
Chrna4	cholinergic receptor, nicotinic, alpha polypeptide 4	−4.01	3.93	63.10
Arsg	arylsulfatase G	4.01	580.93	36.15
Kcnq2	potassium voltage-gated channel, subfamily Q, member 2	−4.00	27.21	436.00
Sfxn1	sideroflexin 1	−3.99	15.50	246.04
Thrb	thyroid hormone receptor beta	−3.91	13.85	208.27
Lmo3	LIM domain only 3	−3.89	12.43	183.70

2.4.4. Comparison between SHAM FEAR vs. SHAM NOT FEAR Groups

In line with PCA highlighting poor segregation between SHAM FEAR vs. SHAM NOT FEAR groups, differential Expression analysis showed only 107 significant DEGs (81 up; 26 down, with reference level set on SHAM NOT FEAR condition) (Figure 6D,E, Table 4). Despite the small number of DEGs, the ORA performed on the obtained DEGs resulted in 129 enriched GO terms (BP: 67; CC: 45; MF: 17) and 3 enriched KEGG terms. The top twenty significant GO terms belonged to BP and CC and suggested the differential involvement of pathways related to synaptic plasticity, in particular related to excitatory synapse, and morphogenesis (Figure 6F) (namely: GO:0009986 Cell surface, GO:0048471 Perinuclear region of cytoplasm, GO:0030424 Axon, GO:0036477 Somato-dendritic compartment, GO:0030425 Dendrite, GO:0097447 Dendritic tree, GO:0043025 Neuronal cell body, GO:0098590 Plasma membrane region, GO:0048812 Neuron projection morphogenesis, GO:0048667 Cell morphogenesis involved in neuron differentiation, GO:0120039 Plasma membrane bounded cell projection morphogenesis, GO:0048858 Cell projection morphogenesis, GO:0032279 Asymmetric synapse, GO:0032990 Cell part morphogenesis, GO:0060076 Excitatory synapse, GO:0005887 Integral component of plasma membrane, GO:0098984 Neuron to neuron synapse, GO:0044456 Synapse part, GO:0097038 Perinuclear endoplasmic reticulum, and GO:0033267 Axon part). KEGG analysis showed modulation of the pathways associated to cAMP signaling (mmu04024), Prostate cancer (mmu05215), and Thyroid hormone signaling (mmu04919).

Table 4. The top 20 DEGs identified by differential expression profiling from RNA extracted by amygdala pyramidal neurons. Comparison between SHAM FEAR vs. SHAM NOT FEAR groups.

Gene Symbol	Gene Name	Fold Change (log2 Scale)	SHAM FEAR adj.mean	SHAM NOT FEAR adj.mean
Gm13574	predicted gene 13574	−3.48	2.78	30.95
Gm49601	predicted gene, 49601	2.45	17.99	3.30
4931406B18Rik	RIKEN cDNA 4931406B18 gene	−2.31	10.68	52.82
D430019H16Rik	RIKEN cDNA D430019H16 gene	−2.30	23.45	115.51
Sphkap	SPHK1 interactor, AKAP domain containing	−2.25	73.14	346.81
Slx4ip	SLX4 interacting protein	1.98	265.56	67.29
Smcr8	Smith-Magenis syndrome chromosome region, candidate 8 homolog (human)	−1.97	21.56	84.55
Olfr1233	olfactory receptor 1233	1.81	403.22	115.33
Hcn2	hyperpolarization-activated, cyclic nucleotide-gated K+ 2	1.80	27.04	7.74
Sumf1	sulfatase modifying factor 1	1.79	859.93	248.23
Cd99l2	CD99 antigen-like 2	1.60	106.02	34.94
Kif6	kinesin family member 6	1.59	182.58	60.68
Plaa	phospholipase A2, activating protein	1.55	185.44	63.29
Clstn3	calsyntenin 3	1.51	344.56	120.72
Adcyap1r1	adenylate cyclase activating polypeptide 1 receptor 1	1.35	235.84	92.81
Ino80	INO80 complex subunit	−1.28	326.26	789.84
Cpe	carboxypeptidase E	1.27	359.63	148.70
Uty	ubiquitously transcribed tetratricopeptide repeat gene, Y chromosome	−1.23	2559.64	6021.69
Vmn2r114	vomeronasal 2, receptor 114	−1.21	448.82	1040.79
Gm13574	predicted gene 13,574	−1.19	720.62	1641.20

2.4.5. DEGs Involved in Learning/Memory and Fear Response in the Comparison between OPTO FEAR vs. SHAM FEAR Groups

Despite the relative pathways were not identified as significantly enriched, we wondered whether genes involved in learning/memory and fear response were significantly modulated in the case of impaired fear extinction caused by the optogenetic stimulation on PrL pyramidal neurons.

By looking at genes with GO annotations related to BP associated to Learning/memory and Fear response, we identified several DEGs between OPTO FEAR vs. SHAM FEAR groups (with reference level set on the SHAM FEAR condition; genes related to Learning/memory: Learning or Memory, 64 over 103 from the gene universe, Learning, 41 over 63, Memory, 25 over 45, Associative Learning, 18 over 23, Long-term memory, 9 over 16, and Visual Learning, 18 over 23; genes related to Fear response: Fear Response and Behavioral Fear Response, 13 over 19) (Figure 7, Supplementary Results 2). This set contains 3 of the top twenty DEGs, all down regulated (with the log2 Fold Change (FC) ranging from −4.24 to −3.23): *Brinp1* (associated to Fear response and Memory), *Bcl2* (associated to Fear response), and *Agt* (associated to Learning) in the comparison between OPTO FEAR and SHAM FEAR groups. We also identified several genes well known to be associated with fear extinction: *ApoE*, *Cacna1* (*Cav1.2*), *Creb1*, *App*, and *Arc*. Most of them appeared to be down regulated in the comparison between OPTO FEAR vs. SHAM FEAR groups, but to a lesser extent (*Apoe* log2FC = −0.33; *Cacna1c* (*Cav1.2*) log2FC = −0.94; *Creb1* log2FC = −0.96), while we noted a mild but significant increase in the expression of *App* (log2FC = +0.41) and *Arc* (log2FC = 0.37) genes. Despite non-being a gene with GO annotations related to BP associated to learning/memory and fear response, we noted that *Homer1*, another gene recently associated to fear memory extinction, appeared significantly down regulated (log2FC = −0.87), in the comparison between OPTO FEAR vs. SHAM FEAR groups. Finally, we also noted the strong down regulation of *Thy1* (log2FC = −2.77) gene.

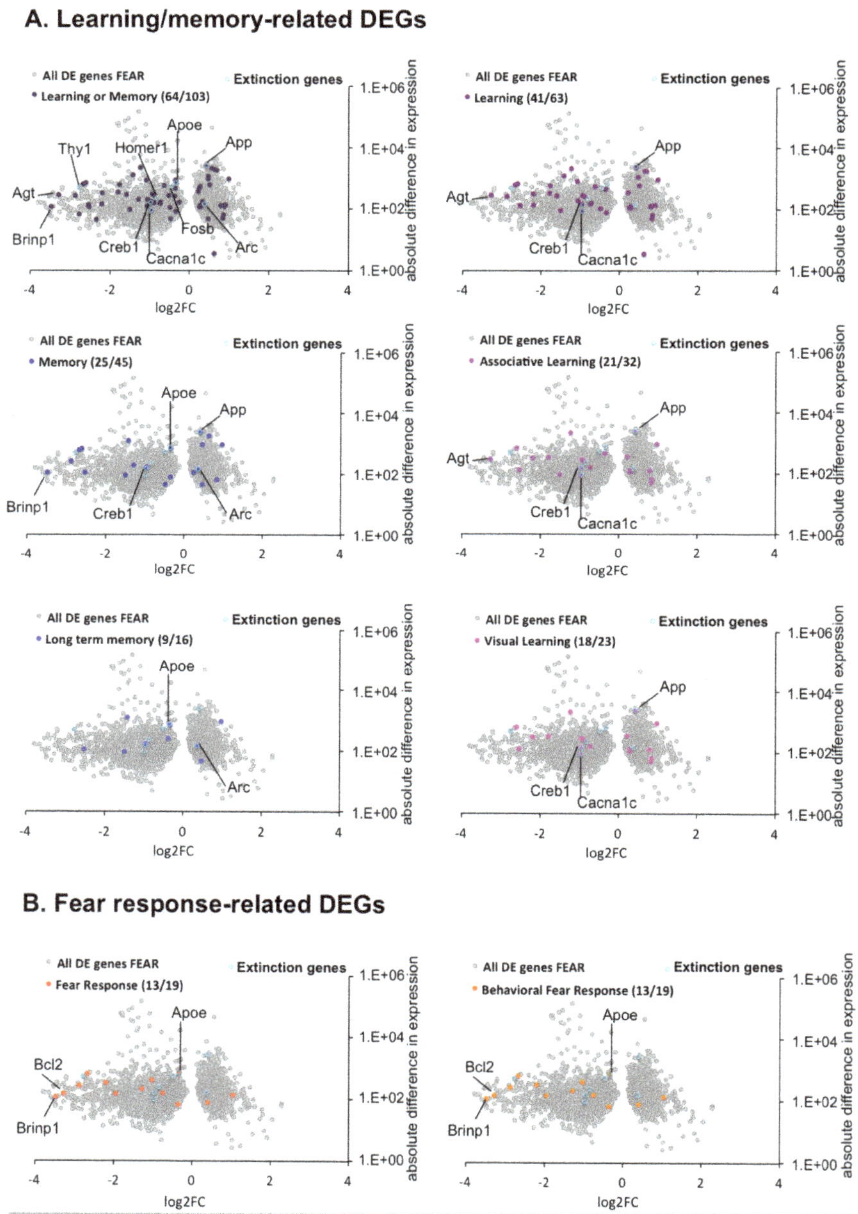

Figure 7. Differentially Expressed Genes (DEGs) with Gene Ontology annotations associated to Learning/memory and Fear response, between OPTO FEAR vs. SHAM FEAR groups. DEGs for each indicated Biological Process (BP) are highlighted as colored dots on the volcano plot representing the total DEGs between OPTO FEAR vs. SHAM FEAR groups, indicated by grey dots. Numbers into brackets indicate the number of DEGs related to each indicated BP over the gene universe. Light blue dots indicate genes described in literature as associated to fear extinction. When DEGs are part of the top twenty DEGs list and/or part of the considered BP the gene name is indicated.

3. Discussion

Starting from the assumption that fear learning and extinction are adaptive processes caused by molecular changes in the amygdala-mPFC neural circuit, in the present research adult mice were submitted to CFC receiving (or not) the aversive US as well as the optogenetic (or sham) stimulations to maintain the activation of the PrL pyramidal neurons and impair fear extinction. At the end of behavioral testing, cellular excitability and excitatory neurotransmission, as well as spinogenesis, were evaluated in PrL pyramidal neurons. In parallel, amygdala pyramidal neurons were sorted to perform cell-type specific RNA sequencing, able to reveal the specific transcriptional signature associated to impaired fear extinction.

All fear-conditioned animals (OPTO FEAR, SHAM FEAR, and No-EX groups) showed consolidation of aversive memory. However, while SHAM FEAR group progressively extinguished fear memory as expected, impaired fear extinction was observed in OPTO FEAR group (Figure 1 and Supplementary Figure S1).

Once reactivated through the re-exposure to the context, the memory of fear conditioning becomes susceptible to be modified [39]. In fact, the re-exposure to the context elicits a series of processes (retrieval, reconsolidation, and extinction), which are governed by "boundary conditions" [39,40]. While extinction refers to the decrement of fear CR that occurs with repeated presentations of CS (no more reinforced with US), and reflects the formation of a new inhibitory memory, reconsolidation is a process whereby previously consolidated memories can be reactivated and again made sensitive to mutate. Notably, the manipulations during or shortly after the period of memory reactivation can influence reconsolidation process. Thus, the present optogenetic stimulations of the PrL pyramidal neurons applied from 3rd min onwards of each day of Extinction phase allowed reducing extinction learning. Although the reconsolidation and extinction are distinct (even if interrelated) processes, distinguishing the role of PrL in the transition between them was out of the scope of the present work.

Interestingly, in fear-conditioned mice the optogenetic stimulation induced a robust increase in evoked firing activity, number of action potentials, EPSC frequency (Figure 2), as well as number and density of dendritic spines in the apical arborizations of PrL pyramidal neurons (Figure 3). Conversely, in not fear-conditioned mice the optogenetic stimulation did not produce any significant change in cellular excitability, evoked firing, EPSC frequency, and spinogenesis (Figures 2 and 3). To better control the effects of learned but not yet extinguished fear, electrophysiological and morphological data of PrL pyramidal neurons from No-EX group were compared with those of the other groups. The findings indicate that No-EX group that had consolidated fear memory, exhibited an excitability of PrL neurons higher than that shown by PrL neurons of SHAM FEAR mice and similar to the excitability shown by OPTO FEAR mice. Furthermore, No-EX mice displayed highest number and density of dendritic spines in apical arborizations of PrL pyramidal neurons. The results demonstrated that while the fear conditioning increased the excitability and morphological properties of PrL pyramidal neurons, the fear extinction decreased them. Optogenetic activation of PrL pyramidal neurons counteracted such reductions, thus leading to impaired extinction.

It has been proposed that the boost of bidirectional amygdala-PrL synaptic transmission leads to changes in the representation of learned CS-US association in mPFC neurons [19,20]. Optogenetic experiments have confirmed that modified synaptic transmission in the amygdala-PrL network during fear learning interferes with long-term fear consolidation [14]. By using ex vivo electrophysiological recordings, combined with optogenetic techniques, it has been demonstrated that fear extinction decreases the efficacy of excitatory transmission from mPFC to the amygdala [21]. In parallel, the amygdala neurons projecting to mPFC exhibit cell-type-specific plasticity during fear extinction [19].

Cell-type-specific transcriptome analysis represents cutting-edge tool to reveal targets useful for understanding and treating fear-related disorders. Differential gene expression and co-expression network analyses identified diverse networks activated or inhibited by fear learning vs. extinction, and upstream regulator analysis and viral vector-manipulations demonstrated that fear extinction is associated with reduced cAMP/Ca^{2+} responsive element binding (CREB) expression [41].

Our study identifies cell-type amygdala pyramidal neuron-specific gene networks (by means of the RNA-sequencing) disclosing pathways associated to adaptive or maladaptive fear extinction and opening innovative possibilities to understand deeper the underlying mechanisms of fear process and its impairment. Here we discuss some of the genes and pathways that result as more relevant.

Optogenetic stimulation of the PrL pyramidal neurons was associated with strong modifications of the amygdala pyramidal neuron transcriptome (Figure 5A,B, Table 1). Bioinformatic analyses revealed that among the top twenty DEGs (resulted down regulated in OPTO FEAR mice when compared with SHAM FEAR mice), there were *Gabra4* (Gamma-aminobutyric acid—GABA—A receptor, subunit alpha 4) and *Kcnh3* (potassium voltage-gated channel, subfamily H, member 3), which are associated with synaptic transmission. Namely, shunting inhibition via GABAA receptors reduces activation of N-methyl-D-aspartate receptors, and impairs long-term potentiation [42], as well as voltage-gated potassium channels control cellular excitability by regulating a variety of neuronal properties, such as inter-spike membrane potential, action potential waveform, and firing frequency [43]. Within the brain, *Kcnh3* is expressed in the cerebral cortex, amygdala, hippocampus, and striatal regions, with specific expression in pyramidal neurons [44].

Other DEGs that we identified as associated with impaired extinction were linked to inflammation processes, such as *Csmd1* [45] and *Bcl2* [46]. Notably, overexpression of *Bcl2* blocks the apoptotic death of the pro-B-lymphocyte cells and neurons [47,48]. Furthermore, in the same comparison OPTO FEAR vs. SHAM FEAR, the top twenty GO terms associated with impaired extinction highlighted the differential involvement of pathways associated with neuronal plasticity and glutamatergic synaptic signaling, resulting in modulation of overall pre- and post-synaptic organization. In parallel, the top twenty KEGG terms were associated with processes related to several neurodegenerative diseases, metabolic pathways, production of lipids and proteins, and thyroid hormone and neurotrophin signaling. Remarkably, fear-associated enrichments have been related with dendritic and post-synaptic processes, while extinction-associated enrichments have been related with cellular metabolism and proliferation [41].

By looking at genes related to Learning/memory or Fear response processes, several DEGs were scored in the comparison between OPTO FEAR vs. SHAM FEAR groups (Figure 7). Among these, 3 DEGs (i.e., *Brinp1*, associated to Fear response and Memory; *Bcl2*, associated to Fear response; *Agt*, associated to Learning) were part of the most impacted genes (top twenty DEGs), and resulted strongly down regulated in OPTO FEAR group. Several genes previously associated with fear extinction in the literature [41,49–54] were present in OPTO FEAR group, as down regulated (*Apoe, Cacna1c, Creb1, Homer1*) or up regulated (*App, Arc*). Although not always the afore-mentioned genes have been associated specifically to amygdala activity, those down regulated have been positively associated with extinction, while those up regulated have been positively associated with fear memory retention.

The photo-stimulation of PrL pyramidal neurons in the presence or absence of fear was differently associated with amygdala pyramidal neuron transcriptome (Figure 6A,B; Table 3). Among the top twenty DEGs, some relevant genes resulted down regulated in OPTO FEAR mice when compared with OPTO NOT FEAR mice. Among these genes, there was *Ngef* (Neuronal guanine nucleotide exchange factor, also known as Ephexin1), an ephrin (Eph) receptor-interacting exchange protein that promotes EphA4 binding and leads to cell morphology changes [55] and reduces spine density [56]. Accordingly, the OPTO FEAR mice (in which *Ngef* is down regulated) exhibited increased number and density of dendritic spines in the apical arborizations. Furthermore, the binding of Eph with its receptors constitutes a molecular link between Eph receptors and actin cytoskeleton and modulates pre-synaptic calcium channel activity [57]. To explore the role of another member of Eph family (EphB2) in memory formation and enhancement, Alapin and colleagues [58] used a photo-activatable EphB2 to activate EphB2 forward signaling in pyramidal neurons of the lateral amygdala. Such a photo-activation during fear learning (but not afterwards) enhances the long-term (but not the short-term) fear response. Accordingly, long-term fear memory is impaired in mice lacking EphB2 forward signaling [58].

Among the top twenty DEGs, another gene associated with actin-binding protein was *Enc1* (ectodermal-neural cortex 1) that resulted down regulated in OPTO FEAR mice. Kim and colleagues [59] showed that expression of *Enc1* induces neuronal process formation, whereas antisense treatment inhibits neurite development. Similarly, *Thy-1* (thymus cell antigen 1) involved in cell–cell interactions, *Ptprg* (protein tyrosine phosphatase, receptor type, G) implicated in the control of cellular proliferation, and *Kcnq2* (potassium voltage-gated channel, subfamily Q, member 2) were down regulated in OPTO FEAR mice. It has been demonstrated that Kcnq2-related proteins are localized on pyramidal neurons, suggesting their pre-synaptic role in action potential propagation and neurotransmitter release [60]. Even *Chrna4* (cholinergic receptor, nicotinic, alpha polypeptide 4) linked to the superfamily of ligand-gated ion channels that mediate fast signal transmission, was down regulated in OPTO FEAR mice. Of note, as an ancillary remark on the extinction process, mice optogenetically stimulated on amygdala cholinergic input during the initial fear learning are more resistant to extinction learning than controls, supporting the role of cholinergic modulation of amygdala circuits in learning and retention of fear memories [61].

Deficit in exploratory behavior and cognitive impairment in learning tasks as well as neuronal death are reported in mutant mice for *Arsg* gene (arylsulfatase G) [62]. Consistently, our analysis shows *Arsg* down regulation in OPTO FEAR mice. Despite the high number (3506) of DEGs, while only 3 GO terms (components of plasma membrane) were enriched, while 169 KEGG terms resulted enriched. The top twenty over-represented KEGG terms were associated with neurodegeneration, metabolism, actin cytoskeleton regulation, and parathyroid hormone regulation.

Optogenetic stimulation of the PrL pyramidal neurons per se, without fear experience, was also associated with a strong down regulation of amygdala gene expression (Figure 5E,F, Table 2). Among DEGs (down regulated in OPTO NOT FEAR mice when compared with SHAM NOT FEAR mice), some genes were implicated in cell proliferation, synaptic activation, and long-term potentiation (such as *Nab1* and *Sos1*), others in inflammation processes (such as *Gpx4* and *Nkap*), with roles in transcriptional repression and RNA splicing and processing (*Nkap*) [63]. Furthermore, in the same comparison the gene universe did not allow enriching many GO terms because of sampling bias correction [38]. Again, the top twenty over-represented KEGG terms were related to neurodegenerative diseases, metabolic pathways, and biosynthesis of protein and unsaturated fatty acids.

Finally, we observed that fear conditioning per se (in the presence of an adaptive extinction), without optogenetic stimulation, was not greatly associated with gene expression in amygdala pyramidal neurons (Figure 6D,E, Table 4), in line with PCA highlighting poor segregation between SHAM FEAR vs. SHAM NOT FEAR groups. This result is in line with those reported by McCullough and colleagues [41] who described a weak separation between the transcriptomes of fear- and not fear-conditioned animals. The authors discussed their findings as result of stress-induced translational changes due to the handling of animals and CS exposure. This hypothesis tested in a separate cohort of mice was confirmed, demonstrating that stress-related genes were similarly regulated in both groups. Thus, it is likely that the signature of associative fear learning was obscured by generalized stress-related changes. Furthermore, while SHAM FEAR group learned the association CS-US, it is not possible to exclude that the SHAM NOT FEAR group encoded the CS (never paired with US). Such a coding might influence or even impede the subsequent acquisition of conditioned associations between the CS and US, as occurs in the case of latent inhibition process. Specifically, latent inhibition refers to the reduced ability to learn the relevance of a stimulus that is paired with an aversive (or positive) condition through classic conditioning if there has been a previous exposure to that stimulus in a neutral context [64,65]. However, the limited segregation between SHAM FEAR and SHAM NOT FEAR gene profiles might also be caused by the time-point of the transcriptomic analyses. In fact, gene profiling of the SHAM FEAR group was performed once the animals had extinguished the fear memories, and fear response was over, although the fear engram was likely still stored in the amygdala-hippocampus-mPFC circuit [66].

Actually, among DEGs present in the comparison between SHAM FEAR vs. SHAM NOT FEAR groups, some genes (such as *Slx4ip* and *Kif6*) were up regulated in SHAM FEAR group and are implicated in DNA/RNA regulation, maintenance, and repair [67,68]. Conversely, the *Ino80* gene resulted down regulated in the same comparison. Notably, it has been reported that the deletion of *Ino80* results in defective cellular proliferation and premature entry into cellular senescence, due to activation of the DNA damage response [69]. Furthermore, other genes associated to metabolism (such as the down regulated *Smcr8* and the up regulated *Clstn3*), neurotransmission (such as the up regulated *Hcn2*), and inflammation (such as the up regulated *Sumf1* and *Plaa*) were found differentially expressed in SHAM FEAR group when compared with SHAM NOT FEAR group. Interestingly, also *Adcyap1r1* (adenylate cyclase activating polypeptide 1 receptor 1) resulted up regulated in SHAM FEAR group. The pituitary adenylate cyclase-activating polypeptide is a hormone that stimulates the secretion of growth hormone, adrenocorticotrophic hormone, catecholamines, and insulin, by its interaction with specific receptors. Interestingly, the methylation of *Adcyap1r1* in peripheral blood has been associated with PTSD, and *Adcyap1r1* mRNA is induced by fear learning [70]. Since DNA methylation of regulatory elements usually acts to repress gene transcription, the findings by Ressler and colleagues [70] indicating that methylation of *Adcyap1r1* is associated with impaired fear extinction (as typically occurring in PTSD) together with the present ones indicating that the up regulation of this gene is associated with efficient fear extinction converge in highlighting that the system pituitary adenylate cyclase-activating polypeptide with its receptors might be an important mediator of abnormal fear responses following trauma [70].

In the same comparison (SHAM FEAR vs. SHAM NOT FEAR), the top twenty GO terms were related to morphogenesis and synaptic plasticity, with the important specificity of the excitatory synapse. KEGG analysis showed modulation of the pathways associated to cAMP signaling, cancer, and thyroid hormone signaling.

Overall, our results show that the optogenetic activation of PrL pyramidal neurons in fear-conditioned mice did not allow the disengagement of the fear matrix, and induced fear extinction deficits mirrored by the increase of cellular excitability, excitatory neurotransmission, and spinogenesis of PrL pyramidal neurons, and by strong modifications of the transcriptome of amygdala pyramidal neurons.

The optogenetic manipulation determined a photo-activation of the pyramidal neurons of layer 5 of PrL cortex projecting to distinct cortical (including IL cortex) and subcortical (including amygdala) targets and allowed deepening the direct (PrL cortex—amygdala) or indirect (PrLcortex—IL cortex—amygdala) crosstalk among structures involved in fear extinction. It is certainly true that the optogenetic stimulation of PrL pyramidal neurons modulated the circuit from mPFC to amygdala, but the effects of PrL optogenetic stimulation were markedly displayed in fear-conditioned animals. This latter observation suggests that the amygdala is the first hub to trigger the fear-associative memory and communicate the fear association to the mPFC.

In optogenetically stimulated fear-conditioned mice the difference in gene expression profiles was characterized by down regulation of genes associated with the synaptic transmission, specifically the inhibitory GABAergic signaling, as well as by differential involvement of pathways associated with neuronal plasticity and glutamatergic signaling.

The present findings indicate that impaired extinction is featured by specific changes of transcriptome that validate previous findings [41], provide targets for future translational research into cell type-specific control of fear extinction, and emphasize the key role of pyramidal neurons belonging to fear matrix. This type of comprehensive cell-type specific analysis may produce an important array of targets potentially useful for diagnosis, treatment, and prevention of fear-related disorders.

Limitations

For sake of clarity, we would like to report some limitations of the present study that will require future and aimed investigations to be dissolved. First of all, the comparison between amygdala pyramidal neuron transcriptomes of naïve, conditioned (only trained without extinction), and optogenetically stimulated fear-conditioned mice would be an interesting deepening of the present study to definitively clarify what is the role of PrL cortex activation during extinction. It is reasonably conceivable that the amygdala gene expression patterns after fear training (without any optogenetic manipulation and before extinction) are similar to those of mice that underwent PrL optogenetic stimulation and thus showed impaired fear extinction. A strong support (even if not the final demonstration) to this interesting hypothesis derives from the evidence that both cellular excitability (EPSC, evoked firing, and rheobase) and morphological (spine number and density) changes are similar in fear-conditioned (only trained without extinction) mice and in PrL optogenetically stimulated mice showing impaired extinction.

Another point to be considered is that in selecting amygdala pyramidal neurons for transcriptome analyses the used methodology, although innovative and fruitful in its results, does not allow distinguishing whether the sorted neurons are directly or indirectly connected with the PrL cortex. Thus, the reported connections between transcriptome changes in amygdala pyramidal neurons and PrL optogenetic stimulation have to be considered associative and not causal.

As final note, we are aware that although the focused laser beam is very well-suited for spatially-localized optogenetic activation of PrL pyramidal neurons, the scarce out-of-focus light could stimulate other photo-activable neurons very close to PrL pyramidal neurons, reducing thus the spatial resolution. However, it is important to notice that the optogenetic stimulation of PrL cortex elicited the typical behavior derived from PrL activation (as the impaired extinction), and no typical behavior derived from IL activation (as the potentiated extinction). Furthermore, it has to be underlined that OPTO NOT FEAR mice showed no change in behavioral, electrophysiological and structural parameters in comparison to SHAM NOT FEAR mice, suggesting that the PrL photo-activation did not elicit per se any unnatural effect.

4. Materials and Methods

4.1. Subjects

Male adult (2.5 month-old) B6.Cg-Tg(Thy1-COP4/EYFP)18Gfng/J (Thy1-COP4) (Jackson Laboratories, Bar Harbor, ME, USA) mice were used in the present research. These transgenic mice express the light-activated ion channel, Channelrhodopsin-2 (ChR2), fused to Yellow Fluorescent Protein (YFP) under the control of the mouse thymus cell antigen 1 (*Thy1*) promoter. The expression of the transgenic ChR2-YFP fusion protein is detected in pyramidal cortical layer 5 neurons, in CA1 and CA3 pyramidal neurons of the hippocampus, amygdala pyramidal neurons, cerebellar mossy fibers, neurons in the thalamus, midbrain and brainstem, and olfactory bulb mitral cells. Transgene-expressing neurons are morphologically and physiologically comparable to non-mutant neurons. The ChR2 functions as a blue light-driven cation channel that depolarizes the cell and elicits action potentials. Thus, illuminating ChR2-expressing neurons with blue light (~470 nm) leads to rapid and reversible photo-stimulation evoking action potential firing/neural activity.

The animals were group-housed (4 mice/cage) with food (Mucedola, Milan, Italy) and water ad libitum, and kept under a 12-h light/dark cycle with the light on at 07:00 h, controlled temperature (22–23 °C) and constant humidity (60 $\pm$ 5%). All experiments took place during the light phase. All efforts were made to minimize animal suffering and to reduce their number, in accordance with the European Directive (Directive 2010/63/EU). The animals assigned to the same experimental group were never siblings.

4.2. Experimental Procedure

Thy1-COP4 mice were unilaterally implanted with a guide cannula on PrL sub-region of right mPFC and then were (or not) fear-conditioned by using the CFC paradigm with the extinction protocol. During the extinction phase of CFC, the mice received optogenetic (OPTO FEAR group, $n = 10$; OPTO NOT FEAR group, $n = 10$) or sham (SHAM FEAR group, $n = 10$; SHAM NOT FEAR group, $n = 10$) stimulations. At the end of the behavioral testing, the animals were sacrificed. Cellular excitability and spine number and density of Thy1-COP4-expressing PrL pyramidal neurons were analyzed. Furthermore, in the same samples in which PrL pyramidal neurons were electrophysiologically recorded, Thy1-COP4-expressing BLA pyramidal neurons were sorted to purify individual cell-specific RNA for transcriptomic analyses. Transcriptome-wide analyses were carried out by RNA-sequencing, after total RNA ultra-low input library preparation and sequencing in PE75 mode on Illumina platform.

To control for the effects of learned but not yet extinguished fear on cellular excitability and spinogenesis of Thy1-COP4-expressing PrL pyramidal neurons, other Thy1-COP4 mice (No-EX group, $n = 6$) were unilaterally implanted with a guide cannula on PrL sub-region of right mPFC and then fear-conditioned by using the CFC paradigm without the extinction protocol. In fact, the day after the Conditioning phase (day 2) the animals were sacrificed and cellular excitability and spine number and density of PrL pyramidal neurons were analyzed.

4.3. Stereotaxic Surgery and Fiber Optic Implantations

All mice were anesthetized by using Zoletil 100 (tiletamine HCl 50 mg/mL + zolazepam HCl 50 mg/mL; Virbac, Milan, Italy) and Rompun 20 (xylazine 20 mg/mL; Bayer S.p.A, Leverkusen, Germany) dissolved in a volume of saline of 4.1 mg/mL and 1.6 mg/mL, respectively and intraperitoneally injected in a volume of 7.3 mL/kg. Mice were mounted onto a stereotaxic frame (David Kopf Instruments, Tujunga, CA, USA) equipped with a mouse adapter and unilaterally implanted with optic fiber (ThorLabs, Newton, NJ, USA) above the PrL part of the right mPFC (AP: +1.8 mm, ML: +0.25 mm, DV: −2.00 mm). The coordinates from bregma were measured according to the atlas of Franklin and Paxinos (1997) and Mouse Brain Atlases (The Mouse Brain Library; www.nervenet.org). Ferrule-terminated implanted optical fibers were secured to the skull using dental acrylic.

Mice were allowed to recover from surgery for 1 week before behavioral testing. During the recovery period, they were habituated to handling and connection of the optic fiber with the optogenetic ferrule. Locations of implanted optical fibers were validated using histology in all experimental mice.

4.4. CFC and In Vivo Optogenetic Stimulations of the PrL Pyramidal Neurons

As previously described [71–73], the CFC was carried out in a soundproof conditioning chamber (50 cm long, 24.5 cm wide, 26.5 cm high) (Ugo Basile, Varese, Italy) made of gray Perspex with a metal grid floor. A video camera placed above the conditioning chamber allowed observing animal behavior. Before the behavioral testing, the mice were handled to connect the optic fiber with the optogenetic ferrule.

As depicted in Figure 1A, on day 1 (Conditioning phase), each mouse was allowed to explore the conditioning chamber for the first 3 min (Baseline). Afterward, only a sample ($n = 26$) received three foot-shocks (0.5 mA, 2.0 s, 1 min inter-shock interval), representing the US. The fear-conditioned animals ($n = 26$) were removed from the conditioning chamber after 1 min from the third foot-shock, while the not fear-conditioned animals ($n = 20$) were removed after 6 min from the insertion in the chamber, to return to their home cages. The employed foot-shock parameters evoked signs of discomfort as freezing, flinching, jumping, and vocalizing.

According to previous reports [71–73] on days 2, 3, 4, 7, and 14 (Extinction phase), a sample of fear-conditioned ($n = 20$) and all not fear-conditioned ($n = 20$) mice were placed again in the conditioning chamber for 6 min (Figure 1A). During the Extinction phase, no shock was delivered. From 3rd min onwards of each day of Extinction phase the mice received three optogenetic or sham stimulations on PrL pyramidal neurons of mPFC, by connecting the optic fiber to a light power source (473 nm; pE2, CoolLED, Andover, UK). Light stimulation parameters were 2s, 20 Hz, 15 ms pulses, 1 pulse every minute, density 14.32–15.91 mW/mm^2 [74]. No light was delivered on sham-stimulated mice. Notably, in vivo optogenetic manipulation of PrL pyramidal neurons was delivered in fear-conditioned mice to maintain fear memories.

The remaining fear-conditioned mice (n = 6) on days 2 were placed again in the conditioning chamber for 6 min (without receiving any optogenetic stimulation) and then sacrificed.

Freezing was recorded by an experimenter blind to the group the animal belonged to and freezing times during the first 3 min for Conditioning (Baseline) and Extinction phases as well as during the entire 6 min of day 14 of Extinction phase were compared among groups.

4.5. Slice Preparation and Electrophysiological Recordings of PrL Pyramidal Neurons

A sample of mice was anesthetized with an overdose of halothane (Sigma-Aldrich, St. Loui, MO, USA) and decapitated on day 2 (n = 3) or day 14 of Extinction phase (n = 5/group). Once removed, the brain was attached with cyanoacrylate glue to a tray and then sectioned into 275 µM-thick coronal slices by means of a vibratome (Leica VT1200s, Wetzlar, Germany). During slicing, brain was maintained in ice-cold artificial cerebrospinal fluid solution (ACSF) containing (in mM): NaCl (126), $NaHCO_3$ (26), KCl (2.5), NaH_2PO_4 (1.25), $MgSO_4$ (2), $CaCl_2$ (2) and glucose (10), gassed with 95% O_2–5% CO_2 (pH 7.4, 300 mOsm). Slices were maintained in the same solution at room temperature (~22 °C) for at least 30 min. Then, a single slice was transferred to a recording chamber mounted on upright infrared microscopy (BX51WI Olympus, Tokyo, Japan) and continuously perfused with oxygenated ACSF (30 °C, 2.5 mL/min) for electrophysiological recordings. Cells were visualized with a 40x water-immersion objective (LumpPlanFI, Olympus, Tokyo, Japan) and by an infrared camera EM-CCD camera (ImagEm, Hamamatsu, Hamamatsu City, Japan). Patch-clamp recordings were made by borosilicate glass pipette (3–5 MΩ) pulled with a micropipette puller (P97, Sutter Instruments, Novato, CA, USA). For current-clamp experiments, the pipette was filled with a solution containing (in mM): $KMeSO_4$ (140), KCL (10), HEPES (10), Mg_2ATP (2) and Na_3GTP (0.4) (pH adjusted to 7.25 with KOH). For voltage-clamp experiments, the intracellular solution contained (in mM): $CsMeSO_3$ (140), EGTA (1), $CsCl_2$, (5.5), $CaCl_2$ (0.1), HEPES (1), $MgCl_2$ (2) Mg-ATP (2) (pH adjusted to 7.3, 290 mOsm). Electrophysiological recordings were acquired by a Multiclamp 700b amplifier, Digidata 1550A, and pClamp 10.4 software (Molecular Devices, San Jos, CA, USA). Signals were digitized at 10 or 20 kHz and filtered at 2 kHz with a low-pass Bessel filter. For voltage-clamp experiments, series resistance was monitored by repeated 5 mV steps. Cells showing an increase of over 20% in series resistance were discarded from statistical analysis. Spontaneous synaptic events were analyzed off-line by using the Minianalysis Program (Synaptosoft Inc., Decatur, GA, USA) and ClampFit 10.2 (Molecular Devices).

4.6. Spine Counting of PrL Pyramidal Neurons

The remaining sample of mice was anesthetized with an overdose of Zoletil (800 mg/kg; Virbac, Milan, Italy) + Rompun (200 mg/kg; Bayer S.p.A, Leverkusen, Germany) dissolved in a volume of saline of 4.1 mg/mL and 1.6 mg/mL respectively, and intraperitoneally injected on day 2 (n = 3) or day 14 of Extinction phase (n = 5/group). The animals were decapitated, the brains were rapidly removed, fixed in 4% paraformaldehyde for 24 h, and then cryoprotected in 30% sucrose solution. The brains were cut on a freezing microtome into 50 µm-thick coronal sections. Sections were collected at the level of PrL region of mPFC (AP: from 2.68 mm to 1.80 mm from bregma) [75] and then mounted onto slides, dehydrated, and coverslipped using Fluoromount (Sigma-Aldrich).

Dendritic spine counts were performed using an optical microscope (Axio Imager M2, Zeiss, Oberkochen, Germany) equipped with a motorized stage and a camera connected to software Neurolucida 2020.1.2 (MicroBright-Field, Williston, VT, USA). Dendrites were traced with spines and images then exported to Neurolucida™ Explorer 2019.2.1 (MicroBright-Field) for spine quantitation.

Due to the difficulty of unequivocally distinguishing filopodia from long thin spines, spine counts included all types of dendritic protrusions ≤4 μm on apical dendrites regardless of their shape or actual function.

Ten dendritic segments (length 20–25 μm) were obtained for each subject of the entire sample. Spine density was calculated by measuring the length of the dendrite segment and counting the number of spines along the segment.

4.7. Amygdala Pyramidal Neuron-Specific RNA Sequencing

4.7.1. Dissociation of Amygdala Tissue for Fluorescence-Activated Cell Sorting (FACS)

On day 14 of Extinction phase, the brains from which PrL pyramidal neurons of mPFC were electrophysiologically recorded were cut to take bilateral amygdala 1-mm punches. Manual and enzymatic dissociations were performed using the Neural Tissue Dissociation Kit (P) (Miltenyi Biotec, Bergisch Gladbach, Germany) with some modifications. Each solution was kept on ice to minimize RNA degradation. Pipette tips were pre-coated in a 0.2 μM filtered 1× PBS-0.5% BSA solution (DPBS without Mg^{2+} and Ca^{2+}, Gibco by Life Technologies, Grand Island, NY, USA; BSA Fraction V (pH 7.0), PanReac AppliChem GmbH, Darmstadt, Germany). Briefly, the amygdala punches were placed on a 35 mm diameter Petri dish, cut into small pieces using a scalpel, and 1 mL of cold Hanks' Balanced Salt Solution without Mg^{2+} and Ca^{2+} (HBBS w/o) (Sigma-Aldrich, St. Louis, MO, USA) was added. The tissue was transferred into a 1.5 mL protein LoBind tube. Additional 1 mL HBBS w/o was used to rinse the dish and added to the 1.5 mL tube. Tissue was centrifuged at 300× *g* for 2 min at room temperature, and the supernatant was carefully aspirated. Then, 975 μL of pre-heated enzyme mix 1 (enzyme P 25 μL, buffer × 950 μL) was added to the tissue, and the 1.5 mL tube was incubated for 15 min at 37 °C under slow, continuous rotation using the MACSmix Tube Rotator (Miltenyi Biotec, Bergisch Gladbach, Germany). Then, 15 μL enzyme mix 2 (enzyme A 5 μL, buffer Y 10 μL) was added to the sample. The sample was gently inverted to mix and mechanically dissociated using the wide-tipped fire-polished Pasteur pipette by pipetting up and down 10 times slowly, followed by a further incubation in the rotator for 10 min at 37 °C under slow rotation. The second round of mechanical dissociation was performed using serially fire-polished filtered-glass Pasteur pipettes with gradual diameter diminution, and pipetting slowly up and down 10 times with each pipette, or as long as until tissue pieces were not yet observable. The sample was again incubated at 37 °C for 10 min using rotator under slow rotation, before being strained through MACS Smart Strainer (70 μm) (Miltenyi Biotec, Bergisch Gladbach, Germany), placed on a 15 mL tube, pre-coated with 0.2 μM filtered 1× PBS-0.5% BSA, adding 8 mL of HBBS with Mg^{2+} and Ca^{2+}. Then, the cell sample was centrifuged at 300× *g* for 10 min at room temperature and the supernatant was completely aspirated and collected into a new 15 mL tube, and centrifuged again at 300× *g* for 10 min at room temperature. The supernatant was again completely aspirated. The pellets obtained from these two centrifugations were pooled into a 1× PBS-0.5% BSA pre-coated SNAP-cap tube containing 1 mL of PBS. Finally, 20U Superase-Inhibitor (Ambion, Invitrogen, ThermoFisher Scientific, Walthem, MA, USA) was added and samples were stored on ice up to sorting.

4.7.2. Cell Sorting and Isolation of Purified Pyramidal Neurons

For the instrument set-up, the samples collected from the amygdala of wild-type YFP-negative mice were used to gate YFP-positive neurons based on forward scatter (FSC) and side scatter (SSC) light scattering and to set YFP negativity. Afterwards, amygdala samples were collected from the Thy1-COP4 mice and stained with 1 μL of propidium iodide (PI) in order to identify dead cells. Pyramidal neurons were then sorted by using the MoFlo Astrios EQ (Beckman Coulter, Brea, CA, USA) and the pyramidal neurons characterized by YFP were collected on the basis of their physical parameters, singlets, PI negative (live cells), and YFP intensity (Figure 8). For initial characterization, samples were collected in PBS and samples examined under a fluorescent microscope to verify correct sorting. Thereafter, cells were sorted directly into ice-cold lysis buffer (Reliaprep RNA Cell Miniprep System, Promega, Fitchburg, WI, USA), mixed by vortexing, kept on ice, and then stored at −80 °C until RNA extraction.

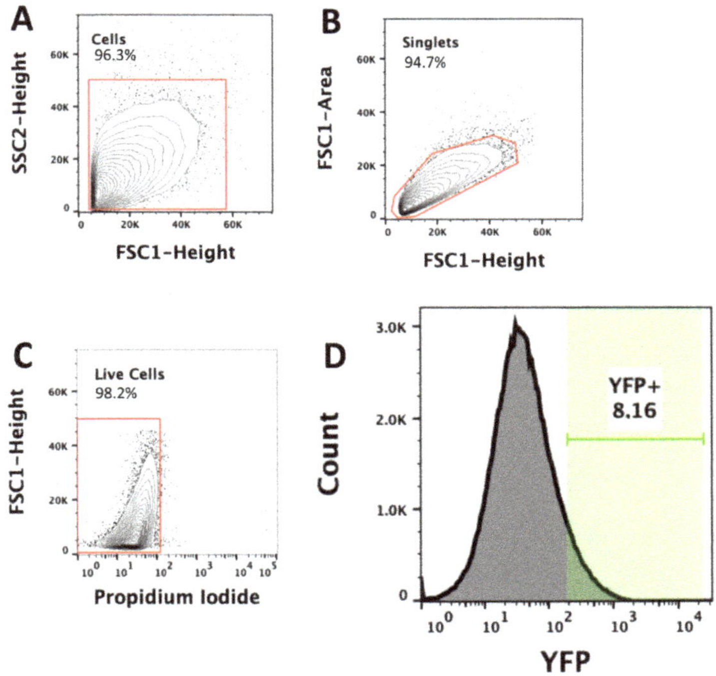

Figure 8. Gating strategy for amygdala pyramidal neurons cell purification. Amygdala pyramidal neurons, excited with a ~488 nm blue laser, were sorted on the basis of their physical parameters (forward scatter, FSC, and side scatter, SSC, light scattering) (**A**), singlets (**B**), propidium iodide negative (live cells) (**C**) and Yellow Fluorescent Protein (YFP) intensity (**D**), and the positive cells were collected. Cells are sorted by high-speed cell sorting (Moflo Astrios EQ).

4.7.3. RNA-Seq Library Preparation

After thawing on ice in presence of additional proteinase K, RNA was isolated according to manufacturer's instructions including on-column DNase treatment. RNA samples were quantified and the quality was tested by Agilent 2100 Bioanalyzer RNA assay (Agilent Technologies, Santa Clara, CA, USA) or Caliper (PerkinElmer, Waltham, MA, USA) (Table 5).

Library preparation and sequencing were performed at IGATechnology (Udine, Italy). At least 3 independent biological replicates were used for each group (Table 5). Each replicate corresponds to the amygdala of a single Thy1-COP4 animal.

Table 5. Sketch of RNA samples extracted by amygdala pyramidal YFP neurons and sequenced in PE75 mode. Only samples used for sequencing, i.e., 14 over the total, are reported. The samples that did not pass quality control were excluded from further analyses.

Conditioning	PrL Stimulation	Sample Name	RNA (ng)	Library (ng)	Reads (M)
FEAR	OPTO	ID_FO1	0.540	0.270	33.954
FEAR	OPTO	ID_FO2	0.846	0.423	38.971
FEAR	OPTO	ID_FO3	0.225	0.112	40.455
FEAR	OPTO	ID_FO4	0.410	0.205	34.162
FEAR	SHAM	ID_FS1	0.140	0.070	34.804
FEAR	SHAM	ID_FS2	2.760	1.380	37.340
FEAR	SHAM	ID_FS3	0.640	0.320	30.547
NOT FEAR	OPTO	ID_NF01	1.770	0.885	30.483
NOT FEAR	OPTO	ID_NF02	0.240	0.120	32.432
NOT FEAR	OPTO	ID_NF03	0.510	0.255	34.756
NOT FEAR	SHAM	ID_NFS1	0.360	0.170	31.432
NOT FEAR	SHAM	ID_NFS2	0.520	0.260	30.438
NOT FEAR	SHAM	ID_NFS3	0.110	0.055	36.824
NOT FEAR	SHAM	ID_NFS4	1.060	0.530	54.741

Libraries were generated from each sample individually, starting from 0.05–1.4 ng of total RNA, using the Ovation SoLo RNA-seq kit for Ultra-low input (NuGEN, Tecan Genomics, Redwood City, CA, USA), following the manufacturer's instructions (library type: fr-second strand). Final libraries were checked with both Qubit 2.0 Fluorometer (Invitrogen by Life technologies, Carlsbad, CA, USA) and Agilent Bioanalyzer DNA assay (Agilent Technologies, Santa Clara, CA, USA) or Caliper (PerkinElmer, Waltham, MA, USA). Libraries were then prepared for sequencing and sequenced on paired-end 2 × 75 bp mode on NextSeq500 (Illumina, San Diego, CA, USA) producing 35.2 MR on average (min 30.4 MR, max 51.7 MR). For the processing of raw data (format conversion and de-multiplexing), Bcl2Fastq 2.20 version of the Illumina pipeline was used. Sequencing data have been deposited in the NCBI Short Read Archive (https://www.ncbi.nlm.nih.gov/geo; GEO accession number GSE162417).

4.7.4. Analysis of RNA Sequencing Data

GRCm38.p6 genome was used to map the reads, and transcript abundances were estimated using Salmon v1.2 [76]. To obtain gene-level count matrices the quantification data were imported using tximport [77]. All further analyses based on these count matrices were performed with the free software R v4.0.2, Bioconductor v3.11 [78], and the package NOISeq v2.31.0 [79]. Differences in RNA composition between samples were corrected by the Trimmed mean of M-values (TMM) normalization [80], and filtered for low counts based on a count per million reads (CPM) criteria. Subsequently, ARSyNseq was used to remove the technical batch effect and NOISeqBIO was used to assess differential gene expression ($q > 0.95$, equivalent to FDR-corrected $p < 0.05$) [79,81]. GO and KEGG pathway analyses were performed by using ClusterProfiler v3.16.1 [82]. The enrichment map method was used to identify functional modules of mutually overlapping gene sets [82,83].

4.8. Statistical Analysis

As regard the behavioral results, a three-factor ANOVA (stimulation × fear × day) or a two-factor ANOVA (group × day) on freezing behavior (measured during 0–3 min of contextual FC) and one-factor ANOVA on freezing behavior (measured during 0–6 min of day 14) were used. Newman-Keuls post-hoc comparisons were applied when permitted.

As regard the morphological results, one-factor ANOVAs on spine number and density were used. Newman-Keuls post-hoc comparisons were applied when permitted.

As regard the electrophysiological results, Pearson r Correlation test (for evoked firing activities and EPSC values) and Mann-Whitney U Test (for rheobase and EPSC values) were used.

As for the behavioral, morphological, and electrophysiological results, all analyses were performed by using Statistica 7.0 for Windows (TIBCO Software Inc., Aliso Viejo, CA, USA) and values of $p = 0.05$ were considered statistically significant.

As regard the transcriptomic results, after TMM normalization and low counts filtering, the resulting genes underwent the downstream analysis. Batch effect correction was applied with ARSyN and a PCA was performed to assess sample clustering based on their expression profiles. Differential expression analysis was performed on Group × Condition design and DEGs were identified using NOISeqBIO, a non-parametric analysis for biological replicates. Significant differentially expressed genes were identified for a $q > 0.95$, equivalent to an FDR-corrected $p < 0.05$. Subsequently, GO and KEGG over-representation analyses were performed and significant pathways were represented by means of enrichment map method to visualize and interpret results.

5. Conclusions

Given the critical role of the pyramidal neurons of amygdala and PrL cortex in fear processing, the characterization of the structural, neurophysiological, and molecular changes of this neuronal population associated to adaptive or maladaptive fear extinction may provide valuable insight for the study of, and therapeutic interventions in, fear-related and psychiatric disorders.

Supplementary Materials: The following are available online at https://www.mdpi.com/1422-0067/22/2/810/s1.

Author Contributions: Conceptualization, D.L., G.S., V.O. and L.P.; Methodology, D.L., G.S., J.G., S.C., L.S., M.D.B., D.C., A.M., C.Z. and L.P.; Software, C.F., A.T., A.P., F.B., N.P.; Validation, all authors; Formal Analysis, D.L., C.F., A.T., A.P., F.B., and N.P.; Investigation, all authors; Resources, D.L. and L.P.; Data Curation, D.L.; Writing—Original Draft Preparation, all authors; Writing—Review & Editing, D.L. and L.P.; Visualization, all authors; Supervision, D.L. and L.P.; Project Administration, D.L.; Funding Acquisition, D.L. and G.S. All authors have read and agreed to the published version of the manuscript.

Funding: This work was supported by Italian Ministry of Health (Grant number GR-2018-12365733 to D.L. and G.S.).

Institutional Review Board Statement: The study was conducted according to the guidelines of the Declaration of Helsinki, and approved by the Institutional Review Board of IRCCS Fondazione Santa Lucia (protocol code 534/2019-PR, date of approval 23 July 2019).

Informed Consent Statement: Not applicable.

Data Availability Statement: All data support the findings of this study are available from the corresponding author upon reasonable request.

Conflicts of Interest: The authors declare no conflict of interest.

References

1. Maren, S.; Fanselow, M.S. The Amygdala and Fear Conditioning: Has the Nut Been Cracked? *Neuron* **1996**, *16*, 237–240.
2. Davis, M. The Role of the Amygdala in Conditioned and Unconditioned Fear and Anxiety. *Amygdala* **2000**, *2*, 213–287.
3. LeDoux, J.E. Emotion Circuits in the Brain. *Annu. Rev. Neurosci.* **2000**, *23*, 155–184. [PubMed]
4. Fanselow, M.S.; Gale, G.D. The Amygdala, Fear, and Memory. *Ann. N. Y. Acad. Sci.* **2003**, *985*, 125–134. [PubMed]
5. Holmes, A.; Singewald, N. Individual Differences in Recovery from Traumatic Fear. *Trends Neurosci.* **2013**, *36*, 23–31. [PubMed]
6. Corcoran, K.A.; Quirk, G.J. Recalling Safety: Cooperative Functions of the Ventromedial Prefrontal Cortex and the Hippocampus in Extinction. *CNS Spectr.* **2007**, *12*, 200–206.
7. Goshen, I.; Brodsky, M.; Prakash, R.; Wallace, J.; Gradinaru, V.; Ramakrishnan, C.; Deisseroth, K. Dynamics of Retrieval Strategies for Remote Memories. *Cell* **2011**, *147*, 678–689. [PubMed]
8. Lacagnina, A.F.; Brockway, E.T.; Crovetti, C.R.; Shue, F.; McCarty, M.J.; Sattler, K.P.; Lim, S.C.; Santos, S.L.; Denny, C.A.; Drew, M.R. Distinct Hippocampal Engrams Control Extinction and Relapse of Fear Memory. *Nat. Neurosci.* **2019**, *22*, 753–761.

9. Tovote, P.; Fadok, J.P.; Lüthi, A. Neuronal Circuits for Fear and Anxiety. *Nat. Rev. Neurosci.* **2015**, *16*, 317–331.
10. Trouche, S.; Sasaki, J.M.; Tu, T.; Reijmers, L.G. Fear Extinction Causes Target-Specific Remodeling of Perisomatic Inhibitory Synapses. *Neuron* **2013**, *80*, 1054–1065.
11. Zhu, H.; Pleil, K.E.; Urban, D.J.; Moy, S.S.; Kash, T.L.; Roth, B.L. Chemogenetic Inactivation of Ventral Hippocampal Glutamatergic Neurons Disrupts Consolidation of Contextual Fear Memory. *Neuropsychopharmacology* **2014**, *39*, 1880–1892. [PubMed]
12. Johansen, J.P.; Hamanaka, H.; Monfils, M.H.; Behnia, R.; Deisseroth, K.; Blair, H.T.; LeDoux, J.E. Optical Activation of Lateral Amygdala Pyramidal Cells Instructs Associative Fear Learning. *Proc. Natl. Acad. Sci. USA* **2010**, *107*, 12692–12697.
13. Tye, K.M.; Prakash, R.; Kim, S.-Y.; Fenno, L.E.; Grosenick, L.; Zarabi, H.; Thompson, K.R.; Gradinaru, V.; Ramakrishnan, C.; Deisseroth, K. Amygdala Circuitry Mediating Reversible and Bidirectional Control of Anxiety. *Nature* **2011**, *471*, 358–362.
14. Klavir, O.; Prigge, M.; Sarel, A.; Paz, R.; Yizhar, O. Manipulating Fear Associations via Optogenetic Modulation of Amygdala Inputs to Prefrontal Cortex. *Nat. Neurosci.* **2017**, *20*, 836–844. [PubMed]
15. Sierra-Mercado, D.; Padilla-Coreano, N.; Quirk, G.J. Dissociable Roles of Prelimbic and Infralimbic Cortices, Ventral Hippocampus, and Basolateral Amygdala in the Expression and Extinction of Conditioned Fear. *Neuropsychopharmacology* **2011**, *36*, 529–538. [PubMed]
16. Beyeler, A.; Eckhardt, C.A.; Tye, K.M. Deciphering memory function with optogenetics. In *Progress in Molecular Biology and Translational Science*; Elsevier: Amsterdam, The Netherlands, 2014; Volume 122, pp. 341–390. ISBN 1877-1173.
17. Kim, H.-S.; Cho, H.-Y.; Augustine, G.J.; Han, J.-H. Selective Control of Fear Expression by Optogenetic Manipulation of Infralimbic Cortex after Extinction. *Neuropsychopharmacology* **2016**, *41*, 1261–1273. [PubMed]
18. Hardt, O.; Nadel, L. Systems Consolidation Revisited, but Not Revised: The Promise and Limits of Optogenetics in the Study of Memory. *Neurosci. Lett.* **2018**, *680*, 54–59.
19. Senn, V.; Wolff, S.B.; Herry, C.; Grenier, F.; Ehrlich, I.; Gründemann, J.; Fadok, J.P.; Müller, C.; Letzkus, J.J.; Lüthi, A. Long-Range Connectivity Defines Behavioral Specificity of Amygdala Neurons. *Neuron* **2014**, *81*, 428–437.
20. Sotres-Bayon, F.; Sierra-Mercado, D.; Pardilla-Delgado, E.; Quirk, G.J. Gating of Fear in Prelimbic Cortex by Hippocampal and Amygdala Inputs. *Neuron* **2012**, *76*, 804–812.
21. Cho, J.-H.; Deisseroth, K.; Bolshakov, V.Y. Synaptic Encoding of Fear Extinction in MPFC-Amygdala Circuits. *Neuron* **2013**, *80*, 1491–1507.
22. Bredy, T.W.; Wu, H.; Crego, C.; Zellhoefer, J.; Sun, Y.E.; Barad, M. Histone Modifications around Individual BDNF Gene Promoters in Prefrontal Cortex Are Associated with Extinction of Conditioned Fear. *Learn. Mem.* **2007**, *14*, 268–276. [CrossRef] [PubMed]
23. Chwang, W.B.; O'Riordan, K.J.; Levenson, J.M.; Sweatt, J.D. ERK/MAPK Regulates Hippocampal Histone Phosphorylation Following Contextual Fear Conditioning. *Learn. Mem.* **2006**, *13*, 322–328. [CrossRef] [PubMed]
24. Miller, C.A.; Sweatt, J.D. Covalent Modification of DNA Regulates Memory Formation. *Neuron* **2007**, *53*, 857–869. [CrossRef] [PubMed]
25. Itzhak, Y.; Anderson, K.L.; Kelley, J.B.; Petkov, M. Histone Acetylation Rescues Contextual Fear Conditioning in NNOS KO Mice and Accelerates Extinction of Cued Fear Conditioning in Wild Type Mice. *Neurobiol. Learn. Mem.* **2012**, *97*, 409–417. [CrossRef] [PubMed]
26. Monsey, M.S.; Ota, K.T.; Akingbade, I.F.; Hong, E.S.; Schafe, G.E. Epigenetic Alterations Are Critical for Fear Memory Consolidation and Synaptic Plasticity in the Lateral Amygdala. *PLoS ONE* **2011**, *6*, e19958. [CrossRef]
27. Stafford, J.M.; Raybuck, J.D.; Ryabinin, A.E.; Lattal, K.M. Increasing Histone Acetylation in the Hippocampus-Infralimbic Network Enhances Fear Extinction. *Biol. Psychiatry* **2012**, *72*, 25–33. [CrossRef]
28. Duke, C.G.; Kennedy, A.J.; Gavin, C.F.; Day, J.J.; Sweatt, J.D. Experience-Dependent Epigenomic Reorganization in the Hippocampus. *Learn. Mem.* **2017**, *24*, 278–288. [CrossRef]
29. Gelernter, J.; Sun, N.; Polimanti, R.; Pietrzak, R.; Levey, D.F.; Bryois, J.; Lu, Q.; Hu, Y.; Li, B.; Radhakrishnan, K. Genome-Wide Association Study of Post-Traumatic Stress Disorder Reexperiencing Symptoms in >165,000 US Veterans. *Nat. Neurosci.* **2019**, *22*, 1394–1401. [CrossRef]
30. Halder, R.; Hennion, M.; Vidal, R.O.; Shomroni, O.; Rahman, R.-U.; Rajput, A.; Centeno, T.P.; Van Bebber, F.; Capece, V.; Vizcaino, J.C.G. DNA Methylation Changes in Plasticity Genes Accompany the Formation and Maintenance of Memory. *Nat. Neurosci.* **2016**, *19*, 102–110. [CrossRef]
31. Ryan, T.J.; Roy, D.S.; Pignatelli, M.; Arons, A.; Tonegawa, S. Engram Cells Retain Memory under Retrograde Amnesia. *Science* **2015**, *348*, 1007–1013. [CrossRef]
32. Yehuda, R.; LeDoux, J. Response Variation Following Trauma: A Translational Neuroscience Approach to Understanding PTSD. *Neuron* **2007**, *56*, 19–32. [CrossRef] [PubMed]
33. Binder, E.B.; Bradley, R.G.; Liu, W.; Epstein, M.P.; Deveau, T.C.; Mercer, K.B.; Tang, Y.; Gillespie, C.F.; Heim, C.M.; Nemeroff, C.B. Association of FKBP5 Polymorphisms and Childhood Abuse with Risk of Posttraumatic Stress Disorder Symptoms in Adults. *JAMA* **2008**, *299*, 1291–1305. [CrossRef] [PubMed]
34. Champagne, F.A. Epigenetic Mechanisms and the Transgenerational Effects of Maternal Care. *Front. Neuroendocrinol.* **2008**, *29*, 386–397. [CrossRef]
35. Franklin, T.B.; Russig, H.; Weiss, I.C.; Gräff, J.; Linder, N.; Michalon, A.; Vizi, S.; Mansuy, I.M. Epigenetic Transmission of the Impact of Early Stress across Generations. *Biol. Psychiatry* **2010**, *68*, 408–415. [CrossRef]

36. Koenen, K.C.; Uddin, M. FKBP5 Polymorphisms Modify the Effects of Childhood Trauma. *Neuropsychopharmacology* **2010**, *35*, 1623–1624. [CrossRef]
37. Yehuda, R.; Bierer, L.M. The Relevance of Epigenetics to PTSD: Implications for the DSM-V. *J. Trauma. Stress* **2009**, *22*, 427–434. [CrossRef]
38. Timmons, J.A.; Szkop, K.J.; Gallagher, I.J. Multiple Sources of Bias Confound Functional Enrichment Analysis of Global-Omics Data. *Genome Biol.* **2015**, *16*, 186. [CrossRef]
39. Alberini, C.M.; LeDoux, J.E. Memory Reconsolidation. *Curr. Biol.* **2013**, *23*, R746–R750. [CrossRef]
40. Vaverková, Z.; Milton, A.L.; Merlo, E. Retrieval-Dependent Mechanisms Affecting Emotional Memory Persistence: Reconsolidation, Extinction, and the Space in Between. *Front. Behav. Neurosci.* **2020**, *14*, 574358. [CrossRef]
41. McCullough, K.M.; Chatzinakos, C.; Hartmann, J.; Missig, G.; Neve, R.L.; Fenster, R.J.; Carlezon, W.A.; Daskalakis, N.P.; Ressler, K.J. Genome-Wide Translational Profiling of Amygdala Crh-Expressing Neurons Reveals Role for CREB in Fear Extinction Learning. *Nat. Commun.* **2020**, *11*, 5180. [CrossRef]
42. Shen, H.; Sabaliauskas, N.; Sherpa, A.; Fenton, A.A.; Stelzer, A.; Aoki, C.; Smith, S.S. A Critical Role for A4βδ GABAA Receptors in Shaping Learning Deficits at Puberty in Mice. *Science* **2010**, *327*, 1515–1518. [CrossRef] [PubMed]
43. Johnston, J.; Forsythe, I.D.; Kopp-Scheinpflug, C. SYMPOSIUM REVIEW: Going Native: Voltage-gated Potassium Channels Controlling Neuronal Excitability. *J. Physiol.* **2010**, *588*, 3187–3200. [CrossRef] [PubMed]
44. Miyake, A.; Mochizuki, S.; Yokoi, H.; Kohda, M.; Furuichi, K. New Ether-à-Go-Go K(+) Channel Family Members Localized in Human Telencephalon. *J. Biol. Chem.* **1999**, *274*, 25018–25025. [CrossRef] [PubMed]
45. Kraus, D.M.; Elliott, G.S.; Chute, H.; Horan, T.; Pfenninger, K.H.; Sanford, S.D.; Foster, S.; Scully, S.; Welcher, A.A.; Holers, V.M. CSMD1 Is a Novel Multiple Domain Complement-Regulatory Protein Highly Expressed in the Central Nervous System and Epithelial Tissues. *J. Immunol.* **2006**, *176*, 4419–4430. [CrossRef] [PubMed]
46. Sentman, C.L.; Shutter, J.R.; Hockenbery, D.; Kanagawa, O.; Korsmeyer, S.J. Bcl-2 Inhibits Multiple Forms of Apoptosis but Not Negative Selection in Thymocytes. *Cell* **1991**, *67*, 879–888. [CrossRef]
47. Williams, G.T. Programmed Cell Death: Apoptosis and Oncogenesis. *Cell* **1991**, *65*, 1097–1098. [CrossRef]
48. Farlie, P.G.; Dringen, R.; Rees, S.M.; Kannourakis, G.; Bernard, O. Bcl-2 Transgene Expression Can Protect Neurons against Developmental and Induced Cell Death. *Proc. Natl. Acad. Sci. USA* **1995**, *92*, 4397–4401. [CrossRef]
49. Clifton, N.E.; Cameron, D.; Trent, S.; Sykes, L.H.; Thomas, K.L.; Hall, J. Hippocampal Regulation of Postsynaptic Density Homer1 by Associative Learning. *Neural Plast.* **2017**, *2017*, 5959182. [CrossRef]
50. Olsen, R.H.; Agam, M.; Davis, M.J.; Raber, J. ApoE Isoform-dependent Deficits in Extinction of Contextual Fear Conditioning. *Genes Brain Behav.* **2012**, *11*, 806–812. [CrossRef]
51. Pardon, M.-C.; Sarmad, S.; Rattray, I.; Bates, T.E.; Scullion, G.A.; Marsden, C.A.; Barrett, D.A.; Lowe, J.; Kendall, D.A. Repeated Novel Cage Exposure-Induced Improvement of Early Alzheimer's-like Cognitive and Amyloid Changes in TASTPM Mice Is Unrelated to Changes in Brain Endocannabinoids Levels. *Neurobiol. Aging* **2009**, *30*, 1099–1113. [CrossRef]
52. Rattray, I.; Scullion, G.A.; Soulby, A.; Kendall, D.A.; Pardon, M.-C. The Occurrence of a Deficit in Contextual Fear Extinction in Adult Amyloid-over-Expressing TASTPM Mice Is Independent of the Strength of Conditioning but Can Be Prevented by Mild Novel Cage Stress. *Behav. Brain Res.* **2009**, *200*, 83–90. [CrossRef] [PubMed]
53. Temme, S.J.; Murphy, G.G. The L-Type Voltage-Gated Calcium Channel CaV1.2 Mediates Fear Extinction and Modulates Synaptic Tone in the Lateral Amygdala. *Learn. Mem.* **2017**, *24*, 580–588. [CrossRef] [PubMed]
54. Trent, S.; Barnes, P.; Hall, J.; Thomas, K.L. AMPA Receptors Control Fear Extinction through an Arc-Dependent Mechanism. *Learn. Mem.* **2017**, *24*, 375–380. [CrossRef] [PubMed]
55. Shamah, S.M.; Lin, M.Z.; Goldberg, J.L.; Estrach, S.; Sahin, M.; Hu, L.; Bazalakova, M.; Neve, R.L.; Corfas, G.; Debant, A.; et al. EphA Receptors Regulate Growth Cone Dynamics through the Novel Guanine Nucleotide Exchange Factor Ephexin. *Cell* **2001**, *105*, 233–244. [CrossRef]
56. Fu, W.-Y.; Chen, Y.; Sahin, M.; Zhao, X.-S.; Shi, L.; Bikoff, J.B.; Lai, K.-O.; Yung, W.-H.; Fu, A.K.; Greenberg, M.E. Cdk5 Regulates EphA4-Mediated Dendritic Spine Retraction through an Ephexin1-Dependent Mechanism. *Nat. Neurosci.* **2007**, *10*, 67–76. [CrossRef]
57. Frank, C.A.; Pielage, J.; Davis, G.W. A Presynaptic Homeostatic Signaling System Composed of the Eph Receptor, Ephexin, Cdc42, and CaV2.1 Calcium Channels. *Neuron* **2009**, *61*, 556–569. [CrossRef]
58. Alapin, J.M.; Dines, M.; Vassiliev, M.; Tamir, T.; Ram, A.; Locke, C.; Yu, J.; Lamprecht, R. Activation of EphB2 Forward Signaling Enhances Memory Consolidation. *Cell Rep.* **2018**, *23*, 2014–2025. [CrossRef]
59. Kim, T.A.; Lim, J.; Ota, S.; Raja, S.; Rogers, R.; Rivnay, B.; Avraham, H.; Avraham, S. NRP/B, a Novel Nuclear Matrix Protein, Associates with P110(RB) and Is Involved in Neuronal Differentiation. *J. Cell Biol.* **1998**, *141*, 553–566. [CrossRef]
60. Cooper, E.C.; Aldape, K.D.; Abosch, A.; Barbaro, N.M.; Berger, M.S.; Peacock, W.S.; Jan, Y.N.; Jan, L.Y. Colocalization and Coassembly of Two Human Brain M-Type Potassium Channel Subunits That Are Mutated in Epilepsy. *Proc. Natl. Acad. Sci. USA* **2000**, *97*, 4914–4919. [CrossRef]
61. Jiang, L.; Kundu, S.; Lederman, J.D.; López-Hernández, G.Y.; Ballinger, E.C.; Wang, S.; Talmage, D.A.; Role, L.W. Cholinergic Signaling Controls Conditioned Fear Behaviors and Enhances Plasticity of Cortical-Amygdala Circuits. *Neuron* **2016**, *90*, 1057–1070. [CrossRef]

62. Kowalewski, B.; Lamanna, W.C.; Lawrence, R.; Damme, M.; Stroobants, S.; Padva, M.; Kalus, I.; Frese, M.-A.; Lübke, T.; Lüllmann-Rauch, R.; et al. Arylsulfatase G Inactivation Causes Loss of Heparan Sulfate 3-O-Sulfatase Activity and Mucopolysaccharidosis in Mice. *Proc. Natl. Acad. Sci. USA* **2012**, *109*, 10310–10315. [CrossRef] [PubMed]
63. Burgute, B.D.; Peche, V.S.; Steckelberg, A.-L.; Glöckner, G.; Gaßen, B.; Gehring, N.H.; Noegel, A.A. NKAP Is a Novel RS-Related Protein That Interacts with RNA and RNA Binding Proteins. *Nucleic Acids Res.* **2014**, *42*, 3177–3193. [CrossRef] [PubMed]
64. Lingawi, N.W.; Holmes, N.M.; Westbrook, R.F.; Laurent, V. The Infralimbic Cortex Encodes Inhibition Irrespective of Motivational Significance. *Neurobiol. Learn. Mem.* **2018**, *150*, 64–74. [CrossRef] [PubMed]
65. Rothenhoefer, K.M.; Stauffer, W.R. Dopamine Signals Learn New Tricks. *Neuron* **2020**, *106*, 11–13. [CrossRef] [PubMed]
66. Bocchio, M.; Nabavi, S.; Capogna, M. Synaptic Plasticity, Engrams, and Network Oscillations in Amygdala Circuits for Storage and Retrieval of Emotional Memories. *Neuron* **2017**, *94*, 731–743. [CrossRef] [PubMed]
67. Li, Y.; Iakoubova, O.A.; Shiffman, D.; Devlin, J.J.; Forrester, J.S.; Superko, H.R. KIF6 Polymorphism as a Predictor of Risk of Coronary Events and of Clinical Event Reduction by Statin Therapy. *Am. J. Cardiol.* **2010**, *106*, 994–998. [CrossRef] [PubMed]
68. Svendsen, J.M.; Smogorzewska, A.; Sowa, M.E.; O'Connell, B.C.; Gygi, S.P.; Elledge, S.J.; Harper, J.W. Mammalian BTBD12/SLX4 Assembles a Holliday Junction Resolvase and Is Required for DNA Repair. *Cell* **2009**, *138*, 63–77. [CrossRef] [PubMed]
69. Min, J.-N.; Tian, Y.; Xiao, Y.; Wu, L.; Li, L.; Chang, S. The MINO80 Chromatin Remodeling Complex Is Required for Efficient Telomere Replication and Maintenance of Genome Stability. *Cell Res.* **2013**, *23*, 1396–1413. [CrossRef]
70. Ressler, K.J.; Mercer, K.B.; Bradley, B.; Jovanovic, T.; Mahan, A.; Kerley, K.; Norrholm, S.D.; Kilaru, V.; Smith, A.K.; Myers, A.J.; et al. Post-Traumatic Stress Disorder Is Associated with PACAP and the PAC1 Receptor. *Nature* **2011**, *470*, 492–497. [CrossRef]
71. Sananbenesi, F.; Fischer, A.; Wang, X.; Schrick, C.; Neve, R.; Radulovic, J.; Tsai, L.-H. A Hippocampal Cdk5 Pathway Regulates Extinction of Contextual Fear. *Nat. Neurosci.* **2007**, *10*, 1012–1019. [CrossRef]
72. Laricchiuta, D.; Centonze, D.; Petrosini, L. Effects of Endocannabinoid and Endovanilloid Systems on Aversive Memory Extinction. *Behav. Brain Res.* **2013**, *256*, 101–107. [CrossRef] [PubMed]
73. Laricchiuta, D.; Saba, L.; De Bartolo, P.; Caioli, S.; Zona, C.; Petrosini, L. Maintenance of Aversive Memories Shown by Fear Extinction-Impaired Phenotypes Is Associated with Increased Activity in the Amygdaloid-Prefrontal Circuit. *Sci. Rep.* **2016**, *6*, 21205. [CrossRef] [PubMed]
74. Jasnow, A.M.; Ehrlich, D.E.; Choi, D.C.; Dabrowska, J.; Bowers, M.E.; McCullough, K.M.; Rainnie, D.G.; Ressler, K.J. Thy1-Expressing Neurons in the Basolateral Amygdala May Mediate Fear Inhibition. *J. Neurosci.* **2013**, *33*, 10396–10404. [CrossRef]
75. Franklin, K.B.J.; Paxinos, G. *The Mouse Brain in Stereotaxic Coordinates*; Academic Press: Cambridge, MA, USA, 1997; p. 186.
76. Patro, R.; Duggal, G.; Love, M.I.; Irizarry, R.A.; Kingsford, C. Salmon Provides Fast and Bias-Aware Quantification of Transcript Expression. *Nat. Methods* **2017**, *14*, 417–419. [CrossRef]
77. Soneson, C.; Love, M.I.; Robinson, M.D. Differential Analyses for RNA-Seq: Transcript-Level Estimates Improve Gene-Level Inferences. *F1000Research* **2015**, *4*, 1521. [CrossRef]
78. Gentleman, R.C.; Carey, V.J.; Bates, D.M.; Bolstad, B.; Dettling, M.; Dudoit, S.; Ellis, B.; Gautier, L.; Ge, Y.; Gentry, J. Bioconductor: Open Software Development for Computational Biology and Bioinformatics. *Genome Biol.* **2004**, *5*, R80. [CrossRef]
79. Tarazona, S.; Furió-Tarí, P.; Turrà, D.; Pietro, A.D.; Nueda, M.J.; Ferrer, A.; Conesa, A. Data Quality Aware Analysis of Differential Expression in RNA-Seq with NOISeq R/Bioc Package. *Nucleic Acids Res.* **2015**, *43*, e140. [CrossRef]
80. Robinson, M.D.; Oshlack, A. A Scaling Normalization Method for Differential Expression Analysis of RNA-Seq Data. *Genome Biol.* **2010**, *11*, R25. [CrossRef]
81. Nueda, M.J.; Ferrer, A.; Conesa, A. ARSyN: A Method for the Identification and Removal of Systematic Noise in Multifactorial Time Course Microarray Experiments. *Biostatistics* **2012**, *13*, 553–566. [CrossRef]
82. Yu, G.; Wang, L.-G.; Han, Y.; He, Q.-Y. ClusterProfiler: An R Package for Comparing Biological Themes among Gene Clusters. *OMICS J. Integr. Biol.* **2012**, *16*, 284–287. [CrossRef]
83. Merico, D.; Isserlin, R.; Stueker, O.; Emili, A.; Bader, G.D. Enrichment Map: A Network-Based Method for Gene-Set Enrichment Visualization and Interpretation. *PLoS ONE* **2010**, *5*, e13984. [CrossRef]

International Journal of
Molecular Sciences

MDPI

Review

Glutamatergic Dysfunction and Synaptic Ultrastructural Alterations in Schizophrenia and Autism Spectrum Disorder: Evidence from Human and Rodent Studies

Ahmed Eltokhi [1,*,†], Andrea Santuy [1,†], Angel Merchan-Perez [2,3] and Rolf Sprengel [4,*]

1 Department of Neurology and Epileptology, Hertie Institute for Clinical Brain Research, Tübingen University, D-72076 Tübingen, Germany; and.santuy@gmail.com
2 Laboratorio Cajal de Circuitos Corticales, Centro de Tecnología Biomédica, Universidad Politécnica de Madrid, 28223 Madrid, Spain; amerchan@fi.upm.es
3 Departamento de Arquitectura y Tecnología de Sistemas Informáticos, Universidad Politécnica de Madrid, 28223 Madrid, Spain
4 Research Group of the Max Planck Institute for Medical Research at the Institute for Anatomy and Cell Biology, Heidelberg University, D-69120 Heidelberg, Germany
* Correspondence: Ahmed.eltokhi@uni-tuebingen.de (A.E.); Rolf.sprengel@mpimf-heidelberg.mpg.de (R.S.)
† These authors contributed equally to this work.

Abstract: The correlation between dysfunction in the glutamatergic system and neuropsychiatric disorders, including schizophrenia and autism spectrum disorder, is undisputed. Both disorders are associated with molecular and ultrastructural alterations that affect synaptic plasticity and thus the molecular and physiological basis of learning and memory. Altered synaptic plasticity, accompanied by changes in protein synthesis and trafficking of postsynaptic proteins, as well as structural modifications of excitatory synapses, are critically involved in the postnatal development of the mammalian nervous system. In this review, we summarize glutamatergic alterations and ultrastructural changes in synapses in schizophrenia and autism spectrum disorder of genetic or drug-related origin, and briefly comment on the possible reversibility of these neuropsychiatric disorders in the light of findings in regular synaptic physiology.

Keywords: glutamatergic system; synaptic ultrastructure; schizophrenia; autism spectrum disorder

check for updates

Citation: Eltokhi, A.; Santuy, A.; Merchan -Perez, A.; Sprengel, R. Glutamatergic Dysfunction and Synaptic Ultrastructural Alterations in Schizophrenia and Autism Spectrum Disorder: Evidence from Human and Rodent Studies. *Int. J. Mol. Sci.* **2021**, *22*, 59. https://dx.doi.org/10.3390/ijms22010059

Received: 29 November 2020
Accepted: 22 December 2020
Published: 23 December 2020

1. The Glutamatergic System

Excitatory neurotransmission in the brain is primarily glutamatergic; glutamatergic neurons in the non-stimulated cerebral cortex consume up to 80% of the total brain metabolic activity [1–3]. For the fast glutamatergic transmission, ionotropic glutamate-gated channels (iGluRs) are recruited. iGluRs are tetrameric, and each of the four subunits has four distinct structural protein domains: an extracellular N-terminal domain, an extracellular ligand-binding domain, a transmembrane channel, and an intracellular C-terminal domain [4]. In total, multiple subunits encoded by 18 different genes can contribute to the formation of iGluRs. Most of these subunits derive from different mRNA splice or pre-mRNA edited variants of genes coding for iGluR subunits, which increase the complexity of the ionotropic glutamatergic neurotransmission system. Based on their main agonist, iGluRs are divided into α-amino-3-hydroxy-5-methylisoxazole-4-propionate receptors (AMPARs), *N*-methyl-D-aspartate receptors (NMDARs), and kainate receptors (KARs). All members of these three receptor families are hetero- or homo-tetramers that are permeable to cations. With the exception of KARs that are only permeable to Na^+, other glutamate-gated ion channels are also permeable to Ca^{2+}. Thus, in response to glutamate activation, they increase the intracellular Ca^{2+} levels, thereby activating Ca^{2+}-dependent intracellular responses upon glutamate stimulation. In the AMPAR family, the Na^+ and Ca^{2+} permeability is usually strictly determined by their subunit composition; only AMPARs that lack the

GluA2 subunit are permeable to Ca^{2+} (CP+AMPARs). CP+AMPARs can contribute to some forms of synaptic plasticity. For the induction of activity-induced synaptic transmission, CP+AMPARs are translocated from extra-postsynaptic to postsynaptic sites in response to intense presynaptic glutamate stimulation. Subsequently, activity-induced incorporated (CP+AMPARs) are replaced by Ca^{2+}-impermeable AMPARs (CP-AMPARs) [5]. In NMDARs, the Ca^{2+} influx is blocked during regular glutamatergic signal transmission at resting membrane potential by a Mg^{2+} ion but is possible after repetitive stimulations that largely depolarize the synaptic membrane and displace the Mg^{2+} ion [6].

The interplay of synaptic iGluR subtypes is thus specialized to regulate the activity-dependent Ca^{2+} influx into the postsynapse, which can lead to either a transient or a long-term alteration of the synaptic efficacy. Glutamate sensing in the synaptic cleft is modified further by the presence of the G-protein-coupled metabotropic glutamate receptors (mGluRs), for slower responses to increased glutamate levels in the synaptic cleft. The mGluRs are classified into three groups based on the receptor structure and ligand sensitivity [7]. Group I (mGluR 1 and 5) is mainly postsynaptic, while groups II (mGluR 2 and 3) and III (mGluR 4, 6, 7, and 8) are primarily presynaptic and modulate the neurotransmitter release [2,8]. Group I works through the activation of phospholipase C, while group II and III mainly operate by decreasing cyclic AMP levels [8]. Their functions are also different: group I potentiates presynaptic glutamate release and postsynaptic NMDAR currents, and groups II and III inhibit presynaptic glutamate release [7].

This molecular complexity of the glutamate sensing system suggests that the glutamate receptors play a major role in the experience-dependent modulation of the CNS. This view was supported by the dependence of the expression of iGluRs and mGluRs on development. Thus, for example, the NMDAR subunit (GluN2A) reaches its adult expression level at postnatal day 12 (P12), while GluN2B mRNA levels fall after P12 in rats [9]; and in young mice, the hippocampal long-term potentiation (LTP), induced by 1×100 Hz tetanic stimulation of CA3 to CA1 synapses, is in part independent of the AMPAR GluA1 subunit [10], which is required for the induction of this LTP form in adult mice [11]. Similarly, the structural organization and numbers of synapses are modified postnatally, demonstrating that the maturation of synaptic transmission and plasticity is accompanied by a structural reorganization of synapses (see below). In this process, glutamatergic postsynaptic scaffolding proteins such as the SH3 and multiple ankyrin domain proteins (SHANKs) appear to play a central role. Similar to the complex glutamatergic receptor system, in mammals, the SHANKs form a huge family of scaffolding proteins encoded by three genes (*SHANK1*, *2*, and *3*) that can express multiple isoforms.

The glutamatergic system plays a central role in neurotransmission and synaptic plasticity and its modulation during development, and it is not surprising that over the past 25 years, substantial pharmacological, genetic, and experimental evidence using genetically modified mice has highlighted the importance of glutamate not only in learning and memory but also in neuropsychiatric disorders such as schizophrenia (SZ). In particular, the numerous iGluR and SHANK genes and their isoforms suggested that the sensitivity to fast signaling of extracellular glutamate is central in neurodevelopment and learning. Mutations introduced by gene targeting verified this hypothesis and provided strong experimental evidence that gene mutations in *Gria1* and the NMDAR subunit 1, *Grin1*, are associated with SZ phenotypes. Mutations that disturb this glutamate-gated NMDAR and AMPAR Ca^{2+} signaling ultimately lead to a dysfunction of neuron-to-neuron communication, neuronal network dynamics, and thus responses to environmental stimuli, as detailed below.

2. The Glutamatergic System in Neuropsychiatric Disorders

2.1. Schizophrenia (SZ)

In humans, SZ is a complex mental disorder characterized by a combination of symptoms including delusions, hallucinations, disorganized speech or behavior, lack of motivation, and cognitive deficits with a severe impact on patients' quality of life and sociability [12–14]. The prevalence of SZ is approximately 1% of the population, and the heritability was calculated to be 79% [15]. As summarized in a recent review, several pharmacological, clinical, and genetic studies could correlate (i) glutamatergic dysfunction in SZ and NMDAR hypofunction in humans, (ii) the schizophrenia-like symptoms due to the autoimmune response against the extracellular domain of NMDARs, (iii) the exacerbation of SZ symptoms in patients administered with glutamatergic receptor antagonists, and (IV) *de novo* copy number variation (CNV) encoding NMDAR subunits and other proteins in the postsynaptic density (PSD) in patients with increased risk of SZ (for a review, see [15]). As summarized in that review, a meta-analytic study reported a significant elevation of glutamate + glutamine (Glx) in glutamatergic transmission in the limbic system, but no significant difference in glutathione—a tripeptide synthesized from glutamate, cysteine, and glycine in proton magnetic resonance spectroscopic imaging (1H-MRSI) studies. Brain imaging and EEG recordings supported the involvement of the glutamatergic system, NMDARs in particular, in SZ (for a review, see [15]). Several studies using postmortem brain tissues from SZ patients showed variable changes in mRNA and protein levels for iGluRs and mGluRs in different brain regions (for reviews, see [15–17]), and in 2012, a study of the Korean population identified *GRIA1* variants, the gene for the AMPAR subunit GluA1, as a SZ risk gene [18].

This strong correlation between genetically based impairments of the fast glutamatergic system and SZ led to several neurocircuitry hypotheses. One main experimentally based hypothesis is that the hypofunction of NMDARs in cortical fast-spiking parvalbumin interneurons leads to changes in cortical network oscillations [19–21], pointing to a communication impairment in the inhibitory and excitatory systems as an underlying mechanism for neuropsychiatric disorders; this was experimentally verified for NMDAR hypomorphic, NMDAR knockout mice [22] and conditional NMDAR mutant mice. In mice lacking the NMDAR, specifically in parvalbumin-expressing neurons, the hippocampal–prefrontal coherence was altered, indicating a disturbed excitatory-inhibitory balance [19,23,24]. In mice with genetic depleted CP+AMPARs, the SZ-like phenotype was consistently confirmed in several studies [25–27] and the SZ phenotype was correlated with the loss of NMDAR-dependent LTP in hippocampal CA1 cells in adults [11,28] (for a review of genetic SZ mouse models with glutamate receptor deficiencies, see [29]).

In summary, data from clinical, pharmacological, and genetic studies strongly implicate the glutamatergic system, and in particular the Ca^{2+} signaling of NMDARs and CP+AMPARs, as the site of many of the abnormalities of brain processes that typically occur in SZ.

2.2. Autism Spectrum Disorder (ASD)

Another prominent neuropsychiatric disorder that is genetically inherited or caused by a *de novo* gene variant is ASD. Unlike SZ, ASD has a clear neurodevelopmental component. The symptoms typically appear before 3 years of age and are characterized by reduced social interactions, limited interest in communication, and repetitive patterns of behavior [30]. ASD is usually associated with other neuropsychiatric disorders, including, but not limited to, intellectual disability, anxiety, and attention-deficit hyperactivity disorder [31]. The heritability is between 70–90%, with a prevalence of around 1.5% in developed countries [32–34]. The high burden of ASD on society is further increased by the fact that its pathophysiology is largely unclear, and that effective therapies for the core symptoms of the disorder are not yet available [35].

The dysfunction of the glutamatergic system has been central to studies of neurotransmitter involvement in ASD. Both hyper- and hypo-glutamate models have been proposed on the basis of a variety of factors, such as ASD phenotypes, patient populations, experimental methods, and the brain regions studied [36–38]. The hyperglutamate theory was supported by the increased level of serum and plasma glutamate in children and adults with ASD [39–46] (for a review and meta-analysis, see [47,48]). Moreover, in the valproic acid (VPA)-induced ASD animal model, an upregulation of the GluN2A and GluN2B subunits of the NMDARs was observed with a corresponding activity-dependent long-term enhancement of synaptic transmission (LTP) [49], and an AMPAR antagonist restored the social behavior [50]. Several other studies confirmed the therapeutic effectiveness of iGluR antagonists, including topiramate (an antagonist of AMPARs/KARs) [51,52], memantine and amantadine (NMDAR antagonists) [53–56], and acamprosate (antagonist of NMDARs and mGluR5 [57,58]) in patients with ASD.

The alternative hypoglutamate theory was based on the dysfunction of the glutamate receptors in ASD and on the pharmacological effects of glutamatergic agonists in rescuing some ASD symptoms. Further evidence for a hypoglutamatergic state in ASD is provided by the therapeutic effects of piracetam, a positive AMPAR modulator [59]. Additionally, in several animal models of ASD, NMDAR signaling appears disrupted, mainly through a hypofunction mediated by the downregulation of the principal NMDAR subunit GluN1 (for a review, see [38]). Interestingly, the NMDAR agonists, e.g., D-cycloserine, which is known to modulate glutamatergic transmission [60], have been shown to improve sociability in patients with ASD [61–63] and in mouse models with an ASD-like phenotype [64–68]. In addition to the rescue of social behavior, D-cycloserine was also reported to be effective in attenuating stereotypic symptoms in adolescents and young adults with ASD [69]. AMPAR-positive allosteric modulators were able to rescue social impairment in *Cntnap2* knockout mice [50]. Various animal models of ASD have revealed alterations in the expression of glutamatergic receptors, along with their functions (for a review, see [38]).

In patients with fragile X syndrome (FXR), a leading genetic cause of autism, a hypofunction of synaptic AMPARs is postulated to be responsible for the intellectual disability (ID) and social–affective symptomatology of FXR patients. Numerous studies of *Fmr1* mouse models showed that AMPAR trafficking to the synapse is impaired in the absence of fragile X mental retardation 1 protein in *Fmr1* KO mice (for reviews, see [70,71]).

Due to the low availability of postmortem human tissues, only a few postmortem studies investigated the expression of iGluRs and mGluRs in ASD patients. These studies revealed several alterations in the expression of glutamatergic receptors in multiple brain regions [72–75] (for a review, see [47]). Genetic studies clearly implicate glutamate receptors in ASD, including NMDARs [76–79], KARs [80–82], AMPARs [83,84], and mGluRs [85] (for a review, see [47]). Many genes expressing molecular components related to the glutamatergic system have been associated with ASD, including *NRXN1*, *2*, and *3* [86–89], *NLGN1*, *3*, and *4* [90–93], *CNTNAP2* [94,95] and *SHANK1*, *2*, and *3* [88,96–100] (for a review, see [101]).

Concerning the glutamate level, several in vivo neuroimaging studies have revealed inconsistent alterations in the levels of glutamate and glutamine in various brain regions, including the cortex and basal ganglia of ASD patients [102–110]. For example, in vivo single-voxel and 1H-MRSI detected hyperglutamatergia (increased glutamatergic metabolites) in the pregenual anterior cingulate cortex in children and adolescents with ASD [102]. 1H-MRSI studies of non-clinical samples found that an increased glutamate/GABA$^+$ ratio in the right hemisphere superior temporal region was correlated with a higher expression of the social disorganization, a shared phenotype within the autistic and schizotypal spectrum [111]. On the other hand, patients with ASD showed significantly lower glutamate concentration in the right anterior cingulate cortex (ACC) [103]. Additionally, translational 1H-MRSI showed a reduced glutamate concentration in the striatum of ASD patients, which was correlated with the severity of social dysfunction, implying that this endophenotype is clinically significant [35]. This reduction was also found in a VPA mouse model of ASD and mice and rats carrying *Nlgn3* mutations, but not in other ASD rodent models [35]. These translational data support the involvement of glutamatergic dysfunction in the corticostriatal pathway in the pathophysiology of ASD.

In summary, studies on humans and rodent models indicate that the glutamatergic system dysfunction via alterations in glutamatergic receptor expression, trafficking, and their synaptic/extrasynaptic localization leads to imbalanced excitatory transmission and alterations in both NMDAR-mediated synaptic development and plasticity and mGluR-mediated signal transduction. All these amendments appear to play a significant role in the pathophysiology of ASD [112].

3. Activity-Induced Modulation of Synaptic Ultrastructure

The study of neurotransmission has led to the conclusion that physiological properties can have a structural correlate, and that different morphological parameters modify the synaptic signal. The main physiological factors altering the structure of synaptic transmission are synaptic plasticity (for reviews, see [113,114]), development [115,116] (for reviews, see [117–119]), and aging [120,121], but other elements such as drug administration have also been shown to affect the synaptic structure [122] (Figure 1). The morphological changes can be directly related to synapses, such as modifications in their density and structure, or affect other organelles involved in synaptic transmission such as mitochondria and the endoplasmic reticulum. Here, we will focus on the changes happening at the level of synapses, briefly describing the changes occurring in the presynaptic side, and focusing on the postsynaptic modifications of the glutamatergic system as well as on the density of synapses. Later, we will describe the relevance of mitochondria in synaptic transmission.

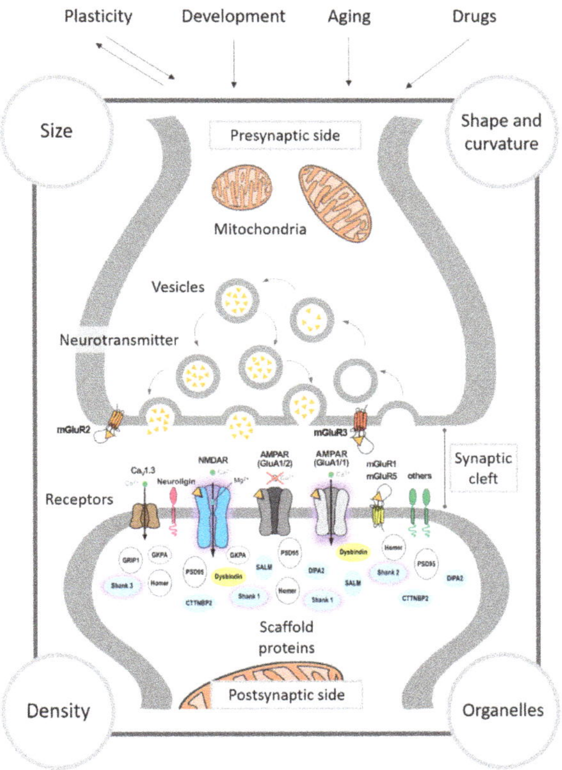

Figure 1. Factors affecting synaptic transmission. Physiological factors that affect synaptic transmission such as synaptic plasticity, development, and aging, and other non-physiological factors such as drug administration, have a synaptic ultrastructural correlate. Morphological changes include the density, size and shape of synapses, or affect organelles involved in synaptic transmision such as mitochondria. The proteins affected in the different animal models that are discussed in this review are colored according to the disorder, yellow for schizophrenia (see Table 1), and blue for autism spectrum disorder (see Table 2). Important key players in ASD and SZ are encircled in purple.

Table 1. Synaptic ultrastructural alterations in schizophrenia. Different studies have assessed the synaptic ultrastructural changes of synaptic transmission in schizophrenia in postmortem human tissues and rodent models (either chemical or genetic). The affected proteins are shown in Figure 1. ↑ stands for increased, ↓ stands for decreased. CA1: Cornu ammonis area 1, CA3: Cornu ammonis area 3, PSD: postsynaptic density.

Schizophrenia	
Humans	
Brain Region	**Ultrastructure Modifications**
Anterior cingulate cortex	↓ density of axospinous synapses and axonal mitochondria [123].
Anterior limbic cortex	↑ density of axospinous and convex synapses; ↓ density of synapses on shafts, flat and concave synapses [124].
Hippocampus CA3	↓ density of axospinous synapses [125,126].
Caudate and putamen	↑ density of in axospinous synapses the caudate matrix and putamen patches [127–129].
Nucleus accumbens	↑ density and ↓ size of axospinous synapses in the core [130].
Rodents	
Rodent Models	**Ultrastructure Modifications**
Ketamine	In posterior cingulate cortex, ↑ thickness and curvature of the synaptic interface; ↑ synaptic cleft width [131].
Dtnbp1$^{-/-}$	In CA1, ↑ vesicle size and thickness of PSD; ↓ vesicles of reserve pool and width of synaptic cleft [132].

Table 2. Synaptic ultrastructural alterations in autism spectrum disorder. Different studies have assessed the synaptic ultrastructural changes of synaptic transmission in autism spectrum disorder. Only one has been done in postmortem human tissue. Rodent models are either chemical or genetic. The affected proteins are shown in Figure 1. ↑stands for increased, ↓stands for decreased. Full names of brain regions, genes, proteins, and drugs are as in the main text and abbreviation list. Cx: cortex, HC: hippocampus, MPFC: medial prefrontal cortex, CA1: Cornu ammonis area 1, PSD: postsynaptic density.

Autism Spectrum Disorder	
Humans	
Brain Region	**Ultrastructure Modifications**
Anterior cingulate cortex	↑ large axons in deep white matter, ↑ small axons in superficial white matter [133].
Rodents	
Rodent Model	**Ultrastructure Modifications**
VPA	In Cx and HC, blurred and thickened synaptic cleft; ↓ synaptic vesicles; altered mitochondrial morphology [134].
Propionic acid	In CA1, few atypically enlarged presynaptic terminals; ↓ density of synaptic vesicles and short active zone [135].
15q11-13 duplication	In somatosensory Cx, ↓size of PSD, spine head volume ↓width, spine neck width ↓density of shaft synapses and mushroom spines; ↑ density of axospinous synapses and filopodial spines [136].
Shank1$^{-/-}$	In CA1, ↓smaller spines and thinner PSD [137].
Shank3B$^{-/-}$	In striatum, ↓ thickness and length of PSD and spine density [138].
Lrfn2$^{-/-}$	In CA1, ↑perforated synapses and synaptic cleft width; ↓PSD length; oddly shaped and spinule-like spines [139].
Cttnbp2$^{-/-}$	In dorsal HC, ↓PSD length and thickness and synaptic vesicle count [140].
Dip2a$^{-/-}$	In Cx, a stubby postsynaptic structure and flattened PSD; defect in spine morphology [141].
Clstn2$^{-/-}$	In MPFC, ↑ density of inhibitory synapses, ↑ negative curved PSD. In ADD, ↑ size of perforated PSD; ↑ density of synaptic vesicles. In HC, ↓ density of synaptic vesicles [142].
Vrk3$^{-/-}$	In CA1, ↓ PSD length and thickness [143].
Fmr1$^{-/-}$	For a review reporting the spine phenotypes in different brain regions, see [144]. In CA1, ↑ diameter of secondary dendrites and dendritic spine density; ↑ mature dendritic spines and ↑ mature postsynaptic densities [145]. In the primary motor Cx, normal density but ↑ turnover rate of dendritic spines [146].

3.1. The Presynapse

Regarding the presynaptic ultrastructure, it was shown in early studies on *Aplysia* that the presynaptic compartment of synapses could be remodeled, leading to an increase in the number, size, and vesicle complement of the active zones during long-term memory formation [147]. Later studies have revealed similar mechanisms in other species, such as *Drosophila*, where the readily releasable pool of vesicles can be dynamically modulated during plasticity [148,149], or in rodents, where the regulation of presynaptic scaffold proteins affects the availability of vesicles [150], and these can be replenished at different rates [151].

3.2. The Synaptic Cleft

It has been proposed, based on computer simulations, that changes in the width of the synaptic cleft alter synaptic function [152]. However, very little experimental research has been done to date on this matter. Glebov et al. (2016) showed a slight reduction in the width of the synaptic cleft after silencing neurons [153], and similar changes have been described in a disease model in which synaptic function is altered [132]. On the other hand, other models of synaptic malfunction exhibited opposite effects [131,139], and further research is needed.

3.3. The Postsynapse

Structural changes have been widely described in mammalian postsynapses, mainly in the glutamatergic system, in both the neocortex and hippocampus.

3.3.1. Synaptic Size

The size of the synaptic junction has been described as a structural correlate of its function. The amplitude of synaptic currents correlates with the volume of dendritic spines [154,155]. In turn, dendritic spine volume strongly correlates with synaptic size [156]. Release probability also scales with synaptic size [157], as well as synapse stability over time, thus, larger synapses survive longer than smaller ones irrespective of synaptic activity [158]. It has been proposed that large synapses represent physical traces of long-term memory, while smaller synapses would be preferential sites for LTP induction [159,160], which has been also supported by simulation studies [160,161]. However, the size of synapses undergoes significant spontaneous changes [162], challenging the notion that they could be stable traces. Thus, new theories have arisen moving the focus of memory storage from individual synapses to network connections [163,164]. In fact, synapses are not distributed in two groups with clearly different sizes. Instead, the sizes of synaptic junctions follow a continuous distribution with a single peak and a long tail to the right, which fits a log-normal distribution both in the neocortex and hippocampus [165–168]. If, as commented above, different functions are performed by synapses of different sizes, there would be a continuous transition between the two types. It is interesting to note that other synaptic parameters, such as the amplitude of unitary excitatory postsynaptic potentials (EPSPs) [169,170] and spike transmission probability [171], also follow log-normal distributions (for a review, see [172]). Model experiments also suggest that larger synapses would not only evoke larger responses but would also be more homogeneous and reliable than smaller ones (i.e., their stochastic variability is reduced when compared to smaller synapses) [173,174].

In any case, the amplitude of postsynaptic response does not depend only on the size of the synapse but also the morphology of dendritic spines [175,176], as well as the geometry of postsynaptic dendrites [177,178]. Another important factor is the concentration of postsynaptic receptors. In the somatosensory cortex of rats, synapses of different sizes express a constant density of AMPARs, so the larger the surface of the PSD, the higher the absolute number of AMPARs [179]. In the hippocampus, however, different types of synapses have different AMPAR content, and the number of AMPARs scales with the synaptic size with different slopes [180]. The fact that NMDARs tend to be more concentrated in smaller synapses [179,181] increases the complexity of the relationship between synaptic size and function. Therefore, although it is clear that the synaptic size plays an important role in synaptic behavior, other factors, including dendritic spine volume, the geometry of the postsynaptic element, or the density of postsynaptic receptors, must also be considered.

3.3.2. Shape of Synapses

Another morphological trace related to synaptic transmission is the shape of the synaptic junctions. Synapses can show a wide range of morphologies that have been classified into four main types: macular (when they are disk-shaped), perforated (with one or more holes in the PSD), horseshoe (when the perimeter is tortuous and horseshoe-shaped with an indentation), and fragmented (when the PSD is divided into two or more fragments) [182–185]. The different shapes also imply differences in the size of synapses: macular synapses have a smaller mean area than more complex morphologies, although their distributions greatly overlap [168].

The role that the shape of synapses plays in synaptic transmission is still under debate. Some findings support the notion that complex morphologies correspond to more active synapses; for example, it has been shown that in the hippocampus, there is an increase in the proportion of fragmented synapses after the induction of LTP [186,187]. It has also been shown that the shape of synapses might affect the proportion of postsynaptic receptors; for example, in the stratum radiatum of the hippocampal CA1 of adult rats, perforated synapses show a higher amount of AMPARs and NMDARs than nonperforated ones. This suggests that perforated synapses may evoke larger postsynaptic responses, and hence contribute to the enhancement of synaptic transmission associated with some forms of synaptic plasticity [188]. However, these differences may only correspond to the different sizes and might be independent of the shape of the synaptic junction.

3.3.3. Curvature of the Synaptic Apposition Surfaces

In the early 1980s, synapses were classified, based on the curvature of their synaptic junction, as positive (when it was curved into the presynaptic terminal, in posterior studies referred to as convex), flat, and negative (when it was curved towards the postsynaptic side—later referred to as concave) [189]. This study described a shift towards more positive junctions as an indication of the increasing maturity of the synapses, but they also related the degree of curvature to the use of the synapse: junctions with pronounced negative curvatures would be non-functional, while flat synapses would be more active [189]. This theory was supported by the fact that the administration of barbiturates changed the curvature of synapses towards more curved shapes [190]. Later studies have shown that not only a variety of anesthetics [191] but also other parameters influence the curvature of synapses in healthy individuals, such as age [185,192] and nutrition [193], making it difficult to draw a conclusion about its relevance. The latest studies performed in rodent hippocampus suggested that changes in synaptic curvature may influence synaptic efficacy associated with long-term depression and LTP [194]. These changes could be due to the increased fusion of vesicles that would result in a growth of the postsynaptic membrane, or to changes in the cytoskeleton caused by the Ca^{2+} influx (for a detailed description of the possible mechanisms responsible for the changes in curvature, see [195]). Furthermore, another study showed that synaptic stimulation increased the curvature of the PSD, making it more convex [196], however, the characterization of synaptic curvature in this study (as in [189,190,193]) was performed in single sections instead of 3D reconstructions, which could be misleading, since the curvature of a single synapse can change from one section to another [195].

3.4. Synaptic Density

In the cortex, most synapses (90–98%) are established in the neuropil [197], which is composed of dendrites, axons, and glial processes. Of these, most are excitatory synapses mainly located on dendritic spines, while the rest correspond to inhibitory synapses, mainly established on the dendritic shafts [167,198,199].

The general view is that each spine establishes one synapse. However, some spines present multiple synapses [167,200–203], while others show a clear lack of them [204]. The presence of multiple synapses per dendritic spine [205–207], as well as a general increase in the total number of synapses [201], has been proposed to be linked to long-term memory formation. The fact that analyzing spines is much easier, due to their larger size, than studying synapses, and the correspondence between the volume of spines and excitatory synaptic strength have led to spines being considered a good structural criterion for excitatory synapses. Different studies have shown an increase in spine number in a variety of conditions, for example, in the mammalian cortex after the exposure to an enriched environment [208]; in the hippocampus after spatial training [209], running [210], and learning [211]; and in the brain of chicks after passive avoidance training [212]. Although an increase in spine density seems to be a trend after brain stimulation, its significance for activity must be considered carefully since, as stated above, changes in spines do not necessarily reflect changes in synapses.

3.5. Mitochondria

Mitochondria play an important role in synaptic transmission by providing most of the energy required by neurons and by acting as Ca^{2+} buffers [213–216]. The ATP provided by mitochondria at the synapse is used for maintaining the resting membrane potential, reversing the ion gradient after action and postsynaptic potentials, and for G-protein signaling, neurotransmitter recycling, and vesicle cycling [217]. Mitochondria are closely associated with synapses [218] and are transported along dendrites and axons to the regions of the cell that require a high energetic supply [219–225]. In the somatosensory cortex, the volume fraction of mitochondria located in axons and dendrites correlates with the local density of synapses, while the volume fraction of mitochondria located in non-synaptic processes does not [226]. The highest density of synapses is found in layer IV, a recipient of thalamic afferents [167]. Interestingly, layer IV also shows the highest density of mitochondria [226], which is associated with the synaptic vesicle pools to ensure the efficacy of the thalamic transmission [227]. This recruitment occurs through variations in the ADP/ATP ratio, but also other factors such as the activation of glutamate receptors [228] and changes in the concentration of Ca^{2+} affect mitochondrial mobility [214,215,222]. As Ca^{2+} buffers, mitochondria play a fundamental role in protecting the terminals by taking up cytosolic Ca^{2+} during repetitive stimulation [229,230]. Synapses and mitochondria are connected even further since there are proteins that regulate the mitochondrial distribution and also influence the formation and maintenance of spines and synapses, e.g., B-cell lymphoma extra large (Bcl-xL), which leads to more mitochondria at synapses and an increase in the synapse number and size when overexpressed [231] (for a review, see [216]).

4. Excitatory Synaptic Ultrastructure Alterations in Neuropsychiatric Disorders

The ultrastructural plastic organization of synapses affects the transmission, thus determining the mechanisms of adaptive or pathological behavior [124]. For example, as discussed before, the size of the active zone and the form of synaptic contacts correlate with the functional activity and maturity of the synapse. Therefore, the analysis of the synaptic ultrastructural alterations is important for the assessment of the functional state of synapses in different neuropsychiatric disorders.

4.1. Schizophrenia

Several studies identified synaptic ultrastructural alterations as evidence for glutamatergic dysfunctions in SZ patients (Table 1 and Figure 1) (for a recent review, see [232]). These alterations may depend on differences in the stage of illness, medication status, and brain regions. In the ACC, which plays a vital role in attention, executive functions and cognitive tasks [233], response conflict [234,235], error rate [236,237], and emotional processing [238], the total synaptic density was found to be decreased by 28% in individuals with SZ compared to healthy controls [123]. However, different types of synapses showed different alterations. The density of the glutamatergic axospinous synapses was reduced by 30%, while synapses on shafts showed no difference in SZ patients. The same study also revealed a selective decrease in the density of mitochondria in axon terminals forming excitatory axospinous synapses, specifically in layer III of the ACC [123], suggesting a decrease in cortical synaptic efficiency which can impact the cognitive functions controlled by the ACC activity.

In layer II of the anterior limbic cortex, a 128% increase in the density of total axospinous synapses and a decrease of 40% of synapses on the dendritic shaft were identified in SZ subjects without determining whether the decreased numbers are in excitatory or inhibitory synapses. These results suggest a disruption in the formation of connections in the neuronal assemblages of the anterior limbic cortex or a defective elimination of synapses during early adolescence, when synaptic pruning occurs by discarding weak and redundant synaptic connections, and strengthening the remaining synapses [239,240]. Regarding the structure of synapses, the same study revealed an increase of 14% in the convex synapses in SZ cases with a decrease in flat and concave synapses by 11% and 3%, respectively [124].

In the hippocampus, two electron microscopy (EM) studies morphometric studies reported a decreased number of glutamatergic axospinous synapses between the mossy fiber axon terminals and branched dendritic spines of pyramidal neurons in the CA3 region in SZ subjects. These results reflect a decrease in the efficacy of mossy fiber synapses in the CA3 hippocampal region, which can impair cognitive maintenance [125,126].

In the caudate nucleus and putamen, as parts of the dorsal striatum that process motor, cognitive, and limbic functions, a selective increase in the glutamatergic synaptic density was reported in SZ patients, suggesting an enhanced cortical excitatory input, as the striatum receives mainly excitatory inputs from the cortex and shows major GABAergic output [127]. The caudate nucleus and putamen are composed of patches embedded in a larger matrix [241]. The matrix compartment preferentially receives inputs from the motor and somatosensory cortices [242], as well as from the dorsolateral prefrontal cortex [243] which processes cognition, including working memory [244]. On the other hand, the patch compartment receives input from the limbic system [128].

The glutamatergic axospinous synaptic density was elevated in both the caudate matrix and the putamen patches in SZ cases on typical antipsychotic drugs [128]. However, the increase in total and excitatory synapses in the patch (but not the matrix) was confined to the treatment-resistant group, while the treatment respondent group had normal levels of synapses except for an increase in the density of glutamatergic axodendritic synapses in both the patch and matrix [129]. There is therefore evidence of altered glutamatergic activity in SZ with an association between striatal function, structure, and treatment response.

The nucleus accumbens acts as a central hub for integrating signals from several regions associated with SZ, including the prefrontal cortex, hippocampus, amygdala, and thalamus [245]. It can be divided into the core and shell. In SZ patients, the core showed an increased density of excitatory axospinous synapses, although they have smaller PSDs. In contrast, the shell did not present any difference from healthy brains. Similarly, mitochondrial density in the nucleus accumbens did not show any alterations in SZ. The frequency of large elaborate multi-perforated synapses was equally found in axospinous synapses and synapses on dendritic shafts and was not altered in SZ [130]. These large synapses play a role in the complex interconnectivity of the nucleus accumbens, which makes it a unique feature of the human brain [130,232].

Despite the presence of genetic rodent models with iGluR or mGluR depletions or mutations described as having an SZ-like phenotype, these studies mainly focused on the electrophysiological alterations (for a review, see [246]) with no synaptic ultrastructural details available except for one pharmacological and one genetic-induced rodent model of SZ. In a ketamine-induced SZ rat model, a reduction in the thickness and curvature of the synaptic interface and an increase in the synaptic cleft width in the posterior cingulate cortex were found, which could be rescued by the administration of vinpocetine, a nootropic phosphodiesterase-1 inhibitor [131]. In the sandy mouse line that harbors a spontaneously occurring deletion in the *Dtnbp1* gene and expresses no dysbindin protein, although the overall appearance of presynaptic terminals and spines in the hippocampal CA1 glutamatergic synapses was normal, quantitative analysis revealed a shift of the vesicle distribution to a ~10% larger size, with a normal count of docked vesicles, a reduced count of reserve pool vesicles, and a decreased width of the synaptic cleft [132]. The glutamatergic synapses in sandy mice also exhibited an increase in the thickness of PSDs.

The inconsistent synaptic ultrastructural alterations between human patients and rodent models as well as between the different rodent models of SZ point to the severe heterogeneity of this complex disorder. Moreover, it may indicate the limitation of the rodents to modulate such complex disorders. Indeed, on the behavioral level, some of the symptoms of SZ, such as auditory hallucinations and delusions, have not yet been modeled due to the difficulty of finding a correlate in animals. In contrast, deficits in sensory processing have proven more amenable to modeling in rodents, including sensorimotor gating. Therefore, face validity is vital to signify that a rodent model can recapitulate important anatomical, biochemical, neuropathological, or behavioral features of a neuropsychiatric disorder. Additionally, before generating a new rodent model, it is important to achieve a construct validation by recreating the same mutation in the gene that was found in the patient to dissect the heterogeneous state of these disorders.

In summary, human postmortem and rodent studies have revealed several glutamatergic synaptic ultrastructural alterations related to SZ, confirming the notion of synaptic involvement in SZ (for a summary of synaptic ultrastructural alterations in SZ, see Table 1 and Figure 1).

4.2. Autism Spectrum Disorder

In ASD, efforts to carry out comparative postmortem brain studies in patients were hindered by poor tissue preservation and small sample sizes. To our knowledge, only one EM study investigating the brains of ASD patients was performed, with a small sample size of adult brains from five ASD patients and four controls. This study did not investigate the synaptic structure but rather focused on the connectivity of myelinated axons in the prefrontal cortex, revealing significantly fewer large axons in the deep white matter below the ACC and much smaller axons in the superficial white matter of the same region, suggesting increased local connectivity and decreased long-range connectivity [133].

Data on synaptic ultrastructure changes are available only for several pharmacological and genetic-induced rodent models of ASD, which document the heterogeneity of ASD (Table 2 and Figure 1). The synaptic ultrastructural alterations in two examples of the pharmacologically induced rodent models of ASD will be described in detail. An ASD rat model of prenatal exposure to VPA exhibited a blurred and thickened structure of the synaptic cleft without clearly marked pre-and postsynaptic membranes. Synaptic vesicles were found to be greatly reduced or completely absent in the presynaptic terminals in the cerebral cortex and hippocampus, with altered mitochondrial shapes and fused mitochondrial cristae and membrane [134]. In a propionic acid mouse model of ASD, multiple synaptic alterations were observed in the CA1 region of the hippocampus, including the presence of some atypically enlarged presynaptic terminals with a reduced density of synaptic vesicles and short active zones [135].

For the genetic-associated ASD rodent models, studying a mouse model with a human 15q11-13 chromosomal duplication revealed that the size of the PSD, the spine head volume, and spine head and spine neck widths were reduced in layer II/III of the somatosensory cortex, indicating a poor maturation of spines. The reported tendencies towards the increased density of spine synapses, reduced densities of shaft synapses with a trend towards increased fractions of filopodia, thin spines, and reduced fractions of mushroom spines support the idea of an altered balance between excitatory and inhibitory synapses in this model [136].

The SHANK protein family comprises a group of postsynaptic scaffolding proteins that play a vital role in the formation, organization, and signaling of glutamatergic synapses [64], acting as the central organizer of the postsynaptic proteins. Since mutations of all three members of the *SHANK* gene family are associated with ASD, the ultrastructural alteration in *Shank* mutant mice was expected to give information on the functional impairment of the network and to provide a link to the alterations found in SZ, in particular *SHANK2* variants associated with SZ [247]. The deletion of *Shank1* in mice displaying an ASD-like phenotype leads to smaller spines, thinner PSDs, and a weak synaptic transmission in the hippocampal CA1 region [137]. In *Shank3B*$^{-/-}$ mice revealing an ASD-like phenotype, the thickness and length of the PSD in the striatum were decreased, with a reduction in the spine density accompanied by a reduced excitatory synaptic transmission, pointing to the disruption of glutamatergic signaling [138]. In another ultrastructural study in the hippocampal CA1 field in *Shank3*-deficient mice, the presence of synaptic pathology at different developmental stages (5 weeks and 3 months of age) was assessed. *Shank3*$^{+/-}$ heterozygote mice had significantly more perforated synapses at 5 weeks than at 3 months of age and significantly more than 5-week-old controls, indicating that the ultrastructural morphological alterations affecting synaptic structure may occur in an age-dependent manner in *Shank3*-deficient mice [248].

LRFN2/SALM1 is a PSD-95-interacting synapse adhesion molecule [249]. In the *Lrfn2*-deficient mouse model of ASD, the CA1 stratum radiatum of the dorsal hippocampus revealed an increased ratio of perforated synapses to total excitatory synapses. Moreover, the synapses had a shorter PSD length but a similar thickness and a wider synaptic cleft. Furthermore, an unstable spine structure was suggested by the presence of oddly shaped, spinule-like spines and perforated PSDs [139].

The ASD-associated cortactin binding protein 2 (CTTNBP2) is known to regulate the subcellular distribution of synaptic proteins, such as cortactin, thereby controlling dendritic spine formation and maintenance [250]. In the *Cttnbp2*$^{-/-}$ mouse model of ASD, in all regions of the dorsal hippocampus, these mice showed a reduction in the length and thickness of PSDs, the count of presynaptic vesicles, and the ratio of vesicle number to PSD length [140].

The ASD candidate gene *DIP2A* (disconnected-interacting protein homolog 2 A) encodes for a protein that is localized to dendritic spines in excitatory neurons [141]. The deletion of *Dip2a* in mice exhibited a postsynaptic structure with a stubby appearance and flattened PSD compared to an enlarged postsynaptic terminal with a distinct neck and abundant PSDs in control mice [141]. These results were accompanied by a defect in spine morphology and synaptic transmission, which may be mediated by PSD size and glutamate receptor dysfunction.

Calsyntenin-2 (CLSTN2) is a synaptic protein that belongs to the superfamily of cadherins and has an important function in learning and memory [251]. In the *Clstn2*$^{-/-}$ mouse model of ASD, the synaptic ultrastructural analysis revealed a reduction in the density of inhibitory synapses both in the medial prefrontal cortex (MPFC) and the hippocampus, as well as a general increase in negatively curved PSDs. The MPFC also showed a reduction in the length of perforated synapses as well as in the width of the synaptic cleft PSD, presenting a surprisingly increased number of synaptic vesicles. In contrast, the presynaptic area of the hippocampal neurons was not changed in *Clstn2* knockout mice, but the number of synaptic vesicles was significantly reduced [142].

In addition to the aforementioned synaptic proteins, nuclear proteins associated with ASD can also affect the synaptic ultrastructure, e.g., vaccinia-related kinases (VRKs) that play a major role in cell signaling, cell cycle progression, apoptosis, and neuronal development. In the hippocampal CA1 region of a VRK3-deficient mouse model of ASD, prominent reductions in PSD length and thickness were revealed [143].

Adding another layer of complexity, a mouse model of Phelan–McDermid syndrome causing autistic phenotypes with a deficiency of mitogen-activated protein kinase 8 interacting protein 2 (MAPK8IP2/IB2), which plays an important role in regulating the ratio of AMPARs to NMDARs at glutamate synapses, revealed no synaptic ultrastructure alterations of cerebellar glutamatergic synapses and featured normal synaptic clefts and postsynaptic densities and abundant presynaptic vesicles [252]. However, despite the normal ultrastructure, a larger NMDAR-mediated current and enhanced intrinsic excitability and LTP were revealed in this mouse model [253].

Although altered dendritic spine morphology is a hallmark of FXS, the degree of spine abnormalities observed in the *Fmr1* KO mouse model is variable (for a review reporting the spine phenotypes in different brain regions of *Fmr1* KO, see [144]). Additionally, in more recent studies, a significant increase in the diameter of secondary dendrites, an increase in dendritic spine density, a decrease in mature dendritic spines and less mature postsynaptic densities were revealed in the hippocampal CA1 region of adult $Fmr1^{-/-}$ mice [145]. In the primary motor cortex of $Fmr1^{-/-}$ mice, a normal density but higher turnover rate of dendritic spines were shown [146].

In summary, several pieces of evidence from rodent studies have revealed heterogeneous synaptic ultrastructural alterations related to ASD, supporting the hypothesis that ASD is, in part, the consequence of a developmental synaptopathy (for a summary of synaptic ultrastructural alterations in ASD, see Table 2 and Figure 1). Improving the quality of postmortem tissue preparation from ASD patients by the development of advanced methods can confirm the synaptic ultrastructural changes revealed in rodent models of ASD and pave the way for dissecting the heterogeneity of the disorder. Moreover, it will facilitate finding ultrastructural alterations that are common with other neuropsychiatric disorders and contribute to their pathophysiology.

5. Conclusions

Numerous experimental and EM-imaging methods have shown that in genetically and pharmacologically caused SZ and ASD, glutamate concentration, glutamatergic receptor expression, and the ultrastructure of glutamatergic synapses are differentially affected in different brain regions, leading to alterations in synaptic signal transmission. Various molecular, physiological, and structural changes of synapses have already been shown to occur during brain development or learning. Those studies identified the glutamatergic neurotransmission as an essential mechanism for an activity-dependent modulation of synapses and the formation (learning) or stabilization (memory) of neural networks. For those network changes, the Ca^{2+} influx at postsynaptic NMDARs and AMPARs and the release of intracellular Ca^{2+} at the activated synapse are essential to initiate the required modification of the synapses that experienced activation through depolarization of the postsynaptic membrane. If the intracellular Ca^{2+} level rises too far, neuronal death will result [254]. Thus, the amount and type of Ca^{2+} increase after glutamate stimulation determines whether and which molecular changes will be activated in the neuron. This could be associated with structural changes in the synapse(s) and increased/decreased excitability. In SZ and ASD, the glutamatergic signal transduction in neuronal networks that controls social and other behaviors is dysregulated. These structural alterations can have a clear impact on synaptic neurotransmission and plasticity, which lead to neuronal circuit defects. The maintenance of cognition and normal behaviors is heavily dependent on the precise formation of mature neuronal circuits. For example, the increase in glutamatergic synaptic density can cause hyper-excitability of cortical circuits consistent with multiple lines of evidence that implicate imbalances in excitatory and inhibitory activity as a shared pathophysiological mechanism in SZ and ASD. Additionally, as synaptic size and shape can have an effect on the number of AMPARs and NMDARs, key components of learning and memory, the alteration of these factors can explain the cognition impairment and memory dysfunction in these neuropsychiatric disorders. However, to understand the different features of both neuropsychiatric disorders and their individual heterogeneity, a more detailed picture of the neuronal network and more comparative ultrastructural information are required. Importantly, structural and functional studies on SZ and ASD also show that the glutamatergic synapses do not appear to lose their plasticity during maturation and learning. In rodent models and human patients, the impaired glutamatergic synaptic plasticity and social behavior can be at least partially normalized genetically or by compensatory medication, as indicated by the rescue of the social impairment via restoration of Shank3 expression in adult *Shank3* KO mice [255] and by the treatment of adult *Shank2* KO mice with an NMDAR agonist, D-cycloserine [65]. Additionally, both experimental and medical interventions were associated with normalized social behavior but also improved learning abilities in adult animals and patients. Detailed synaptic ultrastructural analysis of AMPAR and NMDAR KO mice are necessary to demonstrate that changes in synaptic structures are associated with SZ and ASD phenotypes recognized in AMPAR and NMDAR KO mice, as noticed in patients and the other mouse models for ASD and SZ; in particular, the loss of NMDAR in hippocampal CA3 to CA1 synapses and GluA1 receptor plasticity is associated with impairments in decision making, which depends on the recognition and correct evaluation of environmental stimuli but is not involved in the long-term memory formation [256,257].

Since astrocytes in tripartite synapses also influence synaptic transmission and react to inflowing and intracellularly released Ca^{2+} signals, the role of astrocytes and their mitochondria in neuropsychiatric diseases should be investigated.

Author Contributions: A.E. conceptualized the idea of the review. A.E., A.S., A.M.-P., and R.S. wrote the review. A.E. and A.S. prepared the figure and tables. R.S. provided the receptor cartoons for Figure 1. All authors have read and agreed to the published version of the manuscript.

Funding: A.E. was supported by the German Research Foundation (DFG, Research Unit FOR-2715, grant HE8155/1-1). A.S. received support from the Alexander von Humboldt Foundation, and R.S. received support from the Ingeborg Ständer Foundation.

Institutional Review Board Statement: Not applicable.

Informed Consent Statement: Not applicable.

Data Availability Statement: Not applicable.

Acknowledgments: We would like to sincerely thank John Wray for his critical proofreading, Ulrike Hedrich for her comments on the manuscript, and Shaimaa Madbouly for her constant support.

Conflicts of Interest: The authors declare no conflict of interest.

Abbreviations

1-H-MRSI, Proton magnetic resonance spectroscopic imaging; **ACC**, Anterior cingulate cortex; **AMPAR**, α-amino-3-hydroxy-5-methylisoxazole-4-propionate receptor; **ASD**, Autism spectrum disorder; **Bcl-xL**, B-cell lymphoma extra large; **CA1**, Cornu ammonis area 1; **CA2**, Cornu ammonis area 2; **CA3**, Cornu ammonis area 3; ***Clstn2***, Mouse gene for calsyntenin-2; **CNV**, Copy number variation; ***CNTNAP2***, Human gene for contactin-associated protein-like 2; **CP-AMPAR**, Ca^{2+}-impermeable AMPA receptors; **CP+AMPAR**, Ca^{2+}-permeable AMPA receptors; **CTTNBP2**, Cortactin-binding protein 2; **Cx**, Cortex; **DIP2A**, Disconnected-interacting protein homolog 2A; ***DIP2A*** Human gene for disconnected-interacting protein homolog 2 A; **EPSP**, Excitatory postsynaptic potential; **FXR**, Fragile X syndrome; **GluA1**, Glutamate receptor GluA1 receptor subunit 1; **GluN1**, Glutamate receptor NMDA receptor subunit 1; **GluN2A,B,C,D**, Glutamate receptor NMDA receptor subunit 2A,B,C,D; **GluA1,2,3,4**, Glutamate receptor AMPA receptor subunit A,B,C,D; ***Gria1***, Mouse gene for GluA1; ***Grin1***, Mouse gene for GluN1; **iGluR**, Ionotropic glutamate-gated receptor; **ID**, Intellectual disability; **HC**, Hippocampus; **KAR**— Kainate receptor; ***LRFN2/SALM1***, Synaptic adhesion-like molecule/leucine-rich repeat and fibronectin type III domain containing; **LTP**, Long-term potentiation; **mGluR**, Metabotropic glutamate receptor; **MMRRC**, Mutant Mouse Resource and Research Centers by NIH; **MPFC**, Medial prefrontal cortex; ***NLGN***, Human gene for neurolegin; ***NMDAR***, *N*-methyl-D-aspartate receptor; ***NRXN***, Human gene for neurexin; **P12**, Postnatal day 12; **PSD**, Postsynaptic density; **SH3**, Src-homology domain; **SHANKs**, SH3 and multiple ankyrin domain proteins; ***SHANK1, 2, 3***, Human SHANK1, 2, and 3 genes; ***Shank1, 2, 3***, Mouse genes for SHANK1, 2, and 3 proteins; **SZ**, Schizophrenia; **VPA**, Valproic acid; **VRK**, Vaccinia-related kinase.

Mouse lines

Clstn2$^{-/-}$, global depletion Clstn2—not listed—a CRISPR-Cas9-edited KO mouse line is available at MMRRC: Clstn2$^{em1(IMPC)J}$ (MGI:5766210);

Cttnbp2$^{-/-}$, global depletion of Cttnbp2—not listed—a CRISPR-Cas9-edited KO mouse line is available: Cttnbp2$^{tm1a(KOMP)Wtsi}$ (MGI:4455588);

Dip2a$^{-/-}$, global depletion of Dip3a—not listed—a CRISPR-Cas9-edited KO mouse line is available: Dip2a$^{em\#Ywz}$ (MGI:5787957);

Dtnbp1$^{-/-}$, DtnbP1 (floxed mice B6(Cg)-Dtnbp1tm1c(EUCOMM)Hmgu/DtlJ are available at Jackson labs Stock 030840);

Fmr$^{-/-}$, global Fmr depletion: Fmr1^{tm1Cgr} (MGI:1857169) or Fmr1^{tm2Cgr} (MGI:245108); ***Gria1***$^{-/-}$, global GluA1 depletion: Gria1^{tm1Rsp} (MGI:2178057);

Grin1$^{neo/neo}$, hypomorophic GluN1 expression: Grin1^{tm1Bhk} (MGI:1928280);

Grin1$^{f/f}$***/PV-Cre***$^{-/-}$, GluN1 depletion in parvalbumin-positive neurons: Grin1^{tm1Rsp} (MGI:3611337 (MGI:3590684);

IB2$^{-/-}$, global depletion of Mapk81P/Ib2: Mapk8ip2$^{tm1.1Gol}$ (MGI:4867721);

Lrfn2$^{-/-}$, global depletion of Lrfn2/Salami—not listed—a similar 2lox mouse strain is available from Jackson labs: B6J(Cg)-Lrfn2$^{tm1.2Csbd}$/Mmjax (MMRRC Stock No: 50502-JAX);

Shank1$^{-/-}$, global Shank1 depletion: Shank1^{tm1Shng} (MGI:3762757); ***Shank3B***$^{-/-}$, global depletion of the Shank3B isoform: Shank3^{tm2Gfng} (MGI:4949738);

Vrk3$^{-/-}$, global depletion of Vrk3—not listed—a CRISPR-Cas9-edited KO mouse line is available at MMRRC: VRK3em1$^{(IMPC)MBP/MmUCD}$ (MGI: 63152674).

References

1. Rothman, D.L.; Behar, K.L.; Hyder, F.; Shulman, R.G. In vivo NMR studies of the glutamate neurotransmitter flux and neuroenergetics: Implications for brain function. *Annu. Rev. Physiol.* **2003**, *65*, 401–427. [CrossRef] [PubMed]
2. Howes, O.; McCutcheon, R.; Stone, J. Glutamate and dopamine in schizophrenia: An update for the 21st century. *J. Psychopharmacol.* **2015**, *29*, 97–115. [CrossRef] [PubMed]
3. Stone, J.M. Glutamatergic antipsychotic drugs: A new dawn in the treatment of schizophrenia? *Ther. Adv. Psychopharmacol.* **2011**, *1*, 5–18. [CrossRef] [PubMed]
4. Traynelis, S.F.; Wollmuth, L.P.; McBain, C.J.; Menniti, F.S.; Vance, K.M.; Ogden, K.K.; Hansen, K.B.; Yuan, H.; Myers, S.J.; Dingledine, R. Glutamate receptor ion channels: Structure, regulation, and function. *Pharmacol. Rev.* **2010**, *62*, 405–496. [CrossRef]
5. Park, P.; Kang, H.; Sanderson, T.M.; Bortolotto, Z.A.; Georgiou, J.; Zhuo, M.; Kaang, B.-K.; Collingridge, G.L. On the Role of Calcium-Permeable AMPARs in Long-Term Potentiation and Synaptic Tagging in the Rodent Hippocampus. *Front. Synaptic Neurosci.* **2019**, *11*. [CrossRef]
6. Nicoll, R.A.; Malenka, R.C. Contrasting properties of two forms of long-term potentiation in the hippocampus. *Nature* **1995**, *377*, 115–118. [CrossRef]
7. Niswender, C.M.; Conn, P.J. Metabotropic glutamate receptors: Physiology, pharmacology, and disease. *Annu. Rev. Pharmacol. Toxicol.* **2010**, *50*, 295–322. [CrossRef]
8. Kew, J.N.; Kemp, J.A. Ionotropic and metabotropic glutamate receptor structure and pharmacology. *Psychopharmacology* **2005**, *179*, 4–29. [CrossRef]
9. Monyer, H.; Burnashev, N.; Laurie, D.J.; Sakmann, B.; Seeburg, P.H. Developmental and regional expression in the rat brain and functional properties of four NMDA receptors. *Neuron* **1994**, *12*, 529–540. [CrossRef]
10. Jensen, V.; Kaiser, K.M.; Borchardt, T.; Adelmann, G.; Rozov, A.; Burnashev, N.; Brix, C.; Frotscher, M.; Andersen, P.; Hvalby, O.; et al. A juvenile form of postsynaptic hippocampal long-term potentiation in mice deficient for the AMPA receptor subunit GluR-A. *J. Physiol.* **2003**, *553*, 843–856. [CrossRef]
11. Zamanillo, D.; Sprengel, R.; Hvalby, O.; Jensen, V.; Burnashev, N.; Rozov, A.; Kaiser, K.M.; Köster, H.J.; Borchardt, T.; Worley, P.; et al. Importance of AMPA receptors for hippocampal synaptic plasticity but not for spatial learning. *Science* **1999**, *284*, 1805–1811. [CrossRef] [PubMed]
12. Barch, D.M.; Ceaser, A. Cognition in schizophrenia: Core psychological and neural mechanisms. *Trends Cogn. Sci.* **2012**, *16*, 27–34. [CrossRef] [PubMed]
13. Insel, T.; Cuthbert, B.; Garvey, M.; Heinssen, R.; Pine, D.S.; Quinn, K.; Sanislow, C.; Wang, P. Research domain criteria (RDoC): Toward a new classification framework for research on mental disorders. *Am. J. Psychiatry* **2010**, *167*, 748–751. [CrossRef] [PubMed]
14. Scoriels, L.; Barnett, J.H.; Soma, P.K.; Sahakian, B.J.; Jones, P.B. Effects of modafinil on cognitive functions in first episode psychosis. *Psychopharmacology* **2012**, *220*, 249–258. [CrossRef]
15. Uno, Y.; Coyle, J.T. Glutamate hypothesis in schizophrenia. *Psychiatry Clin. Neurosci.* **2019**, *73*, 204–215. [CrossRef]
16. Catts, V.S.; Lai, Y.L.; Weickert, C.S.; Weickert, T.W.; Catts, S.V. A quantitative review of the postmortem evidence for decreased cortical *N*-methyl-D-aspartate receptor expression levels in schizophrenia: How can we link molecular abnormalities to mismatch negativity deficits? *Biol. Psychol.* **2016**, *116*, 57–67. [CrossRef]

17. Kristiansen, L.V.; Huerta, I.; Beneyto, M.; Meador-Woodruff, J.H. NMDA receptors and schizophrenia. *Curr. Opin. Pharmacol.* **2007**, *7*, 48–55. [CrossRef]
18. Kang, W.S.; Park, J.K.; Kim, S.K.; Park, H.J.; Lee, S.M.; Song, J.Y.; Chung, J.H.; Kim, J.W. Genetic variants of GRIA1 are associated with susceptibility to schizophrenia in Korean population. *Mol. Biol. Rep.* **2012**, *39*, 10697–10703. [CrossRef]
19. Jadi, M.P.; Behrens, M.M.; Sejnowski, T.J. Abnormal Gamma Oscillations in *N*-methyl-D-aspartate Receptor Hypofunction Models of Schizophrenia. *Biol. Psychiatry* **2016**, *79*, 716–726. [CrossRef]
20. Lisman, J.E.; Coyle, J.T.; Green, R.W.; Javitt, D.C.; Benes, F.M.; Heckers, S.; Grace, A.A. Circuit-based framework for understanding neurotransmitter and risk gene interactions in schizophrenia. *Trends Neurosci.* **2008**, *31*, 234–242. [CrossRef]
21. Uhlhaas, P.J.; Singer, W. High-frequency oscillations and the neurobiology of schizophrenia. *Dialogues Clin. Neurosci.* **2013**, *15*, 301–313. [CrossRef] [PubMed]
22. Mohn, A.R.; Gainetdinov, R.R.; Caron, M.G.; Koller, B.H. Mice with reduced NMDA receptor expression display behaviors related to schizophrenia. *Cell* **1999**, *98*, 427–436. [CrossRef]
23. Bygrave, A.M.; Kilonzo, K.; Kullmann, D.M.; Bannerman, D.M.; Kätzel, D. Can *N*-methyl-D-aspartate Receptor Hypofunction in Schizophrenia Be Localized to an Individual Cell Type? *Front. Psychiatry* **2019**, *10*. [CrossRef] [PubMed]
24. Bygrave, A.M.; Masiulis, S.; Nicholson, E.; Berkemann, M.; Barkus, C.; Sprengel, R.; Harrison, P.J.; Kullmann, D.M.; Bannerman, D.M.; Kätzel, D. Knockout of NMDA-receptors from parvalbumin interneurons sensitizes to schizophrenia-related deficits induced by MK-801. *Transl. Psychiatry* **2016**, *6*, e778. [CrossRef] [PubMed]
25. Fitzgerald, P.; Barkus, C.; Feyder, M.; Wiedholz, L.; Chen, Y.-C.; Karlsson, R.-M.; Machado-Vieira, R.; Graybeal, C.; Sharp, T.; Zarate, C.; et al. Does gene deletion of AMPA GluA1 phenocopy features of schizoaffective disorder? *Neurobiol. Dis.* **2010**, *40*, 608–621. [CrossRef] [PubMed]
26. Barkus, C.; Feyder, M.; Graybeal, C.; Wright, T.; Wiedholz, L.; Izquierdo, A.; Kiselycznyk, C.; Schmitt, W.; Sanderson, D.J.; Rawlins, J.N.; et al. Do GluA1 knockout mice exhibit behavioral abnormalities relevant to the negative or cognitive symptoms of schizophrenia and schizoaffective disorder? *Neuropharmacology* **2012**, *62*, 1263–1272. [CrossRef] [PubMed]
27. Sanderson, D.J.; Lee, A.; Sprengel, R.; Seeburg, P.H.; Harrison, P.J.; Bannerman, D.M. Altered balance of excitatory and inhibitory learning in a genetically modified mouse model of glutamatergic dysfunction relevant to schizophrenia. *Sci. Rep.* **2017**, *7*, 1765. [CrossRef]
28. Wiedholz, L.M.; Owens, W.A.; Horton, R.E.; Feyder, M.; Karlsson, R.M.; Hefner, K.; Sprengel, R.; Celikel, T.; Daws, L.C.; Holmes, A. Mice lacking the AMPA GluR1 receptor exhibit striatal hyperdopaminergia and 'schizophrenia-related' behaviors. *Mol. Psychiatry* **2008**, *13*, 631–640. [CrossRef]
29. Inta, D.; Monyer, H.; Sprengel, R.; Meyer-Lindenberg, A.; Gass, P. Mice with genetically altered glutamate receptors as models of schizophrenia: A comprehensive review. *Neurosci. Biobehav. Rev.* **2009**, *34*, 285–294. [CrossRef]
30. Lehman, J.F. *The Diagnostic and Statistical Manual of Mental Disorders*; American Psychiatric Association: Washington, DC, USA, 2000.
31. Fernell, E. Further studies of GABA and Glutamate imbalances in autism are important challenges for future research. *Acta Paediatr.* **2019**, *108*, 200–201. [CrossRef]
32. Bailey, A.; Le Couteur, A.; Gottesman, I.; Bolton, P.; Simonoff, E.; Yuzda, E.; Rutter, M. Autism as a strongly genetic disorder: Evidence from a British twin study. *Psychol. Med.* **1995**, *25*, 63–77. [CrossRef] [PubMed]
33. Prevalence of autism spectrum disorder among children aged 8 years—Autism and developmental disabilities monitoring network, 11 sites, United States, 2010. *MMWR Surveill. Summ.* **2014**, *63*, 1–21.
34. Lyall, K.; Croen, L.; Daniels, J.; Fallin, M.D.; Ladd-Acosta, C.; Lee, B.K.; Park, B.Y.; Snyder, N.W.; Schendel, D.; Volk, H.; et al. The Changing Epidemiology of Autism Spectrum Disorders. *Annu. Rev. Public Health* **2017**, *38*, 81–102. [CrossRef]
35. Horder, J.; Petrinovic, M.M.; Mendez, M.A.; Bruns, A.; Takumi, T.; Spooren, W.; Barker, G.J.; Künnecke, B.; Murphy, D.G. Glutamate and GABA in autism spectrum disorder-a translational magnetic resonance spectroscopy study in man and rodent models. *Transl. Psychiatry.* **2018**, *25*, 106. [CrossRef]
36. Carlsson, M.L. Hypothesis: Is infantile autism a hypoglutamatergic disorder? Relevance of glutamate—Serotonin interactions for pharmacotherapy. *J. Neural. Transm.* **1998**, *105*, 525–535. [CrossRef]
37. Fatemi, S.H. The hyperglutamatergic hypothesis of autism. *Prog. Neuropsychopharmacol. Biol. Psychiatry* **2008**, *32*, 911. [CrossRef]
38. Uzunova, G.; Hollander, E.; Shepherd, J. The role of ionotropic glutamate receptors in childhood neurodevelopmental disorders: Autism spectrum disorders and fragile x syndrome. *Curr. Neuropharmacol.* **2014**, *12*, 71–98. [CrossRef]
39. Moreno, H.; Borjas, L.; Arrieta, A.; Saez, L.; Prassad, A.; Estevez, J.; Bonilla, E. Clinical heterogeneity of the autistic syndrome: A study of 60 families. *Investig. Clin.* **1992**, *33*, 13–31.
40. Moreno-Fuenmayor, H.; Borjas, L.; Arrieta, A.; Valera, V.; Socorro-Candanoza, L. Plasma excitatory amino acids in autism. *Investig. Clin.* **1996**, *37*, 113–128.
41. Aldred, S.; Moore, K.M.; Fitzgerald, M.; Waring, R.H. Plasma amino acid levels in children with autism and their families. *J. Autism Dev. Disord.* **2003**, *33*, 93–97. [CrossRef]
42. Shinohe, A.; Hashimoto, K.; Nakamura, K.; Tsujii, M.; Iwata, Y.; Tsuchiya, K.J.; Sekine, Y.; Suda, S.; Suzuki, K.; Sugihara, G.; et al. Increased serum levels of glutamate in adult patients with autism. *Prog. Neuropsychopharmacol. Biol. Psychiatry* **2006**, *30*, 1472–1477. [CrossRef] [PubMed]

43. Tirouvanziam, R.; Obukhanych, T.V.; Laval, J.; Aronov, P.A.; Libove, R.; Banerjee, A.G.; Parker, K.J.; O'Hara, R.; Herzenberg, L.A.; Herzenberg, L.A.; et al. Distinct plasma profile of polar neutral amino acids, leucine, and glutamate in children with Autism Spectrum Disorders. *J. Autism Dev. Disord.* **2012**, *42*, 827–836. [CrossRef] [PubMed]
44. Shimmura, C.; Suda, S.; Tsuchiya, K.J.; Hashimoto, K.; Ohno, K.; Matsuzaki, H.; Iwata, K.; Matsumoto, K.; Wakuda, T.; Kameno, Y.; et al. Alteration of plasma glutamate and glutamine levels in children with high-functioning autism. *PLoS ONE* **2011**, *6*, e25340. [CrossRef] [PubMed]
45. Hassan, T.H.; Abdelrahman, H.M.; Abdel Fattah, N.R.; El-Masry, N.M.; Hashim, H.M.; El-Gerby, K.M.; Abdel Fattah, N.R. Blood and brain glutamate levels in children with autistic disorder. *Res. Autism Spectr. Disord.* **2013**, *7*, 541–548. [CrossRef]
46. Khalifa, D.; Shahin, O.; Salem, D.; Raafat, O. Serum glutamate was elevated in children aged 3–10 years with autism spectrum disorders when they were compared with controls. *Acta Paediatr.* **2019**, *108*, 295–299. [CrossRef]
47. Rojas, D.C. The role of glutamate and its receptors in autism and the use of glutamate receptor antagonists in treatment. *J. Neural. Transm.* **2014**, *121*, 891–905. [CrossRef]
48. Zheng, Z.; Zhu, T.; Qu, Y.; Mu, D. Blood Glutamate Levels in Autism Spectrum Disorder: A Systematic Review and Meta-Analysis. *PLoS ONE* **2016**, *11*, e0158688. [CrossRef]
49. Rinaldi, T.; Kulangara, K.; Antoniello, K.; Markram, H. Elevated NMDA receptor levels and enhanced postsynaptic long-term potentiation induced by prenatal exposure to valproic acid. *Proc. Natl. Acad. Sci. USA* **2007**, *104*, 13501–13506. [CrossRef]
50. Kim, J.W.; Park, K.; Kang, R.J.; Gonzales, E.L.T.; Kim, D.G.; Oh, H.A.; Seung, H.; Ko, M.J.; Kwon, K.J.; Kim, K.C.; et al. Pharmacological modulation of AMPA receptor rescues social impairments in animal models of autism. *Neuropsychopharmacology* **2019**, *44*, 314–323. [CrossRef]
51. Rezaei, V.; Mohammadi, M.R.; Ghanizadeh, A.; Sahraian, A.; Tabrizi, M.; Rezazadeh, S.A.; Akhondzadeh, S. Double-blind, placebo-controlled trial of risperidone plus topiramate in children with autistic disorder. *Prog. Neuropsychopharmacol. Biol. Psychiatry* **2010**, *34*, 1269–1272. [CrossRef]
52. Doyle, C.A.; McDougle, C.J. Pharmacologic treatments for the behavioral symptoms associated with autism spectrum disorders across the lifespan. *Dialogues Clin. Neurosci.* **2012**, *14*, 263–279.
53. Erickson, C.A.; Posey, D.J.; Stigler, K.A.; Mullett, J.; Katschke, A.R.; McDougle, C.J. A retrospective study of memantine in children and adolescents with pervasive developmental disorders. *Psychopharmacology* **2007**, *191*, 141–147. [CrossRef] [PubMed]
54. Owley, T.; Salt, J.; Guter, S.; Grieve, A.; Walton, L.; Ayuyao, N.; Leventhal, B.L.; Cook, E.H., Jr. A prospective, open-label trial of memantine in the treatment of cognitive, behavioral, and memory dysfunction in pervasive developmental disorders. *J. Child. Adolesc. Psychopharmacol.* **2006**, *16*, 517–524. [CrossRef] [PubMed]
55. Chez, M.G.; Burton, Q.; Dowling, T.; Chang, M.; Khanna, P.; Kramer, C. Memantine as adjunctive therapy in children diagnosed with autistic spectrum disorders: An observation of initial clinical response and maintenance tolerability. *J. Child. Neurol.* **2007**, *22*, 574–579. [CrossRef] [PubMed]
56. King, B.H.; Wright, D.M.; Handen, B.L.; Sikich, L.; Zimmerman, A.W.; McMahon, W.; Cantwell, E.; Davanzo, P.A.; Dourish, C. T.; Dykens, E.M.; et al. Double-blind, placebo-controlled study of amantadine hydrochloride in the treatment of children with autistic disorder. *J. Am. Acad. Child. Adolesc. Psychiatry* **2001**, *40*, 658–665. [CrossRef] [PubMed]
57. Erickson, C.A.; Early, M.; Stigler, K.A.; Wink, L.K.; Mullett, J.E.; McDougle, C.J. An open-label naturalistic pilot study of acamprosate in youth with autistic disorder. *J. Child. Adolesc. Psychopharmacol.* **2011**, *21*, 565–569. [CrossRef] [PubMed]
58. Erickson, C.A.; Wink, L.K.; Ray, B.; Early, M.C.; Stiegelmeyer, E.; Mathieu-Frasier, L.; Patrick, V.; Lahiri, D.K.; McDougle, C.J. Impact of acamprosate on behavior and brain-derived neurotrophic factor: An open-label study in youth with fragile X syndrome. *Psychopharmacology* **2013**, *228*, 75–84. [CrossRef]
59. Akhondzadeh, S.; Tajdar, H.; Mohammadi, M.R.; Mohammadi, M.; Nouroozinejad, G.H.; Shabstari, O.L.; Ghelichnia, H.A. A double-blind placebo controlled trial of piracetam added to risperidone in patients with autistic disorder. *Child. Psychiatry Hum. Dev.* **2008**, *39*, 237–245. [CrossRef]
60. Henter, I.D.; de Sousa, R.T.; Zarate, C.A., Jr. Glutamatergic Modulators in Depression. *Harv Rev. Psychiatry* **2018**, *26*, 307–319. [CrossRef]
61. Posey, D.J.; Kem, D.L.; Swiezy, N.B.; Sweeten, T.L.; Wiegand, R.E.; McDougle, C.J. A Pilot Study of d-Cycloserine in Subjects With Autistic Disorder. *Am. J. Psychiatry* **2004**, *161*, 2115–2117. [CrossRef]
62. Urbano, M.; Okwara, L.; Manser, P.; Hartmann, K.; Deutsch, S.I. A trial of d-cycloserine to treat the social deficit in older adolescents and young adults with autism spectrum disorders. *J. Neuropsychiatry Clin. Neurosci.* **2015**, *27*, 133–138. [CrossRef]
63. Wink, L.K.; Minshawi, N.F.; Shaffer, R.C.; Plawecki, M.H.; Posey, D.J.; Horn, P.S.; Adams, R.; Pedapati, E.V.; Schaefer, T.L.; McDougle, C.J.; et al. d-Cycloserine enhances durability of social skills training in autism spectrum disorder. *Mol. Autism.* **2017**, *8*, 2. [CrossRef] [PubMed]
64. Eltokhi, A.; Rappold, G.; Sprengel, R. Distinct Phenotypes of Shank2 Mouse Models Reflect Neuropsychiatric Spectrum Disorders of Human Patients With SHANK2 Variants. *Front. Mol. Neurosci.* **2018**, *11*. [CrossRef] [PubMed]
65. Won, H.; Lee, H.R.; Gee, H.Y.; Mah, W.; Kim, J.I.; Lee, J.; Ha, S.; Chung, C.; Jung, E.S.; Cho, Y.S.; et al. Autistic-like social behaviour in Shank2-mutant mice improved by restoring NMDA receptor function. *Nature* **2012**, *486*, 261–265. [CrossRef]
66. Burket, J.A.; Benson, A.D.; Tang, A.H.; Deutsch, S.I. D-Cycloserine improves sociability in the BTBR T+ Itpr3tf/J mouse model of autism spectrum disorders with altered Ras/Raf/ERK1/2 signaling. *Brain Res. Bull.* **2013**, *96*, 62–70. [CrossRef]

67. Um, S.M.; Ha, S.; Lee, H.; Kim, J.; Kim, K.; Shin, W.; Cho, Y.S.; Roh, J.D.; Kang, J.; Yoo, T.; et al. NGL-2 Deletion Leads to Autistic-like Behaviors Responsive to NMDAR Modulation. *Cell Rep.* **2018**, *23*, 3839–3851. [CrossRef]
68. Schoch, H.; Kreibich, A.S.; Ferri, S.L.; White, R.S.; Bohorquez, D.; Banerjee, A.; Port, R.G.; Dow, H.C.; Cordero, L.; Pallathra, A.A.; et al. Sociability Deficits and Altered Amygdala Circuits in Mice Lacking Pcdh10, an Autism Associated Gene. *Biol. Psychiatry* **2017**, *81*, 193–202. [CrossRef]
69. Urbano, M.; Okwara, L.; Manser, P.; Hartmann, K.; Herndon, A.; Deutsch, S.I. A trial of D-cycloserine to treat stereotypies in older adolescents and young adults with autism spectrum disorder. *Clin. Neuropharmacol.* **2014**, *37*, 69–72. [CrossRef]
70. Sidorov, M.S.; Auerbach, B.D.; Bear, M.F. Fragile X mental retardation protein and synaptic plasticity. *Mol. Brain* **2013**, *6*, 15. [CrossRef]
71. Cheng, G.R.; Li, X.Y.; Xiang, Y.D.; Liu, D.; McClintock, S.M.; Zeng, Y. The implication of AMPA receptor in synaptic plasticity impairment and intellectual disability in fragile X syndrome. *Physiol. Res.* **2017**, *66*, 715–727. [CrossRef]
72. Purcell, A.E.; Jeon, O.H.; Zimmerman, A.W.; Blue, M.E.; Pevsner, J. Postmortem brain abnormalities of the glutamate neurotransmitter system in autism. *Neurology* **2001**, *57*, 1618–1628. [CrossRef] [PubMed]
73. Fatemi, S.H.; Folsom, T.D.; Kneeland, R.E.; Liesch, S.B. Metabotropic glutamate receptor 5 upregulation in children with autism is associated with underexpression of both Fragile X mental retardation protein and GABAA receptor beta 3 in adults with autism. *Anat. Rec.* **2011**, *294*, 1635–1645. [CrossRef]
74. Lohith, T.G.; Osterweil, E.K.; Fujita, M.; Jenko, K.J.; Bear, M.F.; Innis, R.B. Is metabotropic glutamate receptor 5 upregulated in prefrontal cortex in fragile X syndrome? *Mol. Autism* **2013**, *4*, 15. [CrossRef] [PubMed]
75. Blatt, G.J.; Fitzgerald, C.M.; Guptill, J.T.; Booker, A.B.; Kemper, T.L.; Bauman, M.L. Density and distribution of hippocampal neurotransmitter receptors in autism: An autoradiographic study. *J. Autism Dev. Disord.* **2001**, *31*, 537–543. [CrossRef] [PubMed]
76. Tarabeux, J.; Kebir, O.; Gauthier, J.; Hamdan, F.F.; Xiong, L.; Piton, A.; Spiegelman, D.; Henrion, E.; Millet, B.; S2D team; et al. Rare mutations in *N*-methyl-D-aspartate glutamate receptors in autism spectrum disorders and schizophrenia. *Transl. Psychiatry* **2011**, *1*, e55. [CrossRef] [PubMed]
77. Yoo, H.J.; Cho, I.H.; Park, M.; Yang, S.Y.; Kim, S.A. Family based association of GRIN2A and GRIN2B with Korean autism spectrum disorders. *Neurosci. Lett.* **2012**, *512*, 89–93. [CrossRef] [PubMed]
78. O'Roak, B.J.; Vives, L.; Girirajan, S.; Karakoc, E.; Krumm, N.; Coe, B.P.; Levy, R.; Ko, A.; Lee, C.; Smith, J.D.; et al. Sporadic autism exomes reveal a highly interconnected protein network of de novo mutations. *Nature* **2012**, *485*, 246–250. [CrossRef]
79. Barnby, G.; Abbott, A.; Sykes, N.; Morris, A.; Weeks, D.E.; Mott, R.; Lamb, J.; Bailey, A.J.; Monaco, A.P. Candidate-gene screening and association analysis at the autism-susceptibility locus on chromosome 16p: Evidence of association at GRIN2A and ABAT. *Am. J. Hum. Genet.* **2005**, *76*, 950–966. [CrossRef]
80. Jamain, S.; Betancur, C.; Quach, H.; Philippe, A.; Fellous, M.; Giros, B.; Gillberg, C.; Leboyer, M.; Bourgeron, T.; Paris Autism Research International Sibpair, S. Linkage and association of the glutamate receptor 6 gene with autism. *Mol. Psychiatry* **2002**, *7*, 302–310. [CrossRef]
81. Shuang, M.; Liu, J.; Jia, M.X.; Yang, J.Z.; Wu, S.P.; Gong, X.H.; Ling, Y.S.; Ruan, Y.; Yang, X.L.; Zhang, D. Family-based association study between autism and glutamate receptor 6 gene in Chinese Han trios. *Am. J. Med. Genet. B Neuropsychiatr. Genet.* **2004**, *131*, 48–50. [CrossRef]
82. Strutz-Seebohm, N.; Korniychuk, G.; Schwarz, R.; Baltaev, R.; Ursu, O.; Mack, A.; Ma-Högemeier, Z.; Hollmann, M.; Lang, F.; Seebohm, G. Functional Significance of the Kainate Receptor GluR6(M836I) Mutation that is Linked to Autism. *Cell Physiol. Biochem.* **2006**, *18*, 287–294. [CrossRef] [PubMed]
83. Yonan, A.L.; Alarcon, M.; Cheng, R.; Magnusson, P.K.; Spence, S.J.; Palmer, A.A.; Grunn, A.; Juo, S.H.; Terwilliger, J.D.; Liu, J.; et al. A genomewide screen of 345 families for autism-susceptibility loci. *Am. J. Hum. Genet.* **2003**, *73*, 886–897. [CrossRef] [PubMed]
84. Ramanathan, S.; Woodroffe, A.; Flodman, P.L.; Mays, L.Z.; Hanouni, M.; Modahl, C.B.; Steinberg-Epstein, R.; Bocian, M.E.; Spence, M.A.; Smith, M. A case of autism with an interstitial deletion on 4q leading to hemizygosity for genes encoding for glutamine and glycine neurotransmitter receptor sub-units (AMPA 2, GLRA3, GLRB) and neuropeptide receptors NPY1R, NPY5R. *BMC Med. Genet.* **2004**, *5*, 10. [CrossRef] [PubMed]
85. Serajee, F.J.; Zhong, H.; Nabi, R.; Huq, A.H. The metabotropic glutamate receptor 8 gene at 7q31: Partial duplication and possible association with autism. *J. Med. Genet.* **2003**, *40*, e42. [CrossRef] [PubMed]
86. Autism Genome Project, C.; Szatmari, P.; Paterson, A.D.; Zwaigenbaum, L.; Roberts, W.; Brian, J.; Liu, X.Q.; Vincent, J.B.; Skaug, J.L.; Thompson, A.P.; et al. Mapping autism risk loci using genetic linkage and chromosomal rearrangements. *Nat. Genet.* **2007**, *39*, 319–328. [CrossRef]
87. Bremer, A.; Giacobini, M.; Eriksson, M.; Gustavsson, P.; Nordin, V.; Fernell, E.; Gillberg, C.; Nordgren, A.; Uppstromer, A.; Anderlid, B.M.; et al. Copy number variation characteristics in subpopulations of patients with autism spectrum disorders. *Am. J. Med. Genet. B Neuropsychiatr. Genet.* **2011**, *156*, 115–124. [CrossRef]
88. Pinto, D.; Pagnamenta, A.T.; Klei, L.; Anney, R.; Merico, D.; Regan, R.; Conroy, J.; Magalhaes, T.R.; Correia, C.; Abrahams, B.S.; et al. Functional impact of global rare copy number variation in autism spectrum disorders. *Nature* **2010**, *466*, 368–372. [CrossRef]
89. Gauthier, J.; Siddiqui, T.J.; Huashan, P.; Yokomaku, D.; Hamdan, F.F.; Champagne, N.; Lapointe, M.; Spiegelman, D.; Noreau, A.; Lafreniere, R.G.; et al. Truncating mutations in NRXN2 and NRXN1 in autism spectrum disorders and schizophrenia. *Hum. Genet.* **2011**, *130*, 563–573. [CrossRef]

90. Talebizadeh, Z.; Bittel, D.C.; Veatch, O.J.; Butler, M.G.; Takahashi, T.N.; Miles, J.H. Do known mutations in neuroligin genes (NLGN3 and NLGN4) cause autism? *J. Autism Dev. Disord.* **2004**, *34*, 735–736. [CrossRef]
91. Foldy, C.; Malenka, R.C.; Sudhof, T.C. Autism-associated neuroligin-3 mutations commonly disrupt tonic endocannabinoid signaling. *Neuron* **2013**, *78*, 498–509. [CrossRef]
92. Jamain, S.; Quach, H.; Betancur, C.; Rastam, M.; Colineaux, C.; Gillberg, I.C.; Soderstrom, H.; Giros, B.; Leboyer, M.; Gillberg, C.; et al. Mutations of the X-linked genes encoding neuroligins NLGN3 and NLGN4 are associated with autism. *Nat. Genet.* **2003**, *34*, 27–29. [CrossRef] [PubMed]
93. Laumonnier, F.; Bonnet-Brilhault, F.; Gomot, M.; Blanc, R.; David, A.; Moizard, M.P.; Raynaud, M.; Ronce, N.; Lemonnier, E.; Calvas, P.; et al. X-linked mental retardation and autism are associated with a mutation in the NLGN4 gene, a member of the neuroligin family. *Am. J. Hum. Genet.* **2004**, *74*, 552–557. [CrossRef] [PubMed]
94. Alarcon, M.; Abrahams, B.S.; Stone, J.L.; Duvall, J.A.; Perederiy, J.V.; Bomar, J.M.; Sebat, J.; Wigler, M.; Martin, C.L.; Ledbetter, D.H.; et al. Linkage, association, and gene-expression analyses identify CNTNAP2 as an autism-susceptibility gene. *Am. J. Hum. Genet.* **2008**, *82*, 150–159. [CrossRef] [PubMed]
95. Sampath, S.; Bhat, S.; Gupta, S.; O'Connor, A.; West, A.B.; Arking, D.E.; Chakravarti, A. Defining the contribution of CNTNAP2 to autism susceptibility. *PLoS ONE* **2013**, *8*, e77906. [CrossRef]
96. Gauthier, J.; Spiegelman, D.; Piton, A.; Lafreniere, R.G.; Laurent, S.; St-Onge, J.; Lapointe, L.; Hamdan, F.F.; Cossette, P.; Mottron, L.; et al. Novel de novo SHANK3 mutation in autistic patients. *Am. J. Med. Genet. B Neuropsychiatr. Genet.* **2009**, *150B*, 421–424. [CrossRef]
97. Gauthier, J.; Champagne, N.; Lafreniere, R.G.; Xiong, L.; Spiegelman, D.; Brustein, E.; Lapointe, M.; Peng, H.; Cote, M.; Noreau, A.; et al. De novo mutations in the gene encoding the synaptic scaffolding protein SHANK3 in patients ascertained for schizophrenia. *Proc. Natl. Acad. Sci. USA* **2010**, *107*, 7863–7868. [CrossRef]
98. Durand, C.M.; Betancur, C.; Boeckers, T.M.; Bockmann, J.; Chaste, P.; Fauchereau, F.; Nygren, G.; Rastam, M.; Gillberg, I.C.; Anckarsater, H.; et al. Mutations in the gene encoding the synaptic scaffolding protein SHANK3 are associated with autism spectrum disorders. *Nat. Genet.* **2007**, *39*, 25–27. [CrossRef]
99. Sato, D.; Lionel, A.C.; Leblond, C.S.; Prasad, A.; Pinto, D.; Walker, S.; O'Connor, I.; Russell, C.; Drmic, I.E.; Hamdan, F.F.; et al. SHANK1 Deletions in Males with Autism Spectrum Disorder. *Am. J. Hum. Genet.* **2012**, *90*, 879–887. [CrossRef]
100. Guilmatre, A.; Huguet, G.; Delorme, R.; Bourgeron, T. The emerging role of SHANK genes in neuropsychiatric disorders. *Dev. Neurobiol.* **2014**, *74*, 113–122. [CrossRef]
101. Fung, L.K.; Hardan, A.Y. Developing Medications Targeting Glutamatergic Dysfunction in Autism: Progress to Date. *CNS Drugs* **2015**, *29*, 453–463. [CrossRef]
102. Bejjani, A.; O'Neill, J.; Kim, J.A.; Frew, A.J.; Yee, V.W.; Ly, R.; Kitchen, C.; Salamon, N.; McCracken, J.T.; Toga, A.W.; et al. Elevated glutamatergic compounds in pregenual anterior cingulate in pediatric autism spectrum disorder demonstrated by 1H MRS and 1H MRSI. *PLoS ONE* **2012**, *7*, e38786. [CrossRef] [PubMed]
103. Bernardi, S.; Anagnostou, E.; Shen, J.; Kolevzon, A.; Buxbaum, J.D.; Hollander, E.; Hof, P.R.; Fan, J. In vivo 1H-magnetic resonance spectroscopy study of the attentional networks in autism. *Brain Res.* **2011**, *1380*, 198–205. [CrossRef] [PubMed]
104. Brown, M.S.; Singel, D.; Hepburn, S.; Rojas, D.C. Increased glutamate concentration in the auditory cortex of persons with autism and first-degree relatives: A (1)H-MRS study. *Autism Res.* **2013**, *6*, 1–10. [CrossRef] [PubMed]
105. Horder, J.; Lavender, T.; Mendez, M.A.; O'Gorman, R.; Daly, E.; Craig, M.C.; Lythgoe, D.J.; Barker, G.J.; Murphy, D.G. Reduced subcortical glutamate/glutamine in adults with autism spectrum disorders: A [(1)H]MRS study. *Transl. Psychiatry* **2013**, *3*, e279. [CrossRef] [PubMed]
106. Joshi, G.; Biederman, J.; Wozniak, J.; Goldin, R.L.; Crowley, D.; Furtak, S.; Lukas, S.E.; Gonenc, A. Magnetic resonance spectroscopy study of the glutamatergic system in adolescent males with high-functioning autistic disorder: A pilot study at 4T. *Eur. Arch. Psychiatry Clin. Neurosci.* **2013**, *263*, 379–384. [CrossRef]
107. Kubas, B.; Kulak, W.; Sobaniec, W.; Tarasow, E.; Lebkowska, U.; Walecki, J. Metabolite alterations in autistic children: A 1H MR spectroscopy study. *Adv. Med. Sci.* **2012**, *57*, 152–156. [CrossRef]
108. Naaijen, J.; Forde, N.J.; Lythgoe, D.J.; Akkermans, S.E.; Openneer, T.J.; Dietrich, A.; Zwiers, M.P.; Hoekstra, P.J.; Buitelaar, J.K. Fronto-striatal glutamate in children with Tourette's disorder and attention-deficit/hyperactivity disorder. *Neuroimage Clin.* **2017**, *13*, 16–23. [CrossRef]
109. Page, L.A.; Daly, E.; Schmitz, N.; Simmons, A.; Toal, F.; Deeley, Q.; Ambery, F.; McAlonan, G.M.; Murphy, K.C.; Murphy, D.G. In vivo 1H-magnetic resonance spectroscopy study of amygdala-hippocampal and parietal regions in autism. *Am. J. Psychiatry* **2006**, *163*, 2189–2192. [CrossRef]
110. Tebartz van Elst, L.; Maier, S.; Fangmeier, T.; Endres, D.; Mueller, G.T.; Nickel, K.; Ebert, D.; Lange, T.; Hennig, J.; Biscaldi, M.; et al. Disturbed cingulate glutamate metabolism in adults with high-functioning autism spectrum disorder: Evidence in support of the excitatory/inhibitory imbalance hypothesis. *Mol. Psychiatry* **2014**, *19*, 1314–1325. [CrossRef]
111. Ford, T.C.; Nibbs, R.; Crewther, D.P. Increased glutamate/GABA+ ratio in a shared autistic and schizotypal trait phenotype termed Social Disorganisation. *NeuroImage Clin.* **2017**, *16*, 125–131. [CrossRef]
112. Carlson, G. Glutamate receptor dysfunction and drug targets across models of autism spectrum disorders. *Pharmacol. Biochem. Behav.* **2011**, *100*, 850–854. [CrossRef] [PubMed]

113. Citri, A.; Malenka, R.C. Synaptic Plasticity: Multiple Forms, Functions, and Mechanisms. *Neuropsychopharmacology* **2008**, *33*, 18–41. [CrossRef] [PubMed]
114. Yang, Y.; Lu, J.; Zuo, Y. Changes of Synaptic Structures Associated with Learning, Memory and Diseases. *Brain Sci. Adv.* **2018**, *4*, 99–117. [CrossRef]
115. Kano, M.; Watanabe, T. Developmental synapse remodeling in the cerebellum and visual thalamus [version 1; peer review: 2 approved]. *F1000Research* **2019**, *8*. [CrossRef]
116. Robain, O.; Bideau, I.; Farkas, E. Developmental changes of synapses in the cerebellar cortex of the rat. A quantitative analysis. *Brain Res.* **1981**, *206*, 1–8. [CrossRef]
117. Lohmann, C.; Kessels, H.W. The developmental stages of synaptic plasticity. *J. Physiol.* **2014**, *592*, 13–31. [CrossRef]
118. Farhy-Tselnicker, I.; Allen, N.J. Astrocytes, neurons, synapses: A tripartite view on cortical circuit development. *Neural Dev.* **2018**, *13*, 7. [CrossRef]
119. Prokop, A.; Meinertzhagen, I.A. Development and structure of synaptic contacts in Drosophila. *Semin. Cell Dev. Biol.* **2006**, *17*, 20–30. [CrossRef]
120. Benavides-Piccione, R.; Fernaud-Espinosa, I.; Robles, V.; Yuste, R.; DeFelipe, J. Age-based comparison of human dendritic spine structure using complete three-dimensional reconstructions. *Cereb. Cortex* **2013**, *23*, 1798–1810. [CrossRef]
121. Markus, E.J.; Petit, T.L.; LeBoutillier, J.C. Synaptic structural changes during development and aging. *Brain Res.* **1987**, *432*, 239–248. [CrossRef]
122. Russo, S.J.; Dietz, D.M.; Dumitriu, D.; Morrison, J.H.; Malenka, R.C.; Nestler, E.J. The addicted synapse: Mechanisms of synaptic and structural plasticity in nucleus accumbens. *Trends Neurosci.* **2010**, *33*, 267–276. [CrossRef] [PubMed]
123. Roberts, R.C.; Barksdale, K.A.; Roche, J.K.; Lahti, A.C. Decreased synaptic and mitochondrial density in the postmortem anterior cingulate cortex in schizophrenia. *Schizophr. Res.* **2015**, *168*, 543–553. [CrossRef]
124. Aganova, E.A.; Uranova, N.A. Morphometric analysis of synaptic contacts in the anterior limbic cortex in the endogenous psychoses. *Neurosci. Behav. Physiol.* **1992**, *22*, 59–65. [CrossRef] [PubMed]
125. Kolomeets, N.S.; Orlovskaya, D.D.; Rachmanova, V.I.; Uranova, N.A. Ultrastructural alterations in hippocampal mossy fiber synapses in schizophrenia: A postmortem morphometric study. *Synapse* **2005**, *57*, 47–55. [CrossRef] [PubMed]
126. Kolomeets, N.S.; Orlovskaya, D.D.; Uranova, N.A. Decreased numerical density of CA3 hippocampal mossy fiber synapses in schizophrenia. *Synapse* **2007**, *61*, 615–621. [CrossRef] [PubMed]
127. Roberts, R.C.; Roche, J.K.; Conley, R.R. Synaptic differences in the postmortem striatum of subjects with schizophrenia: A stereological ultrastructural analysis. *Synapse* **2005**, *56*, 185–197. [CrossRef]
128. Roberts, R.C.; Roche, J.K.; Conley, R.R. Synaptic differences in the patch matrix compartments of subjects with schizophrenia: A postmortem ultrastructural study of the striatum. *Neurobiol. Dis.* **2005**, *20*, 324–335. [CrossRef]
129. Roberts, R.C.; Roche, R.J.; Somerville, S.M.; Conley, R.R. Ultrastructural distinctions between treatment responders and non-responders in schizophrenia: Postmortem studies of the striatum. In *Mental Illnesses—Evaluation, Treatments, and Implications*; InTech: London, UK, 2012; pp. 261–286.
130. McCollum, L.A.; Walker, C.K.; Roche, J.K.; Roberts, R.C. Elevated Excitatory Input to the Nucleus Accumbens in Schizophrenia: A Postmortem Ultrastructural Study. *Schizophr. Bull.* **2015**, *41*, 1123–1132. [CrossRef]
131. Xu, Y.; Deng, C.; Zheng, Y.; Liu, N.; Fu, B. Applying vinpocetine to reverse synaptic ultrastructure by regulating BDNF-related PSD-95 in alleviating schizophrenia-like deficits in rat. *Compr. Psychiatry* **2019**, *94*, 152122. [CrossRef]
132. Chen, X.W.; Feng, Y.Q.; Hao, C.J.; Guo, X.L.; He, X.; Zhou, Z.Y.; Guo, N.; Huang, H.P.; Xiong, W.; Zheng, H.; et al. DTNBP1, a schizophrenia susceptibility gene, affects kinetics of transmitter release. *J. Cell Biol.* **2008**, *181*, 791–801. [CrossRef]
133. Zikopoulos, B.; Barbas, H. Changes in prefrontal axons may disrupt the network in autism. *J. Neurosci.* **2010**, *30*, 14595–14609. [CrossRef] [PubMed]
134. Gassowska-Dobrowolska, M.; Cieslik, M.; Czapski, G.A.; Jesko, H.; Frontczak-Baniewicz, M.; Gewartowska, M.; Dominiak, A.; Polowy, R.; Filipkowski, R.K.; Babiec, L.; et al. Prenatal Exposure to Valproic Acid Affects Microglia and Synaptic Ultrastructure in a Brain-Region-Specific Manner in Young-Adult Male Rats: Relevance to Autism Spectrum Disorders. *Int. J. Mol. Sci.* **2020**, *21*, 3576. [CrossRef] [PubMed]
135. Lobzhanidze, G.; Japaridze, N.; Lordkipanidze, T.; Rzayev, F.; MacFabe, D.; Zhvania, M. Behavioural and brain ultrastructural changes following the systemic administration of propionic acid in adolescent male rats. Further development of a rodent model of autism. *Int. J. Dev. Neurosci.* **2020**, *80*, 139–156. [CrossRef] [PubMed]
136. Sato, Y.; Okabe, S. Nano-scale analysis of synapse morphology in an autism mouse model with 15q11-13 copy number variation using focused ion beam milling and scanning electron microscopy. *Microscopy* **2019**, *68*, 122–132. [CrossRef] [PubMed]
137. Hung, A.Y.; Futai, K.; Sala, C.; Valtschanoff, J.G.; Ryu, J.; Woodworth, M.A.; Kidd, F.L.; Sung, C.C.; Miyakawa, T.; Bear, M.F.; et al. Smaller dendritic spines, weaker synaptic transmission, but enhanced spatial learning in mice lacking Shank1. *J. Neurosci.* **2008**, *28*, 1697–1708. [CrossRef] [PubMed]
138. Peca, J.; Feliciano, C.; Ting, J.T.; Wang, W.; Wells, M.F.; Venkatraman, T.N.; Lascola, C.D.; Fu, Z.; Feng, G. Shank3 mutant mice display autistic-like behaviours and striatal dysfunction. *Nature* **2011**, *472*, 437–442. [CrossRef]
139. Morimura, N.; Yasuda, H.; Yamaguchi, K.; Katayama, K.I.; Hatayama, M.; Tomioka, N.H.; Odagawa, M.; Kamiya, A.; Iwayama, Y.; Maekawa, M.; et al. Autism-like behaviours and enhanced memory formation and synaptic plasticity in Lrfn2/SALM1-deficient mice. *Nat. Commun.* **2017**, *8*, 15800. [CrossRef]

140. Shih, P.Y.; Hsieh, B.Y.; Lin, M.H.; Huang, T.N.; Tsai, C.Y.; Pong, W.L.; Lee, S.P.; Hsueh, Y.P. CTTNBP2 Controls Synaptic Expression of Zinc-Related Autism-Associated Proteins and Regulates Synapse Formation and Autism-like Behaviors. *Cell Rep.* **2020**, *31*, 107700. [CrossRef]
141. Ma, J.; Zhang, L.Q.; He, Z.X.; He, X.X.; Wang, Y.J.; Jian, Y.L.; Wang, X.; Zhang, B.B.; Su, C.; Lu, J.; et al. Autism candidate gene DIP2A regulates spine morphogenesis via acetylation of cortactin. *PLoS Biol.* **2019**, *17*, e3000461. [CrossRef]
142. Ranneva, S.V.; Maksimov, V.F.; Korostyshevskaja, I.M.; Lipina, T.V. Lack of synaptic protein, calsyntenin-2, impairs morphology of synaptic complexes in mice. *Synapse* **2020**, *74*, e22132. [CrossRef]
143. Kang, M.S.; Choi, T.Y.; Ryu, H.G.; Lee, D.; Lee, S.H.; Choi, S.Y.; Kim, K.T. Autism-like behavior caused by deletion of vaccinia-related kinase 3 is improved by TrkB stimulation. *J. Exp. Med.* **2017**, *214*, 2947–2966. [CrossRef] [PubMed]
144. He, C.X.; Portera-Cailliau, C. The trouble with spines in fragile X syndrome: Density, maturity and plasticity. *Neuroscience* **2013**, *251*, 120–128. [CrossRef] [PubMed]
145. Jawaid, S.; Kidd, G.J.; Wang, J.; Swetlik, C.; Dutta, R.; Trapp, B.D. Alterations in CA1 hippocampal synapses in a mouse model of fragile X syndrome. *Glia* **2018**, *66*, 789–800. [CrossRef]
146. Padmashri, R.; Reiner, B.C.; Suresh, A.; Spartz, E.; Dunaevsky, A. Altered Structural and Functional Synaptic Plasticity with Motor Skill Learning in a Mouse Model of Fragile X Syndrome. *J. Neurosci.* **2013**, *33*, 19715–19723. [CrossRef]
147. Bailey, C.H.; Chen, M. Morphological basis of long-term habituation and sensitization in Aplysia. *Science* **1983**, *220*, 91–93. [CrossRef]
148. Muller, M.; Liu, K.S.; Sigrist, S.J.; Davis, G.W. RIM controls homeostatic plasticity through modulation of the readily-releasable vesicle pool. *J. Neurosci.* **2012**, *32*, 16574–16585. [CrossRef]
149. Goel, P.; Li, X.; Dickman, D. Estimation of the Readily Releasable Synaptic Vesicle Pool at the Drosophila Larval Neuromuscular Junction. *Bio-Protocol* **2019**, *9*. [CrossRef]
150. Lazarevic, V.; Schone, C.; Heine, M.; Gundelfinger, E.D.; Fejtova, A. Extensive remodeling of the presynaptic cytomatrix upon homeostatic adaptation to network activity silencing. *J. Neurosci.* **2011**, *31*, 10189–10200. [CrossRef]
151. Soares, C.; Lee, K.F.H.; Beique, J.C. Metaplasticity at CA1 Synapses by Homeostatic Control of Presynaptic Release Dynamics. *Cell Rep.* **2017**, *21*, 1293–1303. [CrossRef]
152. Savtchenko, L.P.; Rusakov, D.A. The optimal height of the synaptic cleft. *Proc. Natl. Acad. Sci. USA* **2007**, *104*, 1823–1828. [CrossRef]
153. Glebov, O.O.; Cox, S.; Humphreys, L.; Burrone, J. Neuronal activity controls transsynaptic geometry. *Sci. Rep.* **2016**, *6*, 22703. [CrossRef] [PubMed]
154. Matsuzaki, M.; Ellis-Davies, G.C.; Nemoto, T.; Miyashita, Y.; Iino, M.; Kasai, H. Dendritic spine geometry is critical for AMPA receptor expression in hippocampal CA1 pyramidal neurons. *Nat. Neurosci.* **2001**, *4*, 1086–1092. [CrossRef] [PubMed]
155. Kasai, H.; Matsuzaki, M.; Noguchi, J.; Yasumatsu, N.; Nakahara, H. Structure-stability-function relationships of dendritic spines. *Trends Neurosci.* **2003**, *26*, 360–368. [CrossRef]
156. Arellano, J.I.; Benavides-Piccione, R.; Defelipe, J.; Yuste, R. Ultrastructure of dendritic spines: Correlation between synaptic and spine morphologies. *Front. Neurosci.* **2007**, *1*, 131–143. [CrossRef]
157. Holderith, N.; Lorincz, A.; Katona, G.; Rozsa, B.; Kulik, A.; Watanabe, M.; Nusser, Z. Release probability of hippocampal glutamatergic terminals scales with the size of the active zone. *Nat. Neurosci.* **2012**, *15*, 988–997. [CrossRef]
158. Quinn, D.P.; Kolar, A.; Harris, S.A.; Wigerius, M.; Fawcett, J.P.; Krueger, S.R. The Stability of Glutamatergic Synapses Is Independent of Activity Level, but Predicted by Synapse Size. *Front. Cell Neurosci.* **2019**, *13*, 291. [CrossRef]
159. Mayford, M.; Siegelbaum, S.A.; Kandel, E.R. Synapses and memory storage. *Cold Spring Harb. Perspect. Biol.* **2012**, *4*. [CrossRef]
160. Kasai, H.; Fukuda, M.; Watanabe, S.; Hayashi-Takagi, A.; Noguchi, J. Structural dynamics of dendritic spines in memory and cognition. *Trends Neurosci.* **2010**, *33*, 121–129. [CrossRef]
161. Matsuzaki, M.; Honkura, N.; Ellis-Davies, G.C.; Kasai, H. Structural basis of long-term potentiation in single dendritic spines. *Nature* **2004**, *429*, 761–766. [CrossRef]
162. Minerbi, A.; Kahana, R.; Goldfeld, L.; Kaufman, M.; Marom, S.; Ziv, N.E. Long-Term Relationships between Synaptic Tenacity, Synaptic Remodeling, and Network Activity. *PLoS Biol.* **2009**, *7*, e1000136. [CrossRef]
163. Mongillo, G.; Rumpel, S.; Loewenstein, Y. Intrinsic volatility of synaptic connections—A challenge to the synaptic trace theory of memory. *Curr. Opin Neurobiol.* **2017**, *46*, 7–13. [CrossRef] [PubMed]
164. Susman, L.; Brenner, N.; Barak, O. Stable memory with unstable synapses. *Nat. Commun.* **2019**, *10*, 4441. [CrossRef] [PubMed]
165. Merchan-Perez, A.; Rodriguez, J.R.; Gonzalez, S.; Robles, V.; Defelipe, J.; Larranaga, P.; Bielza, C. Three-dimensional spatial distribution of synapses in the neocortex: A dual-beam electron microscopy study. *Cereb. Cortex* **2014**, *24*, 1579–1588. [CrossRef] [PubMed]
166. Santuy, A.; Tomas-Roca, L.; Rodriguez, J.R.; Gonzalez-Soriano, J.; Zhu, F.; Qiu, Z.; Grant, S.G.N.; DeFelipe, J.; Merchan-Perez, A. Estimation of the number of synapses in the hippocampus and brain-wide by volume electron microscopy and genetic labeling. *Sci. Rep.* **2020**, *10*, 14014. [CrossRef] [PubMed]
167. Santuy, A.; Rodriguez, J.R.; DeFelipe, J.; Merchan-Perez, A. Study of the Size and Shape of Synapses in the Juvenile Rat Somatosensory Cortex with 3D Electron Microscopy. *eNeuro* **2018**, *5*. [CrossRef] [PubMed]
168. Santuy, A.; Rodriguez, J.R.; DeFelipe, J.; Merchan-Perez, A. Volume electron microscopy of the distribution of synapses in the neuropil of the juvenile rat somatosensory cortex. *Brain Struct. Funct.* **2018**, *223*, 77–90. [CrossRef] [PubMed]

169. Lefort, S.; Tomm, C.; Floyd Sarria, J.C.; Petersen, C.C. The excitatory neuronal network of the C2 barrel column in mouse primary somatosensory cortex. *Neuron* **2009**, *61*, 301–316. [CrossRef]
170. Song, S.; Sjostrom, P.J.; Reigl, M.; Nelson, S.; Chklovskii, D.B. Highly nonrandom features of synaptic connectivity in local cortical circuits. *PLoS Biol.* **2005**, *3*, e68. [CrossRef]
171. Mizuseki, K.; Buzsaki, G. Preconfigured, skewed distribution of firing rates in the hippocampus and entorhinal cortex. *Cell Rep.* **2013**, *4*, 1010–1021. [CrossRef]
172. Buzsaki, G.; Mizuseki, K. The log-dynamic brain: How skewed distributions affect network operations. *Nat. Rev. Neurosci.* **2014**, *15*, 264–278. [CrossRef]
173. Montes, J.; Pena, J.M.; DeFelipe, J.; Herreras, O.; Merchan-Perez, A. The influence of synaptic size on AMPA receptor activation: A Monte Carlo model. *PLoS ONE* **2015**, *10*, e0130924. [CrossRef] [PubMed]
174. Franks, K.M.; Bartol, T.M., Jr.; Sejnowski, T.J. A Monte Carlo model reveals independent signaling at central glutamatergic synapses. *Biophys. J.* **2002**, *83*, 2333–2348. [CrossRef]
175. Gulledge, A.T.; Carnevale, N.T.; Stuart, G.J. Electrical advantages of dendritic spines. *PLoS ONE* **2012**, *7*, e36007. [CrossRef] [PubMed]
176. Araya, R. Input transformation by dendritic spines of pyramidal neurons. *Front. Neuroanat.* **2014**, *8*, 141. [CrossRef] [PubMed]
177. Major, G.; Larkum, M.E.; Schiller, J. Active properties of neocortical pyramidal neuron dendrites. *Annu. Rev. Neurosci.* **2013**, *36*, 1–24. [CrossRef] [PubMed]
178. Eyal, G.; Mansvelder, H.D.; de Kock, C.P.; Segev, I. Dendrites impact the encoding capabilities of the axon. *J. Neurosci.* **2014**, *34*, 8063–8071. [CrossRef]
179. Kharazia, V.N.; Weinberg, R.J. Immunogold localization of AMPA and NMDA receptors in somatic sensory cortex of albino rat. *J. Comp. Neurol.* **1999**, *412*, 292–302. [CrossRef]
180. Nusser, Z.; Lujan, R.; Laube, G.; Roberts, J.D.; Molnar, E.; Somogyi, P. Cell type and pathway dependence of synaptic AMPA receptor number and variability in the hippocampus. *Neuron* **1998**, *21*, 545–559. [CrossRef]
181. Racca, C.; Stephenson, F.A.; Streit, P.; Roberts, J.D.; Somogyi, P. NMDA receptor content of synapses in stratum radiatum of the hippocampal CA1 area. *J. Neurosci.* **2000**, *20*, 2512–2522. [CrossRef]
182. Calverley, R.K.; Jones, D.G. Determination of the numerical density of perforated synapses in rat neocortex. *Cell Tissue Res.* **1987**, *248*, 399–407. [CrossRef]
183. Geinisman, Y.; Morrell, F.; de Toledo-Morrell, L. Axospinous synapses with segmented postsynaptic densities: A morphologically distinct synaptic subtype contributing to the number of profiles of 'perforated' synapses visualized in random sections. *Brain Res.* **1987**, *423*, 179–188. [CrossRef]
184. Harris, K.M.; Jensen, F.E.; Tsao, B. Three-dimensional structure of dendritic spines and synapses in rat hippocampus (CA1) at postnatal day 15 and adult ages: Implications for the maturation of synaptic physiology and long-term potentiation. *J. Neurosci.* **1992**, *12*, 2685–2705. [CrossRef] [PubMed]
185. Jones, D.G.; Calverley, R.K. Perforated and non-perforated synapses in rat neocortex: Three-dimensional reconstructions. *Brain Res.* **1991**, *556*, 247–258. [CrossRef]
186. Geinisman, Y.; deToledo-Morrell, L.; Morrell, F.; Heller, R.E.; Rossi, M.; Parshall, R.F. Structural synaptic correlate of long-term potentiation: Formation of axospinous synapses with multiple, completely partitioned transmission zones. *Hippocampus* **1993**, *3*, 435–445. [CrossRef]
187. Toni, N.; Buchs, P.A.; Nikonenko, I.; Povilaitite, P.; Parisi, L.; Muller, D. Remodeling of synaptic membranes after induction of long-term potentiation. *J. Neurosci.* **2001**, *21*, 6245–6251. [CrossRef]
188. Ganeshina, O.; Berry, R.W.; Petralia, R.S.; Nicholson, D.A.; Geinisman, Y. Synapses with a segmented, completely partitioned postsynaptic density express more AMPA receptors than other axospinous synaptic junctions. *Neuroscience* **2004**, *125*, 615–623. [CrossRef]
189. Dyson, S.E.; Jones, D.G. Quantitation of terminal parameters and their inter-relationships in maturing central synapses: A perspective for experimental studies. *Brain Res.* **1980**, *183*, 43–59. [CrossRef]
190. Jones, D.G.; Devon, R.M. An ultrastructural study into the effects of pentobarbitone on synaptic organization. *Brain Res.* **1978**, *147*, 47–63. [CrossRef]
191. Oorschot, D.E.; Jones, D.G. A quantitative ultrastructural study of the effects of phenylacetate on synaptic organization in the developing rat cerebral cortex. *Dev. Neurosci.* **1983**, *6*, 45–57. [CrossRef]
192. Calverley, R.K.; Jones, D.G. A serial-section study of perforated synapses in rat neocortex. *Cell Tissue Res.* **1987**, *247*, 565–572. [CrossRef]
193. Jones, D.G.; Dyson, S.E. The influence of protein restriction, rehabilitation and changing nutritional status on synaptic development: A quantitative study in rat brain. *Brain Res.* **1981**, *208*, 97–111. [CrossRef]
194. Medvedev, N.I.; Popov, V.I.; Dallerac, G.; Davies, H.A.; Laroche, S.; Kraev, I.V.; Rodriguez Arellano, J.J.; Doyere, V.; Stewart, M.G. Alterations in synaptic curvature in the dentate gyrus following induction of long-term potentiation, long-term depression, and treatment with the *N*-methyl-D-aspartate receptor antagonist CPP. *Neuroscience* **2010**, *171*, 390–397. [CrossRef] [PubMed]
195. Marrone, D.F.; Petit, T.L. The role of synaptic morphology in neural plasticity: Structural interactions underlying synaptic power. *Brain Res. Brain Res. Rev.* **2002**, *38*, 291–308. [CrossRef]

196. Tao-Cheng, J.H. Stimulation induces gradual increases in the thickness and curvature of postsynaptic density of hippocampal CA1 neurons in slice cultures. *Mol. Brain* **2019**, *12*, 44. [CrossRef]
197. Alonso-Nanclares, L.; Gonzalez-Soriano, J.; Rodriguez, J.R.; DeFelipe, J. Gender differences in human cortical synaptic density. *Proc. Natl. Acad. Sci. USA* **2008**, *105*, 14615–14619. [CrossRef]
198. Beaulieu, C.; Colonnier, M. A laminar analysis of the number of round-asymmetrical and flat-symmetrical synapses on spines, dendritic trunks, and cell bodies in area 17 of the cat. *J. Comp. Neurol.* **1985**, *231*, 180–189. [CrossRef]
199. Micheva, K.D.; Beaulieu, C. Quantitative aspects of synaptogenesis in the rat barrel field cortex with special reference to GABA circuitry. *J. Comp. Neurol.* **1996**, *373*, 340–354. [CrossRef]
200. Bosch, C.; Martinez, A.; Masachs, N.; Teixeira, C.M.; Fernaud, I.; Ulloa, F.; Perez-Martinez, E.; Lois, C.; Comella, J.X.; DeFelipe, J.; et al. FIB/SEM technology and high-throughput 3D reconstruction of dendritic spines and synapses in GFP-labeled adult-generated neurons. *Front. Neuroanat.* **2015**, *9*, 60. [CrossRef]
201. Jones, E.G.; Powell, T.P. Morphological variations in the dendritic spines of the neocortex. *J. Cell Sci.* **1969**, *5*, 509–529.
202. Petrak, L.J.; Harris, K.M.; Kirov, S.A. Synaptogenesis on mature hippocampal dendrites occurs via filopodia and immature spines during blocked synaptic transmission. *J. Comp. Neurol.* **2005**, *484*, 183–190. [CrossRef]
203. Popov, V.I.; Deev, A.A.; Klimenko, O.A.; Kraev I, V.; Kuz'minykh, S.B.; Medvedev, N.I.; Patrushev, I.V.; Popov, R.V.; Rogachevskii, V.V.; Khutsiyan, S.S.; et al. Three-dimensional reconstruction of synapses and dendritic spines in the rat and ground squirrel hippocampus: New structural-functional paradigms for synaptic function. *Neurosci. Behav. Physiol.* **2005**, *35*, 333–341. [CrossRef] [PubMed]
204. Knott, G.W.; Holtmaat, A.; Wilbrecht, L.; Welker, E.; Svoboda, K. Spine growth precedes synapse formation in the adult neocortex in vivo. *Nat. Neurosci.* **2006**, *9*, 1117–1124. [CrossRef] [PubMed]
205. Radwanska, K.; Medvedev, N.I.; Pereira, G.S.; Engmann, O.; Thiede, N.; Moraes, M.F.; Villers, A.; Irvine, E.E.; Maunganidze, N.S.; Pyza, E.M.; et al. Mechanism for long-term memory formation when synaptic strengthening is impaired. *Proc. Natl. Acad. Sci. USA* **2011**, *108*, 18471–18475. [CrossRef]
206. Giese, K.P.; Aziz, W.; Kraev, I.; Stewart, M.G. Generation of multi-innervated dendritic spines as a novel mechanism of long-term memory formation. *Neurobiol. Learn. Mem.* **2015**, *124*, 48–51. [CrossRef] [PubMed]
207. Toni, N.; Buchs, P.A.; Nikonenko, I.; Bron, C.R.; Muller, D. LTP promotes formation of multiple spine synapses between a single axon terminal and a dendrite. *Nature* **1999**, *402*, 421–425. [CrossRef]
208. Kozorovitskiy, Y.; Gross, C.G.; Kopil, C.; Battaglia, L.; McBreen, M.; Stranahan, A.M.; Gould, E. Experience induces structural and biochemical changes in the adult primate brain. *Proc. Natl. Acad. Sci. USA* **2005**, *102*, 17478–17482. [CrossRef] [PubMed]
209. Moser, M.B.; Trommald, M.; Andersen, P. An increase in dendritic spine density on hippocampal CA1 pyramidal cells following spatial learning in adult rats suggests the formation of new synapses. *Proc. Natl. Acad. Sci. USA* **1994**, *91*, 12673–12675. [CrossRef]
210. Brockett, A.T.; LaMarca, E.A.; Gould, E. Physical exercise enhances cognitive flexibility as well as astrocytic and synaptic markers in the medial prefrontal cortex. *PLoS ONE* **2015**, *10*, e0124859. [CrossRef]
211. Leuner, B.; Falduto, J.; Shors, T.J. Associative memory formation increases the observation of dendritic spines in the hippocampus. *J. Neurosci.* **2003**, *23*, 659–665. [CrossRef]
212. Patel, S.N.; Stewart, M.G. Changes in the number and structure of dendritic spines 25 hours after passive avoidance training in the domestic chick, Gallus domesticus. *Brain Res.* **1988**, *449*, 34–46. [CrossRef]
213. Attwell, D.; Laughlin, S.B. An energy budget for signaling in the grey matter of the brain. *J. Cereb. Blood Flow Metab.* **2001**, *21*, 1133–1145. [CrossRef]
214. Harris, J.J.; Jolivet, R.; Attwell, D. Synaptic energy use and supply. *Neuron* **2012**, *75*, 762–777. [CrossRef]
215. Li, Z.; Okamoto, K.; Hayashi, Y.; Sheng, M. The importance of dendritic mitochondria in the morphogenesis and plasticity of spines and synapses. *Cell* **2004**, *119*, 873–887. [CrossRef]
216. MacAskill, A.F.; Atkin, T.A.; Kittler, J.T. Mitochondrial trafficking and the provision of energy and calcium buffering at excitatory synapses. *Eur. J. Neurosci.* **2010**, *32*, 231–240. [CrossRef]
217. Howarth, C.; Gleeson, P.; Attwell, D. Updated energy budgets for neural computation in the neocortex and cerebellum. *J. Cereb. Blood Flow Metab.* **2012**, *32*, 1222–1232. [CrossRef]
218. Rowland, K.C.; Irby, N.K.; Spirou, G.A. Specialized synapse-associated structures within the calyx of Held. *J. Neurosci.* **2000**, *20*, 9135–9144. [CrossRef]
219. Cai, Q.; Sheng, Z.H. Mitochondrial transport and docking in axons. *Exp. Neurol.* **2009**, *218*, 257–267. [CrossRef]
220. Devine, M.J.; Kittler, J.T. Mitochondria at the neuronal presynapse in health and disease. *Nat. Rev. Neurosci.* **2018**, *19*, 63–80. [CrossRef]
221. Chang, D.T.; Honick, A.S.; Reynolds, I.J. Mitochondrial trafficking to synapses in cultured primary cortical neurons. *J. Neurosci.* **2006**, *26*, 7035–7045. [CrossRef]
222. Obashi, K.; Okabe, S. Regulation of mitochondrial dynamics and distribution by synapse position and neuronal activity in the axon. *Eur. J. Neurosci.* **2013**, *38*, 2350–2363. [CrossRef]
223. MacAskill, A.F.; Kittler, J.T. Control of mitochondrial transport and localization in neurons. *Trends Cell Biol.* **2010**, *20*, 102–112. [CrossRef]

224. Takihara, Y.; Inatani, M.; Eto, K.; Inoue, T.; Kreymerman, A.; Miyake, S.; Ueno, S.; Nagaya, M.; Nakanishi, A.; Iwao, K.; et al. In vivo imaging of axonal transport of mitochondria in the diseased and aged mammalian CNS. *Proc. Natl. Acad. Sci. USA* **2015**, *112*, 10515–10520. [CrossRef]
225. Schwarz, T.L. Mitochondrial trafficking in neurons. *Cold Spring Harb. Perspect Biol.* **2013**, *5*. [CrossRef]
226. Santuy, A.; Turegano-Lopez, M.; Rodriguez, J.R.; Alonso-Nanclares, L.; DeFelipe, J.; Merchan-Perez, A. A Quantitative Study on the Distribution of Mitochondria in the Neuropil of the Juvenile Rat Somatosensory Cortex. *Cereb. Cortex* **2018**, *28*, 3673–3684. [CrossRef]
227. Rodriguez-Moreno, J.; Rollenhagen, A.; Arlandis, J.; Santuy, A.; Merchan-Perez, A.; DeFelipe, J.; Lubke, J.H.R.; Clasca, F. Quantitative 3D Ultrastructure of Thalamocortical Synapses from the "Lemniscal" Ventral Posteromedial Nucleus in Mouse Barrel Cortex. *Cereb. Cortex* **2018**, *28*, 3159–3175. [CrossRef]
228. Rintoul, G.L.; Filiano, A.J.; Brocard, J.B.; Kress, G.J.; Reynolds, I.J. Glutamate decreases mitochondrial size and movement in primary forebrain neurons. *J. Neurosci.* **2003**, *23*, 7881–7888. [CrossRef]
229. David, G.; Barrett, E.F. Stimulation-evoked increases in cytosolic [Ca^{2+}] in mouse motor nerve terminals are limited by mitochondrial uptake and are temperature-dependent. *J. Neurosci.* **2000**, *20*, 7290–7296. [CrossRef]
230. David, G.; Barrett, E.F. Mitochondrial Ca^{2+} uptake prevents desynchronization of quantal release and minimizes depletion during repetitive stimulation of mouse motor nerve terminals. *J. Physiol.* **2003**, *548*, 425–438. [CrossRef]
231. Li, H.; Chen, Y.; Jones, A.F.; Sanger, R.H.; Collis, L.P.; Flannery, R.; McNay, E.C.; Yu, T.; Schwarzenbacher, R.; Bossy, B.; et al. Bcl-xL induces Drp1-dependent synapse formation in cultured hippocampal neurons. *Proc. Natl. Acad. Sci. USA* **2008**, *105*, 2169–2174. [CrossRef]
232. Roberts, R.C.; McCollum, L.A.; Schoonover, K.E.; Mabry, S.J.; Roche, J.K.; Lahti, A.C. Ultrastructural evidence for glutamatergic dysregulation in schizophrenia. *Schizophr. Res.* **2020**. [CrossRef]
233. Koski, L.; Petrides, M. Time-related changes in task performance after lesions restricted to the frontal cortex. *Neuropsychologia* **2001**, *39*, 268–281. [CrossRef]
234. Barch, D.M.; Braver, T.S.; Akbudak, E.; Conturo, T.; Ollinger, J.; Snyder, A. Anterior cingulate cortex and response conflict: Effects of response modality and processing domain. *Cereb. Cortex* **2001**, *11*, 837–848. [CrossRef] [PubMed]
235. Kerns, J.G.; Cohen, J.D.; MacDonald, A.W., 3rd; Cho, R.Y.; Stenger, V.A.; Carter, C.S. Anterior cingulate conflict monitoring and adjustments in control. *Science* **2004**, *303*, 1023–1026. [CrossRef] [PubMed]
236. Braver, T.S.; Barch, D.M.; Gray, J.R.; Molfese, D.L.; Snyder, A. Anterior cingulate cortex and response conflict: Effects of frequency, inhibition and errors. *Cereb. Cortex* **2001**, *11*, 825–836. [CrossRef]
237. Mathalon, D.H.; Fedor, M.; Faustman, W.O.; Gray, M.; Askari, N.; Ford, J.M. Response-monitoring dysfunction in schizophrenia: An event-related brain potential study. *J. Abnorm. Psychol.* **2002**, *111*, 22–41. [CrossRef]
238. Bush, G.; Luu, P.; Posner, M.I. Cognitive and emotional influences in anterior cingulate cortex. *Trends Cogn. Sci.* **2000**, *4*, 215–222. [CrossRef]
239. Sowell, E.R.; Thompson, P.M.; Toga, A.W. Mapping Changes in the Human Cortex throughout the Span of Life. *Neuroscientist* **2004**, *10*, 372–392. [CrossRef]
240. Eltokhi, A.; Janmaat, I.E.; Genedi, M.; Haarman, B.C.M.; Sommer, I.E.C. Dysregulation of synaptic pruning as a possible link between intestinal microbiota dysbiosis and neuropsychiatric disorders. *J. Neurosci. Res.* **2020**, *98*, 1335–1369. [CrossRef]
241. Graybiel, A.M.; Ragsdale, C.W., Jr. Histochemically distinct compartments in the striatum of human, monkeys, and cat demonstrated by acetylthiocholinesterase staining. *Proc. Natl. Acad. Sci. USA* **1978**, *75*, 5723–5726. [CrossRef]
242. Flaherty, A.W.; Graybiel, A.M. Two input systems for body representations in the primate striatal matrix: Experimental evidence in the squirrel monkey. *J. Neurosci.* **1993**, *13*, 1120. [CrossRef]
243. Eblen, F.; Graybiel, A.M. Highly restricted origin of prefrontal cortical inputs to striosomes in the macaque monkey. *J. Neurosci.* **1995**, *15*, 5999–6013. [CrossRef] [PubMed]
244. Goldman-Rakic, P.S. The cortical dopamine system: Role in memory and cognition. *Adv. Pharmacol.* **1998**, *42*, 707–711. [CrossRef] [PubMed]
245. Groenewegen, H.J.; Trimble, M. The Ventral Striatum as an Interface Between the Limbic and Motor Systems. *CNS Spectrums* **2007**, *12*, 887–892. [CrossRef] [PubMed]
246. Rosen, A.M.; Spellman, T.; Gordon, J.A. Electrophysiological endophenotypes in rodent models of schizophrenia and psychosis. *Biol. Psychiatry* **2015**, *77*, 1041–1049. [CrossRef] [PubMed]
247. Peykov, S.; Berkel, S.; Schoen, M.; Weiss, K.; Degenhardt, F.; Strohmaier, J.; Weiss, B.; Proepper, C.; Schratt, G.; Nöthen, M.M.; et al. Identification and functional characterization of rare SHANK2 variants in schizophrenia. *Mol. Psychiatry* **2015**, *20*, 1489–1498. [CrossRef] [PubMed]
248. Uppal, N.; Puri, R.; Yuk, F.; Janssen, W.G.; Bozdagi-Gunal, O.; Harony-Nicolas, H.; Dickstein, D.L.; Buxbaum, J.D.; Hof, P.R. Ultrastructural analyses in the hippocampus CA1 field in Shank3-deficient mice. *Mol. Autism* **2015**, *6*, 41. [CrossRef] [PubMed]
249. Lie, E.; Li, Y.; Kim, R.; Kim, E. SALM/Lrfn Family Synaptic Adhesion Molecules. *Front. Mol. Neurosci.* **2018**, *11*, 105. [CrossRef]
250. Shih, P.Y.; Hsieh, B.Y.; Tsai, C.Y.; Lo, C.A.; Chen, B.E.; Hsueh, Y.P. Autism-linked mutations of CTTNBP2 reduce social interaction and impair dendritic spine formation via diverse mechanisms. *Acta Neuropathol. Commun.* **2020**, *8*, 185. [CrossRef] [PubMed]

251. Lipina, T.V.; Prasad, T.; Yokomaku, D.; Luo, L.; Connor, S.A.; Kawabe, H.; Wang, Y.T.; Brose, N.; Roder, J.C.; Craig, A.M. Cognitive Deficits in Calsyntenin-2-deficient Mice Associated with Reduced GABAergic Transmission. *Neuropsychopharmacology* **2016**, *41*, 802–810. [CrossRef]
252. Giza, J.; Urbanski, M.J.; Prestori, F.; Bandyopadhyay, B.; Yam, A.; Friedrich, V.; Kelley, K.; D'Angelo, E.; Goldfarb, M. Behavioral and cerebellar transmission deficits in mice lacking the autism-linked gene islet brain-2. *J. Neurosci.* **2010**, *30*, 14805–14816. [CrossRef]
253. Soda, T.; Mapelli, L.; Locatelli, F.; Botta, L.; Goldfarb, M.; Prestori, F.; D'Angelo, E. Hyperexcitability and Hyperplasticity Disrupt Cerebellar Signal Transfer in the IB2 KO Mouse Model of Autism. *J. Neurosci.* **2019**, *39*, 2383–2397. [CrossRef] [PubMed]
254. Li, V.; Wang, Y.T. Molecular mechanisms of NMDA receptor-mediated excitotoxicity: Implications for neuroprotective therapeutics for stroke. *Neural. Regen. Res.* **2016**, *11*, 1752–1753. [CrossRef] [PubMed]
255. Mei, Y.; Monteiro, P.; Zhou, Y.; Kim, J.A.; Gao, X.; Fu, Z.; Feng, G. Adult restoration of Shank3 expression rescues selective autistic-like phenotypes. *Nature* **2016**, *530*, 481–484. [CrossRef] [PubMed]
256. Reisel, D.; Bannerman, D.M.; Schmitt, W.B.; Deacon, R.M.; Flint, J.; Borchardt, T.; Seeburg, P.H.; Rawlins, J.N. Spatial memory dissociations in mice lacking GluR1. *Nat. Neurosci.* **2002**, *5*, 868–873. [CrossRef] [PubMed]
257. Bannerman, D.M.; Sprengel, R.; Sanderson, D.J.; McHugh, S.B.; Rawlins, J.N.; Monyer, H.; Seeburg, P.H. Hippocampal synaptic plasticity, spatial memory and anxiety. *Nat. Rev. Neurosci.* **2014**, *15*, 181–192. [CrossRef]

International Journal of
Molecular Sciences

Review

Roadmap for Stroke: Challenging the Role of the Neuronal Extracellular Matrix

Ciro De Luca [1], Assunta Virtuoso [1,2], Nicola Maggio [3,4], Sara Izzo [5], Michele Papa [1,6,*] and Anna Maria Colangelo [6,7]

1 Laboratory of Morphology of Neuronal Network, Department of Public Medicine, University of Campania "Luigi Vanvitelli", 80138 Napoli, Italy; ciro.deluca@unicampania.it (C.D.L.); assunta.virtuoso@unicampania.it (A.V.)
2 School of Medicine and Surgery, University of Milano-Bicocca, 20900 Monza, Italy
3 Department of Neurology and Neurosurgery, Sackler Faculty of Medicine, Sagol School of Neuroscience, Tel Aviv University, Tel Aviv 6997801, Israel; Nicola.maggio@sheba.health.gov.il
4 Department of Neurology and the J. Sagol Neuroscience Center, The Chaim Sheba Medical Center, Tel HaShomer, Ramat Gan 5211401, Israel
5 Multidisciplinary Department of Medical-Surgical and Dental Specialties, Plastic Surgery Unit, University of Campania "Luigi Vanvitelli", 80138 Napoli, Italy; sa_izzo@hotmail.it
6 SYSBIO Centre of Systems Biology ISBE.ITALY, University of Milano-Bicocca, 20126 Milano, Italy; annamaria.colangelo@unimib.it
7 Laboratory of Neuroscience "R. Levi-Montalcini", Department of Biotechnology and Biosciences, University of Milano-Bicocca, 20126 Milano, Italy
* Correspondence: michele.papa@unicampania.it

Received: 18 August 2020; Accepted: 8 October 2020; Published: 13 October 2020

Abstract: Stroke is a major challenge in modern medicine and understanding the role of the neuronal extracellular matrix (NECM) in its pathophysiology is fundamental for promoting brain repair. Currently, stroke research is focused on the neurovascular unit (NVU). Impairment of the NVU leads to neuronal loss through post-ischemic and reperfusion injuries, as well as coagulatory and inflammatory processes. The ictal core is produced in a few minutes by the high metabolic demand of the central nervous system. Uncontrolled or prolonged inflammatory response is characterized by leukocyte infiltration of the injured site that is limited by astroglial reaction. The metabolic failure reshapes the NECM through matrix metalloproteinases (MMPs) and novel deposition of structural proteins continues within months of the acute event. These maladaptive reparative processes are responsible for the neurological clinical phenotype. In this review, we aim to provide a systems biology approach to stroke pathophysiology, relating the injury to the NVU with the pervasive metabolic failure, inflammatory response and modifications of the NECM. The available data will be used to build a protein–protein interaction (PPI) map starting with 38 proteins involved in stroke pathophysiology, taking into account the timeline of damage and the co-expression scores of their RNA patterns The application of the proposed network could lead to a more accurate design of translational experiments aiming at improving both the therapy and the rehabilitation processes.

Keywords: stroke; neuronal extracellular matrix; neurovascular unit; matrix metalloproteinases; systems biology; maladaptive plasticity

1. Introduction

As life expectancy increases, vascular diseases are progressively rising as major causes of disability and death [1]. The central nervous system (CNS) is no exception and stroke represents one of the most common and challenging pathologies of recent decades [2]. Stroke, along with neurodegeneration, can have a great impact on personal health and social burden for elderly people. Indeed, the increasing prevalence of neurovascular disorders, alone or in combination with degenerative diseases, dramatically enhances so-called vascular cognitive impairment (VCI) [3].

Various definitions of stroke have been assessed, however, the American Heart Association/American Stroke Association (AHA/ASA) delineated stroke as a CNS infarction within tissues of the brain, spinal cord or retina, due to ischemia [4]. It should be confirmed through radiological, clinical and/or pathological evidence of permanent injury. Inside the AHA/ASA statement are ischemic or hemorrhagic stroke and subarachnoid hemorrhage. The central element to unravel the vascular disease physiopathology is the neurovascular unit (NVU) [5,6]. Understanding the functionality of the NVU is key to a more efficient approach to stroke. The NVU encompasses the endothelium, pericytes, astrocytes, microglia and neurons. These cellular elements of different embryological origin are anatomically and functionally connected to ensure the correct pairing of metabolic supply/catabolic discharge and CNS activity [5]. Oligodendrocytes are not classically involved in the NVU, although recently they have been shown to actively respond to ischemic insult [7]. The CNS vasculature consists of a specialized form of endothelium with tight junctions (TJs) that regulate osmotic exchanges between the blood and parenchyma [8], the so-called blood–brain barrier (BBB). CNS tissue structure is completed by non-cellular elements such as the basal membrane and other specialized forms such as the neuronal extracellular matrix (NECM), playing a central role in physiological conditions and pathological modifications [9,10].

Injury of the NVU induces the immediate disruption of the BBB, not limited to the ischemic area, with edema of the nearby tissue. Hemostatic and inflammatory responses are helpful in self-limiting acute damage, however, it may lead to an amplification of cellular loss and consequent long-term functional damage, particularly considering the massive necrosis induced by inflammatory responses. In the white matter, it causes further damage to myelin, thus propagating further CNS tissue damage via a vicious cycle mechanism [6,9]. Acute damage is caused by the loss of oxygen and nutrient supply, driven by vessel occlusion (e.g., atheroma), rupture or thromboses (due to endothelial dysfunction or embolization). Neuronal loss within the ictal core is the direct consequence of the acute energy failure. However, the permanent injury of the neuronal networks is mainly caused by the maladaptive reparative processes, the end product of a predominantly metabolic correlative damage of the NVU and NECM with the involvement of glial cells and the immune system [6,11]. Indeed, the inflammatory reaction following the injury within the CNS could be highly disruptive and sustained, with the infiltration of macrophages to the site of injury surrounded by astroglial reactive cells [12]

In this review, we propose evidence for cell matrix and NVU interactions in the pathogenesis of stroke and its sequelae. The experimental data call for the creation of protein–protein interaction (PPI) maps which define the importance of both structural (Figure 1) and sequential coupling (Figures 2 and 3) of cellular and extracellular matrix pathways. The maps are designed using the string database platform [13] and are based on 38 selected proteins obtained through critical analysis of the literature. The papers used to create the structural PPI map are presented considering the logical structure of this review and the main experimental studies are presented in Table 1. The sequential coupling is obtained considering stroke pathophysiological data and the observed co-expression of proteins presented in Figure 2. This systems biology approach is essential to pave the way for an improvement in acute therapy and in the rehabilitation process.

Table 1. Synoptic table of the main representative experimental studies analyzed to select 38 proteins used as input for the string database (Figures 1 and 2). NIH: neuro-immune hemostasis, MMPs: matrix metalloproteinases, ADAMTS: a disintegrin and metalloproteinase with thrombospondin motifs, TIMP: tissue inhibitor of metalloproteinases, CSPGs: chondroitin sulfate proteoglycan, HSPGs: heparan sulfate proteoglycan, TnC: tenascin-C, TnR: tenascin-R, tPA: tissue plasminogen activator, GFAP: glial fibrillary acidic protein, ITGAM: integrin subunit alpha M, HMGB1: high-mobility group box 1 protein, TLR: Toll-like receptor, NFκB: nuclear factor-κB.

Research Paper	Analyzed Pathways	Selected Proteins	Model
Qiu J et al., 2010 [14]	Alarmins—MMPs—NIH	HMGB1, TLR4, MMP9	Mouse, in vivo and in vitro
Gu BJ and Wiley JS, 2006 [15]	Alarmins—MMPs—NIH	MMP9, P2X7R, TIMP-1	Human, in vitro
Gao, H et al., 2011 [16]	Alarmins—NIH	HMGB1, ITGAM, NFkB	Mouse, in vitro
Choi MS et al., 2010 [17]	Alarmins—MMPs—NIH	MMP9, P2YR	Rat, in vitro
Manaenko A et al., 2010 [18]	Alarmins—NIH	HSP70, IL1, TNFα, collagen	Mouse, in vivo
Malik R et al., 2018 [19]	NIH—MMPs	NOS, COL4A	Human, clinical
Clausen BH et al., 2008 [20]	NIH—Glial activation	IL1, TNFα, ITGAM	Mouse, in vivo
Botchkina GI et al., 1997 [21]	NIH—Neurotrophins	TNFα, p75NTR	Rat, in vivo
Atangana E et al., 2017 [22]	NIH—Glial activation	P-selectin, Iba1, ITGAM	Mice, in vivo
Weisenburger-Lile D et al., 2019 [23]	NIH—Glial activation	Neutrophil elastase, ITGAM	Human, clinical
Stubbe T et al., 2013 [24]	NIH—Glial activation	ITGAM, Iba1	Mouse, in vivo
Choucry AM et al., 2019 [25]	NIH—Neurotrophins	p75NTR, TrkA	Rat, in vivo
Candelario-jalil E et al., 2011 [26]	NIH—MMPs	MMP2, MMP9	Human, clinical
Cheng T et al., 2006 [27]	NIH—MMPs	MMP2, MMP9, PAR1, Thrombin, tPA, NFkB	Mouse in vivo, Human in vitro
del Zoppo GJ et al., 2012 [28]	NIH—MMPs—Glial activation	MMP2, MMP9, GFAP, ITGAM, COL4A, HSPGs	NonHuman Primate in vivo, mouse in vitro
Mishiro K et al., 2012 [29]	NIH—MMPs	MMP9, tPA	Mouse, in vivo
Chen W et al., 2009 [30]	MMPs—Glial activation	MMP2, MMP9, TIMP1, TIMP2, GFAP	Rat, in vivo
Ye H et al., 2015 [31]	NIH—Glial activation	S100B	Human, meta-analysis
Maysami S et al., 2015 [32]	NIH—Glial activation	CXCL1, CCL3, Iba1	Mice, in vivo
Quattromani MJ et al., 2018a,b [33,34]	NIH—MMPs	MMP2, MMP9, ADAMTS4, TIMP1, tPA, CSPGs, TnC, TnR	Rat and Mouse, in vivo
Matsumoto H et al., 2008 [35]	NIH—Glial activation	Iba1, NG2, ITGAM	Rat, in vivo
Carmichael ST et al., 2005 [36]	MMPs	CSPGs	Rat, in vivo

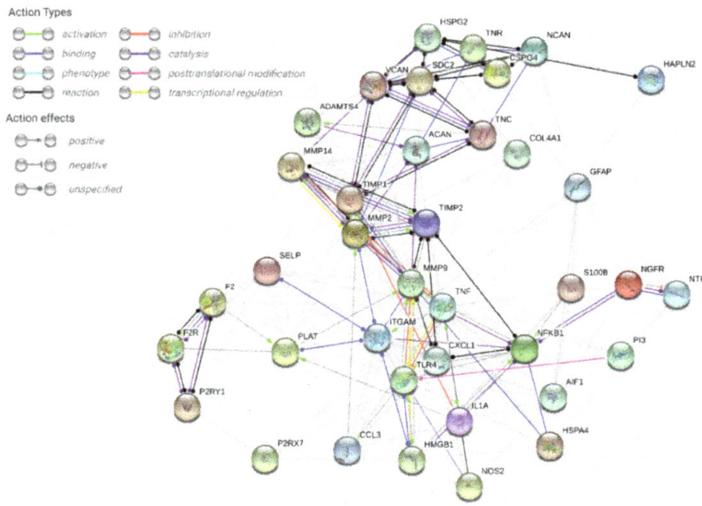

Figure 1. Extracellular/intracellular network analysis within the damaged tissue (string-db.org platform) for protein–protein interactions (PPIs). The proteins related to stroke pathophysiology are clustered and PPIs are defined. This map is difficult to use for the design of further experiments since the interference with any node will apply to every edge at the same time, an atypical phenomenon in biological systems. The map is based on a critical analysis of recent literature (see Table 1 for selected experimental studies). The database query with 38 proteins showed 203 edges with an average node degree of 10.7 (expected number of edges 35, average local clustering coefficient 0.575; PPI enrichment p-value < 10^{-16}). For further details, the map is accessible at this link: [37].

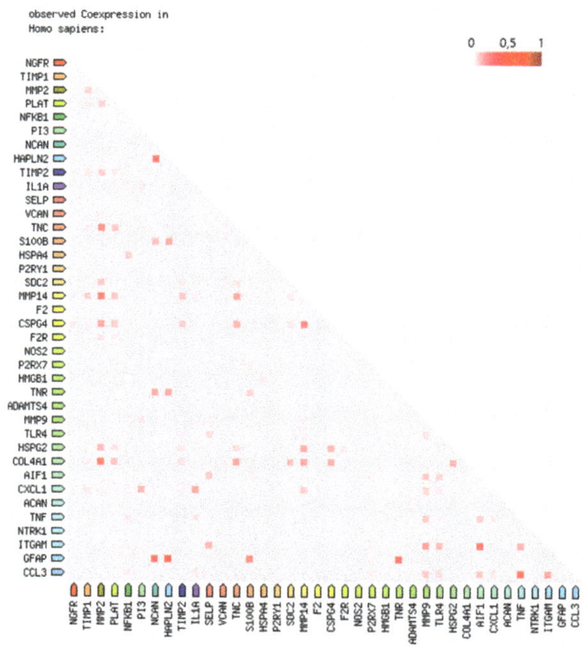

Figure 2. Bidimensional matrix of RNA expression patterns and protein co-regulation. The 38 proteins

proposed in the network map of Figure 1 are listed in a bidimensional matrix to show the co-expression scores (from 0 to 1 on the color visual scale) based on RNA expression patterns and protein co-regulation (string-db.org platform). The regulatory functions are subjected to a precise time-coupled expression. For further details, the matrix is accessible at this link: [38].

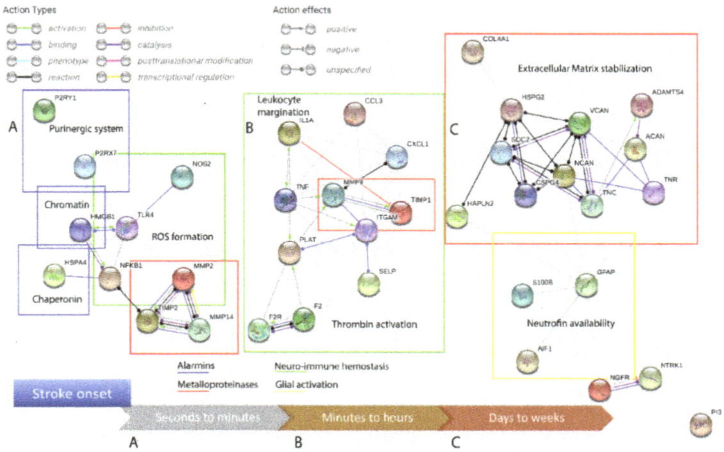

Figure 3. Protein–protein interaction (PPI) maps generated considering the timescale of stroke. Clustering is dependent on time-specific activation, avoiding a direct interaction analysis (presented in Figure 1) and considering the collected literature data and co-expression matrix presented in Figure 2. (**A**) Alarmins are released and constitutive matrix metalloproteinase 2 (MMP2) and purinergic receptors are promptly activated following stroke. (**B**) The inflammasome activates the neuro-immune pathway of cytokines, adhesion molecules, protease receptors and inducible MMPs. (**C**) Leukocytes arrive hours after stroke, act with their enzymes and remain for several days and, while glial activation proceeds, a novel extracellular matrix (NECM) is secreted and reactive gliosis regulates the neurotrophin concentration and receptors.

2. The Brain Energy Failure in Stroke

The interruption of regular perfusion causes an irreversible metabolic imbalance in the territories of the corresponding vasculature, which ultimately causes cell death. The extension and the nature of the vascular insult (e.g., small/large vessel disease, ischemic or hemorrhagic strokes) lead to a variety of parenchymal changes in the CNS, such as damage to cellular and extracellular elements in the ischemic core (i.e., cavitation), as well as a progressive reduction of the CNS tissue with spots of scar tissue and brain atrophy [39]. In the early phases of post-ischemic damage, typically within the first two weeks (before three months, i.e., standard chronic onset timepoint), the majority of patients with clinically diagnosed stroke will have well-defined cavitation at the ischemic core [40]. The cavity is surrounded and limited by glial cells that partially limit the area of damaged tissue (leukocyte infiltration, cellular debris, NECM and signaling molecules) [9,11] with long-lasting inflammation that could perdure more than three months according to injury models of myelinated tissues [41]. Myelin debris in particular should be removed to allow axonal regeneration and to reduce the inflammatory response, as it interferes with the crosstalk between the NVU (i.e., pericytes) and immune system (i.e., macrophages) during tissue repair [42,43]. The volume of the cavitation depends on the size of the afferent vessel and could range from 45 mL (medium sized cavity in a clinically determined stroke) to 150 mL in the case of a malignant middle cerebral artery occlusion [44]. If the damage is superficial, instead of the formation of a cavity, the area could be infiltrated by granulomatous tissue comprehending macrophages and blood vessels where glial cells are obliterated [41]. This macroscopic phenomenon is underlined by distinct molecular steps involving the NVU and NECM elements which interact with CNS resident and blood-derived immune cells [6].

3. Acute Ischemic Damage

The failure to meet the energy demand and counterbalance the osmotic gradient disrupts the organized membrane structures and cell death (necroptosis) [45], a mechanism different from apoptosis (caspase-mediated programmed cell death) or autophagy. The starving cells, lacking a trophic supply and oxygen, release signaling molecules during the death process, such as chaperonins, chromatin proteins and purinergic mediators [46,47]. The early modifications of the NECM generate degraded structural proteins. These molecules are the first to be released as damage-associated molecular patterns (DAMPs). These DAMPs are called alarmins and activate the innate immune response following an acute disruption of the BBB and before the margination of leukocytes [48]. Among the abovementioned molecules, the main factors seem to be high-mobility group box 1 protein (HMGB1), adenosine triphosphate (ATP) and heat-shock protein 70 (HSP70) [14,15,46,49]. Hyaluronic acid (HA), fibronectin and heparan sulfate (HSPGs) or chondroitin sulfate proteoglycans (CSPGs) are the main factors of the NECM [50]. These are endogenous molecules that can activate the immune response within minutes (Figure 3).

3.1. Necroptosis and Chromatin Exposure

HMGB1 is a nuclear protein with transcriptomics functions associated with chromatin structural proteins [51]. The release of the acetylated form of HMGB1 in the extracellular space, following ischemic injury, can activate the inflammatory response, binding to the Toll-like family receptors, particularly types 2 and 4 (TLR2/4) and the receptor for advanced glycation end-products (RAGE), expressed on the resident cells (neurons, astrocytes, microglia) with the secretion of matrix metalloproteinase 9 (MMP9) [14,47]. Indeed, the reshaping of the NECM is an early step, essential in the initial phases of neuroinflammation. Microglia, a scavenging component, express complement receptor 3 (CR3), an integrin $\alpha_M\beta_2$, also known as macrophage antigen complex 1 (MAC1) which can bind HMGB1 and connect the immune response to cell adhesion and phagocytosis [6,16].

3.2. Purinergic System Activation

ATP has been extensively investigated in various damage models as an endogenous factor in maladaptive plasticity both in the acute and chronic rearrangement of the NECM and cellular elements (reactive gliosis), [15,17,52,53]. The activity of extracellular ATP in acute stroke could mediate the migration of microglia, assisted by the rapid secretion of MMP9 [15,17], reinforcing the HMGB1 cascade. The endothelium, neurons, glial cells and leukocytes express purinergic receptors (ionotropic P2Xs and metabotropic P2Ys) which account for the multimodal and rapid signaling function. Microglia–astrocyte signaling with ATP as a neurotransmitter is involved in the inflammatory CNS response with a prominent role of P2X4/7 and P2Y1/6/12 subtypes. ATP can tune the dead or alive response of oligodendrocytes, interacting with P2X7 or P2Y1 receptors, respectively [53]. The cation-selective opening of P2X receptors has an essential role in a faster response to the ligand and in a slower neuromodulatory response that needs to be further clarified [49]. The microglia–astrocyte crosstalk with ATP release and its metabolite adenosine diphosphate (ADP) has been reported in both glutamatergic excitotoxicity and astrocytic feed-forward potentiation in the early phases of reactive gliosis [11,54,55].

3.3. Chaperonine and Stress Response

HSP70 is a chaperonin expressed intracellularly and involved in protein folding or supporting the repair of misfolding processes. It has a role in the proteasome–ubiquitin catabolism of damaged protein [46]. HSP70 is part of a system identified to be stress inducible, with historical reference to heat shock, but currently known to be part of multiple stressor responses (e.g., ischemia, oxidative stress, misfolded protein accumulation) [56]. HSP70 is expressed either as a constitutive or inducible form. Recent studies have demonstrated the role of HSP70 in neuroprotection, considering ischemic and hemorrhagic stroke [46]. HSP70 seems to play a role in both intrinsic and extrinsic apoptotic pathways at multiple levels [46]. It has been shown to reduce vulnerability to specific ischemic insults in astrocytes, preserving higher intracellular ATP and glutathione levels with reduced reactive oxygen species (ROS) accumulation [57]. HSP70 exerts a bivalent role: (i) in anti-inflammatory intracellular pathways and (ii) by immunostimulating extracellular activity [46]. The overexpression of HSP70 reduced the production of tumor necrosis factor-α (TNFα) and IL1 and has been shown to attenuate BBB disruption in a model of hemorrhagic stroke [18,58]. The main HSP70-modulated intracellular systems are inducible nitric oxide synthase (iNOS) expression and nuclear factor-κB (NF-κB) activation [59]. An exonic polymorphism in NOS3 has been recently considered as a locus associated with stroke [19].

NF-κB deregulation probably caused the decreased expression of MMP9 in cultured astrocytes with HSP70 overexpression in an ischemic-like in vitro model [60]. In the same study, an MMP2 reduction was demonstrated, even if its regulation did not depend on NF-κB [60].

On the extracellular side, HSP70 (and other HSPs) can interact with TLR2/4, activate microglia and macrophages and promote lymphocyte differentiation [61–63].

These pathways are deeply connected with (i) neuro-immune hemostasis, (ii) MMP modulation and (iii) reactive gliosis.

4. Neuro-Immune Hemostasis

The essential molecules for transcript formation (HMGB1), cellular energetic metabolism (ATP) and protein-folding assistance (HSP70) constitute the endogenous pathology alarm molecules and damage control systems.

Establishing a network of such proteins in a systems biology model is essential to identify perturbations in metabolic supply, block the energy-consuming activities, save ATP, avoid ROS production and ultimately prompt a fast response to damage in the extracellular medium (Figures 1 and 2). Whether the rescue system is effective or not, it depends on the well-timed resolution of the vascular insult. Nowadays, there are systemic and local treatments to guarantee good tissue reperfusion in ischemic stroke [64], however, the modulation of the intracellular–extracellular alarmins system could increase the time window for efficient therapy and improve the clinical outcome.

The activation of the coagulation cascade and the network involving clot formation, thrombin signaling, immune response and synaptic plasticity has been proposed in several neurological diseases [6]. During stroke, clotting dysfunction is part of the disease itself, with a prominent role in cell survival and the loss of the vital elements in the ischemic core.

4.1. Blood–Brain Barrier Failure

The damage to the BBB structure is the main harm to the NVU during acute stroke, it allows the permeation of blood elements, normally confined to the systemic circulation. These are constituted basically of plasma proteins and immune cells [65]. The DAMP response, on the other hand, produces in a few hours the second phase of the inflammatory process: the production of cytokines such as TNFα and IL1 [20,66]. The astrocytes of the NVU retract their podocytes, the basement membrane is modified by the release of MMPs and the production of cytokines promotes the margination of the first infiltrating leukocytes (neutrophils and monocytes) that act together with the microglia as host defenders.

The activated endothelial cells, in the area of vasogenic edema, express on their membrane cytokine receptors [21,67] and cell adhesion molecules (CAMs) of the immunoglobulin family, such as vascular (V)CAM-1 and platelet (P) and endothelial (E) selectins and integrins (e.g., MAC1), facilitating the rolling, margination and migration of leukocytes in the CNS [22,68].

Indeed, the expression of CAMs has been largely studied as a cell-based phenomenon, but more recent data have shown a direct involvement of NECM molecules, particularly laminin-8/10, fundamentally expressed in the basal lamina of the NVU. It has been shown that in mice lacking laminin-10, a facilitated infiltration of leukocytes with the loss of cadherin junctional support following TNFα administration occurred, while knocking out laminin-8 significantly reduced immune cell margination [69,70].

4.2. Perivascular Space Signaling

Perivascular spaces with low-density collagen type IV, a fundamental constituent of the basement membrane, and laminin are characterized by a prevalent neutrophil infiltration [71]. The mutation of the gene encoding for the α1 chain of collagen type IV, COL4A1, is responsible for a monogenic stroke syndrome (hemorrhagic and/or ischemic small vessel disease) [72], whereas an intron variant of the same gene has been associated with increased risk of multifactorial stroke [19].

The immune cell infiltration (hours to days) of the CNS utilizes both elastases and MMPs to further open the pathway to the second wave of leukocytes: the lymphocytes [23,73]. This process is self-limiting, with a modulatory function principally attributed to T cells, in the tardive phases (Figure 3) of tissue remodeling (weeks to a month timescale) [24,73].

The cytokine shift mediated by the regulatory T cells is typical of this phase, with the production of IL6/10, transforming growth factor β (TGFβ) and possibly neurotrophins such as brain-derived neurotrophic factor (BDNF) and nerve growth factor (NGF) [25]. The stabilization of the lesion is fundamental for neurological signs of partial recovery [74,75].

The schematic description of early to late phases of the immune response needs to be considered as overlapping phenomena playing a pivotal role in NECM remodeling and reactive gliosis (Figure 3) [6,11,76].

5. Matrix Metalloproteinase Modulation

The activation of MMPs is absolutely essential, as it begins during the early phases of stroke pathophysiology. The two main factors are MMP2 and MMP9. MMP2 is constitutively expressed and secreted as a zymogen and can be activated by the formation of a complex involving MMP14 (a membrane-type MMP), tissue inhibitor of metalloproteinase 2 (TIMP 2) and pro-MMP2 itself. MMP14 contains a hydrophobic C-terminal domain and its activity is restricted to the cell membrane [77]. This feature accounts for the spatial definition and limited activation of MMP2 [78]. MMP9 production, on the other hand, is induced during inflammation and its zymogen cleavage is mainly granted by MMP3, activated in neuroinflammatory responses [79]. TIMP1 is the natural inhibitor of MMP9 [80]. TIMP2 has a role as a sensor of damage and confinement of the injury, being able (if not complexed with MMP14) to inhibit MMP2 (Figures 2 and 3) [79,81]. Similarly, the protease thrombin can activate the coagulatory cascade and, once complexed with thrombomodulin and endothelial protein C receptor (EPCR), can mediate the downregulation of coagulatory factors with activated protein C (aPC) [10]. Both MMP2 and MMP9 systems are associated with NVU dysfunction, mainly mediated by the breakdown of claudin-5, occludin and zonula occludens 1 (ZO-1) [26].

These functions are essential in the development of the CNS and in adult brain vascular remodeling following pressor insults, non-occlusive vascular lesions or microvascular recanalization [82–84]. Major ischemic injury causes massive damage to BBB permeability with tissue edema and reperfusion damage; these factors are independent of the extension of the lesion area beyond the volume of the ischemic core [85,86].

The thrombolytic treatment could enhance the activation of already produced MMPs [87,88]. Thrombin or recombinant tissue plasminogen activator (rt-PA)-mediated PAR1 activation contribute to NFκB-induced MMP9 expression [27]. The PAR1-mediated increment of MMP9, particularly through pericytes, could worsen the intracranial hemorrhage clinical outcome [89].

The efficacy of rt-PA has changed the perspective for ischemic stroke patients, however, understanding of the time schedule of NECM modifications could further improve the clinical outcome.

Metalloproteinase Differential Activation

MMPs' timely activation is fundamental in the pathophysiology of stroke. An improved neurological outcome has been shown in an MMP2-deficient mouse model, which presented a smaller hemorrhagic volume and transformation rate upon occlusion [90]. The MMP2 levels were not significantly increased in hemorrhagic transformation, and the infarct area was not affected by MMP2 knockout, which although apparently in contrast with the previous data, further demonstrates its role in the early phases of BBB failure, with a constitutive, scarcely inducible expression and is later replaced by other MMPs (i.e., inducible MMP9) [91].

MMP9 is mainly expressed by resident cells, such as neurons, astrocytes, microglia and endothelial cells [92]. Indeed, MMP9 seems to be involved in the late phases of parenchymal modifications. MMP9 is also expressed by infiltrating leukocytes, but these cells are not considered as the main source [28]. Within the first 24 h of stroke onset, MMP3 activation (via proteolytic cleavage) appears to affect MMP9 levels [79]. In vivo studies with a pan-inhibitor of MMPs (GM6001) showed interesting results, preserving the TJ proteins [29,30]. MMPs are essential in adaptive plasticity phenomena (neuroblast formation, revascularization, synaptic plasticity) [93,94]. MMPs' selective and limited inhibition/activation seems to be the real translational target. Indeed, a potential detrimental effect of pan-MMP inhibition 7 days after cerebral ischemia has already been shown, with the blockage of physiologic reparative phenomena [93].

NVU components (mainly pericytes and endothelial cells) have been shown to be protected by another MMP inhibitor in a focal embolic ischemia model that avoids pericyte contraction and laminin degradation [95].

MMPs and a disintegrin and metalloproteinase with thrombospondin motifs (ADAMTS) are the major NECM modifiers (Figure 2) implicated in the regulatory functions both in health and diseases [9]. Their function is not restricted to the processing of the NECM, they modulate synaptic plasticity. It has been reported that the local inhibition of MMP9 through endogenous inhibitors, hypothermia and synthetic compounds could help in reducing the inflammatory process and improve the outcome in the acute phase [96]. In the subacute phase, MMP9 blockade could be harmful as it has been reported to play a role in the migration of neural progenitors of the subventricular zone (SVZ), suggesting a stimulating role for adult neurogenesis [94]. Moreover, ADAMTS4 has been considered useful in CSPG degradation and axonal regeneration [97].

6. Reactive Gliosis and NECM Deposition

The proliferative response of glial cells to an ischemic event is stimulated by cytokines [11]. An ongoing registry is evaluating cytokines in stroke, based on the previous finding of the S100B (an astrocyte marker) association with neurological outcome and infarct size [31,98]. The upregulation of glial fibrillary acidic protein (GFAP) expression has been generally used as a reactive astrocyte biomarker throughout various models of maladaptive CNS plasticity in CNS diseases [11,52,99,100]. Astroglial response functionally confines the damage with the recruitment of scavenger cells as peripheral macrophages to remove the debris of the acute damage, through the secretion of chemokines, classified based on the first two amino-terminal cysteine (C) sequences as adjacent (CC motif) or separated by another amino acid (CXC). CC ligand 3 (CCL3) and CXC ligand 1/2 (CXCL1/2) have been identified as the main chemokines involved (Figure 1) [32,101,102]. The damaged tissue is replaced by a newly formed NECM and by the proliferation of non-neuronal cells, particularly immune cells and glia (astrocytes and microglia) with the production of tenascins (Tn), proteoglycans and collagen (i.e., fibrous tissue) that rapidly aid tissue repair but definitively inhibit neural plasticity (potentially avoiding maladaptive rewiring) through the scar [103].

6.1. Astroglial Adaptive Plasticity

The subsequent propagation of the necrotic inflammatory response is limited by the surrounding protection of the glial cells. Modeling compromised scar formation showed increased neuronal dysfunction with combinatory death and demyelination in both ischemic and traumatic injury [104,105].

The glial signaling regulates the immune response, organizes the structural reshape of the NECM and avoids maladaptive neuronal plasticity. The leukocyte invasion is indeed regulated by CSPG and tenascin-C (TnC) expression [106,107]. Both these constituents and astrocytic-produced keratan sulfate proteoglycans (KSPGs) are regulatory molecules for neural plasticity and axonal outgrowth [108]. The second Tn expressed in the CNS (TnR) is instead expressed in the perineural nets (PNNs), representing another specialized form of NECM [92]. At this time, rigorous debate concerns if and how it could be beneficial for neural plasticity and limiting scar formation, and if it could play a role in rescuing post-lesional axonal sprouting. Even though targeting astrocytic proliferation and NECM production or NECM degradation has been proposed in various models of injury, [80,82,95,100,109] affirm that it is not always true that astrogliosis "per se" inhibits axonal sprouting and, moreover, it has been reported that astrogliosis has anti-inflammatory and anti-edema activity. The release of CSPGs and TnC may not be detrimental and may help axon regeneration and guidance, while the blockage of MMPs could be useful to increase neurotrophin levels, as demonstrated in injury models [11,100,110].

A recent model has been proposed focusing on the PNNs, a quadruple knockout mouse for brevican, neurocan and TnC/R [33]. It has shed light on NECM modifications in damaged tissue and in the functionally associated parenchyma. The results were controversial, with higher intrahemispheric connectivity and worse behavioral outcomes in the early phases of stroke, further contributing to the idea that a timely modulation is required [33].

6.2. CSPG/HSPG Expression

The NECM has a structural storage function for signaling molecules and growth factors [9,92]. During brain development, HSPGs are necessary for lineage differentiation, cell positioning and neural wiring [111,112]. An important role in the differential expression of CSPGs and HSPGs may provide dynamic path-finding for axonal wiring, with HSPGs enhancing and CSPGs restricting the outgrowth [113].

The CSPGs overexpressed in the late phases of stroke are neurocan, brevican, phosphacan and NG2 [35,114]. The latter is the marker of oligodendrocyte precursor cells (OPCs), which are a putative functional reservoir of multipotent cells, involved in the remyelination of damaged circuitry and supporting axonal outgrowth as the only non-neuronal cell that can form synapses with neurons [115]. OPCs contribute to NECM remodeling by upregulating versican in the damaged area [35]. CSPGs in the PNNs far from the lesion are reduced as a result of the plastic remodeling of the entire system following the injury [36].

A CSPG/HSPG ratio, similar to the developmental processes, has been observed in ischemic models and could be potentially targeted in order to modify neuronal plasticity [113]. The physiological role of the scar could be to inhibit the neural plasticity from the perilesional area into the ischemic core with a certain grade of synaptic plasticity in the ischemic penumbra and neighboring intact tissue [116]. The reactive microglia, astrocytes and leukocytes persist in the ischemic core and are responsible for the structure of the stroke-damaged tissue, including the NECM remodeling. However, the data available were totally produced, except for a few, in rodent animal models typically used in stroke research. We must consider that these animals show very little white matter in their hemispheres, therefore, a vascular injury such as obstruction of the middle cerebral artery leads to gray matter injury, with very little myelin involvement and self-limiting inflammation. To date, despite the growing amount of data, the correct understanding of the role of each element in the clinical outcome following vascular damage is very poor. The systems biology approach could provide designs for conceptual maps with the simultaneous analysis of multiple factors (Figure 1). The benefit of a new methodological approach aimed at assessing the overall molecular networks involved in a specific disease is needed to overcome the old attitude limited to a single cellular or structural element as the key to understanding the pathophysiology of a disease [117]. The therapeutic approach to these models could allow us to overcome the inhibition/enhancement dualism, focused on a single path (valid for monogenic diseases), as this strategy is very poor if applied to the multifactorial pathophysiology of sporadic stroke (Figure 2).

7. Designing the Network: The Time Variable

The modulation of the NECM should be investigated considering its cellular components and how these molecules act as intracellular signal transducers. To the same extent, it is impossible to investigate diseases clearly showing NVU impairment without analyzing its components and blood-derived infiltrating cells. All these factors have their prominent roles on the stage and are mutually influenced by a specific proximity and temporal activation/inhibition (Figure 3). The structural network (Figure 1) needs to be improved, adding the time variable to the same hub connections (Figure 3). In these networks, no targets are defined to block maladaptive phenomena and/or improve adaptive plasticity. The validation of the available data requires adequate experiments based on well-designed networks. These must consider the progression of the injury site and the timing of the reparative processes.

The scaffold of the NECM is mainly composed of HA, proteoglycans, collagen and Tns and, although many of these elements act as signaling molecules in precise phases (development, injury), they are expressed as constitutive elements in the normal adult brain, with physiological turnover but consistent stability [92]. Signaling factors with gradient expression or short-range efficacy are embedded among these functional nets. Indeed, many of these signaling molecules are strictly connected to the abovementioned structural elements and their regulators are found in the same environment [92].

8. Future Perspectives and Therapies

Novel experiments proposed on the basis of the designed rational networks could be helpful to elucidate the role of the NECM in the pathophysiology of stroke. Another limitation to overcome is the current knowledge of both ischemic and hemorrhagic stroke founded on various models (atherogenic, embolic, hypertensive) without discrimination of the involved pathways. BBB failure, on the other hand, could be used to redirect pharmacological treatment to selectively deliver anti-inflammatory drugs to the damaged and edematous tissue.

8.1. Metalloproteinases and Proteoglycans

There is one ongoing double-blind, placebo-controlled clinical trial evaluating minocycline (as an MMP9 inhibitor) and molecular hydrogen (an antioxidant) in stroke [118]. Their protective role could prevent ischemia/reperfusion injury. Moreover, minocycline has been showed to be effective in inhibiting poly(ADP-ribose) polymerase. This trial is a proof of concept for the results of a study that showed a possible role of minocycline in lowering MMP9 plasma levels following rt-PA treatment of stroke [119].

Another clinical trial [120] will also consider MMP9 levels as a possible predictive factor for hemorrhagic transformation of ischemic stroke treated with rt-PA.

Efforts to translate the data on NECM remodeling in stroke recovery is represented by the use of HSPG mimetics that, administered intravenously in an acute ischemic stroke model, demonstrated a good spatial localization and within 6 h of the event, reduced functional deficits and increased plasticity phenomena (neo-angiogenesis, neurogenesis). An open study, the first in humans, to assess the clinical safety of an HSPG mimetic (OTR4132) in anterior circulation acute ischemic stroke of patients re-vascularized with both rt-PA or endovascular thrombectomy, will be conducted [121].

8.2. Engineering Bioscaffolds

The inflammation post trauma or post stroke needs to be the first contained/eliminated issue before an implant is considered. However, the chance to replace the lost NECM using engineered biomaterial has been developed with a top-down approach of modified natural bioscaffolds to achieve a promising bottom-up hydrogel synthesis [122]. There are various technical issues to overcome to realize a translational success in this field, although there are promising results in animal models [122]. In particular, experimental data are based on rodent models of stroke and are limited to brain tissue (with higher gray/white matter ratio than humans) and could be reconsidered to favor nervous tissue richer in white matter, although this has other limitations [123,124].

One of the main biological challenges is represented by the intrinsic connectivity of the CNS. The replacement of novel neurons, infiltrating the bioscaffold, needs to be coupled with a robust newly formed and functional network. The restoration of complex cognitive functions can be achieved with the correct coupling of axonal growth and synapse formation both in local circuitry and with long-distance connections. Hydrogels with HA and vascular endothelial growth factor (VEGF) seem to be able to replicate angiogenesis and axonal networks in the infarct area, similar to the contralateral hemisphere [122]. Studies on axonogenesis, substance deliveries and cell influence on the newly formed tissue have been considered [125–127], but a lack of knowledge on the intricate network of biological, behavioral and experience-dependent processes halts the organization into a functional brain tissue. The major risk is the mere substitution of scar tissue with an organized but useless, or even detrimental, cell/matrix mass. Finally, the role of the environment is crucial for the development of a functional network, as already demonstrated by the impact of early physiotherapy in post-stroke recovery that could partly act on the remodeling of the NECM. It has been shown that an enriched environment could decrease PNNs around neurons and modulate the transcription levels of NECM-modifying proteins in the somatosensory cortex [34].

9. Conclusions

The treatment of stroke has so far been focused on limiting acute damage at the onset and securing the hemodynamic stabilization of the patient to reduce the loss of CNS tissue.

This approach, with rt-PA and mechanical thrombectomy approved for acute ischemic stroke, represented a major contribution and, together with early rehabilitation, significantly improved clinical outcomes. However, this success is limited to selected patients with rescuable brain tissue (within a time window or with significant ischemic core/penumbra mismatch). Hemorrhagic stroke without coagulatory diseases or aneurysmatic ruptures can be clinically modified solely with the strict control of blood pressure values. The actual paradigm is indeed to avoid further damage. We are unaware of how to affect disease progression depending on these variables in the post-acute phase. The focus of further research should be on the cellular capability to reduce the neuroinflammatory phenomena and actively reshape the matrix in early reperfusion and beyond. For instance, a patient treated with rt-PA could benefit from the reduction of constitutive MMP (e.g., MMP2) activation or the blockage of the thrombin-mediated PAR1 pathway. Mechanical thrombectomy, on the other hand, could assist in reducing the prominent BBB failure secondary to large vessel diseases, focusing on endothelial activation and leukocyte margination. These pathways, together with induced MMP activity (e.g., MMP9), could be partially shared with hemorrhagic stroke. These speculations may not actually be proven and may need a systems biology approach to better understand the adaptive reparative mechanisms and try to improve the sequalae with an active approach. MMP and cytokine interference, biomaterial scaffolds, growth signals and an enriched environment are slowly being translated into clinical trials.

The therapeutic benefits, however, may be flawed by the incorrect modulation of proteins without the understanding of the complex structural and temporal phases of their adaptive functioning. This could be the reason for the knockout or pan-inhibitory models that have so far provided contradictory results. All the elements analyzed, cellular and molecular, are affected by a single constant problem, the damage to the NVU. This results in an energy deficit that alters the behavior of all components of the CNS, as reported for neurodegenerative diseases [122]. Defining a model that allows an energy rescue of the NVU and the NECM is the real therapeutic goal.

Author Contributions: Conceptualization, C.D.L., A.M.C. and M.P.; writing—original draft preparation, C.D.L., A.V. and S.I.; writing—review and editing, A.M.C., N.M. and M.P.; visualization, C.D.L.; supervision, M.P. All authors have read and agreed to the published version of the manuscript.

Funding: This research received no external funding.

Acknowledgments: This work was supported by grants from the Italian Minister of University and Research (MIUR) (PRIN2007 to A.M.C. and M.P.; SYSBIO-Italian ROADMAP ESFRI Infrastructures to L.A., M.P. and A.M.C.; PRIN 2015-2015TM24JS_009 to M.P.; PRIN2017-2017XJ38A4_003-to M.P.; ALISEI-IVASCOMAR Italian National Cluster to A.M.C.; and grant Dipartimenti di Eccellenza-2017 to the University of Milano-Bicocca Department of Biotechnology and Biosciences); Corbel: EU-H2020 [PID 2354] to A.M.C.

Conflicts of Interest: The authors declare no conflict of interests.

References

1. Benziger, C.P.; Roth, G.A.; Moran, A.E. The Global Burden of Disease Study and the Preventable Burden of NCD. *Glob. Heart* **2016**, *11*, 393–397. [CrossRef]
2. Feigin, V.L.; Forouzanfar, M.H.; Krishnamurthi, R.; Mensah, G.A.; Connor, M.; Bennett, D.A.; Moran, A.E.; Sacco, R.L.; Anderson, L.; Truelsen, T.; et al. Global and regional burden of stroke during 1990–2010: Findings from the Global Burden of Disease Study 2010. *Lancet* **2014**, *383*, 245–254. [CrossRef]
3. O'Brien, J.T.; Thomas, A. Vascular dementia. *Lancet* **2015**, *386*, 1698–1706. [CrossRef]
4. Sacco, R.L.; Kasner, S.E.; Broderick, J.P.; Caplan, L.R.; Connors, J.J.B.; Culebras, A.; Elkind, M.S.V.; George, M.G.; Hamdan, A.D.; Higashida, R.T.; et al. An updated definition of stroke for the 21st century: A statement for healthcare professionals from the American Heart Association/American Stroke Association. *Stroke* **2013**, *44*, 2064–2089. [CrossRef] [PubMed]
5. Del Zoppo, G.J. The neurovascular unit in the setting of stroke. *J. Intern. Med.* **2010**, *267*, 156–171. [CrossRef]
6. De Luca, C.; Colangelo, A.M.; Alberghina, L.; Papa, M. Neuro-Immune Hemostasis: Homeostasis and Diseases in the Central Nervous System. *Front. Cell. Neurosci.* **2018**, *12*, 459. [CrossRef]
7. Michalski, D.; Keck, A.L.; Grosche, J.; Martens, H.; Härtig, W. Immunosignals of Oligodendrocyte Markers and Myelin-Associated Proteins Are Critically Affected after Experimental Stroke in Wild-Type and Alzheimer Modeling Mice of Different Ages. *Front. Cell. Neurosci.* **2018**, *12*, 23. [CrossRef]
8. Bechmann, I.; Galea, I.; Perry, V.H. What is the blood–brain barrier (not)? *Trends Immunol.* **2007**, *28*, 5–11. [CrossRef]
9. De Luca, C.; Papa, M. Matrix Metalloproteinases, Neural Extracellular Matrix, and Central Nervous System Pathology. *Prog. Mol. Biol. Transl. Sci.* **2017**, *148*, 167–202. [CrossRef]
10. De Luca, C.; Virtuoso, A.; Maggio, N.; Papa, M. Neuro-Coagulopathy: Blood Coagulation Factors in Central Nervous System Diseases. *Int. J. Mol. Sci.* **2017**, *18*, 2128. [CrossRef]
11. Papa, M.; De Luca, C.; Petta, F.; Alberghina, L.; Cirillo, G. Astrocyte-neuron interplay in maladaptive plasticity. *Neurosci. Biobehav. Rev.* **2014**, *42*, 35–54. [CrossRef] [PubMed]
12. Kwiecien, J.M.; Dąbrowski, W.; Yaron, J.R.; Zhang, L.; Delaney, K.H.; Lucas, A.R. The Role of Astrogliosis in Formation of the Syrinx in spinal cord injury. *Curr. Neuropharmacol.* **2020**. [CrossRef] [PubMed]
13. STRING: Functional Protein Association Networks. Available online: https://string-db.org/ (accessed on 30 April 2020).
14. Qiu, J.; Xu, J.; Zheng, Y.; Wei, Y.; Zhu, X.; Lo, E.H.; Moskowitz, M.A.; Sims, J.R. High-mobility group box 1 promotes metalloproteinase-9 upregulation through Toll-like receptor 4 after cerebral ischemia. *Stroke* **2010**, *41*, 2077–2082. [CrossRef] [PubMed]
15. Gu, B.J.; Wiley, J.S. Rapid ATP-induced release of matrix metalloproteinase 9 is mediated by the P2X7 receptor. *Blood* **2006**, *107*, 4946–4953. [CrossRef] [PubMed]
16. Gao, H.-M.; Zhou, H.; Zhang, F.; Wilson, B.C.; Kam, W.; Hong, J.-S. HMGB1 acts on microglia Mac1 to mediate chronic neuroinflammation that drives progressive neurodegeneration. *J. Neurosci.* **2011**, *31*, 1081–1092. [CrossRef] [PubMed]
17. Choi, M.S.; Cho, K.S.; Shin, S.M.; Ko, H.M.; Kwon, K.J.; Shin, C.Y.; Ko, K.H. ATP induced microglial cell migration through non-transcriptional activation of matrix metalloproteinase-9. *Arch. Pharm. Res.* **2010**, *33*, 257–265. [CrossRef] [PubMed]

18. Manaenko, A.; Fathali, N.; Chen, H.; Suzuki, H.; Williams, S.; Zhang, J.H.; Tang, J. Heat shock protein 70 upregulation by geldanamycin reduces brain injury in a mouse model of intracerebral hemorrhage. *Neurochem. Int.* **2010**, *57*, 844–850. [CrossRef] [PubMed]
19. Malik, R.; Rannikmäe, K.; Traylor, M.; Georgakis, M.K.; Sargurupremraj, M.; Markus, H.S.; Hopewell, J.C.; Debette, S.; Sudlow, C.L.M.; Dichgans, M.; et al. Genome-wide meta-analysis identifies 3 novel loci associated with stroke. *Ann. Neurol.* **2018**, *84*, 934–939. [CrossRef]
20. Clausen, B.H.; Lambertsen, K.L.; Babcock, A.A.; Holm, T.H.; Dagnaes-Hansen, F.; Finsen, B. Interleukin-1beta and tumor necrosis factor-alpha are expressed by different subsets of microglia and macrophages after ischemic stroke in mice. *J. Neuroinflamm.* **2008**, *5*, 46. [CrossRef]
21. Botchkina, G.I.; Meistrell, M.E., 3rd; Botchkina, I.L.; Tracey, K.J. Expression of TNF and TNF receptors (p55 and p75) in the rat brain after focal cerebral ischemia. *Mol. Med.* **1997**, *3*, 765–781. [CrossRef]
22. Atangana, E.; Schneider, U.C.; Blecharz, K.; Magrini, S.; Wagner, J.; Nieminen-Kelhä, M.; Kremenetskaia, I.; Heppner, F.L.; Engelhardt, B.; Vajkoczy, P. Intravascular Inflammation Triggers Intracerebral Activated Microglia and Contributes to Secondary Brain Injury After Experimental Subarachnoid Hemorrhage (eSAH). *Transl. Stroke Res.* **2017**, *8*, 144–156. [CrossRef] [PubMed]
23. Weisenburger-Lile, D.; Dong, Y.; Yger, M.; Weisenburger, G.; Polara, G.F.; Chaigneau, T.; Ochoa, R.Z.; Marro, B.; Lapergue, B.; Alamowitch, S.; et al. Harmful neutrophil subsets in patients with ischemic stroke: Association with disease severity. *Neurol. Neuroimmunol. Neuroinflamm.* **2019**, *6*, e571. [CrossRef] [PubMed]
24. Stubbe, T.; Ebner, F.; Richter, D.; Engel, O.; Klehmet, J.; Royl, G.; Meisel, A.; Nitsch, R.; Meisel, C.; Brandt, C. Regulatory T cells accumulate and proliferate in the ischemic hemisphere for up to 30 days after MCAO. *J. Cereb. Blood Flow Metab.* **2013**, *33*, 37–47. [CrossRef]
25. Choucry, A.M.; Al-Shorbagy, M.Y.; Attia, A.S.; El-Abhar, H.S. Pharmacological Manipulation of Trk, p75NTR, and NGF Balance Restores Memory Deficit in Global Ischemia/Reperfusion Model in Rats. *J. Mol. Neurosci.* **2019**, *68*, 78–90. [CrossRef]
26. Candelario-Jalil, E.; Thompson, J.; Taheri, S.; Grossetete, M.; Adair, J.C.; Edmonds, E.; Prestopnik, J.; Wills, J.; Rosenberg, G.A. Matrix metalloproteinases are associated with increased blood-brain barrier opening in vascular cognitive impairment. *Stroke* **2011**, *42*, 1345–1350. [CrossRef] [PubMed]
27. Cheng, T.; Petraglia, A.L.; Li, Z.; Thiyagarajan, M.; Zhong, Z.; Wu, Z.; Liu, D.; Maggirwar, S.B.; Deane, R.; Fernández, J.A.; et al. Activated protein C inhibits tissue plasminogen activator-induced brain hemorrhage. *Nat. Med.* **2006**, *12*, 1278–1285. [CrossRef]
28. del Zoppo, G.J.; Frankowski, H.; Gu, Y.-H.; Osada, T.; Kanazawa, M.; Milner, R.; Wang, X.; Hosomi, N.; Mabuchi, T.; Koziol, J.A. Microglial cell activation is a source of metalloproteinase generation during hemorrhagic transformation. *J. Cereb. Blood Flow Metab.* **2012**, *32*, 919–932. [CrossRef]
29. Mishiro, K.; Ishiguro, M.; Suzuki, Y.; Tsuruma, K.; Shimazawa, M.; Hara, H. A broad-spectrum matrix metalloproteinase inhibitor prevents hemorrhagic complications induced by tissue plasminogen activator in mice. *Neuroscience* **2012**, *205*, 39–48. [CrossRef]
30. Chen, W.; Hartman, R.; Ayer, R.; Marcantonio, S.; Kamper, J.; Tang, J.; Zhang, J.H. Matrix metalloproteinases inhibition provides neuroprotection against hypoxia-ischemia in the developing brain. *J. Neurochem.* **2009**, *111*, 726–736. [CrossRef]
31. Ye, H.; Wang, L.; Yang, X.K.; Fan, L.P.; Wang, Y.G.; Guo, L. Serum S100B levels may be associated with cerebral infarction: A meta-analysis. *J. Neurol. Sci.* **2015**, *348*, 81–88. [CrossRef]
32. Maysami, S.; Haley, M.J.; Gorenkova, N.; Krishnan, S.; McColl, B.W.; Lawrence, C.B. Prolonged diet-induced obesity in mice modifies the inflammatory response and leads to worse outcome after stroke. *J. Neuroinflamm.* **2015**, *12*, 140. [CrossRef] [PubMed]
33. Quattromani, M.J.; Hakon, J.; Rauch, U.; Bauer, A.Q.; Wieloch, T. Changes in resting-state functional connectivity after stroke in a mouse brain lacking extracellular matrix components. *Neurobiol. Dis.* **2018**, *112*, 91–105. [CrossRef] [PubMed]
34. Quattromani, M.J.; Pruvost, M.; Guerreiro, C.; Backlund, F.; Englund, E.; Aspberg, A.; Jaworski, T.; Hakon, J.; Ruscher, K.; Kaczmarek, L.; et al. Extracellular Matrix Modulation Is Driven by Experience-Dependent Plasticity During Stroke Recovery. *Mol. Neurobiol.* **2018**, *55*, 2196–2213. [CrossRef] [PubMed]

35. Matsumoto, H.; Kumon, Y.; Watanabe, H.; Ohnishi, T.; Shudou, M.; Chuai, M.; Imai, Y.; Takahashi, H.; Tanaka, J. Accumulation of Macrophage-Like Cells Expressing NG2 Proteoglycan and Iba1 in Ischemic Core of Rat Brain after Transient Middle Cerebral Artery Occlusion. *J. Cereb. Blood Flow Metab.* **2007**, *28*, 149–163. [CrossRef] [PubMed]
36. Carmichael, S.T.; Archibeque, I.; Luke, L.; Nolan, T.; Momiy, J.; Li, S. Growth-associated gene expression after stroke: Evidence for a growth-promoting region in peri-infarct cortex. *Exp. Neurol.* **2005**, *193*, 291–311. [CrossRef] [PubMed]
37. 38 Items (Human) STRING Interaction Network. Available online: https://version-11-0.string-db.org/cgi/network.pl?taskId=9dpPhoCatq5U (accessed on 30 April 2020).
38. 38 Items (Human) STRING Coexpression View. Available online: https://version-11-0.string-db.org/cgi/coexpression_evidence.pl?networkId=K8sRPIMYJ0ZJ (accessed on 30 April 2020).
39. Hankey, G.J. Stroke. *Lancet* **2017**, *389*, 641–654. [CrossRef]
40. Moreau, F.; Patel, S.; Lauzon, M.L.; McCreary, C.R.; Goyal, M.; Frayne, R.; Demchuk, A.M.; Coutts, S.B.; Smith, E.E. Cavitation after acute symptomatic lacunar stroke depends on time, location, and MRI sequence. *Stroke* **2012**, *43*, 1837–1842. [CrossRef]
41. Kwiecien, J.M.; Dabrowski, W.; Dąbrowska-Bouta, B.; Sulkowski, G.; Oakden, W.; Kwiecien-Delaney, C.J.; Yaron, J.R.; Zhang, L.; Schutz, L.; Marzec-Kotarska, B.; et al. Prolonged inflammation leads to ongoing damage after spinal cord injury. *PLoS ONE* **2020**, *15*, e0226584. [CrossRef]
42. McPhail, L.T.; Borisoff, J.F.; Tsang, B.; Hwi, L.P.; Kwiecien, J.M.; Ramer, M.S. Protracted myelin clearance hinders central primary afferent regeneration following dorsal rhizotomy and delayed neurotrophin-3 treatment. *Neurosci. Lett.* **2007**, *411*, 206–211. [CrossRef]
43. Shibahara, T.; Ago, T.; Tachibana, M.; Nakamura, K.; Yamanaka, K.; Kuroda, J.; Wakisaka, Y.; Kitazono, T. Reciprocal Interaction Between Pericytes and Macrophage in Poststroke Tissue Repair and Functional Recovery. *Stroke* **2020**. [CrossRef]
44. Heiss, W.-D. Malignant MCA Infarction: Pathophysiology and Imaging for Early Diagnosis and Management Decisions. *Cereb. Dis.* **2016**, *41*, 1–7. [CrossRef] [PubMed]
45. Jun-Long, H.; Yi, L.; Bao-Lian, Z.; Jia-Si, L.; Ning, Z.; Zhou-Heng, Y.; Xue-Jun, S.; Wen-Wu, L. Necroptosis Signaling Pathways in Stroke: From Mechanisms to Therapies. *Curr. Neuropharmacol.* **2018**, *16*, 1327–1339. [CrossRef] [PubMed]
46. Kim, J.Y.; Han, Y.; Lee, J.E.; Yenari, M.A. The 70-kDa heat shock protein (Hsp70) as a therapeutic target for stroke. *Expert Opin. Ther. Targets* **2018**, *22*, 191–199. [CrossRef] [PubMed]
47. Gadani, S.P.; Walsh, J.T.; Lukens, J.R.; Kipnis, J. Dealing with Danger in the CNS: The Response of the Immune System to Injury. *Neuron* **2015**, *87*, 47–62. [CrossRef] [PubMed]
48. Liesz, A.; Hu, X.; Kleinschnitz, C.; Offner, H. Functional role of regulatory lymphocytes in stroke: Facts and controversies. *Stroke* **2015**, *46*, 1422–1430. [CrossRef]
49. Khakh, B.S.; North, R.A. Neuromodulation by extracellular ATP and P2X receptors in the CNS. *Neuron* **2012**, *76*, 51–69. [CrossRef]
50. Gaudet, A.D.; Popovich, P.G. Extracellular matrix regulation of inflammation in the healthy and injured spinal cord. *Exp. Neurol.* **2014**, *258*, 24–34. [CrossRef]
51. Almeida, R.; Fernández-Justel, J.M.; Santa-María, C.; Cadoret, J.-C.; Cano-Aroca, L.; Lombraña, R.; Herranz, G.; Agresti, A.; Gómez, M. Chromatin conformation regulates the coordination between DNA replication and transcription. *Nat. Commun.* **2018**, *9*, 1590. [CrossRef]
52. Cirillo, G.; Colangelo, A.M.; Berbenni, M.; Ippolito, V.M.; De Luca, C.; Verdesca, F.; Savarese, L.; Alberghina, L.; Maggio, N.; Papa, M. Purinergic Modulation of Spinal Neuroglial Maladaptive Plasticity Following Peripheral Nerve Injury. *Mol. Neurobiol.* **2015**, *52*, 1440–1457. [CrossRef]
53. Colangelo, A.M.; Cirillo, G.; Lavitrano, M.L.; Alberghina, L.; Papa, M. Targeting reactive astrogliosis by novel biotechnological strategies. *Biotechnol. Adv.* **2012**, *30*, 261–271. [CrossRef]
54. Di Virgilio, F.; Ceruti, S.; Bramanti, P.; Abbracchio, M.P. Purinergic signalling in inflammation of the central nervous system. *Trends Neurosci.* **2009**, *32*, 79–87. [CrossRef] [PubMed]
55. Quintas, C.; Vale, N.; Gonçalves, J.; Queiroz, G. Microglia P2Y(13) Receptors Prevent Astrocyte Proliferation Mediated by P2Y(1) Receptors. *Front. Pharm.* **2018**, *9*, 418. [CrossRef] [PubMed]
56. Chen, B.; Feder, M.E.; Kang, L. Evolution of heat-shock protein expression underlying adaptive responses to environmental stress. *Mol. Ecol.* **2018**, *27*, 3040–3054. [CrossRef] [PubMed]

57. Ouyang, Y.-B.; Xu, L.-J.; Sun, Y.-J.; Giffard, R.G. Overexpression of inducible heat shock protein 70 and its mutants in astrocytes is associated with maintenance of mitochondrial physiology during glucose deprivation stress. *Cell Stress Chaperones* **2006**, *11*, 180–186. [CrossRef] [PubMed]
58. Van Molle, W.; Wielockx, B.; Mahieu, T.; Takada, M.; Taniguchi, T.; Sekikawa, K.; Libert, C. HSP70 protects against TNF-induced lethal inflammatory shock. *Immunity* **2002**, *16*, 685–695. [CrossRef]
59. Feinstein, D.L.; Galea, E.; Aquino, D.A.; Li, G.C.; Xu, H.; Reis, D.J. Heat shock protein 70 suppresses astroglial-inducible nitric-oxide synthase expression by decreasing NFkappaB activation. *J. Biol. Chem.* **1996**, *271*, 17724–17732. [CrossRef]
60. Lee, J.E.; Kim, Y.J.; Kim, J.Y.; Lee, W.T.; Yenari, M.A.; Giffard, R.G. The 70 kDa heat shock protein suppresses matrix metalloproteinases in astrocytes. *Neuroreport* **2004**, *15*, 499–502. [CrossRef]
61. Srivastava, P. Roles of heat-shock proteins in innate and adaptive immunity. *Nat. Rev. Immunol.* **2002**, *2*, 185–194. [CrossRef]
62. Asea, A. Heat shock proteins and toll-like receptors. *Handb Exp. Pharm.* **2008**. [CrossRef]
63. Fischer, N.; Haug, M.; Kwok, W.W.; Kalbacher, H.; Wernet, D.; Dannecker, G.E.; Holzer, U. Involvement of CD91 and scavenger receptors in Hsp70-facilitated activation of human antigen-specific CD4+ memory T cells. *Eur. J. Immunol.* **2010**, *40*, 986–997. [CrossRef]
64. Tsivgoulis, G.; Katsanos, A.H.; Schellinger, P.D.; Köhrmann, M.; Varelas, P.; Magoufis, G.; Paciaroni, M.; Caso, V.; Alexandrov, A.W.; Gurol, E.; et al. Successful Reperfusion With Intravenous Thrombolysis Preceding Mechanical Thrombectomy in Large-Vessel Occlusions. *Stroke* **2018**, *49*, 232–235. [CrossRef] [PubMed]
65. Jiang, X.; Andjelkovic, A.V.; Zhu, L.; Yang, T.; Bennett, M.V.L.; Chen, J.; Keep, R.F.; Shi, Y. Blood-brain barrier dysfunction and recovery after ischemic stroke. *Prog. Neurobiol.* **2018**, *163–164*, 144–171. [CrossRef] [PubMed]
66. Murray, K.N.; Parry-Jones, A.R.; Allan, S.M. Interleukin-1 and acute brain injury. *Front. Cell. Neurosci.* **2015**, *9*, 18. [CrossRef] [PubMed]
67. Levada, O.A.; Troyan, A.S. Poststroke Depression Biomarkers: A Narrative Review. *Front. Neurol.* **2018**, *9*, 577. [CrossRef]
68. Yilmaz, G.; Granger, D.N. Cell adhesion molecules and ischemic stroke. *Neurol. Res.* **2008**, *30*, 783–793. [CrossRef] [PubMed]
69. Song, J.; Zhang, X.; Buscher, K.; Wang, Y.; Wang, H.; Di Russo, J.; Li, L.; Lütke-Enking, S.; Zarbock, A.; Stadtmann, A.; et al. Endothelial Basement Membrane Laminin 511 Contributes to Endothelial Junctional Tightness and Thereby Inhibits Leukocyte Transmigration. *Cell Rep.* **2017**, *18*, 1256–1269. [CrossRef] [PubMed]
70. Kenne, E.; Soehnlein, O.; Genové, G.; Rotzius, P.; Eriksson, E.E.; Lindbom, L. Immune cell recruitment to inflammatory loci is impaired in mice deficient in basement membrane protein laminin alpha4. *J. Leukoc. Biol.* **2010**, *88*, 523–528. [CrossRef]
71. Wang, S.; Voisin, M.-B.; Larbi, K.Y.; Dangerfield, J.; Scheiermann, C.; Tran, M.; Maxwell, P.H.; Sorokin, L.; Nourshargh, S. Venular basement membranes contain specific matrix protein low expression regions that act as exit points for emigrating neutrophils. *J. Exp. Med.* **2006**, *203*, 1519–1532. [CrossRef]
72. Tan, R.Y.Y.; Markus, H.S. Monogenic causes of stroke: Now and the future. *J. Neurol.* **2015**, *262*, 2601–2616. [CrossRef]
73. Grønberg, N.V.; Johansen, F.F.; Kristiansen, U.; Hasseldam, H. Leukocyte infiltration in experimental stroke. *J. Neuroinflamm.* **2013**, *10*, 115. [CrossRef]
74. Chan, A.; Yan, J.; Csurhes, P.; Greer, J.; McCombe, P. Circulating brain derived neurotrophic factor (BDNF) and frequency of BDNF positive T cells in peripheral blood in human ischemic stroke: Effect on outcome. *J. Neuroimmunol.* **2015**, *286*, 42–47. [CrossRef] [PubMed]
75. Brait, V.H.; Arumugam, T.V.; Drummond, G.R.; Sobey, C.G. Importance of T lymphocytes in brain injury, immunodeficiency, and recovery after cerebral ischemia. *J. Cereb. Blood Flow Metab.* **2012**, *32*, 598–611. [CrossRef] [PubMed]
76. Hallmann, R.; Zhang, X.; Di Russo, J.; Li, L.; Song, J.; Hannocks, M.-J.; Sorokin, L. The regulation of immune cell trafficking by the extracellular matrix. *Curr. Opin. Cell Biol.* **2015**, *36*, 54–61. [CrossRef] [PubMed]
77. Seiki, M. The cell surface: The stage for matrix metalloproteinase regulation of migration. *Curr. Opin. Cell Biol.* **2002**, *14*, 624–632. [CrossRef]

78. Strongin, A.Y.; Collier, I.; Bannikov, G.; Marmer, B.L.; Grant, G.A.; Goldberg, G.I. Mechanism of cell surface activation of 72-kDa type IV collagenase. Isolation of the activated form of the membrane metalloprotease. *J. Biol. Chem.* **1995**, *270*, 5331–5338. [CrossRef] [PubMed]
79. Klein, T.; Bischoff, R. Physiology and pathophysiology of matrix metalloproteases. *Amino Acids* **2011**, *41*, 271–290. [CrossRef]
80. Vafadari, B.; Salamian, A.; Kaczmarek, L. MMP-9 in translation: From molecule to brain physiology, pathology, and therapy. *J. Neurochem.* **2016**, *139* (Suppl. 2), 91–114. [CrossRef]
81. Rosenberg, G.A.; Estrada, E.Y.; Dencoff, J.E. Matrix metalloproteinases and TIMPs are associated with blood-brain barrier opening after reperfusion in rat brain. *Stroke* **1998**, *29*, 2189–2195. [CrossRef]
82. Song, I.; Dityatev, A. Crosstalk between glia, extracellular matrix and neurons. *Brain Res. Bull.* **2018**, *136*, 101–108. [CrossRef]
83. Vessières, E.; Freidja, M.L.; Loufrani, L.; Fassot, C.; Henrion, D. Flow (shear stress)-mediated remodeling of resistance arteries in diabetes. *Vasc. Pharm.* **2012**, *57*, 173–178. [CrossRef]
84. Lam, C.K.; Yoo, T.; Hiner, B.; Liu, Z.; Grutzendler, J. Embolus extravasation is an alternative mechanism for cerebral microvascular recanalization. *Nature* **2010**, *465*, 478–482. [CrossRef] [PubMed]
85. Mizuma, A.; Yenari, M.A. Anti-Inflammatory Targets for the Treatment of Reperfusion Injury in Stroke. *Front. Neurol.* **2017**, *8*, 467. [CrossRef] [PubMed]
86. Gasche, Y.; Fujimura, M.; Morita-Fujimura, Y.; Copin, J.C.; Kawase, M.; Massengale, J.; Chan, P.H. Early appearance of activated matrix metalloproteinase-9 after focal cerebral ischemia in mice: A possible role in blood-brain barrier dysfunction. *J. Cereb. Blood Flow Metab.* **1999**, *19*, 1020–1028. [CrossRef] [PubMed]
87. Lees, K.R.; Bluhmki, E.; von Kummer, R.; Brott, T.G.; Toni, D.; Grotta, J.C.; Albers, G.W.; Kaste, M.; Marler, J.R.; Hamilton, S.A.; et al. Time to treatment with intravenous alteplase and outcome in stroke: An updated pooled analysis of ECASS, ATLANTIS, NINDS, and EPITHET trials. *Lancet* **2010**, *375*, 1695–1703. [CrossRef]
88. Wang, X.; Lee, S.-R.; Arai, K.; Lee, S.-R.; Tsuji, K.; Rebeck, G.W.; Lo, E.H. Lipoprotein receptor-mediated induction of matrix metalloproteinase by tissue plasminogen activator. *Nat. Med.* **2003**, *9*, 1313–1317. [CrossRef] [PubMed]
89. Machida, T.; Dohgu, S.; Takata, F.; Matsumoto, J.; Kimura, I.; Koga, M.; Nakamoto, K.; Yamauchi, A.; Kataoka, Y. Role of thrombin-PAR1-PKCθ/δ axis in brain pericytes in thrombin-induced MMP-9 production and blood-brain barrier dysfunction in vitro. *Neuroscience* **2017**, *350*, 146–157. [CrossRef]
90. Lu, A.; Suofu, Y.; Guan, F.; Broderick, J.P.; Wagner, K.R.; Clark, J.F. Matrix metalloproteinase-2 deletions protect against hemorrhagic transformation after 1 h of cerebral ischemia and 23 h of reperfusion. *Neuroscience* **2013**, *253*, 361–367. [CrossRef]
91. Asahi, M.; Sumii, T.; Fini, M.E.; Itohara, S.; Lo, E.H. Matrix metalloproteinase 2 gene knockout has no effect on acute brain injury after focal ischemia. *Neuroreport* **2001**, *12*, 3003–3007. [CrossRef]
92. De Luca, C.; Papa, M. Looking Inside the Matrix: Perineuronal Nets in Plasticity, Maladaptive Plasticity and Neurological Disorders. *Neurochem. Res.* **2016**, *41*, 1507–1515. [CrossRef]
93. Zhao, B.-Q.; Wang, S.; Kim, H.-Y.; Storrie, H.; Rosen, B.R.; Mooney, D.J.; Wang, X.; Lo, E.H. Role of matrix metalloproteinases in delayed cortical responses after stroke. *Nat. Med.* **2006**, *12*, 441–445. [CrossRef]
94. Lee, S.-R.; Kim, H.-Y.; Rogowska, J.; Zhao, B.-Q.; Bhide, P.; Parent, J.M.; Lo, E.H. Involvement of matrix metalloproteinase in neuroblast cell migration from the subventricular zone after stroke. *J. Neurosci.* **2006**, *26*, 3491–3495. [CrossRef] [PubMed]
95. Cui, J.; Chen, S.; Zhang, C.; Meng, F.; Wu, W.; Hu, R.; Hadass, O.; Lehmidi, T.; Blair, G.J.; Lee, M.; et al. Inhibition of MMP-9 by a selective gelatinase inhibitor protects neurovasculature from embolic focal cerebral ischemia. *Mol. Neurodegener.* **2012**, *7*, 21. [CrossRef] [PubMed]
96. Chaturvedi, M.; Kaczmarek, L. Mmp-9 inhibition: A therapeutic strategy in ischemic stroke. *Mol. Neurobiol.* **2014**, *49*, 563–573. [CrossRef] [PubMed]
97. Tauchi, R.; Imagama, S.; Natori, T.; Ohgomori, T.; Muramoto, A.; Shinjo, R.; Matsuyama, Y.; Ishiguro, N.; Kadomatsu, K. The endogenous proteoglycan-degrading enzyme ADAMTS-4 promotes functional recovery after spinal cord injury. *J. Neuroinflamm.* **2012**, *9*, 53. [CrossRef] [PubMed]
98. Cytokine Registry Database of Stroke Patients. Available online: https://clinicaltrials.gov/ct2/show/NCT03297827 (accessed on 30 June 2020).

99. De Luca, C.; Savarese, L.; Colangelo, A.M.; Bianco, M.R.; Cirillo, G.; Alberghina, L.; Papa, M. Astrocytes and Microglia-Mediated Immune Response in Maladaptive Plasticity is Differently Modulated by NGF in the Ventral Horn of the Spinal Cord Following Peripheral Nerve Injury. *Cell Mol. Neurobiol.* **2016**, *36*, 37–46. [CrossRef]
100. Cirillo, G.; Colangelo, A.M.; De Luca, C.; Savarese, L.; Barillari, M.R.; Alberghina, L.; Papa, M. Modulation of Matrix Metalloproteinases Activity in the Ventral Horn of the Spinal Cord Re-stores Neuroglial Synaptic Homeostasis and Neurotrophic Support following Peripheral Nerve Injury. *PLoS ONE* **2016**, *11*, e0152750. [CrossRef]
101. Chamorro, Á.; Meisel, A.; Planas, A.M.; Urra, X.; van de Beek, D.; Veltkamp, R. The immunology of acute stroke. *Nat. Rev. Neurol.* **2012**, *8*, 401–410. [CrossRef]
102. Victoria, E.C.G.; de Brito Toscano, E.C.; de Sousa Cardoso, A.C.; da Silva, D.G.; de Miranda, A.S.; da Silva Barcelos, L.; Sugimoto, M.A.; Sousa, L.P.; de Assis Lima, I.V.; de Oliveira, A.C.P.; et al. Knockdown of C-C Chemokine Receptor 5 (CCR5) is Protective Against Cerebral Ischemia and Reperfusion Injury. *Curr. Neurovasc. Res.* **2017**, *14*, 125–131. [CrossRef]
103. Dias, D.O.; Göritz, C. Fibrotic scarring following lesions to the central nervous system. *Matrix Biol.* **2018**, *68–69*, 561–570. [CrossRef]
104. Faulkner, J.R.; Herrmann, J.E.; Woo, M.J.; Tansey, K.E.; Doan, N.B.; Sofroniew, M.V. Reactive astrocytes protect tissue and preserve function after spinal cord injury. *J. Neurosci.* **2004**, *24*, 2143–2155. [CrossRef]
105. Li, L.; Lundkvist, A.; Andersson, D.; Wilhelmsson, U.; Nagai, N.; Pardo, A.C.; Nodin, C.; Ståhlberg, A.; Aprico, K.; Larsson, K.; et al. Protective role of reactive astrocytes in brain ischemia. *J. Cereb. Blood Flow Metab.* **2008**, *28*, 468–481. [CrossRef] [PubMed]
106. Jachetti, E.; Caputo, S.; Mazzoleni, S.; Brambillasca, C.S.; Parigi, S.M.; Grioni, M.; Piras, I.S.; Restuccia, U.; Calcinotto, A.; Freschi, M.; et al. Tenascin-C Protects Cancer Stem-like Cells from Immune Surveillance by Arresting T-cell Activation. *Cancer Res.* **2015**, *75*, 2095–2108. [CrossRef] [PubMed]
107. Miura, R.; Aspberg, A.; Ethell, I.M.; Hagihara, K.; Schnaar, R.L.; Ruoslahti, E.; Yamaguchi, Y. The proteoglycan lectin domain binds sulfated cell surface glycolipids and promotes cell adhesion. *J. Biol. Chem.* **1999**, *274*, 11431–11438. [CrossRef]
108. Zhang, H.; Uchimura, K.; Kadomatsu, K. Brain keratan sulfate and glial scar formation. *Ann. N. Y. Acad. Sci.* **2006**, *1086*, 81–90. [CrossRef] [PubMed]
109. Rosell, A.; Ortega-Aznar, A.; Alvarez-Sabín, J.; Fernández-Cadenas, I.; Ribó, M.; Molina, C.A.; Lo, E.H.; Montaner, J. Increased brain expression of matrix metalloproteinase-9 after ischemic and hemorrhagic human stroke. *Stroke* **2006**, *37*, 1399–1406. [CrossRef]
110. Anderson, M.A.; Burda, J.E.; Ren, Y.; Ao, Y.; O'Shea, T.M.; Kawaguchi, R.; Coppola, G.; Khakh, B.S.; Deming, T.J.; Sofroniew, M.V. Astrocyte scar formation aids central nervous system axon regeneration. *Nature* **2016**, *532*, 195–200. [CrossRef]
111. Yu, C.; Griffiths, L.R.; Haupt, L.M. Exploiting Heparan Sulfate Proteoglycans in Human Neurogenesis-Controlling Lineage Specification and Fate. *Front. Integr. Neurosci.* **2017**, *11*, 28. [CrossRef]
112. Condomitti, G.; de Wit, J. Heparan Sulfate Proteoglycans as Emerging Players in Synaptic Specificity. *Front. Mol. Neurosci.* **2018**, *11*, 14. [CrossRef]
113. Coles, C.H.; Shen, Y.; Tenney, A.P.; Siebold, C.; Sutton, G.C.; Lu, W.; Gallagher, J.T.; Jones, E.Y.; Flanagan, J.G.; Aricescu, A.R. Proteoglycan-specific molecular switch for RPTPσ clustering and neuronal extension. *Science* **2011**, *332*, 484–488. [CrossRef]
114. Jones, L.L.; Margolis, R.U.; Tuszynski, M.H. The chondroitin sulfate proteoglycans neurocan, brevican, phosphacan, and versican are differentially regulated following spinal cord injury. *Exp. Neurol.* **2003**, *182*, 399–411. [CrossRef]
115. Nishiyama, A.; Komitova, M.; Suzuki, R.; Zhu, X. Polydendrocytes (NG2 cells): Multifunctional cells with lineage plasticity. *Nat. Rev. Neurosci.* **2009**, *10*, 9–22. [CrossRef] [PubMed]
116. George, N.; Geller, H.M. Extracellular matrix and traumatic brain injury. *J. Neurosci. Res.* **2018**, *96*, 573–588. [CrossRef] [PubMed]
117. De Luca, C.; Colangelo, A.M.; Virtuoso, A.; Alberghina, L.; Papa, M. Neurons, Glia, Extracellular Matrix and Neurovascular Unit: A Systems Biology Approach to the Complexity of Synaptic Plasticity in Health and Disease. *Int. J. Mol. Sci.* **2020**, *21*, 1539. [CrossRef] [PubMed]
118. Neuroprotection in Acute Ischemic Stroke (H2M). Available online: https://clinicaltrials.gov/ct2/show/NCT03320018 (accessed on 30 June 2020).

119. Switzer, J.A.; Hess, D.C.; Ergul, A.; Waller, J.L.; Machado, L.S.; Portik-Dobos, V.; Pettigrew, L.C.; Clark, W.M.; Fagan, S.C. Matrix metalloproteinase-9 in an exploratory trial of intravenous minocycline for acute ischemic stroke. *Stroke* **2011**, *42*, 2633–2635. [CrossRef]
120. Relationships among FVIII, t-PA/PAI-1, and MMP-9 Levels and Intracranial Hemorrhage Complications after Thrombolysis with Alteplase in Patients with Acute Ischemic Stroke: Protocol for a Multicenter Retrospective Study. Available online: https://clinicaltrials.gov/ct2/show/NCT03733223 (accessed on 30 June 2020).
121. An Open, Study to Assess the Safety of RGTA® (OTR4132) in Patients with Acute Ischemic Stroke (AIS) (MATRISS). Available online: https://clinicaltrials.gov/ct2/show/NCT04083001 (accessed on 30 June 2020).
122. Modo, M.; Badylak, S.F. A roadmap for promoting endogenous in situ tissue restoration using inductive bioscaffolds after acute brain injury. *Brain Res. Bull.* **2019**, *150*, 136–149. [CrossRef]
123. Tsintou, M.; Dalamagkas, K.; Makris, N. Taking central nervous system regenerative therapies to the clinic: Curing rodents versus nonhuman primates versus humans. *Neural Regen. Res.* **2020**, *15*, 425–437. [CrossRef]
124. Melià-Sorolla, M.; Castaño, C.; DeGregorio-Rocasolano, N.; Rodríguez-Esparragoza, L.; Dávalos, A.; Martí-Sistac, O.; Gasull, T. Relevance of Porcine Stroke Models to Bridge the Gap from Pre-Clinical Findings to Clinical Implementation. *Int. J. Mol. Sci.* **2020**, *21*, 6568. [CrossRef]
125. Pawar, K.; Prang, P.; Müller, R.; Caioni, M.; Bogdahn, U.; Kunz, W.; Weidner, N. Intrinsic and extrinsic determinants of central nervous system axon outgrowth into alginate-based anisotropic hydrogels. *Acta Biomater.* **2015**, *27*, 131–139. [CrossRef]
126. Hong, L.T.A.; Kim, Y.-M.; Park, H.H.; Hwang, D.H.; Cui, Y.; Lee, E.M.; Yahn, S.; Lee, J.K.; Song, S.-C.; Kim, B.G. An injectable hydrogel enhances tissue repair after spinal cord injury by promoting extracellular matrix remodeling. *Nat. Commun.* **2017**, *8*, 533. [CrossRef]
127. Kočí, Z.; Výborný, K.; Dubišová, J.; Vacková, I.; Jäger, A.; Lunov, O.; Jiráková, K.; Kubinová, Š. Extracellular Matrix Hydrogel Derived from Human Umbilical Cord as a Scaffold for Neural Tissue Repair and Its Comparison with Extracellular Matrix from Porcine Tissues. *Tissue Eng. Part C Methods* **2017**, *23*, 333–345. [CrossRef]

International Journal of
Molecular Sciences

MDPI

Article

Interleukin-1β Alters Hebbian Synaptic Plasticity in Multiple Sclerosis

Mario Stampanoni Bassi [1], Fabio Buttari [1], Carolina Gabri Nicoletti [2], Francesco Mori [2], Luana Gilio [1], Ilaria Simonelli [3], Nicla De Paolis [1], Girolama Alessandra Marfia [1,2], Roberto Furlan [4], Annamaria Finardi [4], Diego Centonze [1,2,*] and Ennio Iezzi [1]

1 Unit of Neurology & Neurorehabilitation, IRCCS Neuromed, 86077 Pozzilli (IS), Italy; m.stampanonibassi@gmail.com (M.S.B.); fabio.buttari@gmail.com (F.B.); gilio.luana@gmail.com (L.G.); depaolisnicla@gmail.com (N.D.P.); marfia@med.uniroma2.it (G.A.M.); ennio.iezzi@neuromed.it (E.I.)

2 Multiple Sclerosis Clinical and Research Unit, Department of Systems Medicine, University of Rome Tor Vergata, 00133 Rome, Italy; carolgabri@gmail.com (C.G.N.); francesco.mori@uniroma2.it (F.M.)

3 Service of Medical Statistics & Information Technology, Fondazione Fatebenefratelli per la Ricerca e la Formazione Sanitaria e Sociale, 00153 Rome, Italy; ilaria.simonelli@afar.it

4 Clinical Neuroimmunology Unit, Institute of Experimental Neurology, Division of Neuroscience, San Raffaele Scientific Institute, 20132 Milano, Italy; furlan.roberto@hsr.it (R.F.); finardi.annamaria@hsr.it (A.F.)

* Correspondence: centonze@uniroma2.it; Tel.: +39-06-7259-6010; Fax: +39-06-7259-6006

Received: 30 July 2020; Accepted: 21 September 2020; Published: 23 September 2020

check for updates

Abstract: In multiple sclerosis (MS), inflammation alters synaptic transmission and plasticity, negatively influencing the disease course. In the present study, we aimed to explore the influence of the proinflammatory cytokine IL-1β on peculiar features of associative Hebbian synaptic plasticity, such as input specificity, using the paired associative stimulation (PAS). In 33 relapsing remitting-MS patients and 15 healthy controls, PAS was performed on the abductor pollicis brevis (APB) muscle. The effects over the motor hot spot of the APB and abductor digiti minimi (ADM) muscles were tested immediately after PAS and 15 and 30 min later. Intracortical excitability was tested with paired-pulse transcranial magnetic stimulation (TMS). The cerebrospinal fluid (CSF) levels of IL-1β were calculated. In MS patients, PAS failed to induce long-term potentiation (LTP)-like effects in the APB muscle and elicited a paradoxical motor-evoked potential (MEP) increase in the ADM. IL-1β levels were negatively correlated with the LTP-like response in the APB muscle. Moreover, IL-1β levels were associated with synaptic hyperexcitability tested with paired-pulse TMS. Synaptic hyperexcitability caused by IL-1β may critically contribute to alter Hebbian plasticity in MS, inducing a loss of topographic specificity.

Keywords: paired associative stimulation (PAS); multiple sclerosis (MS); interleukin (IL)-1β; synaptic plasticity; long-term potentiation (LTP); transcranial magnetic stimulation (TMS)

---

1. Introduction

Multiple sclerosis (MS) is an immune-mediated disease characterized by inflammation of the central nervous system (CNS), which is associated with demyelinating white matter lesions and neurodegeneration. Relapsing-remitting (RR)-MS is characterized by a very variable disease course, with stable phases alternating with acute relapses. The ability to compensate for ongoing brain damage critically influences the disease course, promoting clinical stability and preventing the accumulation of disability. Synaptic plasticity, and particularly long-term potentiation (LTP), is one of the main physiological mechanisms involved in clinical recovery after brain damage [1,2]. Accordingly, it has been speculated that efficient LTP induction could have a positive influence on MS disease course [3].

Studies in experimental autoimmune encephalomyelitis (EAE), an animal model of MS, and in patients with MS, have shown that CNS inflammation affects synaptic functioning. In particular, specific proinflammatory mediators released by the immune cells influence synaptic transmission and plasticity. Interleukin (IL)-1β is one of the most important proinflammatory cytokines associated with MS pathogenesis [4] and plays a crucial role in regulating neuronal functioning in both physiological and pathological conditions [5–7]. Experimental studies in EAE have demonstrated that IL-1β alters synaptic functioning, promoting synaptic hyperexcitability and glutamate-mediated excitotoxicity [8,9]. Accordingly, raised IL-1β signaling has been associated with worse disease course and increased neurodegeneration in EAE and in patients with MS [10,11].

However, how IL-1β modifies LTP expression has not yet been conclusively clarified. In fact, studies in mice with EAE have demonstrated both impaired hippocampal LTP [12] and pathologically increased LTP [13] in response to IL-1β. Notably, it has been demonstrated that IL-1β induces both increased glutamatergic transmission and defective GABAergic signaling [8,9,14]. Altered LTP induction has also been found in patients with MS using transcranial magnetic stimulation (TMS). In particular, in RR-MS, a paradoxical LTP-like effect has been evidenced in response to continuous theta burst stimulation (cTBS), a protocol inducing long-term depression (LTD), and has been correlated with IL-1β cerebrospinal fluid (CSF) levels [15].

In the present study, to better characterize how IL-1β could affect LTP in human MS, we investigated in a group of RR-MS patients the effect of IL-1β on LTP induction using paired associative stimulation (PAS). This TMS protocol is particularly suitable to investigate the peculiar features of associative Hebbian synaptic plasticity, such as input specificity, which can be particularly altered in MS patients due to the imbalance between excitatory and inhibitory synaptic transmission induced by neuroinflammation.

2. Results

The clinical characteristics of MS patients and healthy controls are shown in Table 1. TMS was well tolerated by all participants, and no adverse effects were reported.

Table 1. Clinical Characteristics of MS Patients and Controls.

		MS (33)	Controls (15)
Sex, females	N (%)	19 (57.6%)	10 (66.7%)
Age, years	mean (SD)	35.51 (9.34)	28.9 (7.4)
Disease duration, months	median (IQR)	13 (8–29)	-
Radiological activity	N (%)	13 (39.4%)	-
EDSS	median (IQR)	1.5 (1–2)	-
IL-1β, pg/mL	median (IQR)	0.1 (0–26.67)	-

Abbreviations: EDSS (expanded disability status scale); IL (interleukin); IQR (interquartile range); MS (multiple sclerosis); SD (standard deviation).

2.1. PAS-Induced LTP-like Plasticity is Altered in RR-MS Patients

Repeated measures analysis of variance (RM-ANOVA) exploring the effects of PAS in MS patients and controls showed a significant MEP size increase after PAS (time effect: F = 12.336; $p < 0.001$), and the effect differed between abductor pollicis brevis (APB) and abductor digiti minimi (ADM) muscles (muscle effect: F = 11.319; $p = 0.002$). The MEP increase after PAS was significantly greater in controls compared with MS patients (interactions time × group effect: F = 3.557; $p = 0.022$) and more evident in APB compared to ADM muscle (interaction time × muscle effect: F = 3.979; $p = 0.014$). Finally, significant differences were found when comparing the PAS response in APB and ADM muscles between controls and MS patients (interaction muscle × group effect: F = 44.854; $p < 0.001$; interaction time × muscle × group effect: F = 15.061; $p < 0.001$).

Post hoc analyses showed that in control subjects, PAS induced a significant LTP-like effect in the APB muscle and had no effect in the ADM muscle. Conversely, MS patients lacked the expected LTP-like effect in the APB and showed a paradoxical motor-evoked potential (MEP) increase in the ADM (Figure 1).

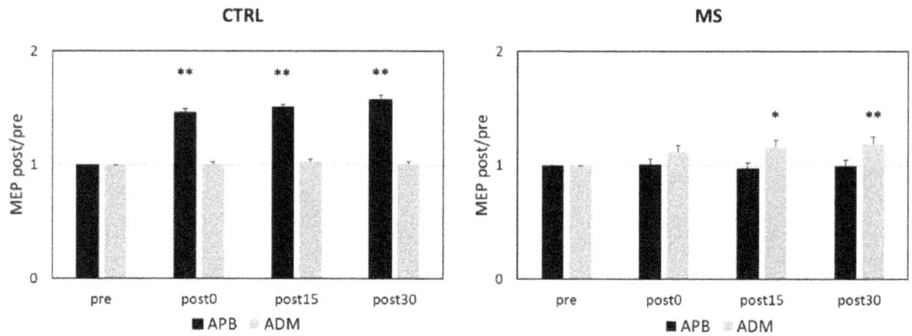

Figure 1. PAS Effects in RR-MS Patients and Controls. Figure 1 legend. In healthy subjects, PAS elicited the expected homosynaptic LTP-like effect in the APB muscle (**left panel**), whereas in MS patients, PAS elicited heterosynaptic LTP in the ADM muscle (**right panel**). * $p \leq 0.05$; ** $p \leq 0.01$. The p values refer to the comparisons between pre and post 0, post 15, and post 30 in each muscle. All p values were adjusted by Benjamini–Hochberg correction. Abbreviations: APB (abductor pollicis brevis); ADM (abductor digiti minimi); MEP (motor-evoked potential); PAS (paired associative stimulation); RR-MS (relapsing-remitting multiple sclerosis).

No significant correlations emerged between PAS effect and the clinical and demographic characteristics explored (age, disease duration, disability; all $p > 0.2$).

2.2. CSF IL-1β Alters PAS Effects

To explore the influence of IL-1β on synaptic plasticity in MS patients, we analyzed the correlation between the CSF levels of this cytokine and the PAS-induced LTP-like effect.

A significant negative correlation emerged between IL-1β CSF concentrations and the amount of LTP-like effect induced by the PAS protocol in the APB muscle at post 0 (Spearman's r (33) = −0.360, $p = 0.040$, B-H adjusted $p = 0.040$) post 15 (Spearman's r (33) = −0.463, $p = 0.007$ B-H adjusted $p = 0.011$), and post 30 (Spearman's r (33) = −0.430, $p = 0.013$, B-H adjusted $p = 0.017$) (Figure 2). Conversely, no significant correlations emerged between the CSF levels of this cytokine and the effect of PAS explored on the ADM.

Finally, a significant negative correlation emerged between IL-1β CSF concentrations and the ratio of the LTP-like effect induced by the PAS protocol on APB and ADM muscles at post 0 (Spearman's r (33) = −0.627, $p < 0.001$, B-H adjusted $p < 0.001$), post 15 (Spearman's r (33) = −0.508, $p = 0.003$, B-H adjusted $p = 0.006$), and post 30 (Spearman's r (33) = −0.499, $p = 0.003$, B-H adjusted $p = 0.006$).

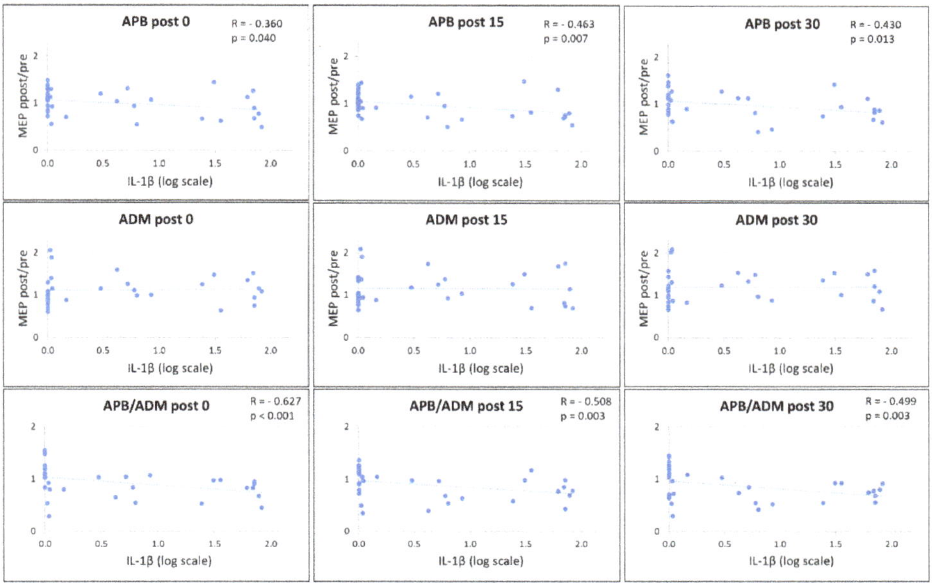

Figure 2. Correlation between IL-1β CSF Levels and PAS Effects. Figure 2 legend. A negative correlation was found between IL-1β CSF concentrations and LTP-like effects in the APB muscle at post 0, post 15, and post 30 (**upper panels**); conversely, no significant correlation emerged with LTP-like effects in the ADM muscle (**middle panels**). A significant negative correlation was found between IL-1β CSF levels and the ratio of the LTP-like effect induced in APB and ADM muscles (**lower panels**). Abbreviations: APB (abductor pollicis brevis); ADM (abductor digiti minimi); CSF (cerebrospinal fluid); IL (interleukin); MEP (motor-evoked potential).

2.3. Detectable IL-1β CSF Levels are Associated with Paradoxical Response to PAS

To further explore the impact of IL-1β CSF concentrations on PAS, we divided MS patients into two groups according to the presence of IL-1β in the CSF. Patients with detectable IL-1β in the CSF (IL-1β positive) amounted to 20, while IL-1β was undetectable in the CSF of 13 patients (IL-1β negative). The clinical characteristics of the two groups are shown in Table 2.

Table 2. Clinical Characteristics of MS Patients According to IL-1β Group.

		IL-1β Negative (13)	**IL-1β Positive (20)**
Sex, females	N (%)	8 (61.5%)	11 (50%)
Age, years	mean (SD)	38 (11.72)	33.9 (7.28)
Disease duration, months	median (IQR)	12 (9–35)	14 (6.5–31.5)
Radiological activity	N (%)	4 (30.8%)	9 (45%)
EDSS	median (IQR)	1.5 (1–2)	1.75 (1–2)

Abbreviations: EDSS (expanded disability status scale); IL (interleukin); IQR (interquartile range); MS (multiple sclerosis); SD (standard deviation).

In MS patients, RM-ANOVA showed that response to PAS in APB and ADM muscles significantly differed according to the presence of IL-1β in the CSF (interactions muscle x group: F = 17.162; $p < 0.001$; interaction time x muscle x group effect: F = 6.144; $p = 0.002$). Post hoc comparisons showed in the IL-1β positive group a significant LTP-like effect in the ADM muscle but not in the APB, while no significant LTP-like effect was observed in the IL-1β negative group on both muscles (Figure 3).

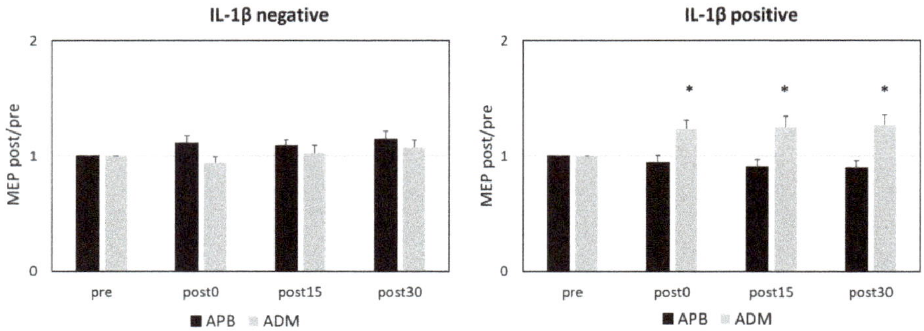

Figure 3. IL-1β CSF Detectability and PAS Effects. Figure 3 legend. In the IL-1β negative group, PAS failed to induce LTP-like effects in both APB and ADM muscles (**left panel**), whereas in the IL-1β positive group, PAS elicited an abnormal LTP in the ADM muscle (**right panel**). * $p \leq 0.05$. The p values refer to the comparisons between pre and post 0, post 15, and post 30 in each muscle. All p values were adjusted by Benjamini–Hochberg correction. Abbreviations: APB (abductor pollicis brevis); ADM (abductor digiti minimi); CSF (cerebrospinal fluid); IL (interleukin); PAS (paired associative stimulation).

2.4. IL-1β and Intracortical Excitability

To explore whether IL-1β affects intracortical excitability, we correlated the CSF concentration of this cytokine with the short-interval intracortical inhibition (SICI) and intracortical facilitation (ICF). A significant correlation emerged between IL-1β and both SICI (Spearman's r (23) = 0.497, $p = 0.016$, B-H adjusted $p = 0.018$) and ICF (Spearman's r (23) = 0.665, $p = 0.001$, B-H adjusted $p = 0.004$) (Figure 4). SICI and ICF did not correlate with the PAS-induced effect on the ADM muscle. Finally, no significant correlations were found between SICI, ICF, and the clinical and demographic characteristics (age, disease duration, disability, and disease activity; all $p > 0.2$).

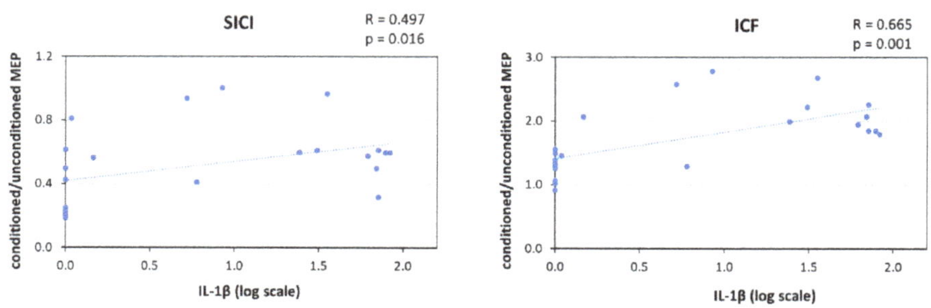

Figure 4. Correlation between IL-1β CSF Levels and Inhibitory/Excitatory Intracortical Transmission. Figure 4 legend. IL-1β CSF levels positively correlated with reduced intracortical inhibition (**left panel**) and increased intracortical facilitation (**right panel**). Abbreviations: CSF (cerebrospinal fluid); ICF (intracortical facilitation); IL (interleukin); MEP (motor-evoked potential); SICI (short-interval intracortical inhibition).

3. Discussion

The PAS protocol significantly changes cortical excitability by combining repeated TMS activation of M1 with peripheral nerve stimulation, separated by specific time intervals [16,17]. This protocol resembles the Hebbian spike-timing-dependent plasticity described in animal experiments where pre- and post-synaptic neuronal activation induces changes in synaptic efficacy [18,19]. Repeated low-frequency electrical stimulation of the median nerve followed 25 ms after by single-pulse TMS over the cortical representation of the contralateral APB muscle is able to increase the amplitude of the MEPs in the APB muscle, persisting up to 60 min. The effect of PAS requires the activity of the N-Methyl-D-Aspartate (NMDA) receptor as shown by the administration of NMDA antagonists [20]. Moreover, PAS effects are bidirectional and strongly governed by temporal rules, as LTD is induced when the TMS pulse follows the electric nerve stimulation at an ISI of 10 ms [21]. Finally, PAS effects are somatotopically specific, being evident only in the muscle receiving homotopical input by afferent stimulation and TMS [20,22]. Therefore, associativity and input specificity of PAS follow the Hebbian rules of synaptic plasticity.

In the present study we found that CSF IL-1β expression is associated with altered PAS25-induced LTP in RR-MS patients. Patients with detectable IL-1β showed both absent LTP in the APB muscle and abnormal LTP expression in the ADM muscle. These results suggest that IL-1β may promote a profound alteration of Hebbian synaptic plasticity mechanisms, characterized by impaired homosynaptic LTP expression and loss of topographic specificity leading to heterosynaptic effects.

The finding that IL-1β CSF detectability is associated with an absent LTP-like effect is in line with previous studies showing that inflammation in MS is associated with altered LTP induction [23,24]. TMS studies investigating the homosynaptic effects of PAS25 in MS patients have shown that elevated CSF levels of the proinflammatory cytokine IL-6 negatively correlated with the LTP-like effects induced by the PAS protocol [24]. Conversely, comparable effects of PAS25 between remitting MS patients and controls have been reported [25].

Notably, inflammation has also been associated with a paradoxical LTP-like response produced by LTD-inducing protocols both in EAE and MS [13,15,26]. IL-1β has been specifically associated with LTP induction in response to low frequency stimulation in mice hippocampal slices [13] and to LTP-like response after cTBS [15]. Lacking LTD-like effects during MS relapses may depend on reduced GABAergic transmission [10,27]. Accordingly, IL-1β has been associated with synaptic hyperexcitability due to enhanced glutamatergic and impaired GABAergic transmission in both EAE and MS, promoting excitotoxic neurodegeneration [10,14]. Interestingly, blocking IL-1β signaling with an IL-1 receptor antagonist normalized synaptic transmission and plasticity [10,15], confirming the specific role of IL-1β as a key player of inflammatory synaptopathy in EAE and MS. While the increased IL-1β levels could explain the lack of PAS-induced homosynaptic LTP, the altered balance between excitatory and inhibitory transmission caused by IL-1β might have altered the topographic specificity of PAS in RR-MS responsible for abnormal heterosynaptic LTP shown here.

Due to the positive-feedback nature of LTP, uncontrolled increases of synaptic efficacy can destabilize neuronal activity, and therefore homeostatic mechanisms are required to constrain synaptic strength in a dynamic physiological range [28]. A compensatory mechanism aimed at preventing excessive synaptic plasticity from developing relies on the Bienenstock–Cooper–Munro theory [29], postulating that the threshold for LTP induction varies as a function of the integrated postsynaptic activity. Accordingly, low levels of postsynaptic activity favour LTP induction whereas high levels block it. Another mechanism involved in the homeostatic control of synaptic activity is synaptic scaling. This form of homeostatic plasticity modifies the expression of α-amino-3-hydroxy-5-methyl-4-isoxazolepropionic acid receptors (AMPARs), inducing increases or decreases of neuronal excitability (upscaling and downscaling, respectively). Unlike Hebbian plasticity, scaling has a negative-feedback nature and lacks input specificity as it involves all synapses in a neuron [30,31]. This mechanism contributes to counterbalance the synaptic instability produced by unrestrained increases in synaptic excitability induced by LTP alone.

Altered spatial specificity of the PAS protocol has been previously reported in other neurological manifestations, in particular in patients with dystonia [22,32], and has been related to a defective neuronal inhibition [33].

Altered mechanisms of homeostatic control of synaptic plasticity have also been demonstrated in EAE and MS. Experimental studies have demonstrated that inflammatory molecules influence the induction and maintenance of synaptic scaling in the brain [34–37]. Accordingly, in EAE, increased post-synaptic glutamatergic excitatory currents have been evidenced before the appearance of clinical manifestations [8]. These alterations have been associated with exacerbated neuronal damage, which can be prevented by the administration of AMPA blockers [8].

In the present study we found that IL-1β CSF levels correlated with enhanced ICF and reduced SICI in RR-MS patients. Therefore, heterosynaptic LTP-like effects on the ADM muscle elicited by PAS might depend on synaptic hyperexcitability associated with raised IL-1β CSF concentrations.

Given the pivotal role of IL-1β in the pathogenesis of MS and in the development of synaptic alterations, anti-IL-1β drugs might have a potential therapeutic role. Although recent reports suggest a possible beneficial effect of anti-IL-1β therapies (i.e., anakinra and canakinumab) in MS [38], current evidence is limited. A clinical trial is ongoing to test the efficacy and safety of anakinra in patients with MS (ClinicalTrials.gov: NCT04025554).

Although various proinflammatory and anti-inflammatory molecules can regulate synaptic plasticity in MS [39], we focused on the proinflammatory cytokine IL-1β which has been previously identified as one of the main determinants of the inflammatory synaptopathy in MS [40]. However, a wider set of CSF molecules should be analyzed to better explore the impact of the inflammatory milieu on synaptic functioning, and additional preclinical investigations are needed to clarify the pathophysiological mechanisms of IL-1β-driven synaptic alterations. Related to this issue, another important feature to address is the contribution of different disease phases in synaptic alterations in MS. In this study, we did not find a significant difference in PAS response between relapsing and remitting patients; however, the presence of radiological activity has been previously associated with synaptic alterations [23,26] and may represent a possible confounding factor.

PAS is particularly suitable for investigating some properties of spike timing-dependent plasticity in humans. Our results have shown that MS synaptic hyperexcitability induced by IL-1β may critically contribute to alter Hebbian plasticity, inducing a loss of topographic specificity.

4. Materials and Methods

4.1. MS Patients

The study involving human subjects was approved by the Ethics Committee of the University Tor Vergata Hospital in Rome, Italy (approval code 123/15, approval date 25 September 2015). All patients gave written informed consent. A group of 33 RR-MS patients admitted to the neurological clinic of University Tor Vergata Hospital participated in the study. MS diagnosis was based clinical, laboratory, and MRI parameters [41]. A group of 15 healthy controls was also included.

All patients underwent clinical examination, MRI scan, CSF withdrawal, and TMS evaluation during hospitalization. Corticosteroids or disease modifying therapies were initiated later if indicated. All patients were asymptomatic in the upper right limb. Disability was assessed using the Expanded Disability Status Scale (EDSS) [42]. Disease duration was calculated as the number of months from disease onset to the time of diagnosis. MRI examination consisted of dual-echo proton density, fast fluid-attenuated inversion recovery, T2-weighted spin-echo images, and pre-contrast and post-contrast T1-weighted spin-echo images.

4.2. CSF Collection and Analysis

After lumbar puncture, CSF was centrifuged and immediately stored at −80 °C until analysed using a Bio-Plex multiplex cytokine assay (Bio-Rad Laboratories, Hercules, CA, USA) according to the manufacturer's instructions. CSF concentrations of IL-1β were calculated according to a standard curve generated for the specific target and expressed as picograms per milliliter (pg/mL).

4.3. Transcranial Magnetic Stimulation

MEPs were elicited through a figure-of-eight coil with an external loop diameter of 70 mm connected to a Magstim 200^2 magnetic stimulator (The Magstim Company Ltd., Whitland, Dyfed, UK). The coil was held tangentially to the scalp with the handle pointing backward and away from the midline at about 45°, in the optimal scalp sites (hot spots) to evoke MEPs in the contralateral APB and ADM muscles. Raw electromyographic signals were recorded with surface electrodes. Responses, sampled at 5 KHz with a CED 1401 A/D laboratory interface (Cambridge Electronic Design, Cambridge, UK), were amplified and filtered (bandpass 20 Hz to 2 kHz) with a Digitimer D360 amplifier (Digitimer Ltd., Welwyn Garden City, Hertfordshire, UK), then recorded by a computer with Signal software (Cambridge Electronic Design).

Resting motor threshold (RMT) was defined as the lowest stimulus intensity able to evoke MEPs at rest with peak-to-peak amplitude of 50 μV in five out of ten consecutive trials. Active motor threshold (AMT) was defined as the lowest intensity able to elicit MEPs of 100 uV in five out of ten consecutive stimuli, during a slight voluntary contraction of the target muscle.

Intracortical excitability was assessed by testing the SICI and ICF with a paired-pulse TMS paradigm. A conditioning stimulus (CS) set at an intensity of 80% AMT was delivered before a test stimulus (TS) with an intensity set to obtain MEPs with peak-to-peak amplitude of 0.5–1.0 mV [43,44]. For SICI, the interstimulus interval (ISI) elapsing between the CS and the TS was 3 ms, whereas for ICF the ISI was 10 ms. Three different conditions (10 test pulses given alone and 10 conditioned pulses for each ISI) were randomly tested. The conditioned MEPs were expressed as a percentage of the unconditioned MEPs.

LTP-like cortical plasticity was explored with the PAS protocol [16]. Median nerve electric shocks were followed by single TMS pulses over the abductor pollicis brevis (APB) muscle hot spot with an ISI of 25 ms. The median nerve was stimulated at the wrist with a constant current stimulator (model DS7A, Digitimer Ltd.) through a pair of surface electrodes (0.2 ms duration, cathode proximal). The intensity of the TMS pulses was set to evoke MEPs of about 0.5–1 mV peak-to-peak amplitude in the APB and in the ADM muscles at baseline. The same intensity was used to elicit MEPs after PAS. The median nerve was stimulated with an intensity set at 300% of the perceptual threshold. Two hundred pairs of electric and magnetic stimuli were repetitively delivered at a rate of 0.25 Hz. Twenty MEPs were recorded from the APB and from the ADM muscles at the baseline, immediately after PAS (post 0) and 15 and 30 min later (post 15 and post 30). At each time interval after PAS, MEP amplitudes were averaged and normalized to the mean baseline amplitude.

4.4. Statistical Analysis

Normality distribution of continuous variables was assessed by Shapiro–Wilk test. Data were expressed as the mean (standard deviation, SD) or, when necessary, the median (interquartile range, IQR). Categorical variables were shown as absolute (n) and relative frequency (%). Logarithmic transformation was applied to reduce the skewness of data distribution and better approximate the normal distribution. Pearson's correlation or, if data were not normally distributed, Spearman's non-parametric correlation, was applied to evaluate possible associations between continuous variables. The relationship between two continuous variables was depicted by a scatter plot.

Possible differences in PAS response between patients and controls were explored with RM-ANOVA with time (pre, post 0, post 15, post 30) and muscle (APB, ADM) as within-subject factors and group (SM, Controls) as between-subject factors. Only for the MS sample was an RM-ANOVA with time (pre, post 0, post 15, post 30) and muscle (APB, ADM) as within-subject factors and MS group (MS IL1-β positive, MS IL-β negative) as between-subject factors performed to assess whether IL-1β CSF detectability influenced the PAS response. For post-hoc comparisons, two-tailed paired sample t tests were conducted to compare pre- vs post- (0, 15, 30) MEP amplitudes. For multiple comparisons, Benjamini–Hochberg correction was applied.

Author Contributions: Conceptualization, M.S.B., F.M., G.A.M., D.C.; methodology, M.S.B., C.G.N., F.M., E.I.; formal analysis, I.S.; investigation, C.G.N., L.G., N.D.P., A.F., R.F.; data curation, L.G.; writing—original draft preparation, M.S.B., E.I.; writing-review and editing, F.B., C.G.N., F.M., L.G., I.S., N.D.P., G.A.M.; R.F., A.F., D.C. supervision, D.C.; funding acquisition, F.B., D.C. All authors have read and agreed to the published version of the manuscript.

Funding: This work was supported by FISM—Fondazione Italiana Sclerosi Multipla—cod. 2019/S/1 awarded to DC; by the Italian Ministry of Health (Ricerca corrente-IRCCS Neuromed awarded to DC; Ricerca Finalizzata 2018, RF-2018-12366144 awarded to DC; Ricerca Finalizzata 2018, GR-2018-12366154 awarded to FB) and by the 5 × 1000 grant from IRCCS Neuromed.

Conflicts of Interest: The authors declare the following potential conflicts of interest with respect to the research, authorship, and/or publication of this article: Fabio Buttari acted as Advisory Board members of Teva and Roche and received honoraria for speaking or consultation fees from Merck Serono, Teva, Biogen Idec, Sanofi, and Novartis and non-financial support from Merck Serono, Teva, Biogen Idec, and Sanofi. Girolama Alessandra Marfia received honoraria for speaking, consultation fees, and travel funding from Roche, Almirall, Bayer Schering, Biogen Idec, Merck Serono, Novartis, Sanofi-Genzyme, Mylan, and Teva. She is the principal investigator in clinical trials for Actelion, Biogen Idec, Merck Serono, Mitsubishi, Novartis, Roche, Sanofi-Genzyme, Teva. Roberto Furlan has received honoraria as a speaker or for research support from Biogen, Novartis, Merck, Roche, Genzyme. Diego Centonze is an Advisory Board member of Almirall, Bayer Schering, Biogen, GW Pharmaceuticals, Merck Serono, Novartis, Roche, Sanofi-Genzyme, and Teva and received honoraria for speaking or consultation fees from Almirall, Bayer Schering, Biogen, GW Pharmaceuticals, Merck Serono, Novartis, Roche, Sanofi-Genzyme, and Teva. He is also the principal investigator in clinical trials for Bayer Schering, Biogen, Merck Serono, Mitsubishi, Novartis, Roche, Sanofi-Genzyme, and Teva. His preclinical and clinical research was supported by grants from Bayer Schering, Biogen Idec, Celgene, Merck Serono, Novartis, Roche, Sanofi-Genzyme, and Teva. The funders had no role in the design of the study; in the collection, analyses, or interpretation of data; in the writing of the manuscript, or in the decision to publish the results. Mario Stampanoni Bassi, Carolina Gabri Nicoletti, Francesco Mori, Luana Gilio, Ilaria Simonelli, Nicla De Paolis, Annamaria Finardi, Ennio Iezzi declare no conflict of interest.

Abbreviations

ADM	abductor digiti minimi
AMPA	α-amino-3-hydroxy-5-methyl-4-isoxazolepropionic acid
AMT	active motor threshold
APB	abductor pollicis brevis
CS	conditioning stimulus
cTBS	continuous theta burst stimulation
EDSS	expanded disability status scale
ICF	intracortical facilitation
IL	interleukin
IQR	interquartile range
ISI	interstimulus interval
iTBS	intermittent theta burst stimulation
LTD	long-term depression
LTP	long-term potentiation
MEP	motor-evoked potential
NMDA	N-Methyl-D-Aspartate
PAS	paired associative stimulation
RMT	resting motor threshold
RR	relapsing–remitting
SD	standard deviation
SICI	short-interval intracortical inhibition
TMS	transcranial magnetic stimulation
TS	test stimulus

References

1. Centonze, D.; Rossi, S.; Tortiglione, A.; Picconi, B.; Prosperetti, C.; De Chiara, V.; Bernardi, G.; Calabresi, P. Synaptic plasticity during recovery from permanent occlusion of the middle cerebral artery. *Neurobiol. Dis.* **2007**, *27*, 44–53. [CrossRef] [PubMed]
2. Mori, F.; Kusayanagi, H.; Nicoletti, C.G.; Weiss, S.; Marciani, M.G.; Centonze, D. Cortical plasticity predicts recovery from relapse in multiple sclerosis. *Mult. Scler. J.* **2014**, *20*, 451–457. [CrossRef]
3. Weiss, S.; Mori, F.; Rossi, S.; Centonze, D. Disability in multiple sclerosis: When synaptic long-term potentiation fails. *Neurosci. Biobehav. Rev.* **2014**, *43*, 88–99. [CrossRef] [PubMed]
4. Lin, C.C.; Edelson, B.T. New Insights into the Role of IL-1β in Experimental Autoimmune Encephalomyelitis and Multiple Sclerosis. *J. Immunol.* **2017**, *198*, 4553–4560. [CrossRef] [PubMed]
5. Coogan, A.N.; O'Neill, L.A.; O'Connor, J.J. The P38 mitogen-activated protein kinase inhibitor SB203580 antagonizes the inhibitory effects of interleukin-1β on long-term potentiation in the rat dentate gyrus in vitro. *Neuroscience* **1999**, *93*, 57–69. [CrossRef]
6. Goshen, I.; Kreisel, T.; Ounallah-Saad, H.; Renbaumb, P.; Zalzstein, Y.; Ben-Hur, T.; Levy-Lahad, E.; Yirmiya, R. A dual role for interleukin-1 in hippocampal-dependent memory processes. *Psychoneuroendocrinology* **2007**, *32*, 1106–1115. [CrossRef]
7. Schmid, A.W.; Lynch, M.A.; Herron, C.E. The effects of IL-1 receptor antagonist on beta amyloid mediated depression of LTP in the rat CA1 in vivo. *Hippocampus* **2009**, *19*, 670–676. [CrossRef]
8. Centonze, D.; Muzio, L.; Rossi, S.; Cavasinni, F.; De Chiara, V.; Bergami, A.; Musella, A.; D'Amelio, M.; Cavallucci, V.; Martorana, A.; et al. Inflammation triggers synaptic alteration and degeneration in experimental autoimmune encephalomyelitis. *J. Neurosci.* **2009**, *29*, 3442–3452. [CrossRef]
9. Rossi, S.; Studer, V.; Motta, C.; De Chiara, V.; Barbieri, F.; Bernardi, G.; Centonze, D. Inflammation inhibits GABA transmission in multiple sclerosis. *Mult. Scler.* **2012**, *18*, 1633–1635. [CrossRef]
10. Rossi, S.; Furlan, R.; De Chiara, V.; Motta, C.; Studer, V.; Mori, F.; Musella, A.; Bergami, A.; Muzio, L.; Bernardi, G.; et al. Interleukin-1β causes synaptic hyperexcitability in multiple sclerosis. *Ann. Neurol.* **2012**, *71*, 76–83. [CrossRef] [PubMed]
11. Rossi, S.; Studer, V.; Motta, C.; Germani, G.; Macchiarulo, G.; Buttari, F.; Mancino, R.; Castelli, M.; De Chiara, V.; Weiss, S.; et al. Cerebrospinal fluid detection of interleukin-1β in phase of remission predicts disease progression in multiple sclerosis. *J. Neuroinflamm.* **2014**, *18*, 11–32. [CrossRef]
12. Di Filippo, M.; Chiasserini, D.; Gardoni, F.; Viviani, B.; Tozzi, A.; Giampà, C.; Costa, C.; Tantucci, M.; Zianni, E.; Boraso, M.; et al. Effects of central and peripheral inflammation on hippocampal synaptic plasticity. *Neurobiol. Dis.* **2013**, *52*, 229–236. [CrossRef] [PubMed]
13. Nisticò, R.; Mango, D.; Mandolesi, G.; Piccinin, S.; Berretta, N.; Pignatelli, M.; Feligioni, M.; Musella, A.; Gentile, A.; Mori, F.; et al. Inflammation subverts hippocampal synaptic plasticity in experimental multiple sclerosis. *PLoS ONE* **2013**, *8*, e54666. [CrossRef]
14. Mandolesi, G.; Grasselli, G.; Musella, A.; Gentile, A.; Musumeci, G.; Sepman, H.; Haji, N.; Fresegna, D.; Bernardi, G.; Centonze, D. GABAergic signaling and connectivity on Purkinje cells are impaired in experimental autoimmune encephalomyelitis. *Neurobiol. Dis.* **2012**, *46*, 414–424. [CrossRef] [PubMed]
15. Mori, F.; Nisticò, R.; Mandolesi, G.; Piccinin, S.; Mango, D.; Kusayanagi, H.; Berretta, N.; Bergami, A.; Gentile, A.; Musella, A.; et al. Interleukin-1β promotes long-term potentiation in patients with multiple sclerosis. *Neuromolecular Med.* **2014**, *16*, 38–51. [CrossRef]
16. Stefan, K.; Kunesch, E.; Cohen, L.G.; Benecke, R.; Classen, J. Induction of plasticity in the human motor cortex by paired associative stimulation. *Brain* **2000**, *123*, 572–584. [CrossRef]
17. Quartarone, A.; Rizzo, V.; Bagnato, S.; Morgante, F.; Sant'Angelo, A.; Girlanda, P.; Siebner, R. Rapid-rate paired associative stimulation of the median nerve and motor cortex can produce long-lasting changes in motor cortical excitability in humans. *J. Physiol.* **2006**, *575*, 657–670. [CrossRef]
18. Wigstrom, H.; Gustafsson, B.; Huang, Y.Y.; Abraham, W.C. Hippocampal long-term potentiation is induced by pairing single afferent volleys with intracellularly injected depolarizing current pulses. *Acta Physiol. Scand.* **1986**, *126*, 317–319. [CrossRef]
19. Bi, G.Q.; Poo, M.M. Synaptic modifications in cultured hippocampal neurons: Dependence on spike timing, synaptic strength, and postsynaptic cell type. *J. Neurosci.* **1998**, *18*, 10464–10472. [CrossRef]

20. Stefan, K.; Kunesch, E.; Benecke, R.; Cohen, L.G.; Classen, J. Mechanisms of enhancement of human motor cortex excitability induced by interventional paired associative stimulation. *J. Physiol.* **2002**, *543*, 699–708. [CrossRef]
21. Wolters, A.; Sandbrink, F.; Schlottmann, A.; Kunesch, E.; Stefan, K.; Cohen, L.G.; Benecke, R.; Classen, J. A temporally asymmetric Hebbian rule governing plasticity in the human motor cortex. *J. Neurophysiol.* **2003**, *89*, 2339–2345. [CrossRef]
22. Quartarone, A.; Bagnato, S.; Rizzo, V.; Siebner, H.R.; Dattola, V.; Scalfari, A.; Morgante, F.; Battaglia, F.; Romano, M.; Girlanda, P. Abnormal associative plasticity of the human motor cortex in writer's cramp. *Brain* **2003**, *126*, 2586–2596. [CrossRef]
23. Mori, F.; Kusayanagi, H.; Buttari, F.; Centini, B.; Monteleone, F.; Nicoletti, C.G.; Bernardi, G.; Di Cantogno, E.V.; Marciani, M.G.; Centonze, D. Early treatment with high-dose interferon beta-1a reverses cognitive and cortical plasticity deficits in multiple sclerosis. *Funct. Neurol.* **2012**, *27*, 163–168.
24. Stampanoni Bassi, M.; Iezzi, E.; Mori, F.; Simonelli, I.; Gilio, L.; Buttari, F.; Sica, F.; De Paolis, N.; Mandolesi, G.; Musella, A.; et al. Interleukin-6 Disrupts Synaptic Plasticity and Impairs Tissue Damage Compensation in Multiple Sclerosis. *Neurorehabil. Neural. Repair* **2019**, *33*, 825–835. [CrossRef]
25. Zeller, D.; Aufm Kampe, K.; Biller, A.; Stefan, K.; Gentner, R.; Schütz, A.; Bartsch, A.; Bendszus, M.; Toyka, K.V.; Rieckmann, P.; et al. Rapid-onset central motor plasticity in multiple sclerosis. *Neurology* **2010**, *74*, 728–735. [CrossRef]
26. Wirsching, I.; Buttmann, M.; Odorfer, T.; Volkmann, J.; Classen, J.; Zeller, D. Altered motor plasticity in an acute relapse of multiple sclerosis. *Eur. J. Neurosci.* **2018**, *47*, 251–257. [CrossRef]
27. Caramia, M.D.; Palmieri, M.G.; Desiato, M.T.; Boffa, L.; Galizia, P.; Rossini, P.M.; Centonze, D.; Bernardi, G. Brain excitability changes in the relapsing and remitting phases of multiple sclerosis: A study with transcranial magnetic stimulation. *Clin. Neurophysiol.* **2004**, *115*, 956–965. [CrossRef]
28. Turrigiano, G.G.; Nelson, S.B. Homeostatic plasticity in the developing nervous system. *Nat. Rev. Neurosci.* **2004**, *5*, 97–107. [CrossRef]
29. Bienenstock, E.L.; Cooper, L.N.; Munro, P.W. Theory for the development of neuron selectivity: Orientation specificity and binocular interaction in visual cortex. *J. Neurosci.* **1982**, *2*, 32–48. [CrossRef]
30. Turrigiano, G.G.; Leslie, K.R.; Desai, N.S.; Rutherford, L.C.; Nelson, S.B. Activity dependent scaling of quantal amplitude in neocortical neurons. *Nature* **1998**, *391*, 892–896. [CrossRef]
31. Lissin, D.V.; Gomperts, S.N.; Carroll, R.C.; Christine, C.W.; Kalman, D.; Kitamura, M.; Hardy, S.; Nicoll, R.A.; Malenka, R.C.; von Zastrow, M. Activity differentially regulates the surface expression of synaptic AMPA and NMDA glutamate receptors. *Proc. Natl. Acad. Sci. USA* **1998**, *95*, 7097–7102. [CrossRef]
32. Weise, D.; Schramm, A.; Stefan, K.; Wolters, A.; Reiners, K.; Naumann, M.; Classen, J. The two sides of associative plasticity in writer's cramp. *Brain* **2006**, *129*, 2709–2721. [CrossRef] [PubMed]
33. Hallett, M. Neurophysiology of dystonia: The role of inhibition. *Neurobiol. Dis.* **2011**, *42*, 177–184. [CrossRef]
34. Stellwagen, D.; Malenka, R.C. Synaptic scaling mediated by glial TNF-α. *Nature* **2006**, *440*, 1054–1059. [CrossRef] [PubMed]
35. Becker, D.; Deller, T.; Vlachos, A. Tumor necrosis factor (TNF)-receptor 1 and 2 mediate homeostatic synaptic plasticity of denervated mouse dentate granule cells. *Sci. Rep.* **2015**, *5*, 12726. [CrossRef] [PubMed]
36. Lewitus, G.M.; Pribiag, H.; Duseja, R.; St-Hilaire, M.; Stellwagen, D. An adaptive role of TNFα in the regulation of striatal synapses. *J. Neurosci.* **2014**, *34*, 6146–6155. [CrossRef] [PubMed]
37. Ren, W.J.; Liu, Y.; Zhou, L.J.; Li, W.; Zhong, Y.; Pang, R.P.; Xin, W.J.; Wei, X.H.; Wang, J.; Zhu, H.Q.; et al. Peripheral nerve injury leads to working memory deficits and dysfunction of the hippocampus by upregulation of TNF-α in rodents. *Neuropsychopharmacology* **2011**, *36*, 979–992. [CrossRef]
38. Ozdogan, H.; Ugurlu, S.; Uygunoglu, U.; Tutuncu, M.; Gul, A.; Akman, G.; Siva, A. The efficacy of anti-IL-1 treatment in three patients with coexisting familial Mediterranean fever and multiple sclerosis. *Mult. Scler. Relat. Disord.* **2020**, *45*, 102332. [CrossRef]
39. Stampanoni Bassi, M.; Mori, F.; Buttari, F.; Marfia, G.A.; Sancesario, A.; Centonze, D.; Iezzi, E. Neurophysiology of synaptic functioning in multiple sclerosis. *Clin. Neurophysiol.* **2017**, *128*, 1148–1157. [CrossRef]
40. Rizzo, F.R.; Musella, A.; De Vito, F.; Fresegna, D.; Bullitta, S.; Vanni, V.; Guadalupi, L.; Stampanoni Bassi, M.; Buttari, F.; Mandolesi, G.; et al. Tumor Necrosis Factor and Interleukin-1β Modulate Synaptic Plasticity during Neuroinflammation. *Neural. Plast.* **2018**, *14*, 8430123. [CrossRef]

41. Polman, C.H.; Reingold, S.C.; Banwell, B.; Clanet, M.; Cohen, J.A.; Filippi, M.; Fujihara, K.; Havrdova, E.; Hutchinson, M.; Kappos, L.; et al. Diagnostic criteria for multiple sclerosis: 2010 revisions to the McDonald criteria. *Ann. Neurol.* **2011**, *69*, 292–302. [CrossRef] [PubMed]
42. Kurtzke, J.F. Rating neurologic impairment in multiple sclerosis: An expanded disability status scale (EDSS). *Neurology* **1983**, *33*, 1444–1452. [CrossRef] [PubMed]
43. Ilic, T.V.; Meintzschel, F.; Cleff, U.; Ruge, D.; Kessler, K.R.; Ziemann, U. Short-interval paired pulse inhibition and facilitation of human motor cortex: The dimension of stimulus intensity. *J. Physiol.* **2002**, *545*, 153–167. [CrossRef] [PubMed]
44. Kujirai, T.; Caramia, M.D.; Rothwell, J.C.; Day, B.L.; Thompson, P.D.; Ferbert, A.; Wroe, S.; Asselman, P.; Marsden, C.D. Corticocortical inhibition in human motor cortex. *J. Physiol.* **1993**, *471*, 501519. [CrossRef]

International Journal of
Molecular Sciences

MDPI

Article

Fingolimod Modulates Dendritic Architecture in a BDNF-Dependent Manner

Abhisarika Patnaik [1], Eleonora Spiombi [2], Angelisa Frasca [2], Nicoletta Landsberger [2], Marta Zagrebelsky [1,*,†] and Martin Korte [1,3,*,†]

1 Zoological Institute, Division of Cellular Neurobiology, TU Braunschweig, D-38106 Braunschweig, Germany; a.patnaik@tu-bs.de

2 Department of Medical Biotechnology and Translational Medicine, University of Milan, 20100 Milan, Italy; eleonora.spiombi@unimi.it (E.S.); Angelisa.frasca@unimi.it (A.F.); nicoletta.landsberger@unimi.it (N.L.)

3 Helmholtz Centre for Infection Research, AG NIND, Inhoffenstr. 7, D-38124 Braunschweig, Germany

* Correspondence: m.zagrebelsky@tu-bs.de (M.Z.); m.korte@tu-bs.de (M.K.)

† These authors contributed equally to this work.

Received: 30 March 2020; Accepted: 23 April 2020; Published: 27 April 2020

Abstract: The brain-derived neurotrophic factor (BDNF) plays crucial roles in both the developing and mature brain. Moreover, alterations in BDNF levels are correlated with the cognitive impairment observed in several neurological diseases. Among the different therapeutic strategies developed to improve endogenous BDNF levels is the administration of the BDNF-inducing drug Fingolimod, an agonist of the sphingosine-1-phosphate receptor. Fingolimod treatment was shown to rescue diverse symptoms associated with several neurological conditions (i.e., Alzheimer disease, Rett syndrome). However, the cellular mechanisms through which Fingolimod mediates its BDNF-dependent therapeutic effects remain unclear. We show that Fingolimod regulates the dendritic architecture, dendritic spine density and morphology of healthy mature primary hippocampal neurons. Moreover, the application of Fingolimod upregulates the expression of activity-related proteins c-Fos and pERK1/2 in these cells. Importantly, we show that BDNF release is required for these actions of Fingolimod. As alterations in neuronal structure underlie cognitive impairment, we tested whether Fingolimod application might prevent the abnormalities in neuronal structure typical of two neurodevelopmental disorders, namely Rett syndrome and Cdk5 deficiency disorder. We found a significant rescue in the neurite architecture of developing cortical neurons from *Mecp2* and *Cdkl5* mutant mice. Our study provides insights into understanding the BDNF-dependent therapeutic actions of Fingolimod.

Keywords: Fingolimod; FTY720; BDNF; primary cultures; dendrites; dendritic spines; Rett syndrome; *Mecp2*; *Cdkl5*

1. Introduction

The secreted growth factor brain derived neurotrophic factor (BDNF) belongs to the Neurotrophin family and is produced by neurons in an activity-dependent manner in all major brain areas [1]. It interacts with the tropomyosin receptor kinase B (TrkB) to accomplish its trophic and plasticity-promoting activities. In the developing central nervous system (CNS), BDNF enhances cell survival, promotes morphogenesis and differentiation, and contributes to the formation and maturation of synapses [2]. In the mature CNS, BDNF is essential for activity-dependent synaptic plasticity, regulating synaptic transmission, and maintaining the mature neuronal architecture, including dendritic spine density, size and shape [3,4], as well as for learning and memory processes [5]. The expression of BDNF is tightly regulated and alterations in BDNF levels and function have been suggested to underlie both the cognitive decline during aging processes as well as the pathogenesis of several neurological and psychiatric disorders [6–8]. While BDNF has been shown to exhibit potent therapeutic effects, its delivery to the brain remains challenging due to its short half-life and poor diffusion across the blood-brain-barrier. Current therapeutic strategies employ viral-mediated *Bdnf* gene delivery or modulation of TrkB activity to promote BDNF-TrkB signaling. Alternatively, the administration of BDNF-inducing drugs has been shown to rescue different neurological diseases characterized by BDNF dysregulation. Of special interest in this context is the sphingosine-1-phosphate analog Fingolimod, an immunomodulatory oral drug used in relapsing-remitting forms of multiple sclerosis [9]. The pro-drug (Fingolimod) diffuses across the blood–brain barrier into the CNS parenchyma where it is phosphorylated into its active form Fingolimod-P and acts as ligand to a subset of Sphingosine-1-phosphate receptors (S1PR1 and 3–5) expressed, among others, also by different brain cells [10]. Interestingly, Fingolimod has been shown to upregulate BDNF mRNA levels and increase BDNF protein release both in vitro and in vivo in an activity- and MAPK-dependent manner [11]. The functional significance of the Fingolimod-induced BDNF release is marked by its ability to improve the cognitive impairment typical of different neurological diseases, e.g., Alzheimer's and Huntington's diseases [12,13]. Furthermore, the rescue of typical symptoms upon Fingolimod treatment was shown to be associated with improved synaptic plasticity and amelioration of dendritic spine loss in the hippocampus of a mouse model for Huntington's disease [13] and prevented spine loss in the cortex during the acute phase of experimental acute encephalitis [14,15].

Rett syndrome (RTT) is an X-linked neurodevelopmental genetic disorder mainly due to loss of the *Mecp2* function resulting at the cellular level in impaired neurite outgrowth and abnormal neuronal architecture [16,17] associated with the reduction of both BDNF mRNA and protein levels [18]. BDNF overexpression, its exogenous application, or TrkB agonistic activation have been shown to restore synaptic plasticity, improve neuronal activity and structural aberrations in *Mecp2* deficient mice, mimicking the human disorder [18–20]. Moreover, Fingolimod administration increased BDNF levels in the cortex, hippocampus, and striatum and significantly improved the motor deficits observed in *Mecp2* mutant mice [11]. Cdkl5 deficiency disorder (CDD), an X-linked atypical variant of RTT is also characterized by impairment in neuronal morphology [21,22] and reduced BDNF mRNA levels [22]. Treatment with a specific TrkB agonist was shown to rescue the structural and functional impairments in a Cdkl5 deficient mouse [23] indicating an important role of the BDNF-TrkB signaling in this context.

While Fingolimod has been established as a promising therapeutic agent for several disease models, whether it is able to modulate the structure of mature, healthy CNS neurons is currently not known. Here, we examine the ability of Fingolimod to modulate dendritic architecture in a BDNF-dependent manner of primary wild type hippocampal neurons. Moreover, in our study, we tested whether Fingolimod application can rescue structural anomalies in cortical neurons of *Mecp2* and *Cdkl5* mutant mice.

2. Results

2.1. *Fingolimod-Phosphate (FTY720-P) Modulates Dendritic Architecture of Mature, Healthy Hippocampal Neurons*

To address a potential role of Fingolimod-phosphate (FTY720-P) in modulating the dendritic architecture of healthy hippocampal neurons, 21 DIV primary hippocampal cultures were treated for 24h with 10 nM FTY720-P or DMSO, as control. The Sholl analysis, performed on feGFP expressing hippocampal neurons, showed a significantly higher dendritic complexity upon FTY720-P treatment when compared to controls (Figure 1A). The increase in dendritic complexity was found to be statistically significant specifically for the portion of the dendritic tree closer to the cell soma (Figure 1A,B; two-way ANOVA $F_{1,4560} = 87.78$, $p < 0.0001$). Accordingly, a significant increase was observed in the total number of dendritic intersections (Figure 1C; two-tailed unpaired t-test: $t = 2.710$, $df = 6$, $p < 0.01$; DMSO: 330.8 ± 13.10 vs, FTY720-P: 383.4 ± 14.20) as well as in the total dendritic lengths of FTY720-P treated neurons compared to controls (Figure 1D; two-tailed unpaired t-test: $t = 2.820$, df = 76, $p < 0.01$; DMSO: 3867 ± 142.9 vs. FTY720-P: 4488 ± 166.0).

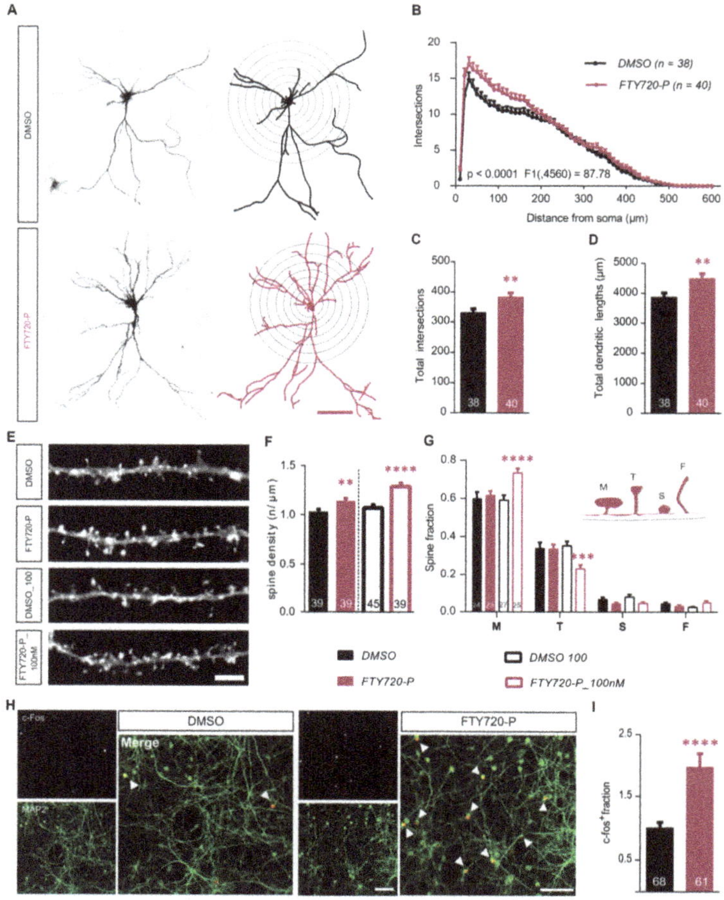

Figure 1. Fingolimod-phosphate (FTY720-P) modulates neuronal morphology in mature hippocampal neurons. FeGFP expressing 21DIV primary hippocampal neurons from healthy wild type BL/6J mice were treated with 10nM FTY720-P for 24h. (**A**) Representative images (left) of feGFP expressing neurons treated with DMSO (above, black) and 10nM FTY720-P (below, magenta) and the relative Neurolucida tracing used to perform the Sholl analysis (right). Scale bar: 100µm. (**B**) Sholl analysis displaying dendritic complexity plotted against distance from the cell soma. F value refers to the DMSO v/s FTY720-P comparison. The graphs show (**C**) the total dendritic complexity and (**D**) total lengths of dendrites for the DMSO (black) and FTY720-P (magenta) treated neurons. (**E**) Representative segments of secondary dendrites from feGFP transfected neurons used to analyze dendritic spine density. Scale bar: 5µm (**F**) Graphical representation of the spine density for neurons treated with 10 (magenta solid) or 100nM (magenta open) FTY720-P their respective DMSO controls (10nM black solid; 100nM black open). (**G**) the graph shows the distribution of different dendritic spine types across all treatment groups. The four different types of spines– mushroom (M), thin (T), stubby (S) and filopodia (F)

as represented in the schematic, were quantified and represented as a fraction of total spines for neurons treated with FYT720P (10 nM magenta solid; 100nM magenta open) or DMSO (10 nM black solid; 100nM black open). (**H**) Images of fields of view (FOV) of DMSO (left) and FTY720-P (right) treated primary hippocampal cultures immunostained for c-Fos (inset: above) and MAP2 (inset: below) with their respective merged images (right). The arrows indicate c-Fos expressing MAP2 positive neurons in the merged picture. Scale bar: 100μm. (**I**) Graph comparing the normalized number of c-Fos$^+$ cells over total MAP2$^+$ neurons between DMSO (black) and FTY720-P (magenta) treated cultures. All plots represent data as mean + SEM. Numbers in the bars indicate total number of neurons or of FOV analyzed, obtained from ≥3 sets of independent experiments. Two-way ANOVA followed by Bonferroni post-hoc test was used in (**B**) and (**G**). For (**C,D**) and (**I**) unpaired Student's *t*-test and for (**F**), one-way ANOVA with Bonferroni post-hoc was used. Significance is denoted as ** $p < 0.01$, *** $p < 0.001$, **** $p < 0.0001$.

Next, to determine if FTY720-P also influences dendritic spines number, dendritic spine density was quantified on secondary dendritic branches of control and FTY720-P treated neurons (Figure 1E). We observed a significant increase (~10%) in the number of dendritic spines per μm in the FTY720-P group (Figure 1E,F; two-tailed unpaired *t*-test: $t = 2.669$, $df = 76$, $p < 0.01$; DMSO: 1.027 ± 0.0283, FTY720-P: 1.137 ± 0.0299). Subsequently, a possible concentration-dependent activity of FTY720-P in regulating dendritic spine density was tested by treating primary hippocampal cultures with a 10-fold higher concentration of FTY720-P (100 nM) or its respective DMSO control. The analysis revealed that, in comparison to the control, the higher drug concentration significantly enhanced spine density by ~20% (Figure 1E,F; two-tailed unpaired t-test: $t = 4.369$, $df = 82$, $p < 0.0001$; DMSO: 1.071 ± 0.0334, FTY720-P_100nM: 1.289 ± 0.0373). Next, a possible concentration-dependent role of FTY720-P was tested in modulating dendritic spine morphology. For this purpose, dendritic spines were structurally classified into four subtypes-mushroom, thin, stubby or filopodia-like (Figure 1G, schematic) based on their length and their head-to-neck ratios [24]. While the neurons treated with lower FTY720-P concentration (10 nM) did not show any difference in their spine-type distribution compared to the relative control neurons, treatment with higher FTY720-P concentration (100 nM) showed a significant increase in the proportion of mushroom-type spines (Figure 1E,G; $p < 0.0001$; DMSO: 0.591 ± 0.027 vs. FTY720-P: 0.735 ± 0.023) associated with a significant decrease in the proportion of thin spines (Figure 1E,G; $p < 0.001$; DMSO: 0.353 ± 0.023 vs. FTY720-P: 0.233 ± 0.021). The fraction of stubby and filopodia-like spines remained unchanged across all treatment groups (Figure 1E,G).

Overall, these observations indicate that, in mature primary hippocampal neurons, a 24-h treatment with FTY720-P results in a mild increase in the dendritic complexity, particularly at the proximal dendrites and in total dendritic length. Moreover, a FTY720-P treatment affects the dendritic spine density and structure in a concentration-dependent manner, with the higher concentration inducing an increase in the proportion of mature dendritic spines.

2.2. *Fingolimod-Phosphate induces the Expression of the Activity-Related Immediate Early Gene c-fos*

FTY720-P has been shown to positively modulate network activity of cultured cortical neurons by promoting excitatory and suppressing inhibitory synaptic transmission signaling via the S1PR1 [11]. Moreover, treatment with FTY720-P induces the phosphorylation of the activity-related kinases ERK1/2 and CREB in primary cortical neurons [11] and FTY720 has been shown to upregulate the activity-related immediate early gene *c-fos* in cultivated cerebellar neurons [25]. Therefore, we next tested whether FTY720-P can also enhance c-Fos protein expression in hippocampal neurons. Therefore, primary hippocampal cultures treated for 24h with FTY720-P or DMSO were immunostained for c-Fos and the number of c-Fos (c-Fos$^+$, Figure 1H, inset above) and MAP2 (MAP2$^+$, Figure 1H, inset below) immunoreactive neurons was quantified. Both treatment groups displayed a clear nuclear localization of c-Fos in neurons, identified by their immune-positivity for MAP2 (Figure 1H). However, in the control group only few neurons were positive for c-Fos in comparison to those treated with FTY720-P (Figure 1H). Quantification of c-Fos$^+$/MAP2$^+$ fraction after FTY720-P treatment revealed a significant, almost two-fold higher number in c-Fos$^+$ neurons than under control conditions (Figure 1I; two-tailed unpaired *t*-test: $t = 4.337$, $df = 127$, $p < 0.0001$; DMSO: 1.000 ± 0.0856, FTY720-P: 1.975 ± 0.2174).

This observation shows the ability of a 24-h treatment with FTY720-P to upregulate c-Fos transcription factor in primary hippocampal neurons.

2.3. Fingolimod-Phosphate (FTY720-P) Mediated Effects are BDNF-Dependent

Several studies have reported the ability of FTY720 and/or FTY720-P to positively modulate the expression and the secretion of BDNF both in vitro and in vivo [11,12,15,26]. Deogracias et al. (2012) showed a time-dependent increase in BDNF mRNA and protein levels with a peak after 24 h FTY720-P application in primary cortical neurons. Interestingly, the increase in BDNF protein was shown to be inversely correlated with the drug concentration [11]. A low dosage of FTY720 was also reported to influence in vitro viability of neural stem cells [27] and recently, also to rescue memory impairment in a mouse model of Alzheimer's disease [28]. Thus, we next tested the activity of a 5-fold lower concentration of FTY720-P in modulating the dendritic architecture of hippocampal neurons in primary hippocampal cultures. DIV21 feGFP expressing neurons were treated for 24 h with 2 nM FTY720-P (FTY720-P_2nM) or DMSO and dendritic complexity was compared using the Sholl analysis (Figure 2A). Interestingly, this low FTY720-P concentration also led to significant alterations in the complexity of proximal dendrites (Figure 2A,B; two-way ANOVA $F_{3,8418} = 33.53$, $p < 0.0001$). Indeed, a significantly higher number of intersections were observed, especially between 20 and 50 μm from the cell body (Figure 2A,B) compared to DMSO treated neurons. Likewise, significantly more total intersections were computed for neurons treated with 2 nM (FTY720-P_2nM) than for the neurons treated with DMSO (Figure 2C; one-way ANOVA $F_{3,138} = 3.469$, $p < 0.05$; DMSO: 243.9 ± 9.940, FTY720-P_2nM: 281.0 ± 9.429).

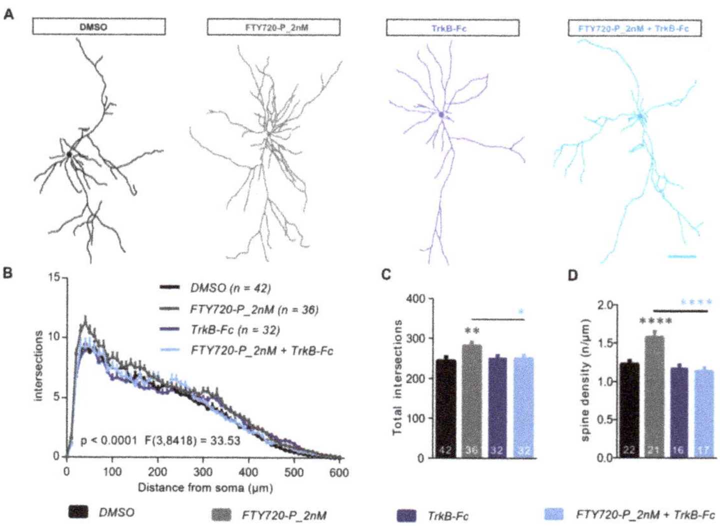

Figure 2. *Cont.*

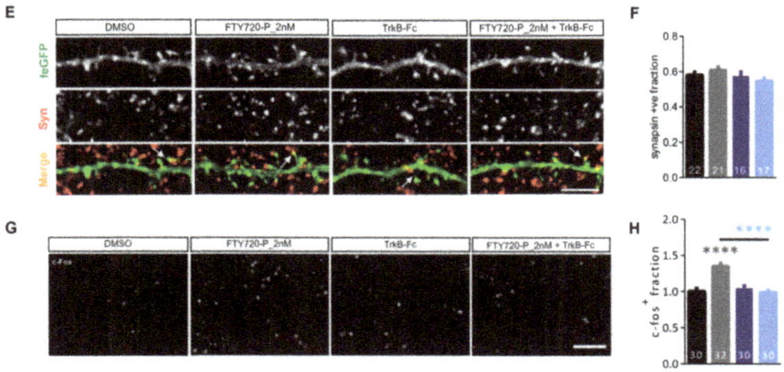

Figure 2. Fingolimod-phosphate (FTY720-P) regulates neuronal architecture in a BDNF-dependent manner. (**A**) Representative Neurolucida tracings, used to perform the Sholl analysis of dendritic complexity for feGFP expressing neurons treated for 24h with: DMSO (black), 2nM FTY720-P (gray), DMSO + TrkBFc (dark blue) and 2nM FTY720-P + TrkBFc (light blue). Scale bar: 100 µm. (**B**) Sholl analysis displayed as number of dendritic intersections against distance from the cell body for DMSO (black), 2nM FTY720-P (gray), DMSO + TrkBFc (dark blue) and 2nM FTY720-P + TrkBFc (light blue) and (**C**) total dendritic complexity of all treatment groups. *F* value in (**B**) refers to comparison between all the 4 treatment groups. (**D**) Dendritic spine densities for DMSO (black), 2nM FTY720-P (gray), DMSO + TrkBFc (dark blue) and 2nM FTY720-P + TrkBFc (light blue) calculated using segments of secondary dendritic branches as shown in (**E**) feGFP panel. The Syn panel displays the corresponding staining of the pre-synaptic marker SynapsinI/II and the merge panel shows the images with overlapping SynapsinI/II puncta (red, pre-synapse) to its matching feGFP dendrite segment (green, post-synapse). The arrows point to coinciding puncta, indicative of mature synapse between the post and pre-synaptic compartments. Scale bar: 5µm. (**F**) The graph compares the fraction of SynapsinI/II positive feGFP labelled spines for DMSO (black), 2nM FTY720-P (gray), DMSO + TrkBFc (dark blue) and 2nM FTY720-P + TrkBFc (light blue). (**G**) Representative fields of view (FOV) of the DMSO, 2nM FTY720-P, DMSO + TrkBFc and 2nM FTY720-P + TrkBFc treated hippocampal cultures stained for c-fos. Scale bar: 100µm. (**H**) Quantification of the proportion of c-Fos expressing neurons represented as normalized fraction for DMSO (black), 2nM FTY720-P (gray), DMSO + TrkBFc (dark blue) and 2nM FTY720-P + TrkBFc (light blue) treated cultures. All graphs represent data as mean + SEM. Numbers in the bars show either total number of neurons or of FOV analyzed, obtained from ≥3 sets of independent experiments. Two-way ANOVA followed by Bonferroni post-hoc test was used in (**B**). For (**C**,**D**,**F**) and (**H**) one-way ANOVA with Bonferroni post-hoc was used. Denotations for significance are * $p < 0.05$, ** $p < 0.01$, **** $p < 0.0001$.

FTY720 has been shown to modulate BDNF levels in a time and concentration-dependent manner. BDNF, which is a critical modulator of structural and functional plasticity in neurons [3], promotes dendritic growth and branching [29,30], and regulates dendritic spine density and morphology in mature neurons [31–33]. Therefore, we asked whether an increase in BDNF release is required for the FTY720-P induced changes in the dendritic architecture. To achieve this aim, DIV21 primary hippocampal cultures were treated with a either FTY720-P_2nM or TrkB receptor bodies (TrkB-Fc) or a combination of the two for 24 h. Treatment with TrkB-Fc alone did not alter dendritic morphology compared to the relative DMSO control as measured using Sholl analysis (Figure 2A,B; two-way ANOVA $F_{3,8418} = 33.53$, $p < 0.0001$). However, when applied in combination with TrkB-Fc, a 24-h treatment with FTY720-P_2nM failed to induce the significant modifications in the dendritic architecture observed upon application of FTY720-P_2nM alone (Figure 2A,B). Indeed, also the increase in total intersections upon FTY720-P_2nM treatment was completely abolished through the TrkB-Fc co-treatment and reduced to control levels (Figure 2C; one-way ANOVA $F_{3,138} = 3.469$, $p < 0.05$; DMSO: 243.9 ± 9.940, FTY720-P_2nM: 281.0 ± 9.429, TrkB-Fc: 251.8 ± 6.460, FTY720-P_2nM+TrkB-Fc: 251.0 ± 8.672). Further, a BDNF-dependent effect of FTY720-P_2nM in regulating the number of dendritic spines was analyzed on secondary dendritic branches of primary hippocampal neurons. While a treatment with TrkB-Fc alone resulted only in a mild decrease in dendritic spine density, a 24-h application of FTY720-P_2nM significantly augmented spine density relative to the control treated neurons (Figure 2E-feGFP, D; one-way ANOVA $F_{3,72} = 13.66$, $p < 0.0001$; DMSO: 1.226 ± 0.0449, FTY720-P_2nM: 1.574 ± 0.0736). This increase was completely prevented when the FTY720-P_2nM was co-applied with TrkB-Fc (Figure 2E-feGFP, D; one-way ANOVA $F_{3,72} = 13.66$, $p < 0.0001$; DMSO: 1.226 ± 0.0449, FTY720-P_2nM: 1.574 ± 0.0736, TrkB-Fc: 1.180 ± 0.0388, FTY720-P_2nM+TrkB-Fc: 1.142 ± 0.0451). To probe whether the increase in dendritic spine density corresponds to an increase in the absolute number of functional synapses, the cultures were immunostained for the pre-synaptic marker SynapsinI/II (Figure 2E-Syn) and its co-localization with the dendritic spines of feGFP transfected neurons was compared between the different treatments (Figure 2E-Merge). Across all treatment groups, the proportion of synapsinI/II positive spines was found to be about ~57% with only a slight increase upon FTY720-P_2nM treatment and a slight decrease upon the combined treatment with FTY720_2nM and TrkB-Fc (Figure 2E-Merge, F; one-way ANOVA $F_{3,72} = 1.398$, $p = 0.2505$; DMSO: 0.5800 ± 0.0168, FTY720-P_2nM: 0.6093 ± 0.0194, TrkB-Fc: 0.5740 ± 0.0267, FTY720-P_2nM+TrkB-Fc: 0.5516 ± 0.0183). These differences, however, did not reach significance (Figure 2F). This constant fraction of SynapsinI/II positive, mature synapses suggest that FTY720-P_2nM indeed increases the absolute number of functional synapses (Figure 2E-Merge, F).

In the CNS, *c-fos* is primarily associated with the regulation of neuronal plasticity [34] and has been shown to be activated downstream of BDNF signaling [33,35]. Thus, we next assessed whether the FTY720-P-induced increase in the proportion of c-Fos positive neurons also depends upon the release of BDNF. Indeed, while treatment with 2nM FTY720-P significantly increased the number of c-Fos positive neurons, combining it with TrkB-Fc completely prevented the enhancement in the number of c-Fos expressing neurons (Figure 2G,H; one-way ANOVA $F_{3,118} = 17.78$, $p < 0.0001$; DMSO: 1.000 ± 0.0414, FTY720-P_2nM: 1.350 ± 0.0454, TrkB-Fc: 1.043 ± 0.0404, FTY720-P_2nM+TrkB-Fc: 0.9985 ± 0.0340).

Taken together, these observations demonstrate that low concentrations of FTY720-P efficiently modulate dendritic architecture and neuronal activation in mature, healthy primary hippocampal neurons. Moreover, the TrkB-Fc co-treatment results indicate a crucial role for the release of BDNF in mediating the effects of FTY720-P.

2.4. Non-Phosphorylated Fingolimod (FTY720) Recapitulates the Dendritic Modifications Induced by the Phosphorylated (FTY720-P) Form

FTY720-P is the active drug form known to interact with its target S1P receptors at membrane surfaces. However, in a clinical scenario the drug is provided in the form of non-phosphorylated FTY720, due to its ability to cross the blood brain barrier and the lipid membrane. Therefore, we next wanted to investigate whether the non-phosphorylated FTY720 is capable of recapitulating the effects we observed using the phosphorylated form. Thus, mature DIV21 hippocampal cultures were treated for 24 h with 10 nM FTY720 or the relative DMSO control. First, the dendritic complexity was analyzed in feGFP expressing neurons using Sholl analysis (Figure 3A,B). The dendritic tree of FTY720 treated neurons was significantly more complex than the one of the DMSO-treated counterparts (Figure 3A,B; two-way ANOVA $F_{1,4636} = 55.64$, $p < 0.0001$). Accordingly, total dendritic complexity (Figure 3B; two-tailed unpaired t-test: $t = 2.277$, $df = 76$, $p < 0.05$; DMSO: 330.8 ± 13.10, FTY720: 373.4 ± 13.29) as well as total dendritic lengths (Figure 3C; two-tailed unpaired t-test: $t = 2.410$, $df = 76$, $p < 0.05$; DMSO: 3867 ± 142.9, FTY720: 4385 ± 159.2) showed a significant increase in the FTY720 treated group when compared to the controls. Finally, the effect of a 24h application of FTY720 was analyzed for dendritic spine density. Also here, FTY720 application led to a significant increase in dendritic spine density compared to the DMSO application (Figure 3D,E; two-tailed unpaired t-test: $t = 3.550$, $df = 77$, $p < 0.001$; DMSO: 1.027 ± 0.0283, FTY720: 1.162 ± 0.0256).

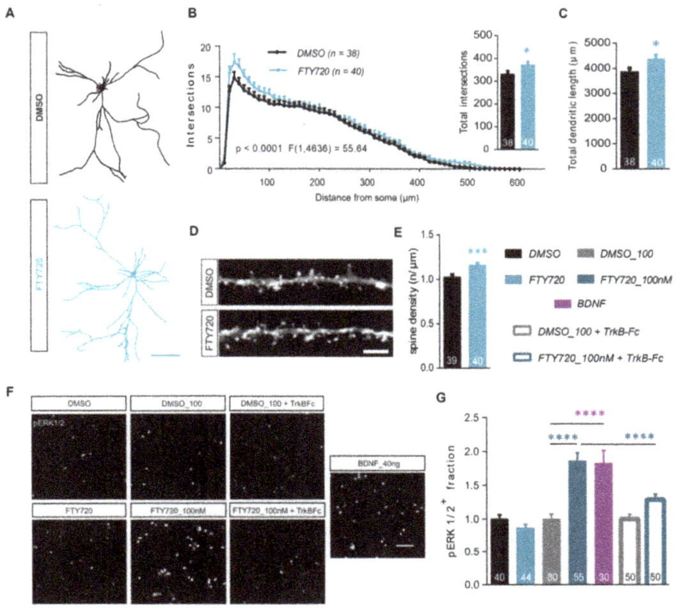

Figure 3. Treatment with the non-phosphorylated Fingolimod (FTY720) modulates neuronal architecture. (**A**)| Representative Neurolucida tracings from feGFP positive hippocampal neurons used for the Sholl analysis from cultures treated either with DMSO or 10nM FTY20 for 24 h. Scale bar: 100μm. (**B**)| Dendritic complexity shown by the number of dendritic intersections plotted against the distance from the soma for DMSO (black) and FTY720 (blue) treated neurons. The *F* value shows the statistical comparison between the two groups. The inset graph represents total dendritic complexity upon treatment with DMSO (black) and FTY720 (blue). (**C**)| Total dendritic length for DMSO (black) and FTY720 (blue) treated neurons. (**D**)| Representative stretches from dendrites of eGFP transfected hippocampal neurons showing dendritic spine protrusions, treated either with DMSO or FTY720 for 24h. Scale bar: 5μm. (**E**)| The graph shows dendritic spine density for DMSO (black) and FTY720 (blue) treated neurons. (**F**)| Representative images of fields of view (FOV) from primary hippocampal cultures stained with anti phospho-ERK1/2 antibody, 30 min post-application of one of the following: DMSO, 10nM FTY720, DMSO_100, 100nM FTY720, DMSO_100 + TrkB-Fc, 100nM FTY720 + TrkB-Fc or 40ng recombinant BDNF protein as a positive control. Scale bar: 100μm. (**G**)| The graph displays the fraction of pERK1/2 expressing neurons relative to the total number of MAP2$^+$ neurons. The data is normalized to the respective controls and compared between the different treatment groups: DMSO (black), 10nM FTY720 (light blue solid), DMSO100 (gray), 100nM FTY720 (dark blue solid), 40ng recombinant BDNF (magenta), DMSO100 + TrkB-Fc (gray open), 100nM FTY720 + TrkB-Fc (dark blue open). All data is plotted as mean + SEM. Numbers in the bars show total number of neurons or FOV analyzed, obtained from ≥3 sets of independent experiments. Two-way ANOVA followed by Bonferroni post-hoc test was used in (**B**). For (**B**) total intersections, (**C**) and (**E**) unpaired Student's t-test and for (**G**) one-way ANOVA with Bonferroni post-hoc was used. Denotations for significance are * $p < 0.05$, *** $p < 0.001$, **** $p < 0.0001$.

As shown by Deogracias et al., (2012) a 30-min FTY720-P treatment upregulated the phosphorylation of extracellular signal-regulated kinase 1/2 (pERK1/2). We tested if this ERK1/2 phosphorylation could also be enhanced by the non-phosphorylated form. Primary hippocampal neurons were treated with 10 nM FTY720 for 30 min and stained using antibodies against pERK1/2 and MAP2. A clear immunofluorescence was observed in the cytoplasm and nucleus of pERK1/2 expressing cells (Figure 3F). The number of cells positive for pERK1/2 (pERK1/2$^+$) were quantified over the total of MAP2$^+$ neurons. While at a concentration of 10 nM, FTY720 did not enhance pERK1/2 expressing cells compared to its DMSO control (Figure 3F,G; one-way ANOVA $F_{6,342} = 18.17$, $p < 0.0001$; DMSO: 1.000 ± 0.0639, FTY720: 0.8692 ± 0.0558), a 10 times higher dose (100 nM), significantly increased the number of pERK1/2$^+$ neurons by ~80% compared to the DMSO treated cultures (Figure 3F,G; one-way ANOVA $F_{6,342} = 18.17$, $p < 0.0001$; DMSO_100: 1.000 ± 0.07469, FTY720_100nM: 1.862 ± 0.1214). it is noteworthy that this increase in the pERK1/2$^+$ fraction was comparable to the one, resulting upon a 30 min application of recombinant BDNF protein (40 ng) (Figure 3F,G; one-way ANOVA $F_{6,342} = 18.17$, $p < 0.0001$; BDNF: 1.825 ± 0.1876). We then probed whether this upregulation of pERK1/2$^+$ after 100 nM FTY720 treatment could be driven by BDNF. Therefore, the 30 min treatment with 100 nM FTY720 was repeated in the presence of TrkB-Fc receptor bodies. TrkB-Fc co-application completely prevented the FTY720-induced increase in the number of pERK1/2 expressing neurons. (Figure 3F,G; one-way ANOVA $F_{6,342} = 18.17$, $p < 0.0001$; DMSO+TrkB-Fc: 1.000 ± 0.07079, FTY720_100nM+TrkB-Fc: 1.291 ± 0.07765).

These observations confirm that the non-phosphorylated FTY720 can indeed successfully induce similar structural alterations in mature hippocampal neurons, as produced by its active phosphorylated form FTY720-P. Additionally, at a higher concentration (100 nM) FTY720 acutely upregulates phosphorylation of ERK1/2 in neurons in a BDNF-dependent fashion.

2.5. Fingolimod-Phosphate (FTY720-P) Completely Rescues the Neurite Defects Observed in Young Mecp2$^{-/y}$ Neurons and Only Partially in Cdkl5$^{-/y}$ Neurons

Our results so far demonstrate the ability of FTY720-P to positively modulate the dendritic architecture of hippocampal neurons in a BDNF-dependent fashion. Therefore, we wondered if the drug could be extended for use onto mouse models for neurodevelopmental diseases characterized by defects in neuronal architecture combined with BDNF deficiency. We selected the *Mecp2$^{-/y}$* transgenic mouse as a model for Rett Syndrome (RTT) caused by the expression of a dysfunctional Mecp2 protein. In addition, *Cdkl5$^{-/y}$* transgenic mice were used as a model for the cyclin-dependent kinase-like 5 (*Cdkl5*) deficiency disorder (CDD) caused by loss of function mutation in the X-linked *Cdkl5*. Both diseases are associated with abnormalities in the neuronal architecture in hippocampal and cortical neurons, including simpler dendritic arbors and reduced dendritic lengths starting already at very early stages during development [16,17,36] [21,22,37,38].

Low-density primary cortical neurons from wild type (WT) or *Mecp2* knockout (*Mecp2$^{-/y}$*) mice were used to test the ability of FTY720-P to rescue the impairment in the neuronal structural development typical of RTT. The DIV1 primary cortical neurons were let develop in the presence of 10 nM FTY720-P or DMSO for 7 days (Figure 4A schematic) before being fixed and stained for MAP2 (Figure 4B). Individual MAP2 positive neurons with well-defined cell bodies and neuritic protrusions were selected to analyze dendritic development using Sholl analysis (Figure 4B,C; two-way ANOVA $F_{3,7826} = 137.4$, $p < 0.0001$). WT cells treated with DMSO or FTY720-P were qualitatively indistinguishable (Figure 4B). In addition, the quantification of their neuritic complexity by Sholl analysis displayed no difference between these two treatment groups (Figure 4C,F). The *Mecp2$^{-/y}$* neurons, on the other hand were remarkably smaller than WT with shorter and fewer neurites (Figure 4B). When compared to WT, *Mecp2$^{-/y}$* neurons were characterized by a significantly lower neuritic complexity as shown by the Sholl analysis (Figure 4C,D). In addition, the total complexity of *Mecp2$^{-/y}$* neurons was significantly lower compared to WT neurons (Figure 4G; one-way ANOVA $F_{3,560} = 43.11$, $p < 0.0001$; WT: 41.02 ± 1.221, *Mecp2$^{-/y}$*: 26.13 ± 0.9886). While WT neurons had an average number of 9.29 ± 0.2442 primary neurites arising from the cell soma (Figure 4H; one-way ANOVA $F_{3,564} = 6.042$, $p < 0.001$), the *Mecp2$^{-/y}$* neurons only had 8.30 ± 0.2507 (Figure 4H). However, *Mecp2$^{-/y}$* neurons that developed for 7 days in the presence of FTY720-P showed longer and more complex neurites than the control-treated counterparts (Figure 4B). This was also true for neurite complexity which significantly increased in the FTY720-P treated group (Figure 4C,E). Moreover, compared to the WT neurons, *Mecp2$^{-/y}$* neurons displayed a decrease of ~37% in total neurite intersections which was completely rescued upon treatment with FTY720-P (Figure 4G;

one-way ANOVA $F_{3,560} = 43.11$, $p < 0.0001$; *Mecp2*$^{-/y}$ + FTY720-P: 38.74 ± 1.098). Also, the average number of primary neurites in the *Mecp2*$^{-/y}$ cultures treated with FTY720-P was restored to WT level (Figure 4H; one-way ANOVA $F_{3,564} = 6.042$, $p < 0.001$; *Mecp2*$^{-/y}$ + FTY720-P: 9.752 ± 0.2274). Lastly, the neurites of *Mecp2*$^{-/y}$ neurons were overall significantly shorter than the WT neurons (Figure 4I) and treatment with FTY720-P completely rescued their total neurite length (Figure 4I; one-way ANOVA $F_{3,560} = 49.24$, $p < 0.0001$; WT: 455.7 ± 12.93, *Mecp2*$^{-/y}$: 279.6 ± 10.69, *Mecp2*$^{-/y}$ + FTY720-P: 417.2 ± 11.79). The FTY720-P treatment in WT cultures did not alter total neurite intersections (Figure 4G; one-way ANOVA $F_{3,560} = 43.11$, $p < 0.0001$; WT: 41.02 ± 1.221, WT+FTY720-P: 42.25 ± 1.307) and length (Figure 4I; one-way ANOVA $F_{3,560} = 49.24$, $p < 0.0001$; WT: 455.7 ± 12.93, WT+FTY720-P: 467.8 ± 14.95).

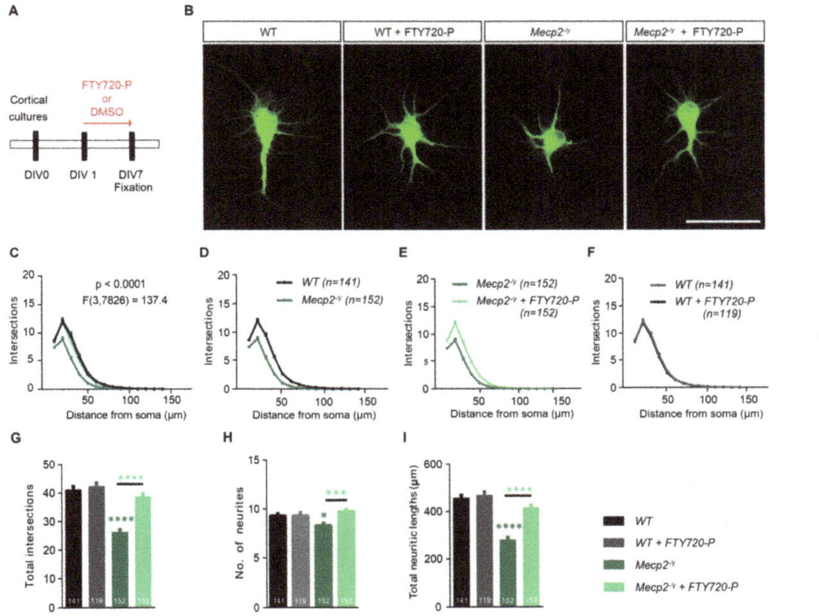

Figure 4. Fingolimod-phosphate (FTY720-P) completely rescues the neuronal architecture in developing *Mecp*$^{-/y}$ cortical neurons. Young DIV7 cortical cultures from *Mecp2*$^{-/y}$ mice or littermate wild type (WT) controls were allowed to develop for one week with either DMSO alone or with 10nM FTY720-P in the growth medium. (**A**) The diagram shows the scheme of treatment from culture preparation on DIV0, treatment from DIV1 to fixation on DIV7. (**B**) Micrographs showing representative fluorescent images of DIV7 WT and *Mecp2*$^{-/y}$ cortical neurons stained for MAP2 and treated with either DMSO or FTY720-P. Scale bar: 50 μm. Neurite complexity plotted as number of neurite intersections against the distance from the soma for: (**C**) all four test groups- WT (black), WT neurons treated with 10nM FTY720-P (gray), *Mecp2*$^{-/y}$ (dark green) and *Mecp2*$^{-/y}$ neurons treated with 10nM FTY720-P (light green). *F* value for comparison between all sets is described with the graph. For better visualization of differences between different pairs, Sholl curves are also plotted separately: (**D**) WT (black) and *Mecp2*$^{-/y}$ (dark green) neurons (**E**) *Mecp2*$^{-/y}$ neurons treated with 10nM FTY720-P (light green) v/s DMSO controls (dark green) and (**F**) WT neurons treated with 10nM FTY720-P (gray) v/s DMSO controls (black). (**G**) Total intersections, (**H**) number of neurites and (**I**) total length of neurites as computed for WT neurons treated with DMSO (black) or FTY720-P (gray) and *Mecp2*$^{-/y}$ treated with DMSO (dark green) or FTY720-P (light green). Data in graphs is plotted as mean + SEM. Numbers in the bars show total number of neurons analyzed for each genotype and treatment, obtained from ≥3 sets of independent experiments. Two-way ANOVA followed by Bonferroni post-hoc test was used in A. For G,H and I one-way ANOVA with Bonferroni post-hoc was used. Denotations for significance are *** $p < 0.001$, **** $p < 0.0001$.

Next, DIV 1 WT or *Cdkl5*$^{-/y}$ cortical cultures were cultured in the presence of 10 nM FTY720-P or DMSO until DIV 7 (Figure 5A). MAP2 positive neurons from each treatment group show cell body with several neurites of different lengths (Figure 5B). WT neurons treated with DMSO alone or with FTY720-P displayed no difference in overall appearance (Figure 5B) and the quantification of their neuritic complexity showed no significant difference both for the Sholl analysis (Figure 5C,F) and their total complexity (Figure 5G; one-way ANOVA $F_{3,669}$ = 10.48, p < 0.0001; WT: 45.51 ± 1.306, WT+FTY720-P: 45.17 ± 1.393). On the other hand, neurons lacking *Cdkl5* exhibited significant structural deficits identifiable by their less complex neurite morphology and shorter neurites (Figure 5B). The Sholl analysis revealed significantly lower complexity for *Cdkl5*$^{-/y}$ neurites in comparison to WT neurons (Figure 5C,D; two-way ANOVA $F_{3,11373}$ = 32.07, p < 0.0001). FTY720-P in the growth medium resulted in a mild, albeit significant increase in the neurite complexity of *Cdkl5*$^{-/y}$ neurons (Figure 5C,E). While total neuritic complexity of FTY720-P treated *Cdkl5*$^{-/y}$ neurons was only slightly higher than control *Cdkl5*$^{-/y}$ neurons (Figure 5G; one-way ANOVA $F_{3,669}$ = 10.48, p < 0.0001; *Cdkl5*$^{-/y}$: 37.18 ± 1.104, *Cdkl5*$^{-/y}$ + FTY720-P: 40.63 ± 1.204), the total neuritic length was significantly enhanced (Figure 5I; one-way ANOVA $F_{3,670}$ = 10.48, p < 0.0001; *Cdkl5*$^{-/y}$: 427.2 ± 12.53, *Cdkl5*$^{-/y}$ + FTY720-P: 478.1 ± 13.89). Although the average number of primary neurites in *Cdkl5*$^{-/y}$ neurons was slightly lower than the WT, it was statistically not significant (Figure 5H; one-way ANOVA $F_{3,669}$ = 1.162, p = 0.3235; WT: 9.086 ± 0.3673, *Cdkl5*$^{-/y}$: 8.804 ± 0.3707, *Cdkl5*$^{-/y}$ + FTY720-P: 9.166 ± 0.4110). WT neurons treated with FTY720-P showed a mild, though not statistically significant increase in the number of primary neurites (Figure 5H; one-way ANOVA $F_{3,669}$ = 1.162, p = 0.3235; WT: 9.086 ± 0.3673, WT+FTY720-P: 9.780 ± 0.3909) and did not show any changes in their total neurite length (Figure 5I; one-way ANOVA $F_{3,670}$ = 10.48, p < 0.0001; WT: 526.1 ± 14.77, WT±FTY720-P: 514.7 ± 15.48).

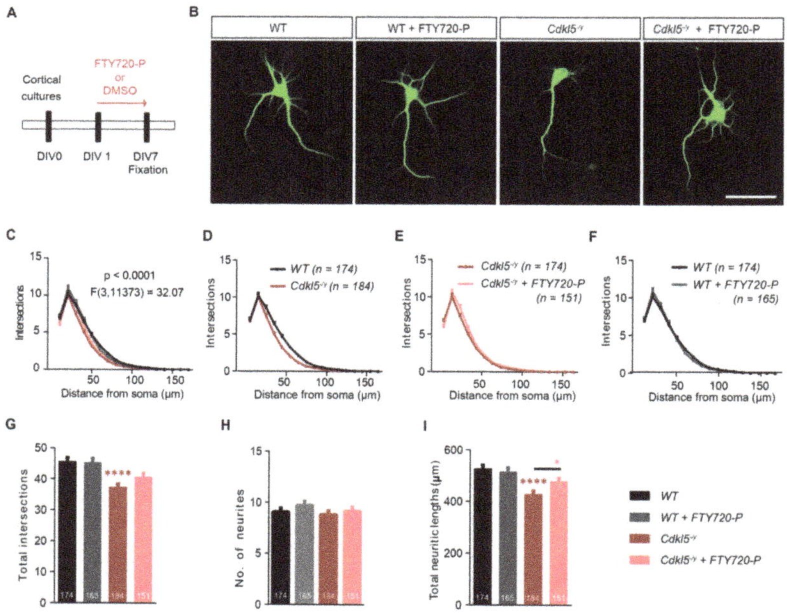

Figure 5. Fingolimod-phosphate (FTY720-P) shows mild rescue in neuronal morphology of young cortical neurons from *Cdkl5*$^{-/y}$ mice. Young DIV7 cortical cultures from mice carrying a *Cdkl5*$^{-/y}$ mutation or littermate wild type (WT) controls, were grown in presence of DMSO alone or 10 nM FTY720-P in

the growth medium. (**A**) The diagram shows the scheme of treatment from culture preparation on DIV0, treatment from DIV1 to fixation on DIV7. (**B**) Micrographs display representative DIV7 MAP2 positive cortical neurons. The WT and *Cdkl5*$^{-/y}$ neurons were both treated either with DMSO or FTY720-P. Scale bar: 50 µm. (**C**) Neurite complexity for both genotypes treated with DMSO and FTY720-P was analyzed and plotted as number of neurite intersections against the distance from the soma. *F* value for comparison between the four sets is described in the graph. The Sholl curve for following genotype and treatment sets are also displayed separately (**D**) WT (black) v/s *Cdkl5*$^{-/y}$ (dark red), (**E**) *Cdkl5*$^{-/y}$ neurons treated with 10nM FTY720-P (rosa) v/s DMSO controls (dark red), (**F**) WT neurons treated with 10nM FTY720-P (gray) v/s DMSO controls (black). (**G**) Total intersections, (**H**) number of neurites and (**I**) total length of neurites as computed for WT neurons treated with DMSO (black) or FTY720-P (gray) and *Cdkl5*$^{-/y}$ neurons treated with DMSO (dark red) or FTY720-P (rosa). Data in graphs is represented as mean + SEM. Numbers in the bars show total number of neurons analyzed for each genotype and treatment, obtained from ≥3 sets of independent experiments. Two-way ANOVA followed by Bonferroni post-hoc test was used in (**A**). For (**G**,**H**) and (**I**) one-way ANOVA with Bonferroni post-hoc was used. Denotations for significance are * $p < 0.05$, **** $p < 0.0001$.

In conclusion, these results illustrate that the deficiency of Mecp2 and Cdkl5 protein respectively leads to severely stunted growth of immature developing cortical neurons as seen by reduction in neurite length and complexity. These defects can be entirely rescued by FTY720-P application into the growth medium for *Mecp2*$^{-/y}$ neurons and only partially for *Cdkl5*$^{-/y}$ neurons.

3. Discussion

Our results describe a role of the immunomodulatory drug Fingolimod in modulating the dendritic architecture of healthy mature primary hippocampal neurons. Indeed, Fingolimod application to hippocampal primary neurons results in a significant increase both in dendritic complexity and in dendritic spine density, associated to a higher proportion of mature, mushroom-like spines. Moreover, we show that a treatment with Fingolimod is able to prevent the structural defects described in hippocampal neurons from the mouse models of two neurodevelopmental disorders, namely Rett syndrome and Cdkl5 deficiency disorder.

3.1. Fingolimod Regulates Neuronal Architecture and the Expression of Activity-Related Genes in Healthy Mature Neurons in a BDNF-Dependent Manner

Fingolimod crosses the blood brain barrier and accumulates within the brain where, binding to S1PRs exerts a series of direct functions on CNS cells including enhanced neuroprotective effects, reduced inflammation and improvement of neuronal pathology in different neurological disorders. Moreover, Fingolimod rescues the cognitive impairment and associated dendritic spine pathology observed in mouse models of Alzheimer and Huntington disease [12–14] indicating that, at least in part, the beneficial effects of Fingolimod may derive from its ability to modulate neuronal architecture. However, very little is currently known about the possible effects of Fingolimod in modulating the dendritic and spine architecture of healthy neurons. We show here that Fingolimod application to primary hippocampal neurons results in increased spine density, associated with a higher proportion of mushroom-like, mature spines identified by their positivity for SynapsinI/II. While application of Fingolimod has been previously shown to promote neurite outgrowth in developing neurons [25], we show that this effect also extends to mature healthy hippocampal neurons, showing an increase in dendritic complexity after treatment. While these observations in vitro do not allow us to conclude that a similar effect could be observed in vivo, they do indicate a role for Fingolimod in modulating dendritic and spine architecture to promote neuronal connectivity within primary hippocampal cultures and are supported by previous reports showing an increase in network activity after Fingolimod application [11].

Several lines of evidence indicate that Fingolimod promotes BDNF synthesis and secretion both in vivo and in vitro [11–13,15]. Moreover, BDNF has been shown to modulate dendritic and spine architecture in an age-dependent manner [3,33]. Here, we show that co-application of the BDNF scavenging TrkB receptor bodies completely prevents the increase in dendritic complexity, spine density and maturation observed upon Fingolimod-P treatment in hippocampal primary neurons indicating the requirement for BDNF secretion in this context. Indeed, several in vitro studies support a role for BDNF signaling in modulating the dendritic architecture during development [3,33,39] and in regulating dendritic spine density and morphology in mature hippocampal neurons [31,32]. We report similar observations on dendritic complexity, spine density and spine morphology of mature hippocampal neurons upon Fingolimod application. It is noteworthy, that although the effects observed upon treatment in mature, healthy neurons are relatively mild they are significant and consistent in their manner of expression. Indeed, Fingolimod specifically increases the complexity of the proximal dendritic tree. A specific effect on proximal dendrites has been previously described in cortical neurons overexpressing BDNF [29] and after exogenous BDNF treatment in cultured hippocampal neurons [30]. However, while in our study Fingolimod affects mature neurons already after 24 h of treatment, the effects of exogenous BDNF application required days [30]. While a clear effect of the exogenous application of BDNF in modulating dendritic spine density in hippocampal neurons could be observed in some studies [31,32], our laboratory failed to reproduce it [33], possibly due to differences in culture conditions. However, in the current study we observe a significant, BDNF-dependent increase in dendritic spine density upon Fingolimod treatment. This observation suggests a higher efficacy of the Fingolimod-induced release of endogenous BDNF compared to its exogenous application, possibly due to a requirement for a specific action at the release site rather that a global effect, as per bath application. On the other hand, it should be noted that previous studies did not observe a regulation of BDNF synthesis upon Fingolimod treatment in vivo [11] indicating the need for further studies.

The effects of BDNF on dendritic spine density have been show to occur in an activity-dependent manner [33] and the positive modulation of dendrite architecture by BDNF occurs via the activity-dependent expression of the cAMP response element-binding protein (CREB) [30]. Interestingly, an increase in the phosphorylation of CREB and of network activity was observed in Fingolimod treated cortical cultures [11] suggesting that the structural changes upon its application, could be mediated via an increase in neuronal activity and CREB phosphorylation. This increased activity is likely since we show a significantly higher number of c-Fos expressing neurons upon Fingolimod application. The immediate early gene *c-fos* is known to be upregulated upon membrane depolarization via CREB phosphorylation and is a commonly used marker for recently activated neurons [40]. Such a sustained increase in c-Fos positive neurons, as seen 24 h after the beginning of Fingolimod treatment could support a positive BDNF-feedback loop and thus contribute to the structural alterations. Indeed, *c-fos* is known to participate in a positive feedback loop downstream of TrkB signaling promoting BDNF expression [41]. Finally, our data show also in hippocampal neurons an increase in the phosphorylation of the extracellular signal-regulated kinase 1/2 (ERK1/2) in treated cultures as previously reported for cortical neurons [11]. The increase in the proportion of pERK1/2$^+$ cells in Fingolimod treated cultures was comparable to the one observed in BDNF treated ones and was, as for c-Fos expression reduced to control levels upon co-treatment with TrkB-Fc receptor bodies again supporting the necessity of BDNF secretion for the Fingolimod induced effects in hippocampal neurons.

3.2. Fingolimod Treatment Rescues the Impaired Neurite Outgrowth in Cortical Neurons in Two X-linked Neurodevelopmental Disorders

While a growing body of evidence supports the benefits of modulating BDNF-TrkB signaling in several neurological conditions, its inability to cross the blood brain barrier as well as the need for temporally and spatially regulated BDNF signaling limit its direct use in therapy. Alternative methods for modulating BDNF levels within specific brain areas include drug-induced BDNF expression. Among these drugs Fingolimod has been successfully used to rescue the cognitive impairment observed in many neurological diseases. However, still, little is known about the cellular mechanisms underlying its beneficial action. Our study adds to this notion, by analyzing the cellular effects of Fingolimod in two X-linked neurodevelopmental disorders, namely Rett syndrome (RTT) and Cdkl5 deficiency disorder (CDD). Both disorders show severe anomalies in neuronal structures both in patients [42] and in the relative mouse models: *Mecp2* [17] or *Cdkl5* knockout mice [37]. Our results confirmed the altered neuronal phenotype in developing cortical neurons derived from *Mecp2*$^{-/y}$ as well as *Cdkl5*$^{-/y}$ mice. The knockout neurons were found to be significantly simpler in complexity and shorter when compared to their relative controls. Fingolimod application completely rescued the impairment in neurite growth in *Mecp2*$^{-/y}$ cortical neurons. While our results cannot exclude a BDNF-independent activity of Fingolimod, it is interesting that the onset of RTT-like phenotypes in *Mecp2* knockouts is associated with the absence of the typical increase in BDNF levels observed postnatally. The *bdnf/Mecp2* double knockout shows exacerbated symptoms and an earlier disease onset [18], indicating a link between the impaired BDNF expression and the RTT pathology. Accordingly, promoting BDNF/TrkB signaling via Fingolimod injection significantly improved motor behavior and the average lifespan of *Mecp2* mutant mice [11]. The current results support the hypothesis that Fingolimod might rescue neuronal structure in *Mecp2*$^{-/y}$ mice, possibly in a BDNF-dependent manner. It is important to note that our findings indicate benefits of Fingolimod by rescuing the typical defects in neuronal architecture at an early developmental stage. However, whether a rescue of the morphological phenotype could be obtained in fully developed neurons in juvenile or adult knockout animals needs further investigation. While application of a TrkB agonist has been shown to rescue both functional and structural impairments in the perirhinal cortex of *Cdkl5*$^{-/y}$ mice [23], little is known about the role of BDNF in the pathology of CDD in contrast to RTT. Our results show that application of Fingolimod results in a partial rescue of the structural impairments observed in *Cdkl5*$^{-/y}$ cortical neurons. While our current results cannot exclude an effect of Fingolimod independent of its ability to promote BDNF release, the partial rescue in *Cdkl5*$^{-/y}$ could be explained by looking at the position of BDNF along the Mecp2 and Cdkl5 downstream signaling pathways. Indeed, while Mecp2 functions upstream of BDNF to regulate its activity-dependent transcription [43], Cdkl5 lies downstream of BDNF-TrkB signaling to activate Rac1 [44]. Thus, while modifying endogenous BDNF levels using Fingolimod may compensate for its reduced levels in *Mecp2*$^{-/y}$ neurons, it may not rescue the effects of the loss of the intermediate signaling protein in *Cdkl5*$^{-/y}$ neurons. The partial rescue observed possibly indicates an indirect, neuroprotective action of Fingolimod in this context.

In summary, the results of our study can be divided in two main independent parts. On one side, our study reports specific BDNF-dependent effects of the immunomodulatory drug Fingolimod in modulating the dendrite architecture of mature, healthy CNS neurons. In addition, our observations add to the understanding regarding the cellular mechanisms underlying the positive effects of a Fingolimod treatment in two highly relevant X-linked neurodevelopmental disorders.

4. Materials and Methods

4.1. Animals

All studies including wild-type (WT) mice were done using C57Bl/6J mice bred in the mouse facility of the TU Braunschweig. The *Mecp2* null mouse strain, originally purchased from Jackson Laboratories (003890 B6.129P2(C)-Mecp2<tm1.1Bird>/J), was backcrossed and maintained on a clean CD1 background [45]. The *Cdkl5* null mouse strain was originally kindly donated by Dr Elisabetta Ciani (University of Bologna). C57BL/6 *Cdkl5* heterozygous females were crossed with CD1 WT male mice for at least ten generations. Timed-pregnant females were generated by overnight crossing wild type or hemizygous CD1 males with *Mecp2*$^{+/-}$ or *Cdkl5*$^{+/-}$ CD1 females. The *Mecp2* and the *Cdkl5* null mutant strains were bred in the laboratory of Prof. Nicoletta Landsberger where the experiments with these mice were performed (University of Milan, Milan). Genotypes were assessed by PCR on genomic DNA purified from tails to distinguish knockouts and littermate WT controls. Mice were housed on a 12 h light/dark cycle in a temperature-controlled environment (21 ± 2 °C) with food and water provided ad libitum. All procedures concerning animals were performed in accordance with the European Union Communities Council Directive (2010/63/EU) and were approved by the animal welfare representative of the TU Braunschweig and the LAVES (Oldenburg, Germany, Az. §4 (02.05) TSchB TU BS) as well as by the Italian Council on Animal Care, the Italian Government decree No. 210/2017 and the San Raffaele Scientific Institutional Animal Care and Use Committee in accordance with the Italian law.

4.2. Reagents and Stock Solutions

Fingolimod hydrochloride: FTY720 (USBiological, Swampscott, MA, USA) and Fingolimod-Phosphate: FTY720-P (Cayman chemical, Ann Arbor, MI, USA), were dissolved in Dimethyl sulfoxide (DMSO, anhydrous; ThermoFisher Scientific,Waltham, MA, USA). FTY720-P required gentle sonication for 2–5 min for dissolution. Stock solution for FTY720 (100mM), FTY720-P (1mM) and control DMSO vials were stored at −20 °C until used. Recombinant Human BDNF protein (R&D systems, Minneapolis, MN, USA) was dissolved in sterile PBS with 0.1% BSA at a concentration of 50 ng/μL. BDNF scavenging TrkB-Fc receptor bodies (Recombinant Human TrkB-Fc Chimera, R&D systems, Minneapolis, MN, USA) were reconstituted in sterile PBS at 50 μg/mL. BDNF and TrkB-Fc were stored at −70 °C until used.

4.3. Primary Hippocampal Cultures

Primary mouse hippocampal and cortical cultures were prepared on embryonic day 18. The pregnant female was rapidly killed by cervical dislocation, the abdomen was sterilized, and the embryos were collected under sterile conditions. Post-decapitation the embryonic brains were quickly isolated and immersed in ice-cold Gey's balanced salt solution supplemented with glucose and with a pH adjusted to 7.3. The hippocampi and/or cortices were dissected, incubated in Trypsin-EDTA (Sigma-Aldrich, St. Louis, MO, USA) at 37 °C for 30 min and further dissociated mechanically using a Pasteur pipette. The cells were re-suspended in Gibco Neurobasal medium (ThermoFisher Scientific, Waltham, MA, USA) supplemented with 2% B27(Invitrogen, Carlsbad, CA, USA) (*v/v*), 10% N2 and 0.5 mM Gibco glutamax (ThermoFisher Scientific, Waltham, MA, USA) and plated at high ($7 \times 10^4/cm^2$) or low density ($3.5 \times 10^4/cm^2$) on 12 mm glass coverslips previously coated with poly-L-lysine (Sigma-Aldrich, St. Louis, MO, USA). The cultures were incubated at 37 °C, 5% CO_2 and 95% O_2 until usage. Medium change was done once a week by replacing 20% of the medium. The C57Bl/6J WT hippocampal cultures were fixed at DIV21 whereas *Mecp2* and *Cdkl5* cortical cultures, at DIV7.

4.4. Transfection and Treatments

At DIV20, mature hippocampal neurons were transfected using Lipofectamine2000 (ThermoFisher Scientific, Waltham, MA, USA) as per manufacturer's instructions. 0.8μg farnesylated-eGFP (Clonetech: pEGFP-F, Mountain View, CA, USA) expressing plasmid were used for each well. Then, 24-h after transfection, the cells were treated with 2, 10, or 100 nM Fingolimod phosphate (FTY720-P) or 10 and 100 nM Fingolimod (FTY720) and the respective DMSO controls diluted into Neurobasal medium without supplements. All FTY720-P concentrations used in the study were in the nM range. Since the physiological levels of Fingolimod crossing the blood brain barrier and accumulating in local neuronal vicinity cannot be clearly estimated, our dosage selection was based on a previous report by Deogracias et al. (2012). They showed nM ranges of FTY720-P to maximally enhance BDNF levels (2-fold) in cortical cultures. This effect however was reduced when the drug concentration reached μM range. In some experimental sets DMSO and FTY720-P (2 nM) treated cultures were co-treated with TrkB-Fc receptor bodies (500 ng/mL). For pERK1/2 experiments, BDNF (40 ng/mL) treatment was used as a positive control. The cells were fixed 24 h or, for pERK 30 min after treatment. *Mecp2* and *Cdkl5* knockout cortical cultures were treated starting at DIV1 by adding 100 μL of fresh Neurobasal medium with supplements containing either 10nM FTY720-P or DMSO to the wells and were fixed at DIV7.

4.5. Immunocytochemistry

The cultured cells were fixed for 10 min in ice cold-Paraformaldehyde (4% in PBS) and washed 3 times with PBS. Permeabilization and blocking of unspecific binding sites was done at room temperature (RT) for 1 h with PBS containing 0.2% Triton X-100 (Sigma-Aldrich, St. Louis, MO, USA) and 1.5% Normal Goat Serum (NGS, ThermoFisher Scientific, Waltham, MA, USA). Next, the cells were incubated overnight with primary antibodies diluted in 1.5% NGS + 0.2% Triton X-100 containing PBS at 4 °C on a rocker. The following primary antibodies were used: mouse anti-MAP2 (1:1000, Sigma-Aldrich, St. Louis, MO, USA)), rabbit anti-cFos (1:10,000, Synaptic systems, Goettingen, NI, Germany), rabbit anti-Phospho-MAPK/Erk1/2 (1:1000, Cell signaling technology, Danvers, MA, USA), rabbit anti- SynapsinI/II (1:1000, Millipore, Burlington, MA, USA). Secondary antibodies were diluted 1:500 in PBS in varying combinations: anti-rabbit IgG, anti-mouse IgG conjugated with appropriate cyanine fluorophores Cy2/Cy3 or Cy5 (Jackson ImmunoResearch Labs, West Grove, PA, USA). The secondary antibody incubation was done for 2h at RT on a rocker. Finally, the cultures were counterstained with 4′,6-diamidino-2-phenylindole (DAPI) diluted 1:1000 in PBS for 10 min. The coverslips were then washed with PBS and mounted onto glass slides using anti fading Fluoro-Gel embedding medium (Electron Microscopy Sciences, Hatfield, PA, USA).

4.6. Image Acquisition and Analysis

Fluorescence microscopy was performed using a Zeiss Axioplan2 microscope equipped with an ApoTome module (Carl Zeiss AG, Oberkochen, BW, Germany), 10× (NA 0.3), 20× (NA 0.8) and 63× (oil, NA 1.4) objectives along with a Zeiss AxioCam MRm camera. For Sholl analysis [46], single feGFP transfected neurons were imaged using a 10× objective. A healthy neuron was identified by a well-defined cell body and absence of irregular membranous protrusions around the soma. Dentate gyrus cells, readily identifiable via their smaller cell body and thicker, shorter dendrites were excluded from the analysis. The neuronal images were subsequently imported into the Neurolucida 6 software (MicroBrightField, Williston, VT, USA), the dendritic trees were traced and thereafter their complexity was analyzed using Neurolucida explorer by quantifying the number of intersecting dendrites at every 10 μm incremental step, starting at the cell body. The total dendritic complexity, as the sum of all dendritic intersections and total dendritic length were also calculated for each cell. Spatially isolated and non-overlapping secondary dendrites were selected for spine density analysis and classification. These were imaged using a 63× (oil, NA 1.4) objective and z-sectioned at 0.5 μm increments using the ApoTome module for structured illumination. Dendritic spine density was quantified by dividing the number of dendritic spines counted with ImageJ (NIH, Bethesda, MD, USA) per the length in μm of the analyzed dendritic segment. Dendritic spines were classified as previously described [24] based on their length (from spine base at dendrite to spine head tip: $Spine_{len}$), head width ($Head_{max}$) and neck width ($Neck_{min}$). These factors were measured manually for individual spines using ImageJ and the spines were sorted into different subtypes based on the following criteria:

$$Spine_{len} \leq 1\mu m \;\&\; Head_{max}/Neck_{min} \leq 1\ \mu m: \textit{Stubby}$$

$$Spine_{len} > 1\mu m \;\&\; Head_{max}/Neck_{min} < 1.5\ \mu m: \textit{Thin}$$

$$Spine_{len} \geq 3\mu m \;\&\; Head_{max}/Neck_{min} \leq 1\ \mu m: \textit{Filopodia}$$

$$Head_{max}/Neck_{min} \geq 1.5\ \mu m: \textit{Mushroom}$$

In some experimental sets, additional immunolabeling for SynapsinI/II was performed. The proportion of the co-localization between immunoreactive SynapsinI/II positive puncta and feGFP-labelled dendritic spines was analyzed using ImageJ (NIH). A feGFP- labelled spine (green) was considered co-localized with SynapsinI/II puncta (red), when a resulting yellow punctum was observed within the same focal plane or two planes above and below. Furthermore, quantification of c-Fos (Cy3) and pERK (Cy3) immunopositivity was done over MAP2+ (Cy2) neurons: neurons were imaged with 10× (NA 0.3) objective under different fluorescent channels. In DIV7, low density cultures, isolated cells were identified based on their MAP2 staining to avoid cells with overlapping neurites. With 20× (NA 0.8) objective, neurons were imaged and analyzed for neurite morphology using Sholl analysis. The analysis was always done by an experimenter blinded for the different genotypes and treatments.

4.7. Data Representation and Statistics

Each treatment was performed in duplicate or triplicate for each experimental set and data from at least three sets of independent experiments were grouped together in Microsoft Excel (Microsoft Corporation, Redmond,WA, USA). For plotting the graphs and performing the appropriate statistical tests, the data sets were imported into Prism 6 (GraphPad Software Inc, San Diego, CA, USA). Unless otherwise mentioned, all data in the graphs are represented as mean +SEM. For comparing only two treatment groups a two-tailed unpaired Student's *t*-test was used. The comparison of more than two experimental groups was done by using a one-way ANOVA followed by a Bonferroni's multiple comparisons post-hoc test. For Sholl analysis measurements, a regular two-way ANOVA was performed followed by a Bonferroni's multiple comparisons post-hoc test. Significance was considered for p value < 0.05.

Author Contributions: Conceptualization, A.P., N.L., M.Z. and M.K.; Methodology, A.F., N.L., A.P., E.S. and M.Z.; Validation, A.F., A.P., E.S. and M.Z.; Formal Analysis, A.P. and M.Z.; Investigation, A.F., A.P., E.S., M.Z.; Resources, N.L., M.K. and M.Z.; Data Curation, A.P. and M.Z.; Writing—Original Draft Preparation, A.P. and M.Z.; Writing—Review & Editing, N.L., M.K., A.P. and M.Z.; Visualization, A.P. and M.Z.; Supervision, N.L., M.K. and M.Z.; Project Administration, N.L., M.K.; Funding Acquisition, N.L., M.K. and M.Z. All authors have read and agreed to the published version of the manuscript.

Funding: This work was supported by DFG Grant (KO 1674/5-1) to M.K. and (ZA 554/5-3) M.Z. and by the German Academic Exchange Service - DAAD (Research Grants for Doctoral Programmes 57299294) to A.P. We are indebted to the Italian Association of parents proRETT research and to "L'albero di Greta" for their support to N.L.

Acknowledgments: We thank Diane Mundil and Heike Kessler for their excellent technical assistance and Maria Fazzari for her contribution.

Conflicts of Interest: The authors declare no conflict of interest. The funders had no role in the design of the study; in the collection, analyses, or interpretation of data; in the writing of the manuscript, or in the decision to publish the results.

References

1. Barde, Y.A.; Edgar, D.; Thoenen, H. Purification of a new neurotrophic factor from mammalian brain. *EMBO J.* **1982**, *1*, 549–553. [PubMed]
2. Cohen-Cory, S.; Kidane, A.H.; Shirkey, N.J.; Marshak, S. Brain-derived neurotrophic factor and the development of structural neuronal connectivity. *Dev. Neurobiol.* **2010**, *70*, 271–288. [CrossRef] [PubMed]
3. Zagrebelsky, M.; Korte, M. Form follows function: BDNF and its involvement in sculpting the function and structure of synapses. *Neuropharmacology* **2014**, *76 Pt C*, 628–638. [CrossRef]
4. Poo, M.M. Neurotrophins as synaptic modulators. *Nat. Rev. Neurosci.* **2001**, *2*, 24–32. [CrossRef] [PubMed]
5. Lu, Y.; Christian, K.; Lu, B. BDNF: A key regulator for protein synthesis-dependent LTP and long-term memory? *Neurobiol. Learn. Mem.* **2008**, *89*, 312–323. [CrossRef] [PubMed]
6. Miranda, M.; Morici, J.F.; Zanoni, M.B.; Bekinschtein, P. Brain-Derived Neurotrophic Factor: A Key Molecule for Memory in the Healthy and the Pathological Brain. *Front. Cell. Neurosci.* **2019**, *13*, 363. [CrossRef]
7. Nagahara, A.H.; Tuszynski, M.H. Potential therapeutic uses of BDNF in neurological and psychiatric disorders. *Nat. Rev. Drug Discov.* **2011**, *10*, 209. [CrossRef]
8. Lima Giacobbo, B.; Doorduin, J.; Klein, H.C.; Dierckx, R.; Bromberg, E.; de Vries, E.F.J. Brain-Derived Neurotrophic Factor in Brain Disorders: Focus on Neuroinflammation. *Mol. Neurobiol.* **2019**, *56*, 3295–3312. [CrossRef]
9. Aktas, O.; Kury, P.; Kieseier, B.; Hartung, H.P. Fingolimod is a potential novel therapy for multiple sclerosis. *Nat. Rev. Neurol.* **2010**, *6*, 373–382. [CrossRef]

10. Healy, L.M.; Antel, J.P. Sphingosine-1-Phosphate Receptors in the Central Nervous and Immune Systems. *Curr. Drug Targets* **2016**, *17*, 1841–1850. [CrossRef]
11. Deogracias, R.; Yazdani, M.; Dekkers, M.P.; Guy, J.; Ionescu, M.C.; Vogt, K.E.; Barde, Y.A. Fingolimod, a sphingosine-1 phosphate receptor modulator, increases BDNF levels and improves symptoms of a mouse model of Rett syndrome. *Proc. Natl. Acad. Sci. USA* **2012**, *109*, 14230–14235. [CrossRef] [PubMed]
12. Fukumoto, K.; Mizoguchi, H.; Takeuchi, H.; Horiuchi, H.; Kawanokuchi, J.; Jin, S.; Mizuno, T.; Suzumura, A. Fingolimod increases brain-derived neurotrophic factor levels and ameliorates amyloid beta-induced memory impairment. *Behav. Brain Res.* **2014**, *268*, 88–93. [CrossRef] [PubMed]
13. Miguez, A.; Garcia-Diaz Barriga, G.; Brito, V.; Straccia, M.; Giralt, A.; Gines, S.; Canals, J.M.; Alberch, J. Fingolimod (FTY720) enhances hippocampal synaptic plasticity and memory in Huntington's disease by preventing p75NTR up-regulation and astrocyte-mediated inflammation. *Hum. Mol. Genet.* **2015**, *24*, 4958–4970. [CrossRef] [PubMed]
14. Hunter, S.F.; Bowen, J.D.; Reder, A.T. The Direct Effects of Fingolimod in the Central Nervous System: Implications for Relapsing Multiple Sclerosis. *CNS Drugs* **2016**, *30*, 135–147. [CrossRef] [PubMed]
15. Smith, P.A.; Schmid, C.; Zurbruegg, S.; Jivkov, M.; Doelemeyer, A.; Theil, D.; Dubost, V.; Beckmann, N. Fingolimod inhibits brain atrophy and promotes brain-derived neurotrophic factor in an animal model of multiple sclerosis. *J. Neuroimmunol.* **2018**, *318*, 103–113. [CrossRef]
16. Kishi, N.; Macklis, J.D. MeCP2 functions largely cell-autonomously, but also non-cell-autonomously, in neuronal maturation and dendritic arborization of cortical pyramidal neurons. *Exp. Neurol.* **2010**, *222*, 51–58. [CrossRef]
17. Baj, G.; Patrizio, A.; Montalbano, A.; Sciancalepore, M.; Tongiorgi, E. Developmental and maintenance defects in Rett syndrome neurons identified by a new mouse staging system in vitro. *Front. Cell. Neurosci.* **2014**, *8*, 18. [CrossRef]
18. Chang, Q.; Khare, G.; Dani, V.; Nelson, S.; Jaenisch, R. The disease progression of Mecp2 mutant mice is affected by the level of BDNF expression. *Neuron* **2006**, *49*, 341–348. [CrossRef]
19. Sampathkumar, C.; Wu, Y.J.; Vadhvani, M.; Trimbuch, T.; Eickholt, B.; Rosenmund, C. Loss of MeCP2 disrupts cell autonomous and autocrine BDNF signaling in mouse glutamatergic neurons. *eLife* **2016**, 5. [CrossRef]
20. Li, W.; Bellot-Saez, A.; Phillips, M.L.; Yang, T.; Longo, F.M.; Pozzo-Miller, L. A small-molecule TrkB ligand restores hippocampal synaptic plasticity and object location memory in Rett syndrome mice. *Dis. Models Mech.* **2017**, *10*, 837–845. [CrossRef]
21. Amendola, E.; Zhan, Y.; Mattucci, C.; Castroflorio, E.; Calcagno, E.; Fuchs, C.; Lonetti, G.; Silingardi, D.; Vyssotski, A.L.; Farley, D.; et al. Mapping pathological phenotypes in a mouse model of CDKL5 disorder. *PLoS ONE* **2014**, *9*, e91613. [CrossRef]
22. Okuda, K.; Takao, K.; Watanabe, A.; Miyakawa, T.; Mizuguchi, M.; Tanaka, T. Comprehensive behavioral analysis of the Cdkl5 knockout mice revealed significant enhancement in anxiety- and fear-related behaviors and impairment in both acquisition and long-term retention of spatial reference memory. *PLoS ONE* **2018**, *13*, e0196587. [CrossRef]
23. Ren, E.; Roncace, V.; Trazzi, S.; Fuchs, C.; Medici, G.; Gennaccaro, L.; Loi, M.; Galvani, G.; Ye, K.; Rimondini, R.; et al. Functional and Structural Impairments in the Perirhinal Cortex of a Mouse Model of CDKL5 Deficiency Disorder Are Rescued by a TrkB Agonist. *Front. Cell. Neurosci.* **2019**, *13*, 169. [CrossRef] [PubMed]
24. Zagrebelsky, M.; Holz, A.; Dechant, G.; Barde, Y.A.; Bonhoeffer, T.; Korte, M. The p75 neurotrophin receptor negatively modulates dendrite complexity and spine density in hippocampal neurons. *J. Neurosci.* **2005**, *25*, 9989–9999. [PubMed]
25. Anastasiadou, S.; Knoll, B. The multiple sclerosis drug fingolimod (FTY720) stimulates neuronal gene expression, axonal growth and regeneration. *Exp. Neurol.* **2016**, *279*, 243–260. [CrossRef] [PubMed]
26. Vidal-Martinez, G.; Vargas-Medrano, J.; Gil-Tommee, C.; Medina, D.; Garza, N.T.; Yang, B.; Segura-Ulate, I.; Dominguez, S.J.; Perez, R.G. FTY720/Fingolimod Reduces Synucleinopathy and Improves Gut Motility in A53T Mice: CONTRIBUTIONS OF PRO-BRAIN-DERIVED NEUROTROPHIC FACTOR (PRO-BDNF) AND MATURE BDNF. *J. Biol. Chem.* **2016**, *291*, 20811–20821. [CrossRef]
27. Zhang, Y.; Li, X.; Ciric, B.; Ma, C.G.; Gran, B.; Rostami, A.; Zhang, G.X. Effect of Fingolimod on Neural Stem Cells: A Novel Mechanism and Broadened Application for Neural Repair. *Mol. Ther. J. Am. Soc. Gene Ther.* **2017**, *25*, 401–415. [CrossRef]

28. Carreras, I.; Aytan, N.; Choi, J.K.; Tognoni, C.M.; Kowall, N.W.; Jenkins, B.G.; Dedeoglu, A. Dual dose-dependent effects of fingolimod in a mouse model of Alzheimer's disease. *Sci. Rep.* **2019**, *9*, 10972. [CrossRef]
29. Horch, H.W.; Kruttgen, A.; Portbury, S.D.; Katz, L.C. Destabilization of cortical dendrites and spines by BDNF. *Neuron* **1999**, *23*, 353–364.
30. Kwon, M.; Fernandez, J.R.; Zegarek, G.F.; Lo, S.B.; Firestein, B.L. BDNF-promoted increases in proximal dendrites occur via CREB-dependent transcriptional regulation of cypin. *J. Neurosci.* **2011**, *31*, 9735–9745. [CrossRef]
31. Tyler, W.J.; Pozzo-Miller, L.D. BDNF enhances quantal neurotransmitter release and increases the number of docked vesicles at the active zones of hippocampal excitatory synapses. *J. Neurosci.* **2001**, *21*, 4249–4258. [PubMed]
32. Ji, Y.; Pang, P.T.; Feng, L.; Lu, B. Cyclic AMP controls BDNF-induced TrkB phosphorylation and dendritic spine formation in mature hippocampal neurons. *Nat. Neurosci.* **2005**, *8*, 164–172. [CrossRef] [PubMed]
33. Kellner, Y.; Godecke, N.; Dierkes, T.; Thieme, N.; Zagrebelsky, M.; Korte, M. The BDNF effects on dendritic spines of mature hippocampal neurons depend on neuronal activity. *Front. Synaptic Neurosci.* **2014**, *6*, 5. [CrossRef] [PubMed]
34. Kaczmarek, L. Expression of c-fos and other genes encoding transcription factors in long-term potentiation. *Behav. Neural Biol.* **1992**, *57*, 263–266. [CrossRef] [PubMed]
35. Gaiddon, C.; Loeffler, J.P.; Larmet, Y. Brain-derived neurotrophic factor stimulates AP-1 and cyclic AMP-responsive element dependent transcriptional activity in central nervous system neurons. *J. Neurochem.* **1996**, *66*, 2279–2286. [CrossRef] [PubMed]
36. Kishi, N.; Macklis, J.D. MECP2 is progressively expressed in post-migratory neurons and is involved in neuronal maturation rather than cell fate decisions. *Mol. Cell Neurosci.* **2004**, *27*, 306–321. [CrossRef]
37. Fuchs, C.; Trazzi, S.; Torricella, R.; Viggiano, R.; De Franceschi, M.; Amendola, E.; Gross, C.; Calza, L.; Bartesaghi, R.; Ciani, E. Loss of CDKL5 impairs survival and dendritic growth of newborn neurons by altering AKT/GSK-3beta signaling. *Neurobiol. Dis.* **2014**, *70*, 53–68. [CrossRef]
38. Fuchs, C.; Gennaccaro, L.; Trazzi, S.; Bastianini, S.; Bettini, S.; Lo Martire, V.; Ren, E.; Medici, G.; Zoccoli, G.; Rimondini, R.; et al. Heterozygous CDKL5 Knockout Female Mice Are a Valuable Animal Model for CDKL5 Disorder. *Neural Plast.* **2018**, *2018*, 9726950. [CrossRef]
39. Horch, H.W.; Katz, L.C. BDNF release from single cells elicits local dendritic growth in nearby neurons. *Nat. Neurosci.* **2002**, *5*, 1177–1184. [CrossRef]
40. Sheng, M.; Greenberg, M.E. The regulation and function of c-fos and other immediate early genes in the nervous system. *Neuron* **1990**, *4*, 477–485. [CrossRef]
41. Tuvikene, J.; Pruunsild, P.; Orav, E.; Esvald, E.E.; Timmusk, T. AP-1 Transcription Factors Mediate BDNF-Positive Feedback Loop in Cortical Neurons. *J. Neurosci.* **2016**, *36*, 1290–1305. [CrossRef] [PubMed]
42. Armstrong, D.; Dunn, J.K.; Antalffy, B.; Trivedi, R. Selective dendritic alterations in the cortex of Rett syndrome. *J. Neuropathol. Exp. Neurol.* **1995**, *54*, 195–201. [CrossRef] [PubMed]
43. Chen, W.G.; Chang, Q.; Lin, Y.; Meissner, A.; West, A.E.; Griffith, E.C.; Jaenisch, R.; Greenberg, M.E. Derepression of BDNF transcription involves calcium-dependent phosphorylation of MeCP2. *Science* **2003**, *302*, 885–889. [CrossRef] [PubMed]
44. Chen, Q.; Zhu, Y.C.; Yu, J.; Miao, S.; Zheng, J.; Xu, L.; Zhou, Y.; Li, D.; Zhang, C.; Tao, J.; et al. CDKL5, a protein associated with rett syndrome, regulates neuronal morphogenesis via Rac1 signaling. *J. Neurosci.* **2010**, *30*, 12777–12786. [CrossRef]
45. Bedogni, F.; Cobolli Gigli, C.; Pozzi, D.; Rossi, R.L.; Scaramuzza, L.; Rossetti, G.; Pagani, M.; Kilstrup-Nielsen, C.; Matteoli, M.; Landsberger, N. Defects During Mecp2 Null Embryonic Cortex Development Precede the Onset of Overt Neurological Symptoms. *Cereb. Cortex* **2016**, *26*, 2517–2529. [CrossRef]
46. Sholl, D.A. Dendritic organization in the neurons of the visual and motor cortices of the cat. *J. Anat.* **1953**, *87*, 387–406.

International Journal of
Molecular Sciences

Article

Acute and Chronic Dopaminergic Depletion Differently Affect Motor Thalamic Function

Giuseppe Di Giovanni [1,2,*], Laura Clara Grandi [3], Ernesto Fedele [4,5], Gergely Orban [3], Agnese Salvadè [3], Wei Song [3], Eleonora Cuboni [1], Alessandro Stefani [6], Alain Kaelin-Lang [3,7,8] and Salvatore Galati [3,8,9,*]

1 Laboratory of Neurophysiology, Department of Physiology and Biochemistry, Faculty of Medicine and Surgery, University of Malta, Msida MSD 2080, Malta; eleonoracuboni@virgilio.it
2 Neuroscience Division, School of Biosciences, Cardiff University, Cardiff CF10 3AX, UK
3 Laboratory for Biomedical Neurosciences, Neurocenter of Southern Switzerland, 6900 Taverne, Switzerland; lauraclaragrandi27@gmail.com (L.C.G.); orbangeri@gmail.com (G.O.); agnese.salvade@eoc.ch (A.S.); song.wei@eoc.ch (W.S.); alain.kaelin@eoc.ch (A.K.-L.)
4 Section of Pharmacology and Toxicology, Department of Pharmacy, Center of Excellence for Biomedical Research, University of Genoa, 16148 Genoa, Italy; fedele@difar.unige.it
5 IRCCS Ospedale Policlinico San Martino, 16132 Genoa, Italy
6 Department of system medicine, Faculty of Medicine and Surgery, University of Rome "Tor Vergata", 00133 Rome, Italy; Stefani@uniroma2.it
7 Medical Faculty, University of Bern, 3008 Bern, Switzerland
8 Faculty of Biomedical Sciences, Università della Svizzera Italiana, 6900 Lugano, Switzerland
9 Center for Movement Disorders, Neurocenter of Southern Switzerland, 6900 Lugano, Switzerland
* Correspondence: giuseppe.digiovanni@um.edu.mt (G.D.G.); salvatore.galati@eoc.ch (S.G)

Received: 9 March 2020; Accepted: 12 April 2020; Published: 15 April 2020

Abstract: The motor thalamus (MTh) plays a crucial role in the basal ganglia (BG)-cortical loop in motor information codification. Despite this, there is limited evidence of MTh functionality in normal and Parkinsonian conditions. To shed light on the functional properties of the MTh, we examined the effects of acute and chronic dopamine (DA) depletion on the neuronal firing of MTh neurons, cortical/MTh interplay and MTh extracellular concentrations of glutamate (GLU) and gamma-aminobutyric acid (GABA) in two states of DA depletion: acute depletion induced by the tetrodotoxin (TTX) and chronic denervation obtained by 6-hydroxydopamine (6-OHDA), both infused into the medial forebrain bundle (MFB) in anesthetized rats. The acute TTX DA depletion caused a clear-cut reduction in MTh neuronal activity without changes in burst content, whereas the chronic 6-OHDA depletion did not modify the firing rate but increased the burst firing. The phase correlation analysis underscored that the 6-OHDA chronic DA depletion affected the MTh-cortical activity coupling compared to the acute TTX-induced DA depletion state. The TTX acute DA depletion caused a clear-cut increase of the MTh GABA concentration and no change of GLU levels. On the other hand, the 6-OHDA-induced chronic DA depletion led to a significant reduction of local GABA and an increase of GLU levels in the MTh. These data show that MTh is affected by DA depletion and support the hypothesis that a rebalancing of MTh in the chronic condition counterbalances the profound alteration arising after acute DA depletion state.

Keywords: electrophysiology; microdialysis; immunohistochemistry; L-DOPA; deep brain stimulation

1. Introduction

The motor thalamus (MTh) plays a pivotal role in motor information processing in both physiological and pathological conditions, being directly connected to both the basal ganglia (BG) and the cortex [1]. In particular, the MTh receives the BG input at the level of the ventral-anterior (VA) and the ventral-medial (VM) nuclei [2]. BG are assumed to inhibit the thalamus under basal conditions because substantia nigra pars reticulata (SNr) and internal globus pallidus (GPi) display high spontaneous spiking rates, releasing GABA in the MTh. However, when the BG network is activated by cortical input, BG output is transiently suppressed and downstream targets, including the MTh, are disinhibited [3,4]. Another source of GABA in the MTh comes from the nucleus reticular thalami (NRT) [5], a thin layer of GABAergic cells adjacent to the relay nuclei of the dorsal thalamus, that we have shown to be able to modulate the oscillatory activity of MTh in dopamine (DA)-depleted animals [6]. On the other hand, the excitatory input to the MTh is classically provided by the cerebellothalamic [7] and corticothalamic synapses [8,9]. Moreover, it has been recently shown that some substantia nigra (SN) neurons project to the NRT and the posterior thalamus, and co-release glutamate (GLU) and DA, while the VA and VM would receive preferential GABAergic SN fibers [10]. GABA, GLU and DA, therefore, may interact when modulating synapse and excitability within the MTh in normal and pathological conditions [11–14]. Although it is well established that the appearance of the cardinal features of Parkinson's disease (PD) is due to the degeneration of DA neurons, with the contextual dysfunction of different neurotransmitter systems and cortical and BG activity [15–17], the evidence of MTh alterations, both at the circuit and cellular level, is scanty and controversial [1,6]. According to the canonical BG schema [18,19], an increase in MTh GABA levels could be the consequence of the DA loss. This agrees with clinical evidence showing that clinical effective sub-thalamic nucleus (STN) deep brain stimulation (DBS) or even levodopa (L-DOPA) in PD patients reversibly reduced the extracellular MTh GABA [20,21]. Nevertheless, a decrease of GABA has been observed in the MTh of 6-OHDA-lesioned rats [22] and certain regions of the thalamus of post-mortem PD brains [23]. Therefore, these discrepancies cannot be only due to the difference in BG circuitry in rats compared to humans.

No changes in GLU concentrations have been described in the thalamuses of PD patients [23], or the striatum or SNr of 6-OHDA-lesioned rats [24]; there was a GLU increase in the entopeduncular nucleus (EPN) [25], in the SNr and the striatum in L-DOPA-induced dyskinesia (LID) 6-OHDA rat [24]. Indeed, AMPA, NMDA and metabotropic GLU receptor antagonists might be a useful target for PD, especially for LID (Cenci et al., 2014).

Finally, the role of the MTh in PD is questioned by the fact that a stereotactic MTh lesion neither worsens Parkinsonian bradykinesia in PD nor regularly causes bradykinesia in patients with essential tremor [26], clearly in contrast to the canonical *box and arrow* schema of BG [18,19].

Thus, the present study was aimed at elucidating the effects of acute and chronic DA depletion, induced respectively by TTX and 6-OHDA [27], on MTh functionality. We, therefore, investigated the electrophysiological changes in the MTh at the level of single neurons from acute and chronic DA-depleted rats. Our main hypothesis was that the MTh is altered differently during rapid changes of DA levels compared to chronic depleted states. This might account for the controversial findings on MTh. Since the BG activity changes in relation to the cortical pattern [28,29], all the recordings were performed during stable cortical slow-wave activity (SWA) induced by urethane anesthesia. This stable cortical pattern allowed us to investigate the coupling of SWA activity with the MTh neuronal activity in acute and chronic DA depletion states. In addition, we measured the extracellular levels of GABA and GLU by microdialysis in the same experimental conditions.

2. Results

2.1. Effects of DA Depletion on the Activity of the Motor Thalamus Neurons

We recorded a total of 82 neurons from the MTh (n = 23 control, n = 25 in 6-OHDA and n = 34 in TTX, Figure 1). We found that the mean firing rate of the recorded MTh neurons was lower after TTX infusion (TTX: 3.69 ± 0.18 spike/s, $p < 0.05$) compared to control group (CTL: 7.12 ± 0.15 spikes/s) and 6-OHDA-lesioned rats (6-OHDA: 8.02 ± 0.29 spikes/s; $p < 0.05$) (Figure 2a). The percentage of spikes in burst in chronic 6-OHDA DA-depleted rats (6-OHDA 76.72 ± 4.18%) was higher than in control

(CTL 59.89 ± 4.22%, $p < 0.05$) and TTX groups (60.53 ± 6.30%, $p < 0.01$) (Figure 2b). Burst length and interspike of burst (Figure 2c,d) were significantly higher in chronic than in acute TTX-mediated DA depletion but were not different from controls (length of burst: 6-OHDA 34.03 ± 3.01 ms, TTX 22.84 ± 2.61 ms, $p < 0.05$; burst interspike: 6-OHDA 15.33 ± 1.65 ms, TTX 10.5 ± 1.29 ms).

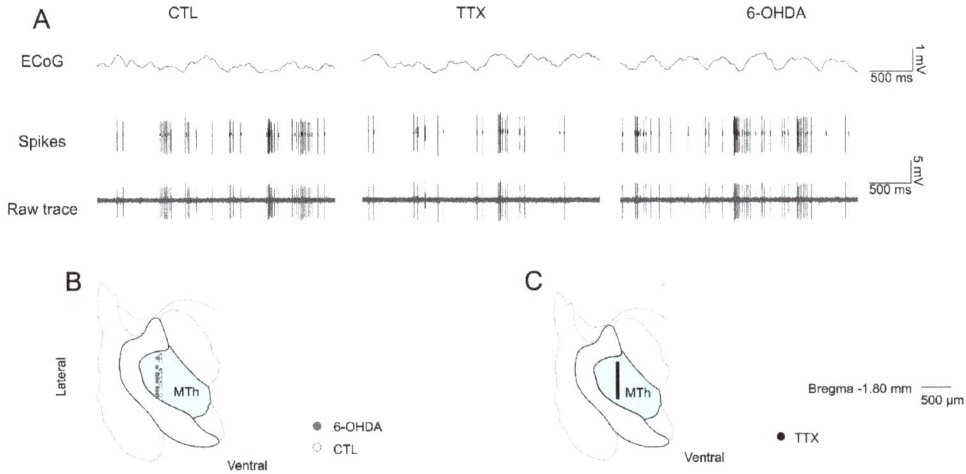

Figure 1. (**A**) Representative electrophysiological recordings of the three groups of animals. From the Table 6. hydroxydopamine (6-OHDA)-lesioned rats, respectively. (**C**) Histological reconstruction of the MTh recording site in tetrodotoxin (TTX)-treated rats (in blue). The black circles represent the recorded neurons in MTh during the TTX-mediated blockade of MFB.

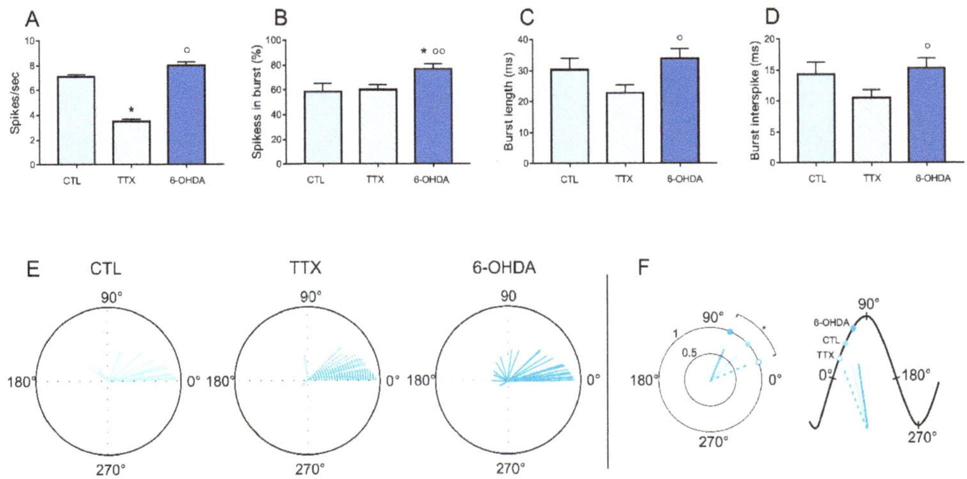

Figure 2. (**A–D**) Firing rate and pattern of the MTh in CTL, during tetrodotoxin (TTX) infusion and in 6-hydroxydopamine (6-OHDA)-lesioned rats. * $p < 0.05$. (**E**,**F**). Phase correlation analysis. (**E**) MTh circular plots of the phase-locked firing of the MTh neurons in control, acute and chronic DA depletion states. The vectors of the preferred firing of individual neurons are shown as lines radiating from the center. The last smaller circular plots show the mean vector for the preferred phases of neurons in each group. (**F**) The histograms on the right show the vector lengths and phase angle comparisons between MTh (blue) in control (dark color), acute (dashed color) and chronic (light color) DA depletion states. On the left, the representation of hypothetical cortical slow oscillation comparing the phase angle and

vector length between MTh (blue) control (dark color), TTX (dashed color) and 6-OHDA (light color) rats. * $p < 0.05$, Kruskal–Wallis test. MTh: motor thalamus; CTL: control; TTX: tetrodotoxin; 6-OHDA: 6-hydroxydopamine. The 0° represents the mid-ascending phase of cortical slow-wave activity (SWA), the 90° represents the cortical SWA peak and the 180° represents the mid-descending phase of cortical SWA.

2.2. Effects of DA Depletion on the Cortical–Thalamic Coupling

We compared the preferred phase angles and the vector lengths in control, acute and chronic DA depletion states. In control rats, the MTh neurons fired preferentially at 42.77 ± 12.36° ($n = 21$, Figure 2e), with a vector length of 0.73 ± 0.06. In acute DA-depleted animals ($n = 34$), the phase was 22.54 ± 6.96° with a vector length of 0.83 ± 0.03, while in the chronic DA depletion state ($n = 22$), it was 79.89 ± 15.98° with a vector length of 0.56 ± 0.07. The vector length, i.e., the strength of corticothalamic coupling, was significantly lower in chronic DA depletion state than in control and in acute DA depletion state ($p < 0.05$). The phase angle was significantly lower in control and acute DA depletion state than that in chronic DA depletion state ($p < 0.05$) (Figure 2e).

2.3. Effects of DA Depletion on the Extracellular Levels of GABA and GLU in the Motor Thalamus

ANOVA analysis followed by Dunnet's multiple comparison test showed a significant effect of TTX infused in the medial forebrain bundle induced an increase in the endogenous GABA extracellular level (215 ± 67.5; $n = 14$; $p < 0.05$) compared to the control (100 ± 7.90, $n = 14$) and post-infusion period (159.2 ± 50.2, $n = 14$) (Figure 3a). No changes in the GLU levels were observed in the MTh after the infusion of TTX (PRE: 100 ± 2.69; TTX: 122.7 ± 24.8, $n = 7$; POST: 123.7 ± 53.9, $n = 14$) (Figure 3b). The extracellular GABA levels in the MTh were significantly lower in 6-OHDA animals than in sham-lesioned animals (100 ± 11.09 vs. 43.54 ± 6.65, $n = 9$; ** $p < 0.01$) (Figure 3c). GLU levels were significantly augmented in the MTh of 6-OHDA rats when compared to those measured in sham-lesioned rats by unpaired student t-test (100 ± 13.75 vs. 276.69 ± 43.52, $n = 9$; ** $p < 0.01$) (Figure 3d) levels.

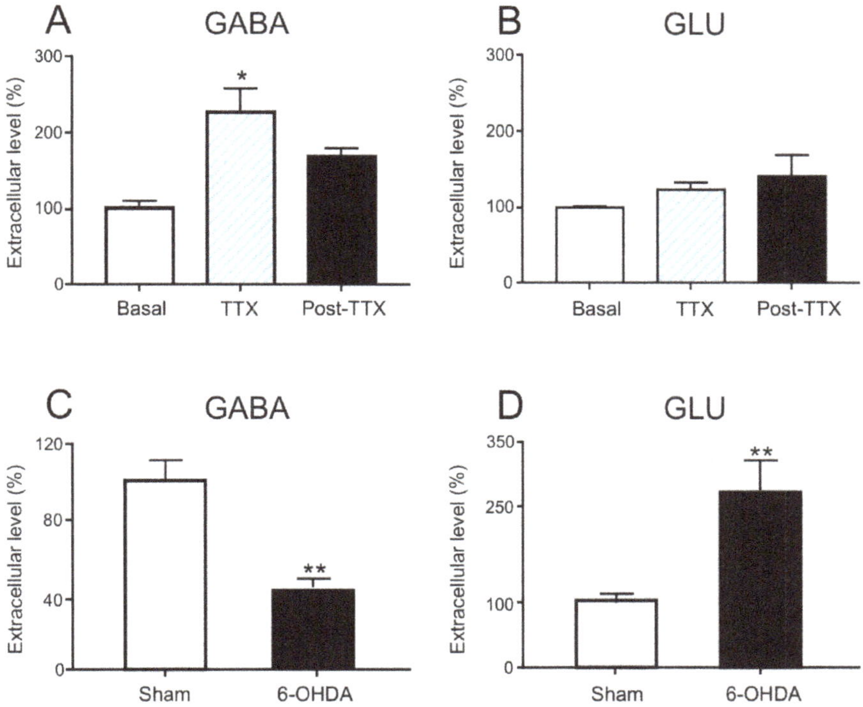

Figure 3. Effect of acute (TTX) and chronic (6-OHDA) dopamine (DA)-depletion on the extracellular levels of endogenous GABA and glutamate (GLU) in the motor thalamus (MTh). TTX infused in the medial forebrain bundle induced an increase in the endogenous GABA extracellular levels (**A**) while did not change the GLU levels in the MTh. Data in (**A**) and (**B**) are expressed as mean ± SEM normalized to the baseline levels (pre TTX infusion) from n = 14 experiments per group; * $p < 0.05$. One-and two-way ANOVA for repeated measures, followed by Dunnet's multiple comparison test. 6-OHDA-lesion infused in the medial forebrain bundle produced, after 2 weeks, a decrease in GABA (**C**) and an increase in GLU extracellular levels in the MTh (**D**). Data in (**C**) and (**D**) are expressed as means ± SEMs normalized to sham-lesioned (n = 9) neurotransmitter levels from n = 9 experiments per group; ** $p < 0.01$, unpaired student t-test.

3. Discussion

We demonstrated that MTh is deeply affected by both acute and chronic DA depletion both in terms of electrophysiological and neurochemical alterations. The main findings of the present study are threefold. Firstly, the results emphasize the different effects of DA deafferentation on MTh, with a decrease in firing rate in TTX-treated animals and an increase of burst firing in 6-OHDA-lesioned rats. Secondly, the results show that DA depletions change, in an opposite fashion, MTh-cortical coupling with the MTh neurons fired preferentially at the beginning of the UP-state in TTX-treated rats compared to the 6-OHDA-lesioned rats, which showed a shift towards the peak of the SWA. Finally, the neurochemical results demonstrated that the extracellular levels of GABA increase and those of GLU do not change in the MTh of TTX-treated animals; in the 6-OHDA-lesioned rats, GABA levels decreased and those of GLU increased.

Despite representing a crucial point in the cortico-BG circuit, only a few studies have investigated the role of the MTh and its properties in PD, and controversial data are present in the literature [1,6]. Furthermore, controversial data are present in the literature which regard MTh electrophysiology and properties in the Parkinsonian state. While some studies show a slight MTh firing rate reduction in awake Parkinsonian monkeys [30–34], others revealed increased [1] or unchanged firing rates [35]. We recently showed that all the EEG frequency bands are altered in TTX and 6-OHDA-depleted states; MTh oscillation changes occur preferentially in acute DA depletion, but not in the chronic state [6].

In the present study, we demonstrated that the TTX-inhibition of the medial forebrain bundle (MFB) (for technical details, see our review [27]) determined an acute increase in GABA content with the reduction of the MTh activity. In rats, the SNr activity is reduced after acute DA depletion caused by TTX [36], in accordance with electrophysiological and biochemical data obtained in PD patients [37]. Thus, the changes in MTh activity have to be due to other GABAergic afferents of the network. For instance, we occasionally recorded NRT neurons during TTX infusion and observed a consistent increment of their activity (unpublished observation). Considering that NRT is a powerful modulator of different high-order thalamic nuclei [38], it is likely that this nucleus tightly controls the MTh as well. Conversely, we observed that the MTh neuronal firing rate returns to control condition in the chronic depletion state induced by 6-OHDA. This could be due to the aberrant decrease in GABA and an increase in GLU levels in the MTh of 6-OHDA-lesioned rats that might compensate for the marked inhibition observed in the acute depletion state. The shift of the excitation/inhibition (E/I) balance toward a net excitation in the MTh of the 6-OHDA lesioned rats might indicate that the plastic changes observed in the striatum and cortex of the PD animal model [39] could also occur at the level of the thalamic area. Surprisingly, GABA levels decrease in the chronic DA-depleted MTh, a finding that cannot be explained by the canonical BG/thalamus/cortex loop [19]. Nevertheless, our data, although surprising, are in line with the findings of the only available in vivo study showing a decrease of GABA in the MTh of rats lesioned by striatal infusion of 6-OHDA [22]. Instead, we are the first to report a sustained GLU increase in the MTh of animal models of PD. The decrease in GABA levels is not trivial since the administration of the vasoactive intestinal peptide (VIP) was effective at reversing the motor deficits and contextually increased thalamic GABA contents [22]. Other areas may be involved in rebalancing the MTh activity in chronic DA depletion, such as hyperactivity of the cortex, likely leading the increased release of GLU that we observed within the MTh of 6-OHDA-lesioned rats. The increase of burst activity in the MTh that we recorded in 6-OHDA conditions might depend on the cortex, which, by acting as the main driver, produces a burst input, leading to a coincident burst of activity in MTh neurons, without necessarily causing LTS bursts [1]. The fact that we did not observe a strong effect on burst activity may depend on the canceling effects of different inputs to the MTh, such as synchronized bursts from the BG. The contextual increase in GLU may also cause the production of nitric oxide and contribute to PD development processes [15,40,41].

However, it is also necessary to underline that, since the lesion in our study was unilateral, we cannot exclude that the contralateral hemisphere could compensate, at least partially, for these changes [42]. Furthermore, a recent tractography study in healthy subjects showed that MTh may receive DAergic projections from the ventral tegmental area that are directed towards the supplementary motor area and primary motor cortex (44).

An interesting finding of this study is the interplay between motor thalamus and cortical activity. The DA depletion influenced mainly the MTh-cortical coupling. The divergence of MTh-SWA coupling in the chronic state of DA depletion further suggests that the MTh has undergone the effect of compensatory mechanisms, changing the cross-talk between MTh and the cortex. In accordance with the SWA pattern, the MTh neurons fired preferentially at the beginning of the UP-state in TTX-treated rats compared to the 6-OHDA-lesioned rats, which showed a shift towards the peak of the SWA. Thus, in 6-OHDA rats, there is a delay in the coupling between neuronal activity and the beginning of cortical SWA

We showed that, following the acute reduction of DA, the MTh immediately changes the activity, but the chronic lack of DA leads other structures to rebalance the MTh activity. It is possible to hypothesize that, inside the complex circuit involved in PD, rapid clinical transitions of the motor and non-motor symptoms could be directly related to rapid changes in MTh activity, as observed during

DBS surgery [20,21]. These hypotheses need further investigation of the MTh activity in PD animal models and PD patients.

Our study presents a number of limitations. For instance, we should keep in mind that our recordings and the position of the microdialysis probe were at ranges of depths in the MTh that correspond to both basal ganglia and cerebellar recipient zones. Therefore, our electrophysiological and neurochemical data may have included values obtained from the two different MTh territories. Moreover, we did not distinguish the two types of MTh neurons during cortical activation by performing the tail-pinch (Nakamura et al., Cerebral cortex, 2014). Further experiments will be required to identify the tonic and burst firing cells in MTh in response to cortical activation, and the eventual changes of their firing characteristics following DA depletion by TTX and 6-OHDA treatment. Another limitation of this study is the use of toxin-induced PD animal models and anesthetized preparations. Despite the usefulness of the animal models, they have limitations that need to be considered when interpreting our findings and translating them to humans [27]. Nevertheless, although rodent models cannot recapitulate all features of PD, they remain a crucial approximation in order to open new avenues.

In conclusion, our data show that in chronic conditions, the firing changes observed in the MTh are relatively mild, considering the profound changes observed in other BG nuclei. This could be due to remodeling and adaptation developed as a consequence of the DA depletion. Indeed, acute TTX induced a strong decrease in the firing rate of MTh neurons. On the other hand, the modification of GLU and GABA are highly significant in 6-OHDA DA-depleted rats and are likely a consequence of changes in the MTh inputs, difficult to explain with the current knowledge of PD pathology. The roles of the MTh in the mechanisms of information processing under physiological and pathological conditions need to be further investigated and clarified.

4. Materials and Methods

4.1. Animals

Most of the electrophysiological, neurochemical and histological procedures were carried out on adult male Sprague-Dawley rats in compliance with Swiss laws on animal experimentation, and all procedures were approved by the Animal Research Committee and the Veterinary Office of the Canton of Ticino, Switzerland (TI-08-2015).

A subset of neurochemical experiments was performed on male Sprague-Dawley rats at the Department of Physiology and Biochemistry at the University of Malta in accordance with the European Union (EU) Directive 2010/63/EU and the Institutional Animal Use and Care Committee (IAUCC) of the University of Malta.

Utmost care was taken to limit the number of rats used and their suffering.

4.2. Pharmacological Blockade of the Medial Forebrain Bundle

The pharmacological blockade of the MFB was performed on adult male Sprague–Dawley rats weighing 399 ± 68.4 g according to our previously described method [6,27]. Briefly, TTX was infused via reverse microdialysis using a probe with a 1 mm dialytic membrane (CMA/11 microdialysis probe, CMA Microdialysis, Stockholm, Sweden), which was positioned according to the following coordinates: 2.5 mm posterior to the bregma, 2 mm lateral to the midline and 8.6 mm below the cortical surface [43]. The probe was perfused with saline solution for 1 h and then with TTX (1 μM) using a syringe pump (CMA/400, CMA Microdialysis) at a flow rate of 1 μL/min. The MTh neuronal activity was recorded and analyzed for at least the next 10 min after the TTX infusion in 27 animals, while the GABA and GLU MTh levels were measured in 14 rats.

4.3. Unilateral 6-OHDA Lesion

We performed standard unilateral 6-OHDA denervation in the left hemisphere [6,27] to obtain chronic DA-depleted animals. The animals (399 ± 36.3 g) were anesthetized with 1.5%–2.5% isoflurane in oxygen and then injected with desipramine (25 mg/kg, i.p.) and pargyline (50 mg/kg, i.p.) to minimize 6-OHDA uptake by noradrenergic neurons and metabolism by MAO, respectively. Then, the animals were mounted on a stereotaxic instrument (Stoelting Co., Wheat Lane, Wood Dale, IL, USA) for injection of the neurotoxin (6-OHDA; Sigma; 8 μg/4 μL of saline solution containing 0.1% of ascorbic acid) in the medial forebrain bundle (MFB; stereotaxic coordinates: 2.5 mm posterior to the bregma, 2 mm lateral to the midline, and 8.6 mm below the cortical surface) [44]. The MTh neuronal activity was recorded an analyzed in 6-OHDA-lesioned rats (n = 8) and control animals (n = 6). The microdialysis experiments were performed in sham animals (n = 9), receiving instead, 4 μL of saline solution containing 0.1% of ascorbic acid infused into the MFB and 6-OHDA-lesioned animals (n = 9). Electrophysiological and microdialysis experiments were performed 21–29 days after the administration of 6-OHDA

4.4. Prerecording Surgery

Rats were anesthetized with urethane (1.4 g/kg, i.p.) (Sigma Chemical Co., St Louis, MO, USA) and mounted on a stereotaxic instrument (Stoelting Co.) according to our standard operating procedure [6,27]. Body temperature was maintained at 37–38 °C with a heating pad (Stoelting Co.). After a local anaesthetic injection (lidocaine 0.5%), a midline scalp incision was made, and the skull was almost completely drilled on the left side. The dura were then spread out to expose the cortical surface.

4.5. Electrophysiological Recordings

We recorded single neurons for the entire dorsoventral track in the same medio-lateral site in the MTh. In particular, we recorded a total of 82 neurons from the MTh (n = 23 control, n = 25 in 6-OHDA and n = 34 in TTX) from 5000 μm from the cortex to a maximum of 5770 μm.

We recorded multiple MTh neurons from TTX infused rats over the 10 min after infusion The MTh neurons (n = 34) recorded after the TTX-mediated blockade of the MFB were localized from 5250 μm to 5870 μm from the first neuronal detection of the cortex.

The MTh neurons recorded were localized at the following coordinates: 1.8 mm posterior and 2.2 medial to bregma. Single unit activity was off-line analyzed using Spike 2 software (Cambridge Electronic Design, Cambridge, UK) and principal component analysis for spike sorting.

The Electrocorticogram (ECoG) was recorded through a screw electrode (Dentorama, 8 mm of total length, 3 mm tip length) placed on the cortical surface above the frontal cortex (3.0 mm anterior of bregma and 2.0 mm lateral to the midline) and referenced against an indifferent screw electrode placed above the cerebellum. Raw ECoG was bandpass filtered (0.1–300 Hz) and amplified (×2000; Neurolog). The ECoG was on-line digitalized at a sample rate of 600 kHz through an analogical/digital interface (Micro1401 mk II, Cambridge Electronic Design, Cambridge, UK) and stored on a computer to identify off-line robust SWA, which will be the object of further analysis. Simultaneously, we collected the single-unit activity from the MTh. The acquisition of the recording was performed after the stabilization of neuronal activity. The recordings were performed using a tungsten electrode (WPI, TM33B01; impedance 0.9–1.2 MΩ). The trace containing the neuron action potentials was amplified (×10,000; Neurolog), bandpass filtered (300–1000 Hz) and sampled at 60 kHz for digital storage on a computer connected to the CED 1401 interface [36]. The neuronal activity was recorded for at least 10 min after the TTX infusion.

Figure 1A shows an example of the recordings with the EEG traces, i.e., SWA (top panel); the discriminated single neurons (middle panel); and the neuronal traces (bottom panel).

4.6. Microdialysis Analysis

An additional series of experiments were conducted on adult rats (312 ± 25 g) to measure GABA and GLU before, during and after the TTX-mediated blockade of MFB and in sham and 6-OHDA-lesioned rats [28]. Microdialysis was performed using concentric microdialysis probes (CMA11 with 1 mm PES membrane, outside diameter 0.6 mm, cut-off 35 kDa, Microbiotech) which were positioned into the right MTh (1.8 mm posterior to the bregma, 2 mm lateral to the midline and 6.2 mm below the cortical surface).

Probes were then connected to a high precision micro-injection pump (CMA/100, CMA 159 Microdialysis) and infused at a flow rate of 5 μL/min with an artificial cerebrospinal fluid containing 160 (in mM): NaCl 145, KCl 3, MgCl2 1, CaCl2 1.26, buffered at pH 7.4 with 2 mM phosphate buffer.

After stabilization (4 h), consecutive samples were collected every 10 min. In the acute depleted animals, TTX was infused in the MFB (see above) at the end of the 3rd sample. All samples were immediately frozen at −80 °C until analyzed. At the end of the experiment, rats were sacrificed, and the correct position of the probe was histologically verified (36).

For the experiments in 6-OHDA and sham-lesioned animals, the microdialysis experiments were performed 21–29 days after the lesion.

The concentrations of GABA and GLU in the samples were determined by HPLC analysis following precolumn derivatization with o-phthalaldehyde and resolution through a C18-reverse phase chromatographic column (Chrompack, Middleburg, The Netherlands) connected to a fluorometric detector (excitation 350 nm; emission 450 nm). Buffers were prepared as follows: solvent A, 0.1 M sodium acetate (pH 5.8): methanol, 80:20; solvent B, 0.1 M sodium acetate (pH 5.8): methanol, 20:80; solvent C, 0.1 M sodium acetate (pH 6.0): methanol, 80:20. Gradient programme: 100% C for 4 min from the initiation of the programme; 90% A and 10% B in 1 min; isocratic step for 2 min; 78% A and 22% B in 2 min; isocratic step 172 for 6.50 min; 66% A and 34% B in 1.10 min; isocratic step for 1.50 min; 42% A and 58% B in 1.10 min; isocratic step for 3.50 min; flow rate 0.9 mL/min (Waters 600MS gradient system). Homoserine was used as an internal standard [21].

4.7. Histological and Microscopic Analysis

At the end of the experiment, the positions of the electrodes (Figure 1b,c) were histologically checked, and the unilateral dopaminergic lesion in each PD rat was confirmed by quantification of tyrosine hydroxylase (TH) immunoreactivity, as described and analyzed in our previous publication [28].

4.8. Statistical Analysis

Concerning the single-unit data, statistical comparisons of firing rates and burst parameters (control vs. 6-OHDA and post-TTX) were conducted using the Kruskal–Wallis test with the Man–Whitney test for post hoc comparisons (asymptotic significance (2-tailed) p-value).

The firing rate comparison was performed taking into account the mean of the first 5 min of recording. The comparison of the burst activity was performed in selected 100-s increments in the presence of stable cortical urethane-induced SWA with a frequency below 3 Hz, as shown elsewhere (45.) SWA is not affected by DA denervation and DAergic treatment [45].

To investigate the correlation between cortex and MTh, we performed the phase correlation analysis, with the calculation of descriptive circular statistics being performed using the CircStat toolbox for MATLAB (2). For the burst analysis, the phase correlation analysis was performed in the same selected 100 s in the presence of stable cortical SWA. For circular analyses and statistical comparisons, we included all neurons from the previous analyses.

For each neuron, the linear phase histograms were analyzed, taking into account the up phase of the SWA, and the spike phase values were used for circular analysis. Then, each neuron was tested for significant SWA phase-locked firing (comparison of degree; Rayleigh's uniformity test). The selected neurons were then used for the analysis of the preferred phase angle, i.e., the mean resultant vector in degree (with 90° corresponding to the peak of the SWA phase), and the vector length (corresponding to the level of dispersion of the firing). The mean resultant vector length of the phase distribution, bound between zero and one, was used to quantify the level of phase-locking around the mean angle (the closer to one, the more grouped the firing; 2).

The differences between the mean of the resultant vector (degree) and the vector lengths of the three groups (Control, TTX and 6-OHDA) of MTh were tested using the Kruskal-Wallis test.

All circular analyses were performed using the degree and the upper peak in SWA oscillations corresponding to a 90° phase and the down peak to a 270° phase.

In all experimental conditions, all neurons were tested for significantly phase-locked firing; i.e., not randomly distributed in relation to the SWA (comparison of degree; Rayleigh's uniformity test) according to the method described by Nakamura and colleagues [2]. Firstly, we analyzed the activity by using the CircStat toolbox for MATLAB. As reported in the text, for each neuron, the linear phase histograms were analyzed, taking into account the up phase of the SWA, and the spike phase values were used for circular analysis. Then, each neuron was tested for significant SWA phase-locked firing; i.e., not-randomly distributed in relation to the SWA (comparison of degree; Rayleigh's uniformity test). To exclude this hypothesis, we performed chi-square tests on the percentage of discarded MTh neurons in the three conditions; i.e., control, acute and chronic DA depletion states. The significance threshold was set at $p < 0.05$ for all analyses.

Author Contributions: Conceptualization, S.G and G.D.G.; methodology G.O., L.C.G., E.F., A.S. (Alessandro Stefani), S.G. and G.D.G.; formal analysis, L.C.G., S.G. and G.D.G.; investigation, G.O., L.C.G., E.F., A.S. (Agnese Salvadè), W.S., E.C., L.C.G., S.G. and G.D.G; resources, S.G. and G.D.G.; data curation, G.O., L.C.G., E.F., A.S. (Agnese Salvadè), S.G. and G.D.G.; writing—original draft preparation, S.G., L.C.G. and G.D.G.; writing—review and editing, E.F., A.S. (Alessandro Stefani), A.K.-L. and G.D.G.; supervision, S.G. and G.D.G.; project administration, S.G. and G.D.G.; funding acquisition, S.G. and G.D.G. All authors have read and agreed to the published version of the manuscript.

Funding: This work was supported by the Parkinson Svizzera and the Fondazione per lo studio delle malattie neurodegenerative delle persone adulte e dell'anziano del Ticino to SG and by the University of Malta, Internal Research Grant Scheme to G.D.G.

Conflicts of Interest: The authors declare no conflict of interest.

Abbreviations

MTh	motor thalamus
DA	dopamine
GLU	glutamate
GABA	*gamma*-aminobutyric acid
TTX	tetrodotoxin
6-OHDA	6-hydroxydopamine
VA	ventral-anterior thalamus
VM	ventral-medial
SNr	substantia nigra pars reticulata
GPi	internal globus pallidus
NRT	nucleus reticular thalami
SN	substantia nigra
substantia	substantia nigra pars compacta
PD	Parkinson's disease
STN	sub-thalamic nucleus
DBS	deep brain stimulation
L-DOPA	levodopa
HT2A	entopeduncular nucleus
EPN	immunohistochemistry
LID	L-DOPA-induced dyskinesia
SWA	slow-wave activity
MFB	medial forebrain bundle

References

1. Bosch-Bouju, C.; Hyland, B.I.; Parr-Brownlie, L.C. Motor thalamus integration of cortical, cerebellar and basal ganglia information: Implications for normal and parkinsonian conditions. *Front. Comput. Neurosci.* **2013**, *7*, 163. [CrossRef]

2. Nakamura, K.C.; Sharott, A.; Magill, P.J. Temporal coupling with cortex distinguishes spontaneous neuronal activities in identified basal ganglia-recipient and cerebellar-recipient zones of the motor thalamus. *Cereb. Cortex* **2014**, *24*, 81–97. [CrossRef] [PubMed]
3. Wichmann, T.; Kliem, M.A. Neuronal activity in the primate substantia nigra pars reticulata during the performance of simple and memory-guided elbow movements. *J. Neurophysiol.* **2004**, *91*, 815–827. [CrossRef]
4. Avila, I.; Parr-Brownlie, L.C.; Brazhnik, E.; Castaneda, E.; Bergstrom, D.A.; Walters, J.R. Beta frequency synchronization in basal ganglia output during rest and walk in a hemiparkinsonian rat. *Exp. Neurol.* **2010**, *221*, 307–319. [CrossRef] [PubMed]
5. Pare, D.; Steriade, M.; Deschenes, M.; Oakson, G. Physiological characteristics of anterior thalamic nuclei, a group devoid of inputs from reticular thalamic nucleus. *J. Neurophysiol.* **1987**, *57*, 1669–1685. [CrossRef] [PubMed]
6. Grandi, L.C.; Kaelin-Lang, A.; Orban, G.; Song, W.; Salvade, A.; Stefani, A.; Di Giovanni, G.; Galati, S. Oscillatory Activity in the Cortex, Motor Thalamus and Nucleus Reticularis Thalami in Acute TTX and Chronic 6-OHDA Dopamine-Depleted Animals. *Front. Neurol.* **2018**, *9*, 663. [CrossRef]
7. Yamamoto, T.; Noda, T.; Miyata, M.; Nishimura, Y. Electrophysiological and morphological studies on thalamic neurons receiving entopedunculo- and cerebello-thalamic projections in the cat. *Brain Res.* **1984**, *301*, 231–242. [CrossRef]
8. Kakei, S.; Na, J.; Shinoda, Y. Thalamic terminal morphology and distribution of single corticothalamic axons originating from layers 5 and 6 of the cat motor cortex. *J. Comp. Neurol.* **2001**, *437*, 170–185. [CrossRef]
9. Kultas-Ilinsky, K.; Sivan-Loukianova, E.; Ilinsky, I.A. Reevaluation of the primary motor cortex connections with the thalamus in primates. *J. Comp. Neurol.* **2003**, *457*, 133–158. [CrossRef]
10. Antal, M.; Beneduce, B.M.; Regehr, W.G. The substantia nigra conveys target-dependent excitatory and inhibitory outputs from the basal ganglia to the thalamus. *J. Neurosci.* **2014**, *34*, 8032–8042. [CrossRef]
11. Yague, J.G.; Cavaccini, A.; Errington, A.C.; Crunelli, V.; Di Giovanni, G. Dopaminergic modulation of tonic but not phasic GABAA-receptor-mediated current in the ventrobasal thalamus of Wistar and GAERS rats. *Exp. Neurol.* **2013**, *247*, 1–7. [CrossRef] [PubMed]
12. Lavin, A.; Grace, A.A. Dopamine modulates the responsivity of mediodorsal thalamic cells recorded in vitro. *J. Neurosci.* **1998**, *18*, 10566–10578. [CrossRef] [PubMed]
13. Carlsson, M.; Carlsson, A. Interactions between glutamatergic and monoaminergic systems within the basal ganglia–implications for schizophrenia and Parkinson's disease. *Trends Neurosci.* **1990**, *13*, 272–276. [CrossRef]
14. Munoz, M.D.; de la Fuente, N.; Sanchez-Capelo, A. TGF-beta/Smad3 Signalling Modulates GABA Neurotransmission: Implications in Parkinson's Disease. *Int. J. Mol. Sci.* **2020**, *21*, 590. [CrossRef] [PubMed]
15. Esposito, E.; Di Matteo, V.; Di Giovanni, G. Death in the substantia nigra: A motor tragedy. *Expert Rev. Neurother.* **2007**, *7*, 677–697. [CrossRef]
16. Hammond, C.; Bergman, H.; Brown, P. Pathological synchronization in Parkinson's disease: Networks, models and treatments. *Trends Neurosci.* **2007**, *30*, 357–364. [CrossRef]
17. Boraud, T.; Bezard, E.; Bioulac, B.; Gross, C.E. From single extracellular unit recording in experimental and human Parkinsonism to the development of a functional concept of the role played by the basal ganglia in motor control. *Prog. Neurobiol.* **2002**, *66*, 265–283. [CrossRef]
18. Albin, R.L.; Young, A.B.; Penney, J.B. The functional anatomy of basal ganglia disorders. *Trends Neurosci.* **1989**, *12*, 366–375. [CrossRef]
19. DeLong, M.R. Primate models of movement disorders of basal ganglia origin. *Trends Neurosci.* **1990**, *13*, 281–285. [CrossRef]
20. Stefani, A.; Fedele, E.; Vitek, J.; Pierantozzi, M.; Galati, S.; Marzetti, F.; Peppe, A.; Bassi, M.S.; Bernardi, G.; Stanzione, P. The clinical efficacy of L-DOPA and STN-DBS share a common marker: Reduced GABA content in the motor thalamus. *Cell Death Dis.* **2011**, *2*, e154–e154. [CrossRef]
21. Stefani, A.; Fedele, E.; Pierantozzi, M.; Galati, S.; Marzetti, F.; Peppe, A.; Pastore, F.S.; Bernardi, G.; Stanzione, P. Reduced GABA Content in the Motor Thalamus during Effective Deep Brain Stimulation of the Subthalamic Nucleus. *Front. Syst. Neurosci.* **2011**, *5*, 17. [CrossRef] [PubMed]

22. Korkmaz, O.T.; Tuncel, N.; Tuncel, M.; Oncu, E.M.; Sahinturk, V.; Celik, M. Vasoactive intestinal peptide (VIP) treatment of Parkinsonian rats increases thalamic gamma-aminobutyric acid (GABA) levels and alters the release of nerve growth factor (NGF) by mast cells. *J. Mol. Neurosci.* **2010**, *41*, 278–287. [CrossRef] [PubMed]
23. Gerlach, M.; Gsell, W.; Kornhuber, J.; Jellinger, K.; Krieger, V.; Pantucek, F.; Vock, R.; Riederer, P. A post mortem study on neurochemical markers of dopaminergic, GABA-ergic and glutamatergic neurons in basal ganglia-thalamocortical circuits in Parkinson syndrome. *Brain Res.* **1996**, *741*, 142–152. [CrossRef]
24. Robelet, S.; Melon, C.; Guillet, B.; Salin, P.; Kerkerian-Le Goff, L. Chronic L-DOPA treatment increases extracellular glutamate levels and GLT1 expression in the basal ganglia in a rat model of Parkinson's disease. *Eur. J. Neurosci.* **2004**, *20*, 1255–1266. [CrossRef]
25. Biggs, C.S.; Starr, M.S. Dopamine and glutamate control each other's release in the basal ganglia: A microdialysis study of the entopeduncular nucleus and substantia nigra. *Neurosci. Biobehav. Rev.* **1997**, *21*, 497–504. [CrossRef]
26. Marsden, C.D.; Obeso, J.A. The functions of the basal ganglia and the paradox of stereotaxic surgery in Parkinson's disease. *Brain A J. Neurol.* **1994**, *117*, 877–897. [CrossRef]
27. Grandi, L.C.; Di Giovanni, G.; Galati, S. Animal models of early-stage Parkinson's disease and acute dopamine deficiency to study compensatory neurodegenerative mechanisms. *J. Neurosci. Methods* **2018**, *308*, 205–218. [CrossRef]
28. Galati, S.; Stanzione, P.; D'Angelo, V.; Fedele, E.; Marzetti, F.; Sancesario, G.; Procopio, T.; Stefani, A. The pharmacological blockade of medial forebrain bundle induces an acute pathological synchronization of the cortico-subthalamic nucleus-globus pallidus pathway. *J. Physiol.* **2009**, *587*, 4405–4423. [CrossRef]
29. Prosperetti, C.; Di Giovanni, G.; Stefani, A.; Moller, J.C.; Galati, S. Acute nigro-striatal blockade alters cortico-striatal encoding: An in vivo electrophysiological study. *Exp. Neurol.* **2013**, *247*, 730–736. [CrossRef]
30. Voloshin, M.; Lukhanina, E.P.; Kolomietz, B.P.; Prokopenko, V.F.; Rodionov, V.A. Electrophysiological investigation of thalamic neuronal mechanisms of motor disorders in parkinsonism: An influence of D2ergic transmission blockade on excitation and inhibition of relay neurons in motor thalamic nuclei of cat. *Neuroscience* **1994**, *62*, 771–781. [CrossRef]
31. Schneider, J.S.; Rothblat, D.S. Alterations in intralaminar and motor thalamic physiology following nigrostriatal dopamine depletion. *Brain Res.* **1996**, *742*, 25–33. [CrossRef]
32. Vitek, J.L. Mechanisms of deep brain stimulation: Excitation or inhibition. *Mov. Disord. Off. J. Mov. Disord. Soc.* **2002**, *17* (Suppl. 3), S69–72. [CrossRef]
33. Ni, Z.G.; Gao, D.M.; Benabid, A.L.; Benazzouz, A. Unilateral lesion of the nigrostriatal pathway induces a transient decrease of firing rate with no change in the firing pattern of neurons of the parafascicular nucleus in the rat. *Neuroscience* **2000**, *101*, 993–999. [CrossRef]
34. Kammermeier, S.; Pittard, D.; Hamada, I.; Wichmann, T. Effects of high-frequency stimulation of the internal pallidal segment on neuronal activity in the thalamus in parkinsonian monkeys. *J. Neurophysiol.* **2016**, *116*, 2869–2881. [CrossRef] [PubMed]
35. Pessiglione, M.; Guehl, D.; Rolland, A.S.; Francois, C.; Hirsch, E.C.; Feger, J.; Tremblay, L. Thalamic neuronal activity in dopamine-depleted primates: Evidence for a loss of functional segregation within basal ganglia circuits. *J. Neurosci.* **2005**, *25*, 1523–1531. [CrossRef] [PubMed]
36. Galati, S.; D'Angelo, V.; Olivola, E.; Marzetti, F.; Di Giovanni, G.; Stanzione, P.; Stefani, A. Acute inactivation of the medial forebrain bundle imposes oscillations in the SNr: A challenge for the 6-OHDA model? *Exp. Neurol.* **2010**, *225*, 294–301. [CrossRef] [PubMed]
37. Galati, S.; Mazzone, P.; Fedele, E.; Pisani, A.; Peppe, A.; Pierantozzi, M.; Brusa, L.; Tropepi, D.; Moschella, V.; Raiteri, M.; et al. Biochemical and electrophysiological changes of substantia nigra pars reticulata driven by subthalamic stimulation in patients with Parkinson's disease. *Eur. J. Neurosci.* **2006**, *23*, 2923–2928. [CrossRef]
38. Crick, F. Function of the thalamic reticular complex: The searchlight hypothesis. *Proc. Natl. Acad. Sci. USA.* **1984**, *81*, 4586–4590. [CrossRef]
39. Calabresi, P.; Mercuri, N.B.; Di Filippo, M. Synaptic plasticity, dopamine and Parkinson's disease: One step ahead. *Brain A J. Neurol.* **2009**, *132*, 285–287. [CrossRef]
40. Di Matteo, V.; Benigno, A.; Pierucci, M.; Giuliano, D.A.; Crescimanno, G.; Esposito, E.; Di Giovanni, G. 7-nitroindazole protects striatal dopaminergic neurons against MPP+-induced degeneration: An in vivo microdialysis study. *Ann. N. Y. Acad. Sci.* **2006**, *1089*, 462–471. [CrossRef]

41. Di Matteo, V.; Pierucci, M.; Benigno, A.; Esposito, E.; Crescimanno, G.; Di Giovanni, G. Critical role of nitric oxide on nicotine-induced hyperactivation of dopaminergic nigrostriatal system: Electrophysiological and neurochemical evidence in rats. *CNS Neurosci. Ther.* **2010**, *16*, 127–136. [CrossRef] [PubMed]
42. Pierucci, M.; Di Matteo, V.; Benigno, A.; Crescimanno, G.; Esposito, E.; Di Giovanni, G. The unilateral nigral lesion induces dramatic bilateral modification on rat brain monoamine neurochemistry. *Ann. N. Y. Acad. Sci.* **2009**, *1155*, 316–323. [CrossRef] [PubMed]
43. Hosp, J.A.; Coenen, V.A.; Rijntjes, M.; Egger, K.; Urbach, H.; Weiller, C.; Reisert, M. Ventral tegmental area connections to motor and sensory cortical fields in humans. *Brain Struct Funct.* **2019**, *224*, 2839–2855. [CrossRef] [PubMed]
44. Paxinos, G.; Watson, C. *The Rat Brain in Stereotaxic Coordinates*, 6th ed.; Academic Press: New York, NY, USA, 2007; p. 456.
45. Galati, S.; Song, W.; Orban, G.; Luft, A.R.; Kaelin-Lang, A. Cortical slow wave activity correlates with striatal synaptic strength in normal but not in Parkinsonian rats. *Exp Neurol.* **2018**, *301*, 50–58. [CrossRef]

MDPI
St. Alban-Anlage 66
4052 Basel
Switzerland
www.mdpi.com

International Journal of Molecular Sciences Editorial Office
E-mail: ijms@mdpi.com
www.mdpi.com/journal/ijms

www.ingramcontent.com/pod-product-compliance
Lightning Source LLC
Chambersburg PA
CBHW040816250926
51850CB00007BA/1184
* 9 7 8 3 0 3 6 5 9 0 4 5 5 *